# 23rd International Conference on Condensed Matter Nuclear Science 2021



Published in the Journal of Condensed Matter Nuclear Science (JCMNS) Volume 36

Online
9 – 11 June 2021

**Editor:**

**Jean-Paul Biberian**

ISBN: 978-1-7138-6328-1

**Printed from e-media with permission by:**

Curran Associates, Inc.
57 Morehouse Lane
Red Hook, NY 12571

**Some format issues inherent in the e-media version may also appear in this print version.**



Printed by Curran Associates, Inc. (2023)

For permission requests, please contact
International Society for Condensed Matter Nuclear Science (ISCMNS)
at the address below.

International Society for Condensed Matter Nuclear Science (ISCMNS)
The Willows, Hobro,
Wolverley, Kidderminster
Worcs DY11 5ST
England

www.iscmns.org
webmaster@iscmns.org

**Additional copies of this publication are available from:**

Curran Associates, Inc.
57 Morehouse Lane
Red Hook, NY 12571 USA
Phone: 845-758-0400
Fax: 845-758-2633
Email: curran@proceedings.com
Web: www.proceedings.com

# JOURNAL OF CONDENSED MATTER NUCLEAR SCIENCE

## Experiments and Methods in Cold Fusion

**Proceedings of the 23rd International Conference on Condensed Matter Nuclear Science, Virtual Conference from Xiamen, China June 9–11, 2021**

**VOLUME 36, November 2022**

# JOURNAL OF CONDENSED MATTER NUCLEAR SCIENCE

Experiments and Methods in Cold Fusion

# Preface

The 23rd International Conference on Condensed Matter Nuclear Science (ICCF-23) was held as a virtual conference, during June 9th-11th, 2021, due to the worldwide COVID-19 pandemic and related travel restrictions. With the full support from the International Advisory Committee, the Local Organizing Committee (LOC) made great efforts to promote the development of the field of Condensed Matter Nuclear Science (CMNS). In order to attract researchers and people who were interested in participating in the conference, the LOC did not charge registration fees. The conference was organized and technically supported by the Innovation Laboratory for Sciences and Technologies of Energy Materials of Fujian Province (IKKEM), the State Key Laboratory of Physical Chemistry of Solid Surfaces (PCOSS), and Xiamen University of China.

The ICCF-23 Conference Chair was Prof. Zhong-Qun Tian of Xiamen University, with vice chair Profs. Kang Shi and Han-Tao Zhou from Xiamen University and Prof. Wu-Shou Zhang from the Institute of Chemistry of the Chinese Academy of Sciences. There were 433 registered participants from 32 countries, including China (193), USA (111), Japan (20), India (16), Italy (14), Russia (11), Canada (9), France (9), Switzerland (7), and UK (6). This was a larger number of participants than any previous ICCF series conference. There were 86 abstracts from 17 countries including 10 plenary, 22 invited, 14 oral and 40 3-min poster presentations. Three sessions expanding from morning, afternoon, and evening in Beijing time, and a one-hour live Q&A session were arranged per day, which had 9 sessions in total in 3 days. This time arrangement was to overcome the large time difference between the host country, China, and the locations of many participants. LOC asked presenters to pre-record their talks and submit the videos and materials ahead of the conference, which helped the flow and timeliness of the event, and also let the participants view all the presentations at their convenience. Forty 3-min poster presentations were grouped and presented repeatedly during the intervals of three sessions. One highlight of this conference was the live Q&A sessions for the deep and broad scientific exchange (see Figure 1). These active, inspiring and fruitful discussions were recorded, and then quickly uploaded and shared with all participants on the website for each day. This fast procedure effectively facilitated participation from different time zones.

ICCF-23 covered a wide range of topics including Heat Production, Transmutations, Radiation and Other Measurements, Electrochemical Experiments, Plasma Experiments, Hot Gas Experiments, Beam Experiments, Instrumentation,

**Figure 1.** Some screenshots during the conference.

Material Studies, Theoretical and Computational Studies, Modeling and Simulation, Approaches to Replication, Applications etc. The detailed program is available at http://ikkem.com/iccf-23_program.php. We must thank Christy L. Frazier for writing a detailed report on "ICCF23 Held Virtually from China" in Issue 157 of *Infinite Energy* in August of 2021. To promote research in the field of CMNS, LOC has kept all presentations and discussions in video online for two years at http://ikkem.com/iccf-23_oralab.php and http://ikkem.com/iccf23_3min.php.

It should be noted that *Science* magazine published the 125 Questions theme that focuses on exploration and discovery in April 2021. "Will cold fusion ever be possible?" was among the 125 questions. It has been 33 years after Martin Fleischmann and Stanley Pons declared the existence of an unknown type of nuclear reaction in a simple electrolysis cell. But this phenomenon cannot be explained by traditional theory, and it is not well supported by the experimental results mainly due to the lack of good reproducibility and the poor yield rate. Based on our research and inspired by the presentations and discussions at ICCF-23, we thought that nuclear reactions in condensed matter could be revisited rigorously and pointedly in four aspects (Figure 2) and further explored by several non-traditional ways as follows.

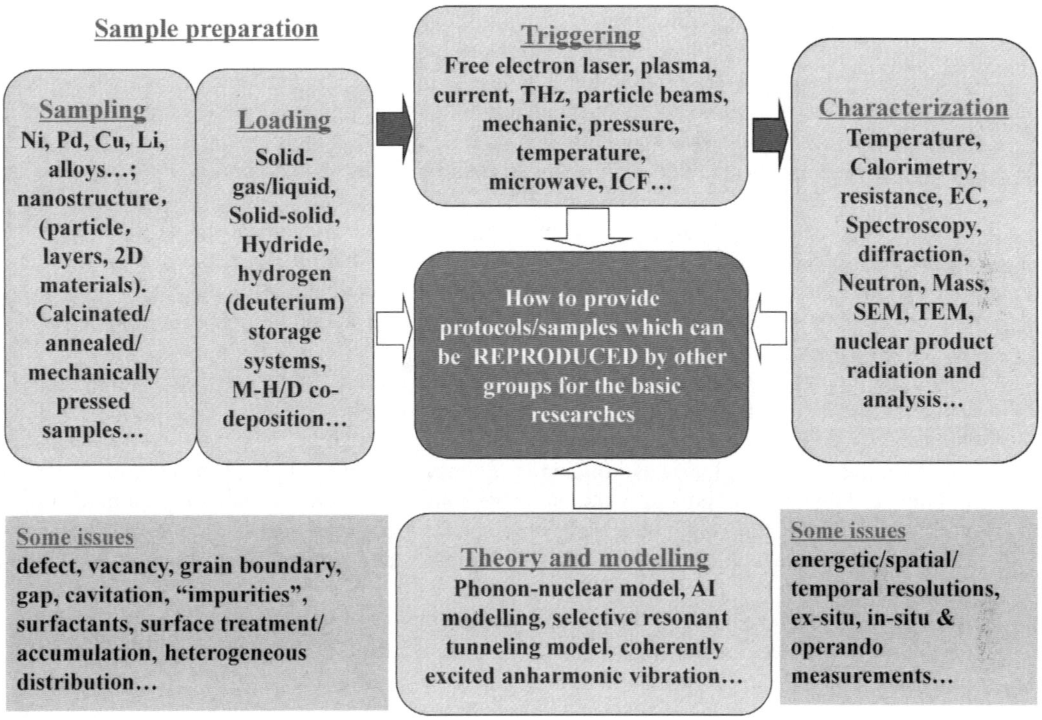


**Figure 2.** Schematic summary of key issues in CMNS research.

1. Sample preparation is a key, as the whole experiment should start from the sample preparation of the material/electrolyte/gas with the same component, purity, surface cleaning and pretreatment, etc. The problem of reproducibility in current research is at least partially due to the irreversible change of lattice/materials after the D/H loading and especially triggering procedures. The defined structure cannot be recreated completely for the next experiment and during long-term cycling because of distortion of the lattice, phase transition or cracking.

2. Nanostructures may have some advantages over bulk materials if the reaction mainly occurs in the region of sub-surface, interface or interphase of several or several tenths of nanometers. Nanoparticles can provide much more active sites. After the experiment, these nanostructures can be easily dissolved by chemicals and then re-synthesized to fresh nanoparticles for the next experiment. Moreover, various core-shell and core-shell-island nanoparticles can be rationally synthesized to implement the desired composition (e.g., Pd, Ni, Cu, Li) and even their quantity ratio as alloys.
3. The small tubes with ultrathin wall, ultrathin wires or nanoparticles of metal or metal hydride could be one of the best systems for the high temperature/pressure studies. For the small nanoparticles, the rationally designed surfactant/layer to protect the particle surface should be used to avoid aggregation of ill-defined structures as well as to promote high loading of D/H.
4. Triggering is essential for observing the reaction to avoid the conflict of the high loading and triggering targets. The sharp increase of temperature by the laser, current, or electro-magnetic pulse triggering is obviously helpful because the absorbed D/H has no time to escape from the loaded material. That most likely results from some processes far from equilibrium in highly D/H loaded metals. More importantly, it can change/distort the material lattice significantly (e.g., deliberately close to the melting point of the metal/alloy) in the transient state for realizing lattice-assisted nuclear reaction. Rapidly raising the temperature close to the melting point of the highly loaded materials could be a way to create a special non-equilibrium state that may promote the reaction effectively.
5. One of the extreme trigger methods could be the utilization of inertial confinement fusion (ICF) facility. The highly D/H loaded metal nanoparticles are filled in the target ball then it is compressed to extremely high densities and temperatures by the initiating laser beams. The sufficiently high power density and extremely high temperature are achieved before the target disassembles. This combined hot and cold fusion method may significantly reduce the reaction threshold of technical parameters for ICF.
6. The highly sensitive measurement of the reaction products and characterization methods with energetic, spatial and temporal resolution should be developed to analyze the defect, vacancy, grain boundary, gap, cavitation, surfactants, heterogeneous distribution, etc. For the comprehensive and deep analysis on all steps (i.e., sample preparation, pretreatment, loading, triggering and reaction), ex-situ, in-situ and operando measurements should be combined synergistically. Surface contamination must be avoided and many characterization methods and tools under high vacuum condition must be developed to extract the weak but useful signal contributed from the surface. The most difficult task could be the operando measurement during the triggering process. The relevant big data analysis and AI may play a key role.
7. Regarding the theoretical and modelling approaches, not only normal condensed matter but also abnormal condensed matter (e.g., 'soft materials' close to the melting point) may support the reaction. It may need to have stimulated surface phonon emission or coherent shaking of surface and/or sub-surface atoms periodically. The localized anharmonic vibrations might be one of the possible ways to realize the localized excited surface phonon, which could be triggered by thermal heating, THz pumping, gas pumping or inflating, etc. The abnormal phenomena may be more distinct when the condensed matter is getting 'soft' in a non-equilibrium state when the condensed matter is input with carefully designed energy flux.
8. One fundamental question could be how to design then construct unique structures and conditions which may lead to the formation of nucleus pairs (or clusters), which could follow the track of the field of superconductors that can form the electron pair.

ICCF-23 was the first online meeting and we tried our best to open it on all aspects as wide as possible, and all the activities were to promote the development in this field. However, we all know that any first attempt in any new experiments cannot be perfect. So, we are committed to maintaining the conference website for future reference: http://ikkem.com/iccf-23.php.

On behalf of the LOC, we sincerely thank all speakers, session chairs, poster presenters, all participants, and in particular all the authors of this special issue for your important contribution and active cooperation. We must thank the organizers IKKEM and PCOSS, and financial sponsors Yushi Energy Group and New Energy Power System, LLC, and the International Advisory Committee for their full support and helpful advice.

**Figure 3.** Group Photo of Chinese speakers and members of the Local Organizing Committee, 11th June 2021, Xiamen, China.

**Chong-En Huang, Ren Hu, Ling-Ni Duan, Kang Shi, and Zhong-Qun Tian**
Innovation Laboratory for Sciences and Technologies of Energy Materials of Fujian Province (IKKEM)
State Key Laboratory of Physical Chemistry of Solid Surfaces (PCOSS)
College of Chemistry and Chemical Engineering
Xiamen University
Xiamen 361005, China

J. Condensed Matter Nucl. Sci. 36 (2022) 1–435
 ISSN 2227-3123

# JOURNAL OF CONDENSED MATTER NUCLEAR SCIENCE

Volume 36 2022

## CONTENTS

RESEARCH ARTICLES

J. Condensed Matter Nucl. Sci. 36 (2022) 1–24

Research Article

# Lattice Energy Converter II: Iron Hydrogen Host Material

F. E. Gordon*

*Inovl Inc., 2787 Curie Place, San Diego, CA, 92122 USA*

H. J. Whitehouse

*Inovl Inc., 6441 Del Paso Ave., San Diego, CA. 92120 USA*

---

**Abstract**

Continuing development of Lattice Energy Converter (LEC) technology has resulted in both experimental and theoretical advancements. Replicated experimental results and analysis for a LEC wherein a codeposited palladium-hydrogen working electrode produced spontaneous and sustained electrical energy attributed to ionizing radiation have been previously reported. Herein is reported the use of a working electrode comprised of codeposited iron-hydrogen from an aqueous solution of $FeCl_2$ which demonstrated similar capabilities to produce spontaneous and sustained electrical energy as well as ionizing radiation. These results also have been replicated. http://ikkem.com/iccf23/PPT/Invited%20Gordon%20ICCF%2023%20LEC%20T5.MP4. This paper updates the presentation at the workshop in honor of Dr. Srinivasan in January 2021 which is available at: https://www.youtube.com/watch?v=J4dzTWY_aWM This paper provides additional analysis that supports the observed experimental results from both presentations.


*Keywords:* Direct energy conversion, Electricity generation, Ionizing radiation

---

## 1. Introduction

As previously reported, the Lattice Energy Converter (LEC) has demonstrated the capability to convert thermal energy in a codeposited palladium-hydrogen lattice into ionizing radiation [1]. It is well known that hydrogen will diffuse into many metals and alloys to form an interstitial metal hydride [2] but it was not known if other metal hydrides would produce results similar to the LEC palladium-hydrogen system.

H. Mehrer [3] discusses the diffusivity of hydrogen in various metals and includes a plot on page 322 of the diffusivity of iron, palladium, and nickel as a function of inverse Kelvin temperature.

As shown, in Fig. 1, the hydrogen diffusivity of iron is approximately two orders of magnitude greater than that of palladium at room temperature and is approximately equal to the hydrogen diffusivity of palladium at its melting point.

*Corresponding author. Email: feg@inovl.com, Ph: +1-858-349-7869

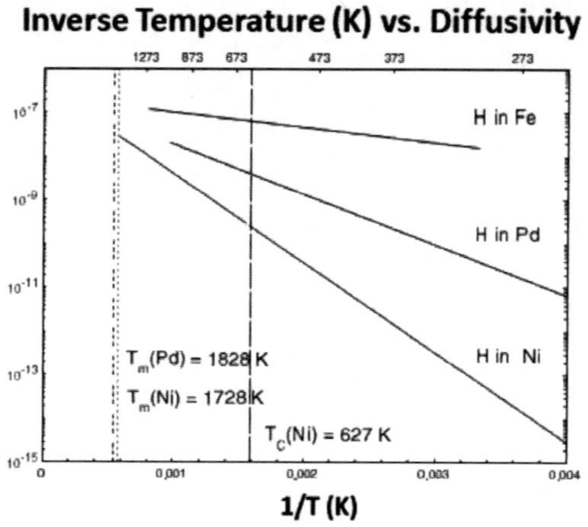


**Figure 1.** Plot showing hydrogen diffusivity of Fe, Pd, and Ni vs. inverse temperature.

Diffusivity describes the ability of a H atom to circulate through the lattice under a relative H concentration gradient. The reader may be surprised to see that diffusivity in Fe is larger than in Pd because it is well known that palladium is quite permeable to hydrogen. This is because the permeability is given by the product of the diffusivity by the hydrogen solubility in the metal. H solubility is much higher in Pd than in Fe and it is the reason why the permeability is so high in Pd in spite of a lower diffusivity.

Although properties such as permeability and solubility may also contribute to LEC cell performance, diffusivity may be an important parameter. The much lower cost and worldwide availability of iron could be important factors in the future economic feasibility of LEC devices. For these reasons, LEC cells were prepared using working electrodes of codeposited iron hydrogen-host-material (HHM) to evaluate iron-hydrogen performance.

## 2. Experimental description

As shown in Fig. 2, LEC cells were prepared where the codeposited Pd-H brass working electrode (WE) as previously described [1], was replaced with a 1/8 inch by 4 inch black iron (low carbon steel) pipe nipple. The iron pipe nipple included a protective coating to prevent oxidation which was removed by sanding. The iron pipe nipple was the cathode in a plating solution of 0.1M $FeCl_2{\cdot}4H_2O$ in distilled $H_2O$ and platinum wires were the anode. The initial plating current was approximately 100 $\mu$A/cm$^2$ for 15 minutes and then increased to approximately 1 mA/cm$^2$ for 15 minutes followed by an increase to approximately 2-3 mA/cm$^2$, for approximately 4 hours, until the codeposited Fe-H was visible and appeared to be uniform. The codeposited iron HHM WE was removed from the plating solution and placed in a 1/2 inch galvanized pipe nipple counter electrode (CE)while avoiding physical contact between the pipe nipples and connected to a digital voltmeter (DVM) to determine if a voltage was produced. If no voltage is measured, additional codeposition of the HHM may be required. The codeposited iron WE is blotted dry and assembled into a LEC cell. A vacuum ($\sim$50 to 75 Torr) is pulled and the cell is refilled with hydrogen gas. Hydrogen gas pressures ranging from several hundred Torr to a few thousand Torr have been successfully tested and other pressures are expected to work.

Multiple Fe-H LEC cells have demonstrated the ability to spontaneously produce a voltage and current. In addition, there have been multiple successful replications of the Pd-H, and/or the Fe-H LEC by several individuals [4], [5],

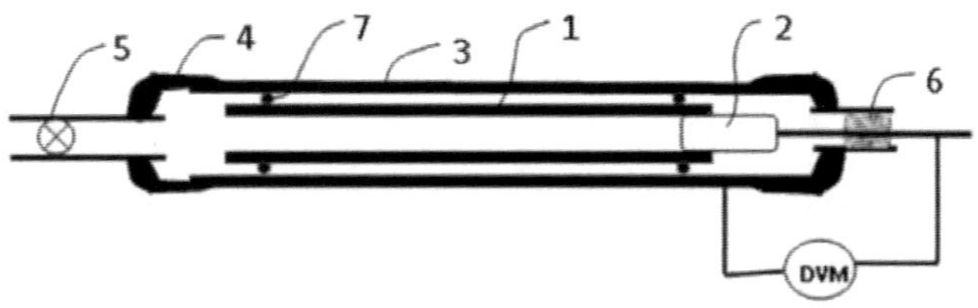


**Figure 2.** LEC cell using iron pipe nipple codeposited with FeH for the working electrode.
1. 1/2 inch by 4 inch iron pipe nipple with id threaded at one end with $5/16$ 24 tap and then codeposited with Fe – H
2. $5/16 \times 24$ set screw with Cu wire brazed on one end and screwed into threaded nipple
3. 1/2 inch by 5.5 inch brass or galvanized pipe nipple (provides different work functions between outer pipe and the inner iron pipe nipple.)
4. $3/2$ inch to $1/4$ inch bushings on each end of the $3/2$ inch nipple
5. $1/4$ inch nipple and valve to evacuate and fill LEC with hydrogen or deuterium gas
6. $1/4$ inch nipple with high temperature epoxy fill to provide electrical insulation
7. Small high temperature Orings ora bead of high temperature epoxy to maintain physical separation between items 1 and 3 while allowing gas to pass between. Note-Orings provide separation and do not seal the gas.

[6], [7]. Similar results have been reported using modified cell designs including hydrogen or deuterium, modified codeposition protocols, modified electrode geometries, as well as the inclusion of additional components between the electrodes. [8], [9], [10]. The collective results provide compelling evidence of the existence of these surprising phenomena.

## 3. Experimental Results

The Fe-H LEC cells have been characterized following the same procedures used for Pd-H LEC cells including tests in air as well as in hydrogen [1]. A typical LEC cell test consists of several steps. The cell is placed in an electric kiln at laboratory temperature that is instrumented with one or more thermocouples. The LEC electrodes are connected to a high input impedance DVM ($\sim$1000 MΩ) with the WE positive and the CE negative. A variable resistance box is connected between the electrodes with a nominal starting resistance of 1 MΩ. The DVM is connected to a digital computer via an optical USB connection and the recording sample rate is set at 2 S/s. The kiln is powered by a variable voltage auto-transformer that is started at a low voltage setting and intermittently raised. LEC voltage and kiln temperature are recorded, and variable resistance load tests are performed occasionally in order to characterize LEC performance, i.e., LEC voltage versus load resistance at selected LEC operating temperatures.

Figure 3 is an example of Fe-H LEC performance in air at approximately 30 °C soon after it was removed from the plating bath. The working electrode was a 1/8 inch by 4-inch black iron pipe nipple that was codeposited with iron from a solution of 0.1 M $FeCl_2$ in $H_2O$.

A test to measure the LEC voltage under various load conditions was conducted immediately after the working electrode was removed from the plating solution and placed in the counter electrode containing air at approximately 30 °C. The performance of this Fe-H LEC is qualitatively similar to that of a Pd-H LEC in that the load current of the LEC behaves as if it were a current source shunted by a variable voltage conductance. More explanations can be found in [1].

The shunt current as a function of load resistance is thus the difference between the load current at low load resistance and the measured load current as a function of resistance. Load power is computed as load voltage squared divided by load resistance or the product of load voltage by load current. Load power is found to be a maximum at intermediate values of load resistance. Figure 4 shows LEC performance for a Pd-H WE comprised in part of a

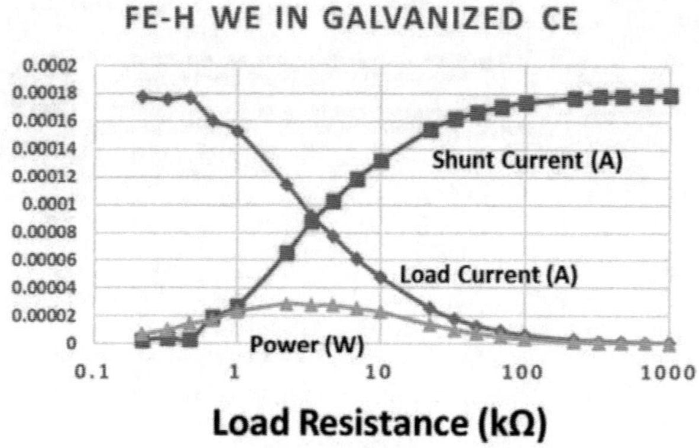


**Figure 3.** Load current, shunt current, and power as a function of resistance.

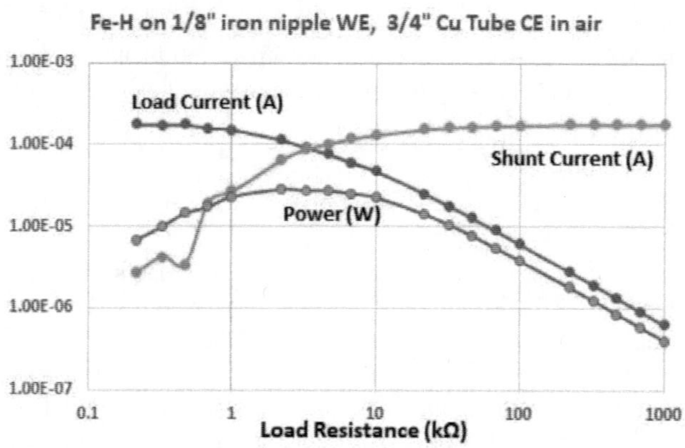


**Figure 4.** Plots of load current, shunt current, and power as a function of load resistance in kΩ.

stainless-steel screen electrode with codeposited Pd HHM and measured shortly after the electrode was removed from the plating solution.

As experiments have demonstrated, a LEC cell develops a spontaneous voltage between the electrodes with no input other than the energy in the lattice of the hydrogen-host-material including the thermal kinetic energy of the hydrogen that is occluded therein. The current-voltage (I-V) characteristics of a LEC cell is non-ohmic in the sense that the current density and the electric field do not go to zero simultaneously where cell current, $I$, is the integral of the current density, $i$, over the surface area of the WE and the cell voltage, $V$, is the integral of the electric field, E, over the separation distance between the electrodes.

JSE Townsend observed [11].

"It is difficult to calculate mathematically the equation of the current-electric-force curve in the general case for parallel plate electrodes, when recombination and diffusion are in progress, as the problem becomes very

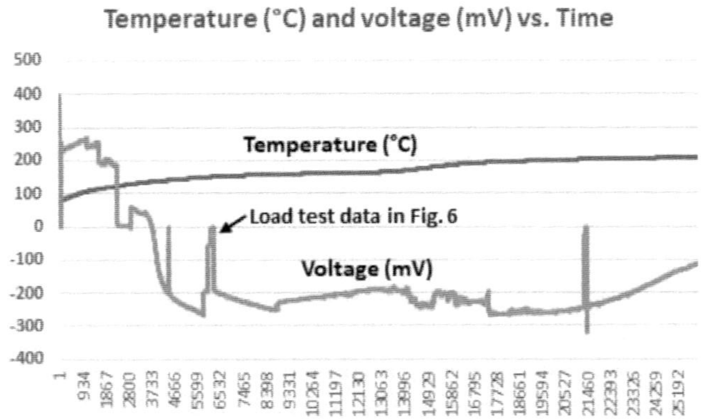


**Figure 5.** Plots of the voltage (mV) under varying load resistances and temperature (°C) as a function of time. Total time for this test was approximately 13 hours.

complicated owing to the fact that the number of ions per cubic centimetre varies at different points of the gas."

Eduard Riecke reported the full set of equations for the conduction of electricity through an ionized gas containing two types of ions, positive and negative, in a parallel-plate condenser in 1903 [12]. An English translation of Riecke is included as Appendix A. He considered the case where there was a space charge in the gas and included the effect of the electric field on the ions, the diffusion of the ions, as well as their recombination to form neutral molecules and then was able to calculate the approximate distribution of the electric field between the electrodes. Using this electric field distribution approximation, the spatial distribution of the ions between the electrodes now can be estimated by integration assuming the densities of positive and negative ions are equal. Using modern notation, the equation giving the current density in terms of ion drift and diffusion can be rewritten as:

$$i/e = (\mu^+ + \mu^-)\boldsymbol{E}n + (\boldsymbol{D}^- - \boldsymbol{D}^+)dn/dx \tag{1}$$

which has a solution of the form, [Appendix B]:

$$\boldsymbol{n(x)} = (1/\boldsymbol{u}(x))[\int^x u(x')g(x')dx' + C], u(x) = \exp[\int^x p(x')dx'] \tag{2}$$

where $p(x) = (\mu^+ + \mu^-)E(x)/(D^- - D^+)$, $g(x) = (i/e)/(D^- - D^+)$, $D^\pm = \mu^\pm k_{eV}T$, $E(x)$ is the electric field within the gas, and $C$ is a constant of integration that can be evaluated using boundary conditions.

As shown, the voltage increased initially when the temperature was less than 100 °C. As the temperature increased, the voltage started to decay and ultimately changed polarity. During the initial phase of the test, load resistances were changed from 3.3 k to 2.2 k to 1.0 kΩ which were selected to be approximately the loads that had produce the maximum power in previous tests. Some of the abrupt changes in voltage during this time are due to the changes in load resistance. After the voltage changed sign, the resistance was set to 1 MΩ except for times when load tests were conducted.

Figure 6 is a plot of a load test of an iron pipe nipple, codeposited with iron and placed in a galvanized CE that was filled with hydrogen gas. The test was conducted as indicated on Fig. 5 when the temperature was approximately 160 °C. Even though the polarity of the voltage has switched, a plot of the absolute values of the variables has the same features as shown in Figs. 3 and 4.

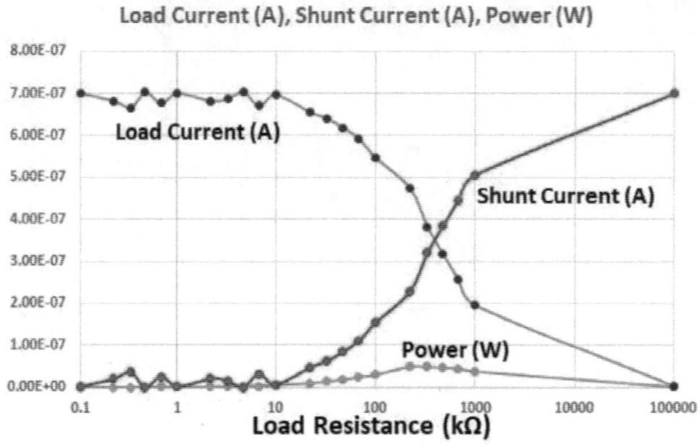


**Figure 6.** Load current, shunt current, and power as a function of resistance.

**Figure 7.** Photograph of the working electrode taken after the temperature test shown in figures 5 and 6.

Figure 7 is a photograph of the working electrode that produced the data plotted in Figs. 5 and 6. The WE is a 1/8 inch black iron pipe nipple by 4 inches long that had codeposited with iron from an aqueous solution of 31% HCl using iron wires as the anode. Prior to the test, the Fe codeposition was lumpy but after the test, multiple unusual features are present. As shown these features appear to be the result of ejecta that produced a "cylindrical tower" that is perpendicular to the surface of the WE and are clearly visible to the naked eye, unlike features reported from Pd-D working electrodes that were a few microns in diameter [13]. The color of the inside surfaces of the ejecta is what would be expected if the Fe had oxidized. This is the first WE that we have observed with these features.

## 4. Possible explanation of LEC voltage generation and its behavior

As currently understood by the authors, LEC voltage generation and the conversion of lattice energy into electrical energy is a multi-step process. Some of these steps are readily explained while others await full explanation. The first step, and perhaps least understood step, is the conversion of the thermal energy in the hydrogen occluded lattice of electrodeposited HHM into some form of radiation. The second step is the conversion of the radiation into ionized gas in fluidic contact with the HHM and heat. The third step, and perhaps better understood step, is the conversion of the ionized gas into voltage and current in an external resistive load impedance. Throughout this multistep process, total energy, including mechanical, chemical, thermal, and nuclear, must be conserved. No 'excess energy' is being produced.

One possibility is that the radiation generation process might be purely mechanical in origin. Thermally generated phonons in the lattice of the HHM interact linearly or nonlinearly to produce regions of high amplitude atomic motion [14], [15] that interacts either directly with the occluded hydrogen atoms within the lattice; or indirectly with the hydrogen atoms within lattice vacancies, super abundant vacancies, or other crystal defects to emit particulate and/or electromagnetic radiation into the gas. Alternatively, the radiation generation might be due to some form of lattice enabled nuclear reaction (LENR) processes that produces particulate and/or electromagnetic radiation. Whatever process or combination of processes responsible for the production of the radiation itself is difficult to measure and determine its form. That radiation is produced is indicated both by the fogging of film [16], [17], the color change of chemically treated paper exposed to the radiation [18], the production of voltage in semiconductor material exposed to the radiation [19], the development of voltage in combinations of materials in contact with hydrogen and/or deuterium gas [9], and by the conduction of electricity through $H_2$, $D_2$, and air exposed to the radiation [20], [21].

The conversion of radiation into ionized gas and heat is a well-documented process. The ionization of gas by Röntgen rays (X-rays) was described by JJ Thomson and E Rutherford in 1896 [22] and by uranium rays in 1899 [23] and it is now known that any energetic charged particle can ionize a gas [24] or water vapor [25]. The minimum energy required to remove the first electron from a gas is called the ionization potential and tables of ionization potential are readily available [26]. The ionization potential for $H_2$ is about 13.6 eV and that for $D_2$ is about 15.5 eV. However, in the process of ionizing a gas, translational and rotational energy also is imparted to the gas. In the case of polyatomic molecules such as $H_2$ and $D_2$, energy also goes into internal bond motion so that the average energy required to ionize gases is greater than their ionization potential. The average energy required to ionize a gas is known as the W-value of the gas and is about 35 to 36 eV for $H_2$ and $D_2$ [27]. Thus, the thermal energy, enthalpy, or heat energy, of a gas is increased by ionization and the total heat energy increase of the gas can be estimated by multiplying the W-value of the gas by the ionization rate per unit volume of the gas and integrating over the volume of the gas being ionized. The calculated thermal energy being produced based on the number of ions in the gas and the current produced, is several orders of magnitude below that which calorimetry can detect. This opens the possibility that the occurrence of LENR may be much more pervasive than calorimetry indicates, and when combined with the self-initiation and self-sustaining capabilities demonstrated by the LEC, suggests a possible connection to the reactions and processes leading to biological and nuclear transmutations. [28] [29].

The generation of the LEC voltage can occur in several ways. One possible way is that charged particles are emitted from the WE with sufficient energy to ionize the gas and while ionizing the gas transit through the gas to the CE and deposit their charge on the counter electrode. This mode of voltage generation is analogous to that of a direct charge nuclear battery [30] [31]. Another method of LEC voltage generation is by the photoelectric effect. Electromagnetic energy emitted from the working electrode causes electrons to be emitted from the CE. The emitted electrons then ionize the gas while behaving as free electrons before they attach to neutral molecules to become negative ions. The emission of the electrons from the CE causes a charge to develop on the electrode and thus a voltage to be generated. Another possibility of LEC voltage generation is that the WE and the CE have different work functions thus generating a Volta effect voltage when the gas between the electrodes of different work functions is ionized. LEC voltage generation also may be caused by the different rates of diffusion of positive and negative ion distributions in the gas. These last two methods will be discussed in the following section.

In 1903 JJ Thomson published the first edition of his treatise entitled Conduction of Electricity through Gases, [32] that expanded on his 1899 paper "On the theory of the conduction of electricity through gases by charged ions," where he proposed a method to estimate the electric field distribution in a conducting gas when diffusion of the ions is neglected. In the same year, 1903, Eduard Riecke [12] at the University of Göttingen published in German his method to estimate by successive approximation the electric field distribution in a conducting gas when the diffusion of the ions is included. An English translation is presented in Appendix A. Measurements of spontaneously conducting LEC cells has shown that LEC cells behave as if they are a current source due to the diffusion of charge concentration within the gas and an internal electric field dependent shunt current caused by the drift of the ions due to the electric field within the gas. Thus, to explain the behavior of LEC cell voltage as a function of both time and temperature, it is necessary to consider both the voltage across the cell and the distribution of charge within the cell.

In 1932 Karl K Darrow explained the spontaneous voltage developed when an ionized gas is contained between two plane-parallel electrodes [33]. A summary of Darrow's and other researchers' contributions are given in [1]. Darrow's insightful contribution explaining the spontaneous voltage produced by a LEC where there are only two types of ions, positive and negative, with different diffusion coefficients is reproduced here with emphasis added.

> "Suppose the ions of the two signs are spread identically through a gas between two walls—by 'identically' I mean that everywhere the concentrations of the two kinds are equal, though their common value varies from place to place—and that this state of affairs is stationary. **Then there must be ionizing rays acting continually on the gas, and also there must be a potential-difference between the walls**. For if there were no field, the negatives would diffuse along the concentration-gradient more rapidly than the positives; there would be a net current; this would result in a depletion of negatives, and an excess of positive charge would arise in the gas, in contradiction with the assumption. But suppose there is a P.D. [Potential Difference] between the walls, in such a sense as to oppose the negatives and pull the positives forward as they stream together down the gradient. A value for this potential-difference can be found, such that the field strength will retard the negatives and encourage the positives just sufficiently to annul the net current aforesaid;"

When a spontaneously conducting LEC is observed over an extended period of time, or when the temperature is slowly changed, the voltage may vary with time or even change sign multiple times. When an externally applied voltage on a LEC is reduced, the current may go to zero before the voltage goes to zero. This behavior is different than that reported in early experiments on the conduction of electricity through gases that were carried out at the Cavendish Laboratory in Cambridge [22], [23], [32] [34] and in later experiments using ionization chambers [35] where the I-V characteristic was observed to be ohmic at low voltage. These unusual electrical phenomena need to be understood in order that they can be either mitigated or exploited in future engineering applications of a LEC.

A cylindrical coaxial LEC with an externally applied voltage is similar to a Gerdien Condenser that is traditionally used to measure the density of ions in a gas. Some Gerdien Condenser applications are to characterize atmospheric pressure plasmas [36] or for atmospheric ion measurements [37]. A plane-parallel electrode LEC with external voltage applied is similar to a plane-parallel ionization chamber [24] while a cylindrical LEC with externally applied voltage is similar to a DC voltage ionization chamber with cylindrical electrodes [38]. Both a Gerdien Condenser and a LEC behave strangely when no voltage is externally applied to the electrodes and although the explanations for the Gerdien Condenser and LEC performance are different, similar phenomena may be responsible for both unusual behaviors.

### 4.1. Contact potential or work function effect

Two possible causes of LEC voltage change with time and temperature are the variation of the work function of the electrodes or variation in the mobility of the ions in the gas due to changes in the molecular or atomic composition of the gas. With regard to the variation of the work function of the electrodes The Radio Corporation of America (RCA) [39] reports that contact potential difference between electrodes of different material may lead to a measured cell voltage when the gas between the electrodes is ionized. They report that

> "The principle involved here goes back to 1897, when Lord Kelvin[12] only a year after the discovery of radioactivity by [Henri] Becquerel, used the radioactivity from uranium to ionize the air between two plates

of dissimilar metals and reported that he had observed a potential between them of up to 1.9 volts. ···
[12] Kelvin, Lord, "Experiments on the Electrical Properties of Uranium", *Edinb. Roy. Soc. Proc.* Vol. 21, p. 417 (1897); also Mattr. and Phy. Papers, Vol. 6, p. 84 (1911). "The Direct Conversion of Radiation into Electrical Energy""

In 1928 *The Electrician* [40] reports "A New Electronic Battery" by Mr. J. Kramer in his presidential address before the Scientific Society of the Birmingham and Midlands Institute.

> "This electronic cell would be similar to a voltaic couple, but the electrolyte would be replaced by radio-active material. It might also be described as a "self-charging" condenser. ···. Turning to the new development, it is found that if the air gap between the dissimilar plates is occupied by radio-active materials a steady e.m.f. is produced."

In 1951 PE Ohmart [41] reports that

> "Experiments have shown that if a cell, made up of two electrochemically dissimilar materials separated by a gas, is connected to a current measuring device, a small continuous current will be caused to flow from the more noble to the more active electrode without an external source of voltage when the separating gas is forcibly ionized by exposure to nuclear radiation. The current thus produced has been found to be dependant upon the nature of the electrodes and their surfaces, the type and flux of the energizing radiation, the type and pressure of the filling gas, and (when excited by gamma radiation) the gamma stopping power of the electrodes."

With respect to the role that contact potential could have on the performance of a LEC, KL Alpin [37] reports that for a Gerdien Condenser

> "The tube material could be significant for two reasons: firstly that of its durability when used in the field, and secondly that of the effect of contact potential. Wåhlin (1986) discusses the effect of the material of the tube, opining that contact potentials cause offsets in the i-V response. This is credible, as two electrodes touching a fluid containing ions do make an electrochemical cell (e.g. Atkins, 1989), even if the electrolyte (air) is dilute. According to Wåhlin (1986), a Gerdien will measure a non-zero current when the bias voltage is zero, due to electrochemical potential at the surface of the tube. Wåhlin (1986) suggests that a steel tube has to be biased at about 0.4 V and the aluminum at 1 V to counteract this. ···
> Atkins P.W. (1989), *Physical Chemistry,* $4^{th}$ edition, Oxford University Press ··· Wåhlin L (1986), *Atmospheric electrostatics,* Research Studies Press, Letchworth"

Consider the variation of the work function or contact potential of the electrodes of a LEC. Although there are listings of work function for many different materials, the work function of a given material is sensitive to the surface properties of the material. During prolonged LEC operation the ionizing radiation as well as the gas ions that transit to the electrodes can change the surface condition of the electrodes. Additionally, the fractional atomic loading of the HHM with atomic ($^1H_1$) or molecular ($^1H_2$) hydrogen or a combination of these gases is a function of the density of WE vacancies and other HHM structural defects such as grain boundaries. Fractional atomic loading of the HHM depends on the over-potential of the ionized hydrogen gas as an electrolyte and the current load voltage of the LEC. The number of HHM lattice vacancies varies exponentially with the inverse kelvin temperature of the HHM in accordance with an Arrhenius equation of the form $N_V = N_0\exp(-E_a/k_{eV}T)$ where $N_0$ is a constant depending on the lattice material, $E_a$ is an activation energy in electron volts (eV), $k_{eV}$ is Boltzmann's constant expressed in eV/K, and $T$ is the HHM temperature in Kelvin. Additionally, the number of grain boundaries increases with increased stress on the HHM due to changes in lattice dimension due to the occlusion of the hydrogen in the HHM lattice structure.

### 4.2. Gas ion mobility effect

There is a second phenomenon that in combination with changes in the work function or contact potential of the electrodes of a LEC may explain the unusual voltage variation of a LEC. This is the variation of the composite mobility of the gas ions due to changes in the composition of the gas during operation. A gas that is being ionized either by interaction with energetic particles or by electromagnetic radiation, *e.g.*, UV, X-rays, or gamma rays, may contain several different types of molecules with different masses as well as free electrons that have not yet attached to neutral molecules to form negative ions. Therefore, following Riecke [11], [12], assume that all ions carry a single charge, $e$, the conduction equation for a LEC can be written for plane-parallel electrodes as

$$i/e = \sum\nolimits_{\mathrm{k}} (v_{\mathrm{dk}}{}^{+} n_{\mathrm{k}}{}^{+} - D_{\mathrm{k}}{}^{+} \mathrm{d}n_{\mathrm{k}}{}^{+}/\mathrm{d}x) + \sum\nolimits_{\ell} (v_{d\ell}{}^{-} n_{\ell}{}^{-} + D_{\ell}{}^{-} \mathrm{d}n_{\ell}{}^{-}/\mathrm{d}x) \quad (3)$$

where $i$ [A·cm$^{-2}$] is the measured cell current density, $v_d{}^{\pm}$ [cm·s$^{-1}$] is the magnitude of the drift velocity vector, $\boldsymbol{v}_d{}^{\pm} = \mu^{\pm}\boldsymbol{E}(x)$, of the ions, $\mu^{\pm}$ [cm$^2$·V$^{-1}$·s$^{-1}$] is their mobility, $E(x)$ [V·cm$^{-1}$] is the electric field strength, $n_k{}^{\pm}(x)$ [cm$^{-3}$] their density, $D^{\pm} = \mu^{\pm} k_{eV} T$ their diffusion coefficient, $k_{eV} = 8.617\mathrm{x}10^{-5}$ eV/K is Boltzmann's constant, $T$ [K] the effective temperature of a gas ion or an electron, $x$ [cm] the distance measured from the anode, $k$ the number of different positive ions, and $\ell$ the number of different negative ions. Note that the effective temperature of a free electrons can be much greater than the temperature of the gas ions which are approximately the ambient temperature of the gas.

## 5. Observations and Conclusions

1) The surprising phenomena produced by a LEC is real.
2) Carefully conducted control experiments using blank electrodes do not produce a voltage and current above the extremely small currents and voltages predicted for conduction through air that has not been ionized except for the naturally occurring ions [7].
3) LEC's are easy to replicate and are not very sensitive to codeposition protocols.
4) Both the origin as well as the nature of the LEC radiation that ionizes the gas is unknown.
5) Possible mechanisms that might produce the initial ionizing radiation at the working electrode include:
   (a) thermally induced vibration of the hydrogen host material's lattice enhanced by nonlinear wave-wave mixing in conjunction with the presence of hydrogen occluded in lattice vacancies near the surface of the hydrogen host material;
   (b) thermally induced interaction between multiple occluded hydrogen atoms contained within a vacancy, or super abundant vacancy, or other crystal defects;
   (c) LENR reactions such as when different nuclear spin orientations are present within a vacancy; or
   (d) a combination of mechanisms.
6) Given that the gas between the electrodes is ionized, LEC properties such as spontaneous and self-sustained production of a voltage and current has been explained.

## 6. Future research and development

Scaling up these experimental results to produce a commercial energy device requires 6-10 orders of magnitude. Possible areas for improvement include:

(1) better metallurgy/materials and plating processes to increase the activity that produces the ionizing radiation;
(2) methods to increase flux such as flowing gas, increased temperature, gas pressure;
(3) cell designs to optimize electrode separation distance and electrode design; and
(4) the application of magnetic and/or electric fields.

## 7. Conclusions

The conduction of electricity in a LEC is significantly different than that conventionally used to describe the conduction of electricity through gases for several reasons [33], [44], [45]. First and foremost is the fact that the source of the conduction is from some form of radiation produced by or within the hydrogen occluded hydrogen-host-material (HHM) of the cell's working electrode (WE). Secondly, the equivalent gas pressure caused by 'over-potential' in the gas near the HHM of the working electrode [43] appears to cause the conduction to not 'saturate' as the voltage across the cell increases.

Thirdly, diffusion of the ions plays a major role in the generation of the spontaneous voltage of a LEC when there is no electrical input to the LEC. Finally, the electrical work function of the material of the counter electrode (CE) also appears to play a role in the conduction of electricity in a LEC [41].

## 8. Appendix A

English translation of Eduard Riecke from the news of the K Society of Sciences in Göttingen. Mathematical-physical class. Issue 4. (1903).

### 8.1. On approximately saturated currents between plane-parallel plates

In two earlier works[1]) I developed the theory of the currents which occur between two concentric spherical surfaces as a result of the ionization of the air contained in the cavity. It was found that the observation of these currents provided a means of determining the reunion [recombination] constant $\alpha$. In considering this possibility, it seemed to me of interest to also subject the case of an air space enclosed between two plane-parallel plates to a somewhat more precise investigation.

#### 8.1.1. *The general equations of the problem*

Let one of two infinite plates, which are parallel to each other at the distance $l$, be positively charged, the other connected to the conductive [grounded to earth]. Let the intensity of the electric field enclosed between the plates be $\mathfrak{F}$. The air in the space between the plates is ionized, for example, by X-rays or radium rays. The ionization strength is $q$, the density of the electric current generated by the force $\mathfrak{F}$ [is] c. The densities of the positive and negative ions are $N^+$ and $N^-$, their absolute mobilities $U$ and $V$, their diffusion coefficients $k^+$ and $k^-$. The coefficient of reunification [or recombination] is $\alpha$, the electrical elementary quantum is denoted by $\varepsilon$, the speed of light by $v$. We establish the normal at some point on the positive plate. We make its base point the zero point, its two, which crosses over to the negative plate, the positive direction of an axis $x$.

Then the following equations result:

$$\left\{\begin{aligned} \frac{d\mathfrak{F}}{dx} &= 4\pi\varepsilon(N^+ - N^-), \\ v\varepsilon\left(UN^+ + VN^-\right)\mathfrak{F} - \varepsilon\frac{d}{dx}\left(k^+N^+ - k^-N^-\right) &= c, \\ q = vU\frac{d(N^+\mathfrak{F})}{dx} - k^+\frac{d^2N^+}{dx^2} &+ \alpha N^+N^-, \\ q = -vV\frac{d(N^-\mathfrak{F})}{dx} - k^-\frac{d^2N^-}{dx^2} &+ \alpha N^+N^-. \end{aligned}\right. \tag{I}$$

We have four equations and we can consider the quantities $\mathfrak{F}$, $q$, $N^+$, $N^-$ as unknowns; it is then assumed that the current density c is determined by observation.

8.1.2. *The saturation current*

If one neglects the influence of reunification and diffusion, one obtains:[1]

$$\begin{cases} \dfrac{d\mathfrak{F}}{dx} = 4\pi\varepsilon(N^+ - N^-), \\ v\varepsilon(UN^+ + VN^-)\mathfrak{F} = c, \\ q = vU\dfrac{d(N^+\mathfrak{F})}{dx} = -vV\dfrac{d(N^-\mathfrak{F})}{dx}. \end{cases} \tag{II}$$

Solving the equations leads to the following formulas:

$$\begin{cases} \mathfrak{F}^2 = \mathfrak{F}_0^2 - \dfrac{8\pi c}{\nu V}x + \dfrac{4\pi(U+V)}{\nu UV}qx^2, \\ N^+\mathfrak{F} = \dfrac{q}{\nu U}x, \quad N^-\mathfrak{F} = \dfrac{c}{\nu\varepsilon V} - \dfrac{q}{\nu V}x, \\ c = \varepsilon ql. \end{cases} \tag{III}$$

For x = 0 it becomes:

$$\mathfrak{F} = \mathfrak{F}_0, \quad N_o^+ = 0, \quad N_o^-\mathfrak{F}_0 = \frac{c}{\nu\varepsilon V} = \frac{q}{\nu V}l.$$

For x = $l$:

$$N_\imath^+\mathfrak{F}_\imath = \frac{q}{\nu U}l, \quad N_\imath^-\mathfrak{F}_\imath = 0.$$

So it is:

$$\varepsilon vVN_0^-\mathfrak{F}_0 = \varepsilon vUN_l^+\mathfrak{F}_l = \mathfrak{c}.$$

Using these values, equations (III) can be reduced to the form:

$$\begin{cases} \mathfrak{F}^2 = \mathfrak{F}_0^2 - 8\pi\varepsilon N_0^-\mathfrak{F}_0 x + 4\pi\varepsilon\left(N_0^-\mathfrak{F}_0 + N_l^+\mathfrak{F}_l\right)\dfrac{x^2}{l}, \\ N^+\mathfrak{F} = N_l^+\mathfrak{F}_l\dfrac{x}{l}, N^-\mathfrak{F} = N_0^-\mathfrak{F}_0\left(1 - \dfrac{x}{l}\right), \\ c = \varepsilon ql. \end{cases} \tag{III$'$}$$

In order to get a more definite idea of the values of the coefficients appearing in these uations, it seems useful to put them as much as possible in numerical form.

We use values for this purpose:

$$U = 1,26 \times 10^{-8}, \quad V = 1,74 \times 10^{-8}, \quad \frac{V}{U} = 1,38.$$

$$\frac{k^+}{U} = \frac{k^-}{V} = 2,33 \times 10^{-6}.$$

$$k^+ = 0,0293, \quad k^- = 0,0405,$$

$$\varepsilon = 4,69 \times 10^{-10}, \quad a = 1,59 \times 10^{-6}.$$

[1]E. Riecke, Nacbr. d. K Gesellsch. d. Wissensch. zu Göttingen, math.-phys. Kl. Heft 1 u. 2. 1903; Ann. d. Phys. 12. P. 52, 1903.

Then it will be:

$$N_0^- \mathfrak{F}_0 = 0,00192 \times ql, \quad N_l^+ \mathfrak{F}_l = 0,00265 \times ql,$$
$$\mathfrak{F}^2 = \mathfrak{F}_0^2 - 0,226 \times 10^{-10} qlx + 0,268 \times 10^{-10} qx^2.$$

It follows from this that the variability of the field strength will in general only be very small. The theory of the saturation current is only applicable if

$$\alpha N^+ N^-, \quad k^+ \frac{d^2 N^+}{dx^2} \text{ and } k^- \frac{d^2 N^-}{dx^2}$$

are very small with respect to $q$.

Now is:

$$\alpha N^+ N^- \mathfrak{F}^2 = \alpha N_l^+ \mathfrak{F}_l N_0^- \mathfrak{F}_0 \frac{x(l-x)}{l^2},$$

thus using the given values:

$$\alpha N^+ N^- = 8,10 \times 10^{-12} \frac{q^2 x(l-x)}{\mathfrak{F}^2}.$$

It must therefore be small [relative to] $q$

$$\frac{\mathfrak{F}^2}{8,10 \times x(l-x)} \times 10^{12}.$$

Also:

$$k^+ N^+ = k^+ N_l^+ \mathfrak{F}_l \frac{x}{\mathfrak{F} l},$$
$$\frac{1}{\mathfrak{F}} = \frac{1}{\mathfrak{F}_0} + 0,113 \times 10^{-10} \frac{qlx}{\mathfrak{F}_0^2} - 0,134 \times 10^{-10} \frac{qx^2}{\mathfrak{F}_0^3}.$$

Consequently:

$$k^+ \frac{d^2 N^+}{dx^2} = \frac{q^2}{\mathfrak{F}_0^3}(1,75l - 6,21x)10^{-15}.$$

The theory of the saturation current assumes that $q$ is small compared to:

$$\frac{\mathfrak{F}_0^3}{1,75l - 6,21x} \times 10^{15}.$$

From the condition that $q$ should be large compared to $k^- \mathrm{d}^2 N^- / \mathrm{d}x^2$ it also follows:

$q$ small compared to

$$\frac{\mathfrak{F}_0^3}{6,21x - 3,78l} \times 10^{15}.$$

## 8.2. Unsaturated current, first approximation

From the quantities that we calculated above according to the theory of the saturation current, we distinguish those that apply in a first approximation for the unsaturated current by means of the index 1. The general equations of the problem then come in the following form:

$$\left\{\begin{aligned} \frac{d\mathfrak{F}_1}{dx} &= 4\pi\varepsilon\left(N_1^+ - N_1^-\right), \\ \upsilon\varepsilon\left(UN_1^+ + VN_1^-\right)\mathfrak{F}_1 &= c + \varepsilon\frac{d}{dx}\left(k^+N^+ - k^-N^-\right), \\ q_1 &= \upsilon U\frac{d\left(N_1^+\mathfrak{F}_1\right)}{dx} - k^+\frac{d^2N^+}{dx^2} + \alpha N^+N^-, \\ q_1 &= -\upsilon V\frac{d\left(N_1^-\mathfrak{F}_1\right)}{dx} - k^-\frac{d^2N^-}{dx^2} + \alpha N^+N^-. \end{aligned}\right. \tag{IV}$$

Solving the equations leads to the following approaches:

$$\begin{aligned} \mathfrak{F}_1^2 = \mathfrak{F}_0^2 + c_1x + &\frac{4\pi\varepsilon(U+V)}{\nu UV}q_1x^2 + 8\pi\varepsilon\frac{k^+}{\nu U}\left(N^+ + N^- - N_0^-\right) \\ &-\frac{8\pi\varepsilon(U+V)}{\nu UV}\alpha\int_0^x\int_0^x N^+N^-dxdx. \\ N_1^+\mathfrak{F}_1 = &\frac{\mathrm{c}}{\nu\varepsilon(U+V)} + \frac{q_1}{\nu U}x + \frac{k^+}{\nu U}\frac{dN^+}{dx} - \frac{\alpha}{\nu U}\int_0^\infty N^+N^-dx \\ &+\frac{V}{8\pi\varepsilon(U+V)}c_1, \\ N_1^-\mathfrak{F}_1 = &\frac{\mathrm{c}}{\nu\varepsilon(U+V)} - \frac{q_1}{\nu V}x + \frac{k^-}{\nu V}\frac{dN^-}{dx} + \frac{\alpha}{\nu V}\int_0^\infty N^+N^-dx \\ &-\frac{U}{8\pi\varepsilon(U+V)}c_1. \end{aligned}$$

The integration constant $c_1$ is determined by the conditions that c must [originate] on the electrode plates.

$$\mathrm{c} = \varepsilon\nu V\left(\mathfrak{F}_1N_1^-\right)_0 + \varepsilon k^-\left(\frac{dN^-}{dx}\right)_0 = \varepsilon\upsilon U\left(\mathfrak{F}_1N_1^+\right)_1 - \varepsilon k^+\left(\frac{dN_1^+}{dx}\right)_1$$

One finds:

$$c_1 = -\frac{8\pi c}{\nu V};$$

the solution of equations (IV) is then given by the following formulas:

$$\begin{cases} q_1 = \frac{c}{\varepsilon l} + \frac{\alpha}{l}\int_0^l N^+N^- dx = q + \frac{\alpha}{l}\int_0^l N^+N^- dx. \\ \mathfrak{F}_1^2 = \mathfrak{F}^2 + \frac{4\pi\epsilon(U+V)}{\nu UV}\alpha\left\{\frac{x^2}{l}\int_0^l N^+N^- dx - 2\int_0^x\int_0^x N^+N^- dxdx\right\} \\ \qquad + 8\pi\varepsilon\frac{k^+}{\nu U}\left(N^+ + N^- - N_0^-\right), \\ N_1^+\mathfrak{F}_1 = N^+\mathfrak{F} + \frac{\alpha}{\nu U}\left\{\frac{x}{l}\int_0^l N^+N^- dx - \int_0^x N^+N^- dx\right\} + \frac{k^+}{\nu U}\frac{dN^+}{dx}, \\ N_1^-\mathfrak{F}_1 = N^-\mathfrak{F} + \frac{\alpha}{\nu V}\left\{\frac{l-x}{l}\int_0^l N^+N^- dx - \int_x^l N^+N^- dx\right\} - \frac{k^-}{\nu V}\frac{dN^-}{dx}. \end{cases} \quad \text{(V)}$$

Using the numerical values we get:

$$\mathfrak{F}_1^2 = \mathfrak{F}^2 + 36,3\times 10^{-24}\frac{q^2}{\mathfrak{F}_0^2}(l-x)^2x^2 + 0,67\times 10^{-15}\times\frac{qx}{\mathfrak{F}_0},$$

$$N_1^+\mathfrak{F}_1 = \overset{+}{N}\mathfrak{F} + 3,56\times 10^{-15}\frac{q^2}{\mathfrak{F}_0^2}(l-2x)(l-x)x + 0,228\times 10^{-6}\frac{q}{\mathfrak{F}_0},$$

$$N_1^-\mathfrak{F}_1 = N^-\mathfrak{F} - 2,55\times 10^{-15}\frac{q^2}{\mathfrak{F}_0^2}(l-2x)(l-x)x + 0,149\times 10^{-6}\frac{q}{\mathfrak{F}_0}.$$

### 8.2.1. *Unsaturated current, second approximation*

In the formulas (I), instead of $N^+$and $N$ in the terms multiplied by $\alpha$ and $k$, we put the previously calculated quantities ${N_1}^+$ and ${N_1}^-$. We then get the formulas to calculate the second approximation:

$$\frac{d\mathfrak{F}_2}{dx} = 4\pi\varepsilon\left(N_2^+ - N_2^-\right),$$

$$v\varepsilon\left(UN_2^+ + VN_2^-\right)\mathfrak{F}_2 = c + \varepsilon\frac{d}{dx}\left(k^+N_1^+ - k^-N_1^+,\right.$$

$$q_2 = vU\frac{d\left(N_2^+\mathfrak{F}_2\right)}{dx} - k^+\frac{d^2N_1^+}{dx^2} + \alpha\overset{+}{N}_1 N_1^-,$$

$$q_2 = -vV\frac{d\left(N_2^-\mathfrak{F}_2\right)}{dx} - k^-\frac{d^2N_1^-}{dx^2} + \alpha N_1^-N_1^-.$$

The solution of the equations gives:

$$\begin{cases} q_2 = q + \frac{\alpha}{l}\int_0^z N_1^+ N_1^- dx \\ \mathfrak{F}_2^2 = \mathfrak{F}_0^2 - \frac{8\pi c}{\nu V}x + \frac{4\pi\varepsilon(U+V)}{\nu UV}q_2 x^2 \\ \quad + 8\pi\varepsilon\frac{k^+}{\nu U}\left(N_1^+ + N_1^- - N_{10}^+ - N_{10}^-\right) \\ \quad - \frac{8\pi\varepsilon(U+V)}{\nu UV}\alpha\int_0^x\int_0^x N_1^+ N_1^- dx dx \\ N_2^+\mathfrak{F}_2 = \frac{q_2}{\upsilon U}x + \frac{k^+}{\nu U}\frac{dN_1^+}{dx} - \frac{\alpha}{\upsilon U}\int_0^x N_1^+ N_1^- dx. \\ N_2^-\mathfrak{F}_2 = \frac{q_2}{\upsilon V}(l-x) - \frac{k^-}{\nu V}\frac{dN_1^-}{dx} - \frac{\alpha}{\upsilon V}\int_x^l N_1^+ N_1^- dx. \end{cases} \tag{VI}$$

If in these formulas, instead of $q_2, N_1^+$ and $N_1^-$, one substitutes the values that follow from the previous one, the result is complicated expressions. We shall only elaborate a little further on the first of the equations, since it is linked to the possibility of determining the value of the constant $\alpha$ of the reunification from observations of unsaturated currents.

To do this, we bring the equation to the form:

$$\varepsilon l_{q_2} - c = \alpha\varepsilon\int_0^l N_1^+ N_1^- dx.$$

Let's use the abbreviation:

$$\frac{x}{l}\int_0^l N^+N^- dx - \int_x^l N^+N^- dx = \mathfrak{X}_1, \quad \frac{l-x}{l}\int_0^l N^+N^- dx - \int_x^l N^+N^- dx = \mathfrak{X}_2,$$
$$\frac{x^2}{l}\int_0^l N^+N^- dx - 2\int_0^x\int_0^x N^+N^- dx dx = \Xi$$

so [this] becomes:

$$N_1^+\mathfrak{F}_1 = N^+\mathfrak{F} + \frac{\alpha}{\nu U}\mathfrak{X}_1 - \frac{k^+}{\nu U}\cdot\frac{dN^+}{dx},$$
$$N_1^-\mathfrak{F}_1 = N^-\mathfrak{F} + \frac{\alpha}{\nu V}\mathfrak{X}_2 - \frac{k^-}{\nu U}\cdot\frac{dN^-}{dx},$$
$$\mathfrak{F}_1^2 = \mathfrak{F}^2 + \frac{4\pi\varepsilon(U+V)}{\nu UV}\alpha\Xi + 8\pi\varepsilon\frac{k^+}{\nu U}\left(N^+ + N^- - N_0^-\right).$$

If you use these values to calculate the integral $\int_0^l N_1^+ N_2^- \mathrm{d}x$, you get the equation for $\alpha$:

$$\varepsilon l q_2 - c = \alpha\varepsilon\left[\int_0^l N^+N^- dx + \frac{k^+}{\nu U}\left\{\int_0^l \frac{1}{\mathfrak{F}}\left(N^-\frac{dN^+}{dx} - N^+\frac{dN^-}{dx}\right)dx - 8\pi\varepsilon\int_0^1 \frac{N^+N^-\left(N^+ + N^- - N_0^-\right)}{\mathfrak{F}^2}dx\right\}\right] + \alpha^2\varepsilon\frac{1}{\nu UV}\left[\int_0^l \frac{VN^-\mathfrak{X}_1 + UN^+\mathfrak{X}_2}{\mathfrak{F}}dx - 4\pi\varepsilon\left(U+V\right)\int_0^l \frac{N^-N^+\Xi}{\mathfrak{F}^2}dx\right].$$

The numerical calculation gives:

$$\varepsilon l q_2 - \mathrm{c} = \alpha\varepsilon \times 0,85 \times 10^{-6}\frac{q^2l^3}{\mathfrak{F}_0^2}\left\{\begin{matrix} 1 + 4,65 \times 10^{-4}\dfrac{1}{\mathfrak{F}_0 l} \\ -3,34 \times 10^{-16}\dfrac{ql}{\mathfrak{F}_0^3}\end{matrix}\right\} + \alpha^2\varepsilon \times 0,143 \times 10^{-12}\frac{q^3l^5}{\mathfrak{F}_0^4}\left\{1 - 0,81 \times 10^{-12}\frac{ql^2}{\mathfrak{F}_0^2}\right\}.$$

You can roughly write:

$$\varepsilon l q_2 - \mathrm{c} = \alpha\varepsilon \times 0,85 \times 10^{-6}\frac{q^2l^3}{\mathfrak{F}_0^2} + \alpha^2\varepsilon \times 0,143 \times 10^{-12}\frac{q^3l^5}{\mathfrak{F}_0^4}.$$

Here is:

$$c = \varepsilon l q.$$

The equation therefore comes in the simpler form:

$$l\left(q_2 - q\right) = \alpha \times 0,85 \times 10^{-6} \times \frac{q^2l^3}{\mathfrak{F}_0^2} + \alpha^2 \times 0,143 \times 10^{-12} \times \frac{q^3l^5}{\mathfrak{F}_0^4}.$$

If one uses this equation to calculate $\alpha$, the result is:

$$\alpha = 1,18 \times 10^6 \times \frac{\mathfrak{F}_0^2\left(q_2 - q\right)}{l^2q^2}\left\{1 - 0,20\frac{q_2 - q}{q}\right\}.$$

So the value of $\alpha$ becomes smaller if the quadratic term is also taken into account.

If one introduces instead of $q_2$ and $q$ the dimensions $\mathfrak{C}$ of the saturation current and $c$ of the actually observed current, then:

$$\alpha = 5,52 \times 10^{-4} \times \frac{(\mathfrak{C} - c)\mathfrak{F}_0^2}{lc^2}\left(1 - 0,20\frac{\mathfrak{C} - c}{c}\right).$$

(Received 4 August 1903.)

Note: Partial translation prepared by HJ Whitehouse from original text and equations in German using Google Translate and reviewed by Melita Hein, San Diego.

## 9. Appendix B: LEC Current Behavior and Ion Distribution

### 9.1. Introduction

Much helpful information to understand the behavior of LEC cells can be learned from the work on the conduction of electricity through gases by the late Sir JJ Thomson and other researchers at the Cavendish Laboratory at Cambridge University near the turn of the 19th century [22], [23], [34]. However, the various different configurations and uses of $LEC$ cells result in fundamental differences in performance than those reported by early authors. One of their primary interests was to understand the processes that take place when gas is ionized. Thomson makes the following assumptions [34] using the CGS-ESU system of units (emphasis added):

"Let $n_1, n_2$ be respectively the number of positive and negative ions per unit volume at a place fixed by the coordinate $x$, let $q$ be the number of positive or negative ions produced in unit time per unit volume at this point by the ionizing agent; let $X$ be the electric intensity at this point, $k_1, k_2$ the velocities of the positive and negative ions under unit electric intensity, so that the velocities of those ions at this point are respectively $k_1X, k_2X$; let $e$ be the charge on an ion. The volume density of the electrification, supposed due entirely to the presence of the ions, is $(n_1 - n_2)\, e$; hence we have

$$\mathrm{d}X/\mathrm{d}x = 4\pi\,(n_1 - n_2)\, e \tag{1}$$

If $i$ is the current through unit area of the gas, **and if we neglect any motion of the ions except that caused by the electric field**, we have

$$n_1 e k_1 X + n_2 e k_2 X = i \tag{2}$$

Thomson continues by combining equations (1) and (2) and solving for $n_1\mathrm{e}$ and $n_2 e$. Then he observes that in a steady ionization state, ion losses by recombination must be balanced by ionization (emphasis added):

"We shall suppose that the number of positive or negative ions which recombine in unit volume in unit time is $\alpha n_1 n_2$ : this is the rate at which unit volume is losing positive and negative ions in consequence of recombination; in consequence of ionization it is gaining them at the rate $q$, and in consequence of the motion of the ions under the electric force it is losing positive ions at the rate $\mathrm{d}\,(n_1 k_1 X)\,/dx$ and negative ones at the rate $-\mathrm{d}\,(n_2 k_2 X)\,/\mathrm{d}x$. The diffusion of the ions causes unit volume to lose positive and negative ions at the rates $D_1\, \mathrm{d}^2\mathrm{n}_1/\mathrm{d}x^2, -D_2\, \mathrm{d}^2\mathrm{n}_2/\mathrm{d}x^2$, where $D_1$ and $D_2$ are the coefficients of diffusion of the positive and negative ions. **Unless the electric field is very weak the motion of the ions by diffusion is, except in quite exceptional cases, insignificant in comparison with that under the electric field.**"

The authors believe that it is the diffusion of the ions that is the primary source of the spontaneous 'open circuit' voltage of a LEC as well as the spontaneous 'short circuit' current measured at low values of load resistance and thus 'very weak' electric fields. Although diffusion is most evident for the spontaneous conduction mode of LEC operation it nevertheless still plays a role in determining the I-V characteristics of LECs with multiple counter electrodes (CE) as well as two-electrode LECs operated in the impressed voltage or current mode.

### 9.2. Current Saturation during Impressed Conduction

An important characteristic of the conduction of electricity through gases is the rate of gas ionization $q$. This is traditionally accomplished by measuring the current density, $i$, at voltages sufficiently high that the magnitude of the drift velocities, $v_\mathrm{d}$, of the ions, i.e., $k_1X, k_2X$ in CGS ESU or $\mu_1 E, \mu_2 E$ in CGSA units, are sufficiently high that the ions drift to the electrodes in a time short enough so that there is a minimum loss of ion charge by ion-ion recombination. JJ Thomson and E Rutherford [22] report that:

"for a given intensity of radiation the current through the gas does not exceed a certain maximum value whatever the electromotive force may be, the current gets, as it were, 'saturated.'"

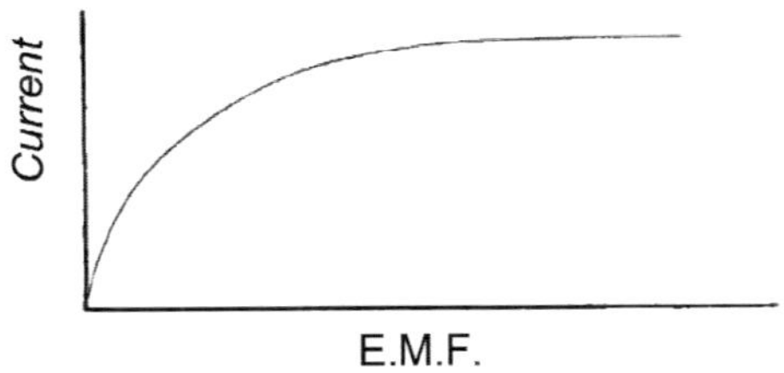


**Figure B.1.** Current saturation curve adapted from JJ Thomson [34].

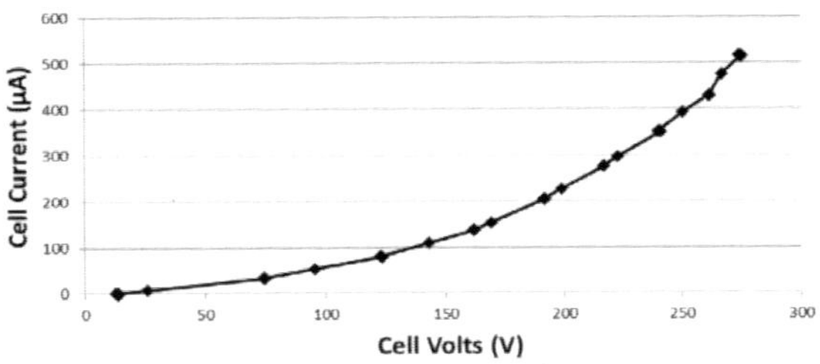


**Figure B.2.** Impressed current-voltage (I-V) curve for a Pd − D cell with 6 mm electrode separation.

An example current voltage (I-V) saturation curve similar to that in [22] is shown in Fig. B.1. This curve clearly shows an 'ohmic' behavior at low voltage, i.e., current goes to zero linearly as the electromotive force (EMF) goes to zero.

However, when a two-electrode LEC cell, i.e., one working electrode (WE) and one counter electrode (CE), is operated in the impressed current/voltage mode the measured I-V curve is remarkably different. Figure B.2 shows the measured (I-V) curve for a Pd-D LEC cell reported in [1],

Note, that the current instead of saturating as the voltage is increased, the current increases rapidly and shows no indication of saturation at a voltage across the cell of approximately 250 V for a cell with an electrode separation of approximately 6 mm. Analysis of the conduction data for this cell showed that the measured current increased as a cubic polynomial in cell voltage. This difference in performance is thought to be caused by the significant differences in the way that the gas is being ionized. When the cube root polynomial of the cell current is plotted versus the cell voltage the result is an approximately straight line that upon extrapolation to zero voltage does not pass through the origin of the graph. This is due to the presence of ion diffusion in addition to ion drift and ion-ion recombination. In fact, it was this observation that lead to the discovery of spontaneous LEC conduction that was reported in [1].

Only when the full set of equations that include the diffusion term are examined, such as those published by Eduard Riecke in 1903, is it seen that a LEC can have 'non-ohmic' behavior at low voltage. It is the diffusion term that introduces the 'non-ohmic' behavior, i.e., current does not necessarily go to zero when the electric field goes to zero. An English translation of Riecke's University of Göttingen is presented in Appendix A. This German language paper includes the complete set of 1-dimensional equations, including diffusion, for a gas composed of both positive and negative ions.

A potential explanation of the difference in the behavior of the ionized-gas conduction experiments by early researchers at the Cavendish Laboratory and experiments performed by the authors is in the structure and preparation of the working electrode (WE) of the LEC. In all Inovl experiments the WE is comprised in part of hydrogen-host-

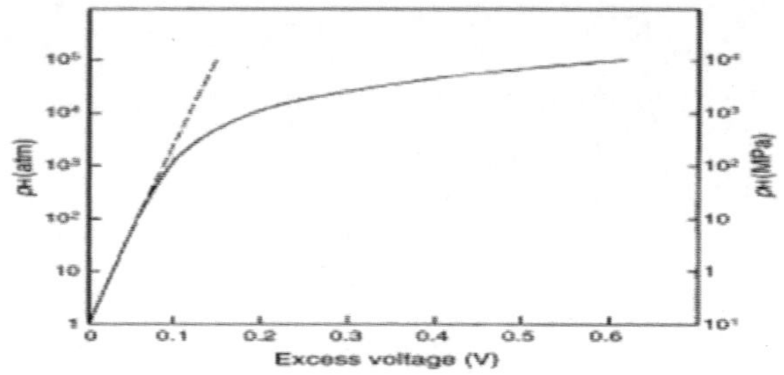


**Figure B.3.** Equivalent gas pressure versus excess voltage during electrolysis.

material (HHM), i.e., black Pd or black Fe, occluded with atomic hydrogen during electrochemical HHM codeposition performed in light water $(H_2O)$. In addition to occluding hydrogen in the HHM during electrodeposition, Fukai [2] also has shown that a large number of super-abundant vacancies (SAV) are introduced in the electrodeposited metal. In most 2-electrode impressed conduction experiments the gas in fluidic contact with both the WE and the counter electrode (CE) is hydrogen $\left({}^1H_2\right)$ although for the experiment described in Fig. B.2. the gas was deuterium $\left({}^2H_2\right)$. Since ionized gas acts as electrolyte during conduction there is a potential difference developed in the gas which increases as the voltage across the cell increases. This causes an increase in the excess or 'over-potential' near the surface of the WE as a function of the voltage across the cell. Fukai [43] has shown that a small increase of 'over-potential' causes a large equivalent gas pressure, e.g., several thousand atm, due to fugacity at the cathode during aqueous electrodeposition or electrolysis. Figure B.3 is a plot of equivalent gas pressure versus excess or over-potential from Fukai [43]. Thus, the presence of numerous SAV that can receive atomic hydrogen and the large equivalent hydrogen gas pressure to fill the SAV may account to the remarkably different I-V characteristics, e.g., convex for Inovl versus concave for Cavendish, as well as the lack of saturation for the Inovl impressed conduction experiments.

## 9.3. Current Saturation during Spontaneous Conduction

When a 2-electrode LEC is operating in the spontaneous voltage mode then diffusion and ion-ion recombination are the dominant conduction controlling factors rather than excess or 'overpotential' since the voltage across the cell and thus the electric field strength within the gas are low relative to their values during impressed current/voltage LEC operation. Figures B.4 and B.5 show the I-V (Note: V is not shown) curve for a freshly prepared Fe-H LEC operating in air at NTP. In addition to the I-V curve calculated shunt current and power in the load resistance also are shown.

## 9.4. Ion Distribution Equations

Consider first analyzing the performance of a LEC under the condition of an impressed voltage. To simplify the analysis, assume that the electrical properties of the electrodes are the same, i.e., their work functions are the same, Also assume that the current density through the gas, $i$, is the same everywhere over the surface of the electrodes, that the density of ions, $n^+$and $n^-$, only depends on the electrode separation variable, $x$ or $r$, that the positive and negative ion mobilities are effective measured values, i.e., the weighted harmonic means of the different ion-type mobilities, and that the electric field can be estimated everywhere throughout the gas. Additionally, following the practice of early authors, let the equations be written in CGS-ESU, i.e., current density is in statAmps, voltage is in statVolts, the unit charge, $e$, is approximately $4.8032 \times 10^{-10}$ statCoulombs or esu, and the ions carry only one charge per ion.

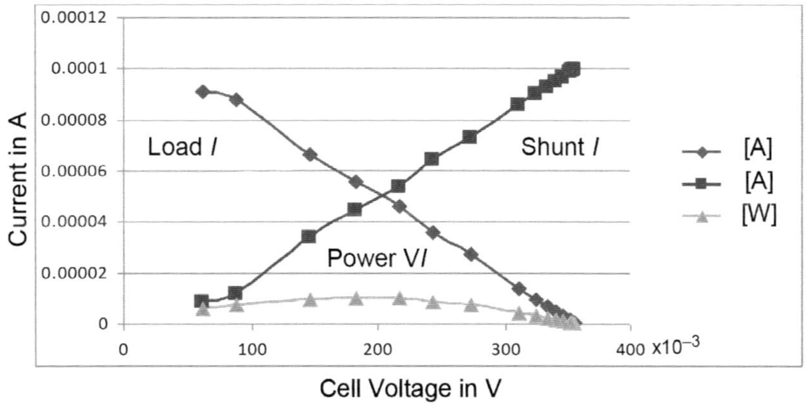


**Figure B.4.** Fe-H $1\frac{1}{4}$ "black Fe WE and 3/4" Cu CE using Air at NTP.

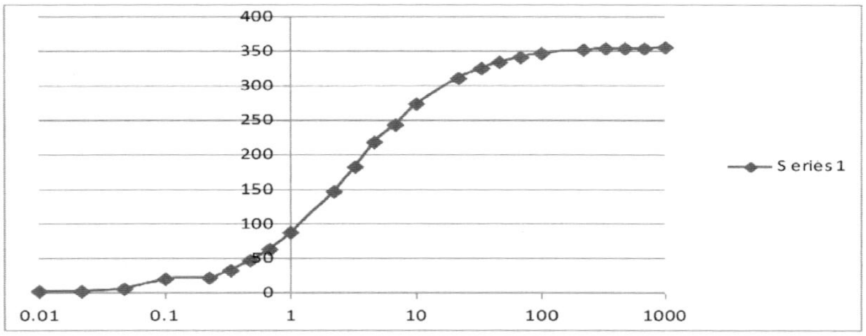


**Figure B.5.** Cell voltage (mV) versus load resistance in k$\Omega$ for Fe-H cell using air.

Making these assumptions, the general equations describing the conduction can be written, following Riecke but using modern notation, as:

$$\mathrm{d}E/\mathrm{d}x = 4\pi e\left(n^+ - n^-\right), \tag{B1a}$$

$$e\left(\mu^+ n^+ + \mu^- n^-\right)E - ed\left(D^+ n^+ - D^- n^-\right)/\mathrm{d}x = i, \tag{B1b}$$

$$q = \mu^+ \mathrm{d}\left(n^+ E\right)/\mathrm{d}x - D^+ \mathrm{d}^2 n^+/\mathrm{d}x^2 + \alpha n^+ n^-, \tag{B1c}$$

$$q = -\mu^- \mathrm{d}\left(n^- E\right)/\mathrm{d}x - D^- \mathrm{d}^2 n^-/\mathrm{d}x^2 + \alpha n^+ n^-. \tag{B1d}$$

Using CGSA, MKSA or the SI system of units published in 1960 and used in [36], [37], and [38] equation (B1a) becomes equation (B1a′)

$$\mathrm{d}E/\mathrm{d}x = e(n_+ - n_-)/\varepsilon \tag{B1a′}$$

where $\varepsilon$ is the permittivity of the gas and the value of the elementary charge is $e := 1.602176634 \times 10^{-19}$C since 2019.

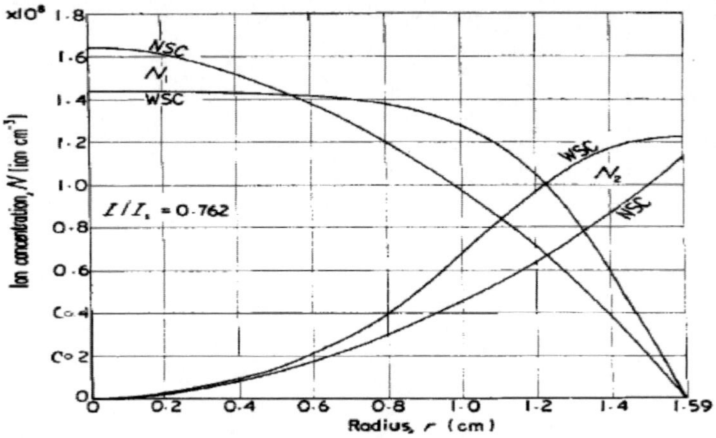


**Figure B.6.** Ion distribution in a cylindrical chamber with radii $a = 0.0159$ cm and $b = 1.59$ cm for air at 1 atm at a current saturation of 0.762 for no space-charge (NSC) and space charge (WSC) [46].

Further, it should be noted that the measured variables of a LEC are the total current, $I$, and the voltage, $V$, between the electrodes. These quantities are related to the corresponding variables, $i$, and $E$ as integrals, i.e., $I = \int_S i\,\mathrm{d}S$ and $V = \int E(x)\,\mathrm{d}x$ where d$S$ is an element of electrode surface area and $x$ is the variable representing the separation distance between the electrodes, *e.g.*, for coaxial cylindrical electrodes of radius $r_a \leq r \leq r_b$ then $V = \int_{ra}^{rb} E(r)\,\mathrm{d}r$. Also, it is important to observe that even if the electric field were identically zero everywhere within the gas, $E(x) = 0$, the current density $i$ would not be zero unless $d(D_+n_+ - D_-n_-)/\mathrm{d}x$ vanishes also. This means that the conductivity of the gas is not necessarily ohmic in the sense that $I \neq GV$ were $G$ is the conductance of the gas.

Darrow assumes that there is only one type of positive ion and one type of negative ion at that these ions are present in equal numbers [33]. Then equation (B1b) can be rewritten as

$$i/e = (\mu_+ + \mu_-)\,En + (D_- - D_+)\,\mathrm{d}n/\mathrm{d}x \tag{B2}$$

which has a solution of the form [42]:

$$n(x) = (1/u(x))\left[\int^x u\,(x')\,g\,(x')\,\mathrm{d}x' + \mathrm{C}\right], u(x) = \exp\left[\int^x p\,(x')\,\mathrm{d}x'\right] \tag{B3}$$

where $p(x) = (\mu_+ + \mu_-)\,E(\mathrm{x})/\,(D_- - D_+)\,, g(\mathrm{x}) = (i/e)/\,(D_- - D_+)\,, D_\pm = \mu_\pm k_{\mathrm{eV}}T$, $E(x)$ is the electric field within the gas, and C is a constant of integration.

The equations for the distribution of charge, neglecting diffusion but including recombination,have been developed by Townsend in 1915 for the case of plane-parallel electrode geometry on page 72 [11]. However,the equations are not evaluated to produce a graph of the ion charge-density as a function of position between the electrodes. R Rosen and EP George in 1975, also neglecting diffusion but including recombination, have evàluated the equations for the distribution of charge for both the cases of plane-parallel and cylindrical electrode geometry.Theirevaluation for cylindrical geometry is reproduced as Fig. B.6 for the case of electrode radii $a = 0.059$ cm and $b = 1.59$ cm and air at 1 atm.

## 10. Conclusions

For several reasons the conduction of electricity in a LEC is significantly different than that conventionally used to describe the conduction of electricity through gases [33], [44], [45]. First and foremost is the fact that the source of the conduction is from some form of radiation produced by or within the hydrogen occluded hydrogen-host-material (HHM) of the cell'sworking electrode (WE). Second, the equivalent gas pressure caused by 'over-potential' in thegas near the HHM of theworking electrode [43] appears to cause the conduction to not 'saturate' as the voltage across the cell increases. Third, diffusion of the ions plays a major role in the generation of the spontaneous voltage of a LEC when there is no electrical input to the LEC. Finally, the electrical work function of the material of the counter electrode (CE) also appears to play a role in the conduction of electricity in a LEC [41].

## References


[1] F. E. Gordon, H. J. Whitehouse, Lattice Energy Converter, JCMNS, (2021).

[2] Y. Fukai, The Metal-Hydrogen System, second edn, (Springer, 2005).

[3] H. Mehrer, Diffusion in Solids, (Springer, 2007).

[4] J. P. Biberian, IWAHLM presentation (2021). Also available from https://www.youtube.com/watch?v=fUsKv1af1DQ&t=2129s

[5] A. Smith, IWAHLM (a) paper and (b) presentation (2021). https://www.lenr-forum.com/attachment/18421-assisi-iwahlm-2021-prsentation-pdf/, https://www.lenr-forum.com/attachment/18471-iwahlm-assisi-2021-latest-pdf/

[6] A. Erickson, private communication (2020).

[7] J. Stevenson, Replication Report (2021). https://www.lenr-forum.com/attachment/17858-lec-replication-report-pdf/

[8] F. David, IWAHLM presentation (2021). Also available from https://www.researchgate.net/publication/354191227_DIRECT_CONVERSION_REPLICATIONS_F_David_14th_IWAHLM_29_august-1_september_2021_Assisi_Italy

[9] T. K. Keene et al., US 9,472,812 B2, (18 Oct. 2016).

[10] BioSearch, A Study in the Direct Creation of Electricity from the Interaction of Palladium with Hydrogen/Deuterium Gas, BioSearch, Sarasota, FL USA, 2020.

[11] J. S. E. Townsend, Electricity in Gases, (Oxford at the Clarendon Press, 1915).

[12] E. Riecke, Über näherungsweise gesättigte Ströme zwischen plan-parallelen Platten, [About approximately saturated currents between plane-parallel plates], (a) Nachrichten von der Gesellschaft Wissenschaften zu Göttingen [News from the Society of Sciences in Göttingen] (1903). Available at https://gdz.sub.uni-goettingen.de/id/PPN252457811_1903 (b) *Ann. d. Phys.* (1903).

[13] S. Szpak et al., Evidence of Nuclear Reactions in the Pd Lattice, Naturwissenschaften 00 (Short Communication) (2005) 1-4.

[14] M. R. Staker, A model and simulation of lattice vibrations in a superabundant vacancy phase of palladium–deuterium, Modelling Simul. Mater. Sci. Eng. 28(**??**) (2020). Also available from https://doi.org/10.1088/1361-651X/ab9994

[15] (a) H. S. Bhat, E. Afshari. Nonlinear constructive interference in electrical lattices, Phys. Rev. E. 77 (2008). (b) P.G. Kevrekidis, Nonlinear Waves in Lattices: Past, Present, Future, IMA Journal of Applied Mathematics (2010) Also available from https://arxiv.org/abs/1009.3178

[16] S. Szpak, et al., On the behavior of the cathodically polarized Pd-D system: Search for emanating radiation, Physics Letters A 210 (1996) 382-390.

[17] R. K. Rout, et al., Reproducible, anomalous emissions from palladium deuteride/hydride. Fusion Technol., 30 (1996) 273-. Also available from http://www.lenr-canr.org/acrobat/RoutRKreproducib.pdf

[18] F. David, J. Giles, Possible production of atomic deuterium by palladium cathode, Materials Research Innovations 2008 VOL 12 NO 4 p172 Also available from URL https://www.researchgate.net/publication/249844377

[19] F. David, J. Giles, (a) Alternatives to Calorimetry, J. Condensed Matter Nucl. Sci. 31 (2020) 53–61, (b) F. David, J. Giles, Self-Polarisation of Fusion Diodes: From Excess Energy to Energy ICCF-14 2008. Also available from https://www.researchgate.net/publication/265312308_Self-Polarisation_of_Fusion_Diodes_From_Excess_Energy_to_Energy

[20] F. Gordon, H. Whitehouse, Gaseous-Phase Ionizing Radiation Generator. US 10,841,989 B2 (17 November 2020).

[21] F. Gordon, H. Whitehouse, Lattice Energy Conversion Device, patent pending

[22] J. J. Thomson, E. Rutherford, XL. On the Passage of Electricity through Gases exposed to Röntgen Rays, Phil. Mag. S5 42 (1896) 392-407.

[23] J.J. Thomson, XIX. On the theory of the conduction of electricity through gases by charged ions, Phil. Mag. S5 47:286 (1899) 253-268. Also available from https://zenodo.org/record/2226317#.YFFzOriZegc

[24] B. B. Rossi, H. H. Staub, Ionization Chambers and Counters, (McGraw-Hill Book Co., 1949).

[25] N. N. Das Gupta, S. K. Ghosh. A Report on the Wilson Cloud Chamber and its Applications in Physics". Reviews of Modern Physics. 18(2): 225–365.

[26] Anonymous, Periodic Trends- Ionization Energy. (2021, June 19). Available from https://chem.libretexts.org/@go/page/53708

[27] H. Bichsel *et al.*, W Values for Gases: Experimental Data and Suggested Values, *J. Int. Commission on Radiation Units and Measurements*, **os16**(2), (1979), 18–32.

[28] J. P. Biberian, Fusion in All Its Forms, Cold Fusion, ITER, Alchemy, Biological Transmutations, in French (2012), English translation (Infinite Energy Press, 2015).

[29] T. Mizuno, Nuclear Transmutation: The Reality of Cold Fusion, in Japanese 1997, English translation, (Infinite Energy Press, 1998).

[30] M. A. Prelas et al., A Review of Nuclear Batteries, Progress in Nuclear Energy, 75, (2014) 117–148. Also available from https://www.researchgate.net/publication/262341758

[31] M. G. Spencer, High power direct energy conversion by nuclear batteries, Appl. Phys. Rev. 6, (2019).

[32] J..J. Thomson, The Discharge of Electricity Through Gases. (Archibald Constable & Co., Westminster, UK, 1898), reprinted (Charles Scribner's Sons, New York, 1903)

[33] K. K. Darrow, Electrical Phenomena in Gases, (The Williams & Wilkins Co. Baltimore, MD, 1932).

[34] J. J. Thomson, Conduction of Electricity Through Gases, (Cambridge at the University Press, 1906). Also available from https://archive.org/details/conductionofele00thomuoft/mode/2up

[35] P. A. Tate, P. C. East, The Saturation Curve in Ionization Chambers, Report No. 355, Defence Research Chemical Laboratories, (1961) Ottawa Canada.

[36] M.C.C. Lacdan, Development of Gerdien condenser for Atmospheric Pressure Plasmas, Ph.D thesis, Doshisha University, Kyoto, Japan, 2017.

[37] K. L. Aplin, Instrumentation for Atmospheric Ion Measurements, Ph.D thesis, University of Reading, Reading, UK, 2000.

[38] D Novkovic, Numerical solutions of differential equations of a cylindrical ionization chamber, Phys. Med. Biol. 41 (1996) 725–741.

[39] Radio Corporation of America, The Direct Conversion of Radiation into Electrical Energy, RB 3 1955.Available at http://www.one-electron.com/Archives/RCA/RCA-RB/RCA-RB.html

[40] J. V. Kramer, A New Electronic Battery, The Electrician, 93 (1924) 411.

[41] P. F. Ohmart, A Method of Producing an Electric Current From Radioactivity, J. Appl. Phys. 22 (1951) 1504–.

[42] W. E. Boyce, R. C. DiPrima, Elementary Differential Equations and Boundary Value Problems $3^{rd}$ edn. (1977) John Wiley and Sons.

[43] Y. Fukai, The structure and phase diagram of M–H systems at high chemical potentials—High pressure and electrochemical synthesis, Journal of Alloys and Compounds 404–406 (2005) 7–15.

[44] J.J. Thomson, G.P. Thomson, (a)Conduction of Electricity Through Gases, $3^{rd}$ edn, vol I, (Cambridge University Press, 1928), (b) Conduction of Electricity Through Gases, $3^{rd}$ edn, vol II, (Cambridge University Press, 1933).

[45] L. B. Loeb, (a) Fundamental Processes of Electrical Discharge in Gases, John Wiley and Sons (1939), (b) Basic Processes of Gaseous Electronics, (University of California Press, 1955).

[46] R. Rosen, E. P. George, Ion Distributions in Plane and Cylindrical Chambers, *Phys. Med.* Biol, 20(6) (1975) 990–1002.

J. Condensed Matter Nucl. Sci. 36 (2022) 25–29

Research Article

# Mitigating the Loss of Irreplaceable LENR Research Records

Thomas W. Grimshaw*

*Ph.D. LENRGY, LLC 3571 Far West Blvd, PMB 102, Austin, TX 78731, USA*

**Abstract**

An initiative is underway to document the records of cold fusion researchers while the information is still available. Now more than 30 years after the 1989 LENR announcement, many of these investigators are leaving the field. Projects under the LENR Research Documentation Initiative (LRDI) generally begin with a site visit where information is collected and interviews are conducted. Provision is made for securing the records, and a project report is prepared. More than 20 projects including over 25 participants and collaborators have been performed or are underway. The LRDI includes a major outreach effort to ensure participation and promote the overall standing of cold fusion. Future plans for the initiative call for additional participants and more in-depth coverage of research documented in previous projects. An integrated LRDI report including summaries of the individual projects is being prepared. Documentation of the large body of LENR research is essential both for the field and for humankind generally. Capturing the records enhances the prospects of the field by demonstrating the large amount of the information and the credibility of the researchers. The future of humanity depends on developing an inexhaustible supply of clean and inexpensive energy. LENR is one of the best candidates currently available to meet this need.


*Keywords:* LENR Research, Research Documentation, Cold Fusion Records

## 1. Introduction

When cold fusion was announced by Martin Fleischmann and Stanley Pons in 1989, it seemed possible that the needs of humankind for inexpensive, clean and inexhaustible energy would be met for the foreseeable future [1], [2]. However, for a variety of reasons related to politics in society and within science, cold fusion (now often referred to as low energy nuclear reactions, LENR) was rejected by mainstream science. Consequently, the potential energy benefits of LENR were not rigorously pursued as a top priority. Nevertheless, a significant number of reputable scientists worldwide continued research in the phenomenon at the margins of mainstream science. Cold fusion did not die out like other rejected scientific claims. The principal issues of the field continue to be adequate reproducibility and an acceptable explanation, which is not uncommon for scientific discoveries.

Many of the LENR researchers began their work in the early months and years after the 1989 announcement. Because of its pariah status in the scientific world, few new investigators were attracted to the field. The determined

*Corresponding author: thomaswgrimshaw@gmail.com; Ph: 512.784.1078.

© 2022 ICCF. All rights reserved. ISSN 2227-3123

researchers who continued have developed a large body of evidence for the reality of the phenomenon so that LENR may yet fulfill its promise as a major new source of energy. But now more than 30 years after the announcement, many of these investigators are leaving the field because of retirement, health issues and death.

The loss of their research records would be a tragedy not only for the LENR field, but also for all of humankind, given the continued promise of LENR as an energy source. The LENR Research Documentation Initiative (LRDI) is underway at LENRGY, LCC [3] to help mitigate the loss of these records. Its objectives are to preserve the records while they are still available and protect the information for future analysis as more understanding of LENR is gained.

## 2. Methods

The procedure of the LRDI [4] is straightforward (Figure 1). The collected records are supplemented with recorded interviews, and a research timeline is prepared. Arrangements are made for records preservation, and draft and final reports are prepared. Opportunities for additional work are normally included in the report and are pursued in a subsequent phase in many cases. In most projects, one or more site visits are made to gather information and conduct the recorded interviews. As progress is made, the components of the record are described in a series of memos. The collection of memos then serves as the basis for the project report.

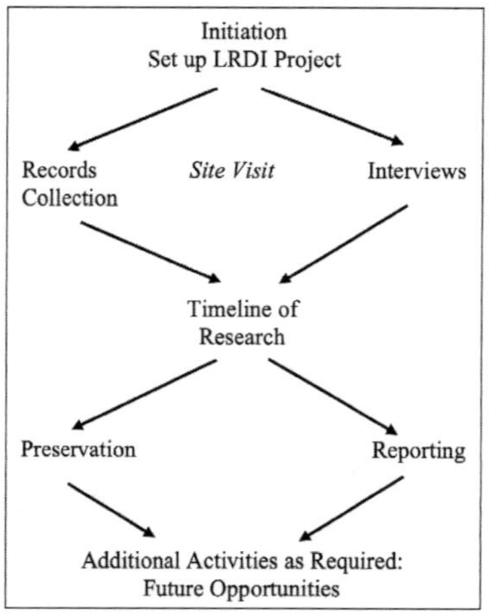


**Figure 1.** LRDI Procedure.

## 3. Projects and Participants

The LRDI has been privileged to engage some of the longest-standing and most prominent investigators in the LENR field. More than 20 projects are underway or have been performed under the umbrella of the LRDI [5], beginning with a pilot project with Dr. Edmund Storms [6]. The projects have included more than 25 participants and collaborators. They are listed in Table 1 generally in order of participation in the LRDI.

The methods of LRDI projects described above were mostly developed in a pilot project with Dr. Ed Storms and have been refined in subsequent projects. Given the thoroughness of his investigations and his comprehensive records, Dr. Storms proved to be a nearly ideal candidate for the pilot project. He is retired from Los Alamos National

**Table 1.** LRDI Participants and Collaborators

| | | |
|---|---|---|
| Storms, E. | Miles, M. | Beaudette, C. |
| Claytor, T. | Miley, G. | Mossier-Boss, P. |
| Fowler, M. | Tanzella, F. | Gordon, F. |
| Pease, D. | Imam, A. | Carat, R. |
| El-Boher, A. | Little, S. | Kowalski, L. |
| Hubler, G. | Little, S. | Lomax, A. |
| Nagel, D. | Passell, T. | Biberian, J, |
| Srinivasan, M. | Forsley, L. | Fox, H. |
| Letts, D. | Gluck, P. | University of Utah Marriott Library |
| | | SRI International |

Laboratory and has been a foremost LENR researcher going back to the 1989 announcement. He has written two books on the subject [7], [8]. He received the Preparata Medal, one of the most prestigious in the field, in 2005. Eight phases of his cold fusion work were identified in the Storms LENR Research Documentation Project. The results of the pilot project were presented as a poster at ICCF-21 in 2018 [9], and other LRDI projects followed soon afterward.

## 4. Project Components

As might be expected, the types and amounts of available research records are unique for each LRDI participant. In many cases, a LENR laboratory is available and is documented, including the experimental methods and equipment utilized. Typical project components are papers, presentations and unpublished reports, lab notebooks, and other hardcopy and the electronic files. Most participants have a library of LENR-relevent books, magazines and papers, often including proceedings of the International Conferences on Cold Fusion (ICCFs), as well as recorded media, such as DVD tapes and CDs. As noted above, a memo is prepared describing each component.

## 5. Future Project Opportunities

In many LRDI projects, more information is identified than can be included in the first set of memos and the initial report. These opportunities are noted in the report. For some participants, more work – including more site visits – is performed, and additional versions of the report are prepared.

## 6. Outreach

Several measures have been taken to "get the word out" on the LRDI and its objectives and methods. After the Storms pilot project was presented as an ICCF-21 poster in 2018, a presentation on methods and participants was made at the MIT CF/LANR workshop in early 2019 [10]. A comprehensive article appeared in Infinite Energy magazine in 2020 [11]. LRDI updates have been provided at ICCF-22 in 2019 [12], ICCF-23 in 2021 [13] and ICCF-24 in 2022 [14]. A 2022 paper has also been published in the JCMNS [15] on the LENR career of Dr. Mahadeva Srinivasan, one of the earliest and most productive researchers in the field. That paper is derived from the Srinivasan LENR Research Documentation Project, for which a report [16] had been prepared in 2020, not long before his death.

## 7. Why Must the Loss of LENR Research Records Be Mitigated?

Documentation and preservation of LENR research is important both to the field specifically and to humankind more generally. The cumulative body of the LENR research records provides ample evidence of the reality of the phenomenon. This evidence in turn establishes the basis for financial support of the field for conducting experiments and developing an explanation. The records manifest the depth, breadth and quality of the work done in the field, which demonstrates the credibility of the participants.

The case for cold fusion is strengthened by recording as much of the evidence for its existence as possible. As more understanding of the phenomenon is gained, these previous records may be re-analyzed to help address the primary issues of insufficient reproducibility and inadequate explanation.

The benefit of lessons learned in the cold fusion case is another reason for careful documentation of past research. Given LENR's potential energy benefits, the obvious question of why it was rejected within a short time after it was announced demands better answers. Answers to this question may provide guidance on how to deal more constructively with revolutionary scientific claims in the future and protect the interests of humankind.

To the extent that it is able to displace current energy sources like fossil fuels and nuclear power, LENR will go far in ensuring the long-term habitability of the earth. The need of humankind for unlimited, cheap and clean sources of energy can hardly be overstated. Measures like documentation and preservation of the records will help LENR meet this urgent need and must be undertaken to enhance the long-term prospects of humanity.

The LENR research records must also be documented and preserved for proper recognition of the investigators who have pursued it in the face of rejection, and even ridicule, from the scientific establishment. Many of them have persisted despite damage to their professional careers as they pursued the phenomenon both for the advancement of science and for possible financial gain. Particularly if and when the potential cold fusion benefits are finally realized, these determined pioneers in this new scientific field will deserve the recognition and gratitude of all of humanity.

## 8. Where to from Here?

An integrated report that includes a chapter for each LRDI participant is underway. Provision is made for review of each chapter by its participant (if available) in order to meet confidentiality provisions of the initiative. At the same time, more participants may be added, and more work on previous projects may be undertaken as identified in the future opportunities section of the reports. Solutions are being evaluated for a central repository for the LRDI information to ensure both adequate security and easy access by the participants.

## Acknowledgments

Thanks go in particular to the LRDI participants for their willingness to make their records available and their efforts in support of their projects. Dr. Ed Storms has been not only helpful and cooperative in the pilot project, but also has been a long-standing mentor for Dr. Grimshaw's work in the LENR field. Dr. David Nagel has also been a strong supporter of the LRDI as well as a co-author of LENR papers, presentations and articles.

Gratitude is expressed to the Anthropocene Institute [17] for financial support for LRDI expenses. The LRDI is being performed by LENRGY, LLC, which is based in Austin Texas. Thanks are extended to the Energy Institute of The University of Texas at Austin for providing a "home" organization before then.

## References

[1] Lemonick, M., 1989. Fusion Illusion? Two Obscure Chemists Stir Up a Fascinating Controversy in the Lab, but New Tests Challenge Their Hopes of Creating Limitless Energy. Time Magazine, May 8.
[2] Begley, S., 1989. The Race for Fusion. Newsweek Magazine, May 8.
[3] LENRGY, LLC. lenrgyllc.com
[4] Grimshaw, T., 2019. Collection, Organization and Documentation of LENR Research Results: Guideline. Unpublished Report. January.
[5] Grimshaw, T., 2022. LENR Research Documentation Initiative: Status Report and Future Directions, Update 4. Unpublished Report. Update of February 2021 (Update 4) in Preparation.

[6] Grimshaw, T. and E. Storms, 2018, Documentation of Dr. Edmund Storms' 29 Years of Cold Fusion Research - Experiments, Explanations, and Related Scientific Contributions: Energy Institute, University of Texas, Unpublished Draft Summary Report.

[7] Storms, E.K., 2007. The Science of Low Energy Nuclear Reaction: A Comprehensive Compilation of Evidence and Explanations about Cold Fusion: Singapore, World Scientific Publishing.

[8] Storms, E.K., 2014. The Explanation of Low Energy Nuclear Reaction: an Examination of the Relationship Between Observation and Explanation. Concord, NH. In?nite Energy Press.

[9] Grimshaw, T., and E. Storms. 2018. Documentation of Dr. Ed Storms' 29-Year LENR Research Career. Poster for 21st International Conference on Cold Fusion (ICCF-21), Fort Collins, Colorado. June.

[10] Grimshaw, T., 2019. LENR Research Documentation Initiative: Objectives, Procedure, Participants, Future Opportunities. Presentation and Poster at CF/LANR Colloquium at MIT, Cambridge, MA. March. https://theworld.com/?mica/2019colloq.html

[11] Grimshaw, T., 2020. Documenting Cold Fusion Research: Preserving a Vital Asset for Humankind. Infinite Energy, Issue 150, March/April.

[12] Grimshaw, T., 2019. LENR Research Documentation Initiative: Progress in Methods and Participants. Presentation and Poster at 22nd International Conference on Cold Fusion (ICCF-22), Assisi, Italy. September. https://iscmns.org/iccf22/

[13] Grimshaw, T., 2021. Mitigating the Loss of Irreplaceable LENR Research Records: Initiative Update. Virtual Poster Presentation at 23rd International Conference on Cold Fusion (ICCF-23), Xiamen, China, June. http://ikkem.com/iccf-23.php

[14] Grimshaw, T., 2022. LENR Research Documentation: What Have We Learned So Far? Presentation and Poster at 24th International Conference on Cold Fusion (ICCF-24), Mountain View, California, July. https://www.iccf24.org/

[15] Grimshaw, T., 2022. Dr. Mahadeva Srinivasan: Research Pioneer in the LENR Field. Jour. Condensed Matter Nuclear Science. Vol. 35, p. 21-29.

[16] Srinivasan, M., and T. Grimshaw, 2020. Documentation of Dr. Mahadeva Srinivasan's Cold Fusion Research. LENRGY Unpublished Report. Volume 1, Third Draft Report, June. Volume 2, Collection of Papers and Related Documents, June 2018.

[17] Anthropocene Institute. https://www.anthropoceneinstitute.com/

J. Condensed Matter Nucl. Sci. 36 (2022) 30–37

Research Article

# A Possible Heuristic Explanation of Exotic Vacuum Objects (EVO's, Charge Clusters)

Graham K. Hubler*

*The Medical School, University of Missouri, Columbia, MO USA*

**Abstract**

In early 90's, Ken Shoulders was granted 5 patents on Exotic Vacuum Objects (EVO) claiming that they were **a new form of matter**. He produced many monographs about them and suggested they were the physics that explained cold fusion. In Ken Shoulders words, EVO's are, "Highly organized, **micron-sized clusters of electrons**, having soliton behavior, with electron populations density on the order of *Avogadro's number per* $cm^3$ (A typical $2\mu$m EVO has a population of $10^{11}$ to $10^{13}$ electrons). When interacted with solid material, these charge clusters perform a low-energy phase transformation type of atomic disruption that **liquefies the lattice** and propels the material to a **high velocity** without apparent signs of conventional heating. Using an ordinary thermal interpretation, a thermal gradient for bulk material greater than **26,000 degrees C** per micrometer would be required to achieve these effects". This paper presents lessons from thin film deposition methods like Vacuum Arc, Pulsed Electron beam, Pulsed Laser whose commonality with EVO generation is pulse energy impingement on a target. Rather than the hypothesis of a "new form of matter" as an explanation of EVO's, it is hypothesized that generation of a micro shaped-charge, in analogy with explosively formed shaped-charge munitions, can explain the characteristics of surfaces that were struck by EVO's. This hypothesis reproduces the effects that are underlined in the text above.


*Keywords:* Exotic vacuum objects, EVO, clusters of electrons, pulsed electron beams, vacuum arc

## 1. Introduction

In early 90's, Ken Shoulders was granted 5 patents on Exotic Vacuum Objects claiming that they were **a new form of matter**. He produced many monographs about them and suggested they were the physics that explained cold fusion [1].

In Ken Shoulders words, EVO's are, "Highly organized, **micron-sized clusters of electrons**, having soliton behavior, with electron population density on the order of *Avogadro's number per* $cm^3$. When interacted with solid material, these charge clusters perform a low-energy phase transformation type of atomic disruption that **liquefies the lattice** and propels the material to a **high velocity** without apparent signs of conventional heating." Ken Shoulders named at first these objects *Electron Validum* (EV) and later EVOs. Other researchers call the same phenomenon Condensed Plasmoids (CP) and make similar claims of their properties [2]–[5].

*Corresponding author: hublerg@umsystem.edu, 6635 Prestwick Drive, Highland, MD 20777

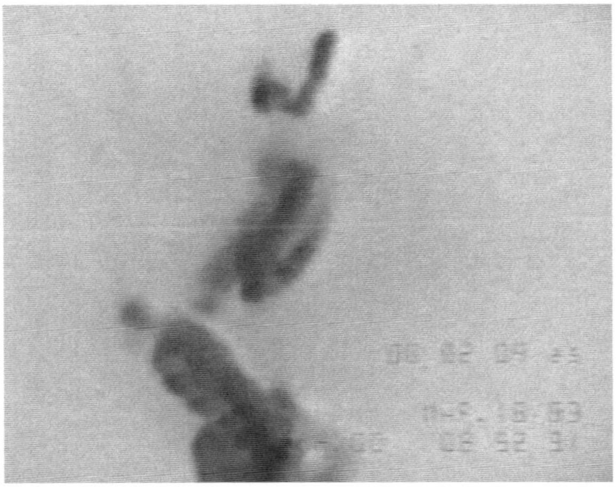

**Figure 1.** Pin hole high speed camera photo of an EVO in flight.

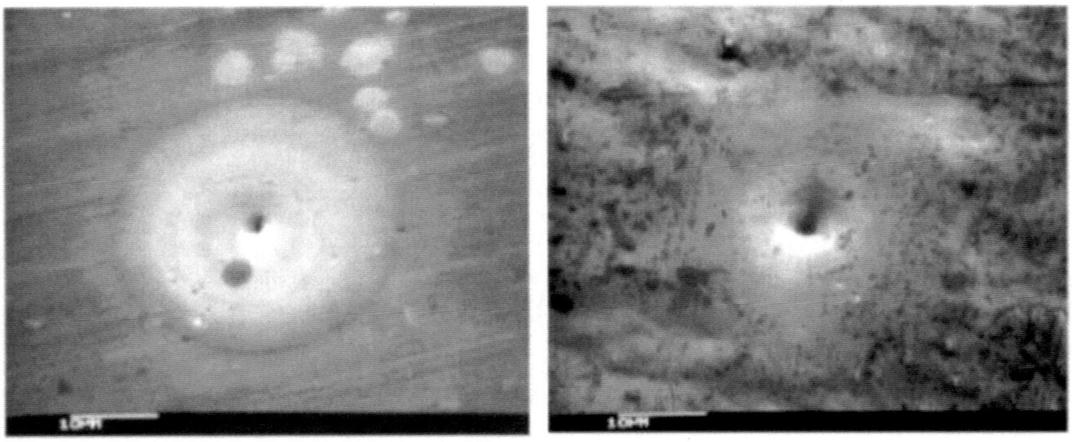

**Figure 2.** The entrance and exit sides of a 6 $\mu$m thick Al foil that was struck by an EVO leaving a $\sim$2 $\mu$m diameter hole [5].

Presented here are lessons from the thin film deposition methods Vacuum Arc, Pulsed Electron Beam, Pulsed Laser whose commonality with EVO generation is directed energy pulses impinging on a solid target. Rather than the hypothesis of a "new form of matter" as an explanation of EVO's, it is hypothesized that generation of a micro shaped-charge, in analogy with explosively formed shaped-charge munitions, can explain many of the characteristics of target surfaces and substrate surfaces that were struck by EVO's. This hypothesis reproduces the effects that are underlined in the text above.

## 2. Discussion

Figure 1 shows high speed photography of an EVO showing $\sim$2 $\mu$m diameter elongated features connected together like sausage links pointing in different directions [4], [5]. Figure 2 shows the entrance and exit sides of a 6 $\mu$m thick

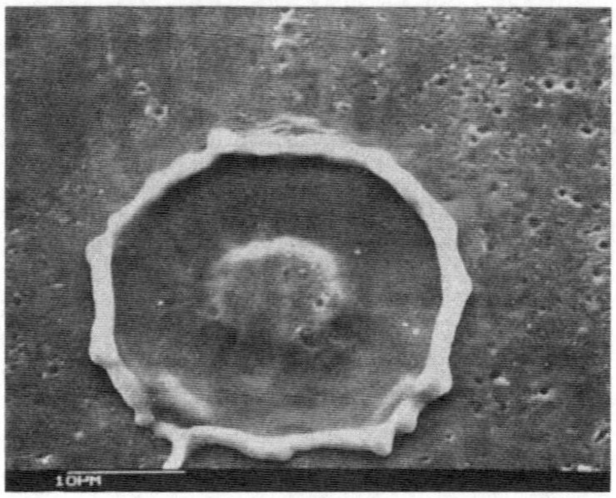


**Figure 3.** Circular feature left by charged cluster strike on a surface [1].

Al foil that was struck by an EVO leaving a $\sim$2 $\mu$m diameter hole [5]. An interpretation of Ken Shoulders is: "Using an ordinary thermal interpretation, a thermal gradient for bulk material greater than 26,000 degrees C per micrometer would be required to achieve these effects. This suggests that 26,000C would be the temperature required to melt a clean hole right through material with a melting point of 2,600 degrees C (i.e., SiC, not shown)".

Rather than postulate a new form of matter, let's first approach the problem using known physics from thin film deposition processes where pulsed lasers (PLD), pulsed electrons (PED) and vacuum arc (VAD) are employed to melt and/or vaporize a target and the subsequent vapor moves in a plume to coat a substrate. There is extensive literature on finite element and computational fluid dynamic simulations of these processes [6]–[12]. In PLD, typically 1 J/cm$^2$, 10-100 ns UV laser pulse atomizes a $\sim$50 $\mu$m diameter 0.2 $\mu$m deep slug of material [11], [12]. The ionized plasma expands into the space above the target by the high-pressure region expanding into the low-pressure region that reaches plume velocities of $\sim$1000 m/s in a highly directional expansion perpendicular to the target. PLD works well for opaque materials. In PED on an insulator [6, 7, 10], typically a 1 kA, 100 ns pulse of 15 keV electrons melts and vaporizes a $\sim$mm diameter, $\sim$2 $\mu$m deep slug of material that expands into space, greatly aided by Coulomb explosion due to the milli-Coulombs of charge deposited in the material, reaching plume velocities of up to $\sim$10,000 m/s [10]. In VAD [8], the process is similar to PED but with less control, and both PED and VAD work well on any material including insulators. In PED and VAD, the cathode is shaped to support a relatively large area impingement on the target in order to maximize the volume of material transferred to the target for overall efficiency of the deposition process. Figure 3 shows an unusual circular pattern left on a surface stuck by an EVO [4]. But Fig. 7 in Ref. [7] shows a similar feature on a Cu PED anode and Fig. 2 in Ref [9] also shows similar features for an insulating VAD anode. Other features of EVOs appear in numerous references regarding research and simulations of pulsed electron beam deposition [6], [7], [10].

Figure 4 is a schematic diagram of Ken Shoulders EVO generator [1]. A 10 to 20 KeV pulse of electrons is emitted by the sharp cathode at 12a and an EVO emerges and is caused to strike a secondary substrate at various angles or is photographed as in Fig. 1. This set-up to produce EVO's has several modifications making it distinct from the deposition processes. These are:

- The cathode is very sharp, making the electron beam small in diameter
- The cathode is placed very close to the target
- The target is a "high quality" insulator like quartz

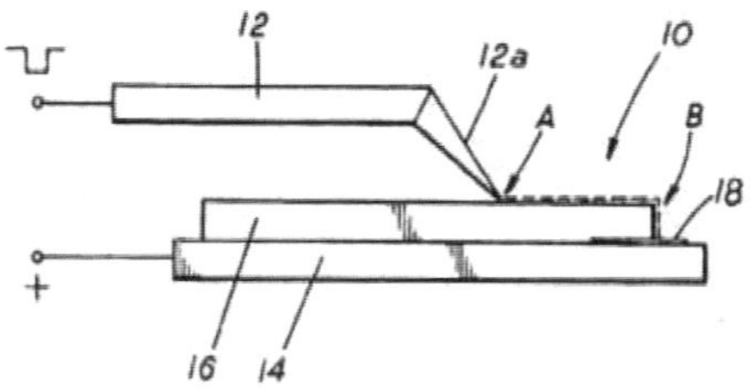


**Figure 4.** Schematic representation of EVO generator. A is a point cathode, 16 is thin insulating material, 14 is a conducting ground plane. A negative pulse is applied at A [1].

**Figure 5.** Lichtenberg Figure (Linac Tree, Wikipedia).

- The target is thin, and placed on a conducting ground plane
- The EVO's are observed in single pulse experiments versus high pulse rate for film deposition.

Figure 4 in Ref. [8] shows a series of snapshots of a VAD discharge on Cu from which plume velocity of 10,000 m/s can be estimated. From snapshots in Fig. 8 of Ref [10] for a PED discharge on an insulator, a plume velocity of 10,000 m/s can also be estimated.

To illustrate an effect of a mC of charge in a solid, Fig. 5 presents photos of a Linac Tree (Lichtenberg figure) [13]. The lightning-like pattern is produced by a milli-Coulomb dose of 60 MeV electrons incident across the entire face of a Lucite block. Putting your body near the block makes your hair stand on end. On the bottom of the block a grounded nail is struck with a hammer and the shock causes the trapped electrons to avalanche to the nail, damaging the Lucite block, leaving behind the beautiful Linac Tree. Question: What happens when the same milli-Coulomb of charge enters an insulator to a $10^{10}$ smaller volume of a $\sim$2 $\mu$m diameter, 2 $\mu$m thick cylinder of material?

A final example to flush out the landscape of the effects of directed energy impingement on solids, I describe the shaped-explosive charge anti-armor munition as it may be directly related to EVO's. Fig. 6 shows a schematic representation of a shaped charge munition [14]. The high explosive (HE) melts the copper liner beginning at the Vee. As the detonation progresses the copper liner is expelled as an elongated slug of melted copper with a velocity of 10,000 m/s. Figure 7 is high speed photography of typical stages of a shaped charge. As a frame of reference – explosive front velocity of C4 high explosive is 1,680 m/s. The appearance of the slug is similar to an EVO in flight. The Cu slug penetrates thick steel armor by ***plastic flow***, not melting. Figure 8 shows how a shaped charge penetrated

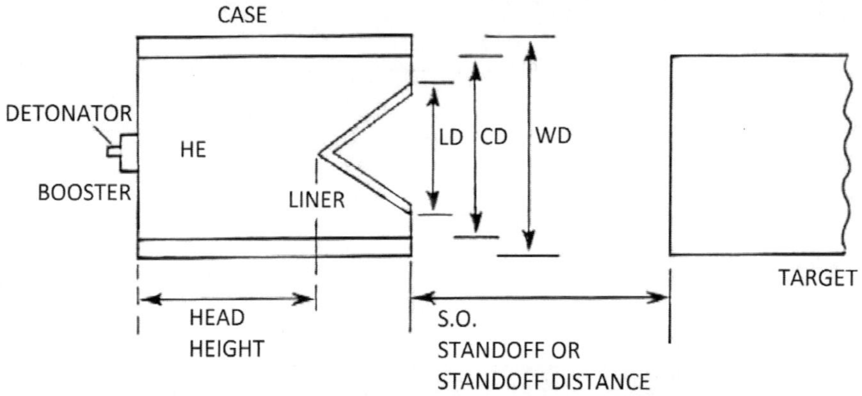


**Figure 6.** Schematic representation of a shaped-charge munition [13].

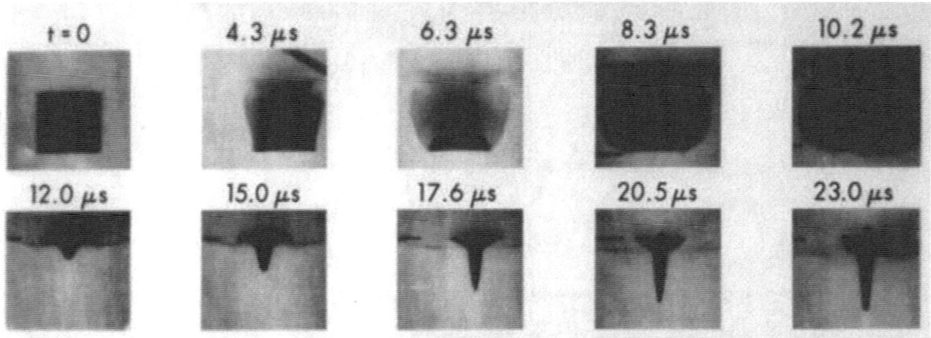


**Figure 7.** High speed photographs of detonation stages of a typical shaped charge [14].

a 10 cm thick armor steel plate. The pressure of the copper slug travelling at 10,000 m/s exceeds the Hugoniot elastic limit (HEL) where a solid flows like a liquid [15].

Now let's examine the EVO generator depicted in Fig. 4. The cathode is a sharp point placed close to the target so the area of impingement on the insulating target is small. Figure 9 shows a heuristic (common sense) version of what may occur when 15 kV, 100 ns, $\sim$mC pulse of electrons impinges on the target. In $\sim$100 ns, electrons stop in the insulator, depositing their several Joules of energy. In $<1$ $\mu$s, A slug of material melts that is $\sim$2 $\mu$m deep, $\sim$2 $\mu$m in diameter. The image charge induced in the nearby underlying ground plane aids in tight diameter control of the electrons as they penetrate over the pulse length time. This feature is not present in PED and VAD due to the thick insulating target used for deposition. The electrons spread laterally during the pulse to a greater extent than in the EVO target due to the repulsion of previously embedded charge. In $\sim$1 $\mu$s, the melted material is expelled from the surface by a combination of Coulomb explosion of trapped charge, and push from vaporized material.

Underneath the melted slug (due to energy loss of electrons, subsurface is hotter that the surface), and lateral pressure from the strain in unheated material surrounding the melt, the slug elongates, reaching a velocity of 10,000 m/s as it separates from the insulator, leaving a crater behind. As point of reference, a 5$\mu$m x 5$\mu$m x 5$\mu$m volume of

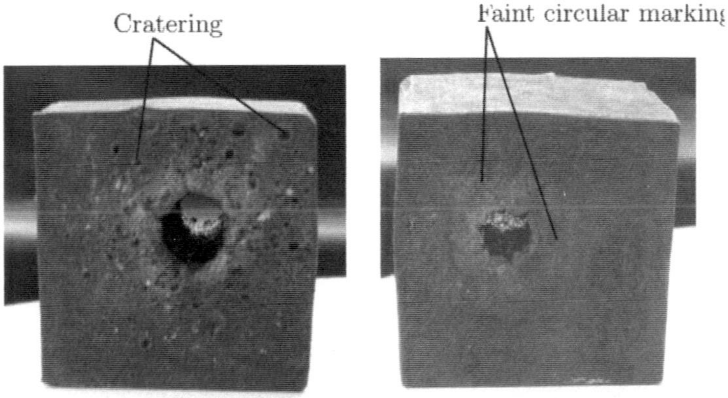


(a) The entry wound of the uppermost block of the penetrated target.

(b) The exit wound of the uppermost block of the penetrated target.

**Figure 8.** Photographs of hole in steel armor plate caused be strike of shaped charge munition [14].

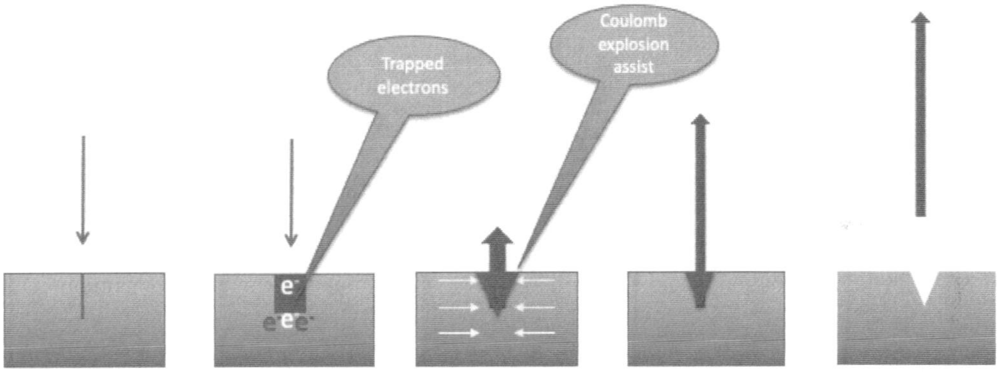


**Figure 9.** Schematic representation of strike of electron pulse on an insulator (see text).

insulator material corresponds to $\sim 10^{-10}$ grams. Figure 10 is a schematic of the 10,000 m/s liquid slug penetrating a target by hydrodynamic flow, leaving a clean hole behind. Again, the micro shaped-charge penetrates targets by plastic flow, not melting. The slugs will leave carved out tracks when launched parallel to surfaces, and the motion of the slugs are influenced by electric/magnetic fields due to small mass and the fact that they are charged. Another mechanism according to simulations can be that the electrons emerge perpendicular to the surface with a velocity of 100,000 m/s due to Coulomb explosion and the electrons drag the melted material behind them at velocities up to 30 km/s [10]. Figures 2 and 8 show the similarities in the ability of high velocity slugs to penetrate materials.

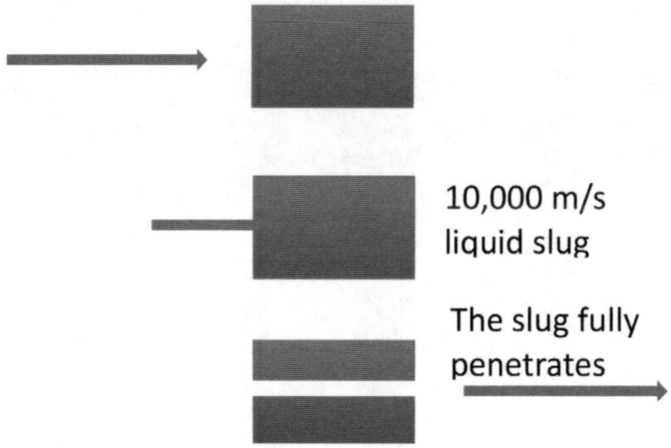


**Figure 10.** Schematic representation of strike of micro liquid slug of anode material on a solid (see text).

## 3. Summary

In conclusion, the characteristics of EVO's or charged clusters presented in the literature were compared to the characteristics of pulsed deposition systems and shaped-charge munitions. EVO's are generated with a single 10-20 kV, $\sim$100 ns electron pulse on a high quality, thin insulator on a ground plane. The sharp point cathode, thin insulator on a ground plane, and close proximity of the cathode and anode seem to be the unique features of the generator as compared to deposition pulses of electrons and photons. In the latter case the cathode spot is large by comparison to maximize the amount of material deposited with each pulse. It is suggested that EVO's are not a new form of matter (dense ball of electrons with diameter of microns containing $10^{23}$ electrons per cubic centimeter) but a manifestation of known thermomechanical and electromagnetic mechanisms acting on the thermal insulator, a small portion of which is rapidly heated that then generates a micro shaped charge. Many of the attributes of EVO's or charged clusters can be explained by this model.

Thermomechanical/electromagnetic modelling would be very useful to validate or invalidate the present hypothesis.

## References

[1] See U.S. Patents by K.R. Shoulders. 5,018,180 (1991) - 5,054,046 (1991) - 5,054,047 (1991) - 5,123,039 (1992), - 5,148,461 (1992).

[2] Lutz Jaitner, "The Physics of Condensed Plasmoids and LENR", The Physics of Condensed Plasmoids (CPs) and Low-Energy Nuclear Reactions (LENR) 1-101, February 2020, lutz.jaitner t-online de, www.condensed-plasmoids.com.

[3] A history of EVO's and charged cluster research may be found at, https://condensed-plasmoids.com/history.html.

[4] Ken Shoulders produced numerous unrefereed monographs that may be accessed by searching Ken Shoulders EVO; What's an EVO? by Ken; or visit http://blog.hasslberger.com/2007/10/ken_shoulders_evos_exotic_vacu.html

[5] K.R. Shoulders, S. Shoulders, "observations on the role of charged clusters in nuclear reactions", J. New Energy, **1**(3) (1996) 111.

[6] Muddassir Ali, "Modeling of plasma dynamics during pulsed electron beam ablation", Ph.D. Thesis, Laurentian University, Sudbury, Ontario, Canada (2017).

[7] Ying Qin, Chuang Dong, Xiaogang Wang, Shengzhi Hao, Aimin Wu, Jianxin Zou, Yue Liu, "Temperature profile and crater formation induced in high-current pulsed electron beam processing", J. Vac. Sci Tech. A **21**, (2003) 1934; https://doi.org/10.1116/1.1619417.

[8] Zhipeng Zhou, Andreas Kyritsakis, Zhenxing Wang, Yi Li, Yingsan Geng, Flyura Djurabekova, "Direct observation of vacuum arc evolution with nanosecond resolution", Sci. Reports **9** (2019) 7814; https://doi.org/10.1038/s41598-019-44191-6 1.

[9] Peng Zhang, Tungwai Leo Ngai, Zhi Ding, Yuanyuan Li, "Erosion craters on Ti3SiC2 anode", Phys. Letters **A 378** (2014) 2417–2422.

[10] M. Nistor1, N.B .Mandache, J. Perrière, "Pulsed electron beam deposition of oxide thin films", J. Phys. D: Appl. Phys. **41** (2008) 165205; doi:10.1088/0022-3727/41/16/165205.

[11] S.S. Harilal, C.V. Bindhu, M.S. Tillack, F. Najmabadi, and A.C. Gaeris, "Internal structure and expansion dynamics of laser ablation plumes into ambient gases", J. Appl. Phys. **93**, (2003) 2380.

[12] D.B. Chrisey, G.K. Hubler, *"Pulsed Laser Deposition of Thin Films"*, Book Editors, (Wiley Interscience, NY, 1994).

[13] An internet browser search will reveal many descriptive sites for Lichtenberg figures (Linac Tree).

[14] William Walters, "Introduction to Shaped Charges", Army Research Laboratory Report ARL-SR-150, March 2007.

[15] Y. Ashuach, Z. Rosenberg, E. Dekel, "On the spall strengths and Hugoniot elastic limits of some strong ceramics", AIP Conference Proceedings **955** (2007) 473. https://doi.org/10.1063/1.2833106

J. Condensed Matter Nucl. Sci. 36 (2022) 38–47

Research Article

# Long Term Anomalous Heat from 9 nm Pd Nanoparticles in an Electrochemical Cell

Graham K. Hubler* and Kavita K. Katti

*Sidney Kimmel Institute for Nuclear Renaissance (SKINR), University of Missouri, Columbia, MO USA*
*Institute of Green Nanotechnology; Medical School, University of Missouri, Columbia, MO 65212 USA*

Dennis Pease, Arik El Boher, Orchideh Azzizi, and Jinghao He

*Sidney Kimmel Institute for Nuclear Renaissance (SKINR), University of Missouri, Columbia, MO USA*

Kattesh V. Katti

*Institute of Green Nanotechnology; Medical School, University of Missouri, Columbia, MO 65212 USA*

**Abstract**

We report 58 days of 120 milliwatts anomalous excess heat in a Fleischman-Pons type open electrochemical cell which integrates to 610,000 Joules of liberated energy, including 4 days of anomalous heat after electrolysis power was shut off. The Pd foil cathode was placed in tension and 9 nm Pd nanoparticles were deposited on the cathode in situ at the onset of the experimental run.


*Keywords:* Fleischman-pons, anomalous heat, nano particles, Pd foil, heat after electrolysis

## 1. Introduction

There have been numerous papers on the measurement of anomalous heat during the electrolysis with $D_2O$ on Pd cathodes since the first reports by Fleischman and Pons [1]. We describe this heat as the anomalous heat effect (AHE) rather than cold fusion (CF) or low energy nuclear reactions (LENR) since the underlying mechanism has not been discovered nor is there incontrovertible evidence that a nuclear process is involved. Since Arata first reported the AHE using gas loading methods on Pd nanoparticles at elevated temperature [2], many researchers shifted to performing variations of elevated temperature gas loading on Pd and Ni nanoparticle containing powders with considerable success [3]–[6]. Here we report for the first time the use of Pd nanoparticles in the electrolysis method of exploring the AHE that resulted in the appearance of the AHE with higher probability than for bulk Pd cathodes based upon SKINR data from over 1000 runs on Pd foils that seldom produced excess heat.

*Corresponding author: 6635 Prestwick Drive, Highland, MD 20777, hublerg@umsystem.edu

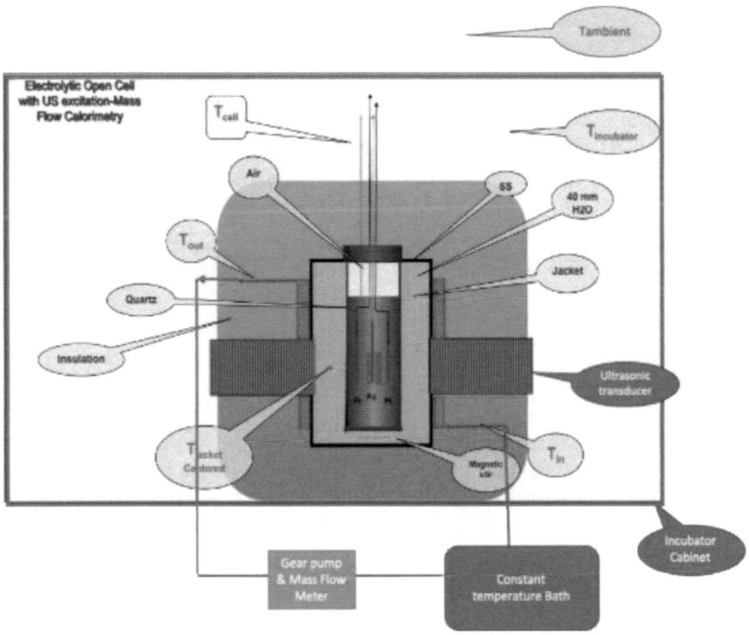


**Figure 1.** Schematic representation of calorimeter system US4 at SKINR.

In this effort, 9 different calorimeters in the SKINR lab were employed including mass flow (1), $^4$He tight mass flow (1), Isoperibolic (6) and differential (1). The latter was constructed by SKINR and the mass flow and isoperibolic calorimeters were legacy calorimeters from Energetics, LTD. Definitive results were obtained with the open cell mass flow calorimeter and will be described in this report. Various less robust results were obtained in the other calorimeters and will be included later in the discussion.

## 2. Experimental

Figure 1 is a schematic view of the open cell mass flow calorimeter designated US4 [7]. This calorimeter (built by Energetics, Inc., modified by SKINR to provide vertical tension on cathode) is designed to bathe the cathode in ultrasound stimulation using 4 ultrasound transducers placed equidistant around the cell centered on the cathode vertically. The ultrasound transducers were not used in this experiment. A unique feature of this calorimeter is a water jacket surrounding the cell containing the electrolyte. The water jacket served to transmit the ultrasound acoustic waves (when used) through the water jacket to the walls of the quartz cell, and then into the electrolyte and then to the cathode. Energetics obtained modest heat results using this method of stimulation [7] and this mass flow calorimeter proved to be a reliable and yield accurate heat data in our previous AHE studies. In the present experiment, the ultrasound transducers were not used as we were interested in just using the mass flow calorimeter to test the nanoparticle inspired cathodes. Also implemented in this cell was spring actuated constant vertical tension of 4.8E-5 Newtons/m$^2$ (1/3 the Pd yield strength of 180 N/m$^2$) on the 8cm x 0.8cm x 100$\mu$m, 0.9995 purity Pd foil cathode. When a Pd foil cathode is loaded with H, the change in volume between the alpha and beta phases causes the cathode to expand and crinkle and become misshapen. The spring loaded tension was used to allow the cathode to expand and therefore minimize

**Figure 2.** Photos of US4 calorimeter; a) anode-cathode (non-tension version) assembly and electrochemical cell, b) SS chamber, c) assembled calorimeter in incubator cabinet.

the deformation of the cathode. It was hoped that his feature would prevent detachment of Pd nanoparticles deposited on the foil surface.

Surrounding the water jacket were Cu coils that formed the mass flow portion of the calorimeter and the cell, water jacket and coils were placed inside a stainless-steel (SS) cylindrical chamber. The SS chamber was covered in insulation and placed inside an incubator shown in Fig. 2. Outside of the incubator was a constant temperature bath that fed water to the inlet of the mass flow coil and the heated water returned through a gear pump and mass flow sensor to the constant temperature bath.

There were six PT100 thermal sensors in the system shown in Fig. 1: water input temperature to the mass flow coil (Tin), water outlet temperature from the mass flow coil (Tout), temperature in the water jacket centered on the cathode (Tjack), cell electrolyte temperature centered on the cathode (Tcell), temperature in the incubator cabinet (Tcab) and the ambient temperature outside of the cabinet (Tamb). The thermal power measured by the mass flow system was Power = Mass Flow x (Tout-Tin) where the product is read directly as Watts. The anodes were 4 Pt wires wound around Teflon supports as shown in Fig. 2. Several electrolytes were used, depending on the test being conducted; 0.1M LiOD in $H_2O$, 0.1M LiOD in $D_2O$, 0.1M LiOH in $D_2O$. For the long term run that is the primary focus of this report, the electrolyte was 0.1M LiOD in $H_2O$.

The 4 point-probe technique was used to measure the cathode resistance in situ. The electrolysis current and voltage were measured at the sampling rate of 50,000 Hz, averaged, and reported at 1 Hz. The 6 temperature sensors, cathode resistance and mass flow sensor were all reported at 1 Hz. Also shown in Fig. 1 is a magnetic stir in the bottom of the water jacket. This was implemented by Energetics as necessary to obtain robust mass flow calorimetry operation. The stir inserts energy into the calorimeter that must be accounted for. Energetics found that the stir caused a 0.786 C positive offset in Tout, Tjack, Tcell and that this offset was stable over a period of years. Tin is not influenced by the magnetic stir. Therefore, to obtain accurate heat output from the calorimeter, 0.786 C was subtracted in the software from Tout (Tcell, Tjack were not corrected) in order to make Tin-Tout a precise measure of heat output.

Pd nanoparticles (NP) were synthesized in the U. Missouri Medical School following protocols described in the U.S. patent 20170009366 entitled, "*EGCG Stabilized Pd nanoparticles, method for making and electrochemical cell*" [8]. Initially, 9 nm diameter NP were emphasized and during the experimental program sizes down to 5 nm diameter were studied. The size distribution at any size was $\sim\pm1.7$ nm. A mass of 379 mg of Pd nanoparticles were added to electrolyte consisting of 150 ml of 0.1M LiOD in $H_2O$ just prior to starting the experiments, turning the clear electrolyte to a black color. The electrolyte became transparent within 2 days from time electrolysis began as Pd nanoparticles plated out on cathode (as determined by supporting experiments outside of a calorimeter where the electrolyte could be observed). Figure 3 is a TEM image of a typical 9 nm batch of Pd NP.

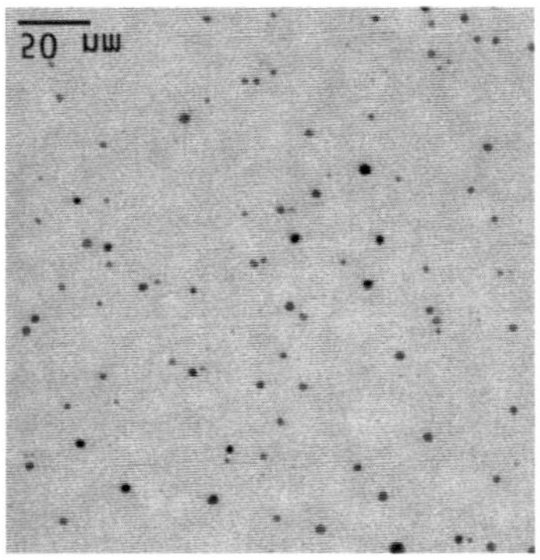


**Figure 3.** Transmission electron micrograph (TEM) image of nominal 9 nm Pd nanoparticles. Scale marker is 20 nm.

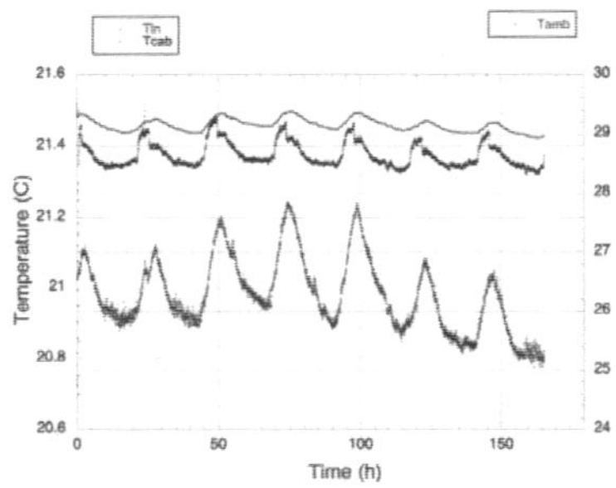


**Figure 4.** Tin, Tcab, Tamb versus time showing effects of daily room temperature variation.

## 3. Initial Test Data

Figure 4 displays Tin, Tcab and Tamb for 166 hours. There is very poor summertime temperature control in the room as Tamb shows swings of 2.7 C. The incubator cabinet reduces the swing to less than 0.2 C. Note that the cabinet and constant temperature bath were regulated to about 21.4 C which is ~5C less than ambient temperature so that the cabinet and constant temperature bath only use cooling cycles. Temperature variations in Tin are further reduced by the constant temperature bath to 0.07 C. All three sensors are highly correlated in time.

Figure 5 displays 166 hours of test data for a Pt cathode, showing the variation of temperatures with input electrical power. During the test, the electrical power was applied for open-cell electrolysis in a constant current mode and the input power was computed using densely sampled electrical currents and voltages during the electrolysis of 0.1M LiOD in $H_2O$ with a thermal neutral potential correction of 1.482 V. With a mixed electrolyte, the precise value to assign for the thermoneutral potential is unknown so the value for $H_2O$ was used as a lower limit to the correct value. Plotted are Tcell-Tin, Tjack-Tin and Tout-Tin, where Tin is delay shifted by 1.9 hours optimized by least squares fitting. This is due to the fact that the constant temperature bath is correlated with Tamb while Tcell, Tjack, and Tout are correlated with Tcab, and it takes 1.9 hours for Tin to respond to Tcab. A benefit of mass flow calorimetry is that the heat signal is a differential of Tin and Tout so that the daily temperature variations are largely removed when Tin is subtracted from Tout as can be seen in Fig. 5. The sensors for Tcell, Tjack and Tout are highly correlated in time and Tout and Tjack are so highly correlated that Tjack could serves as a secondary Tout for measurement of heat output. Note that Tcell is larger than Tjack as it should be since the heat is generated at the cathode. Tjack lags Tcell in time by 6 minutes

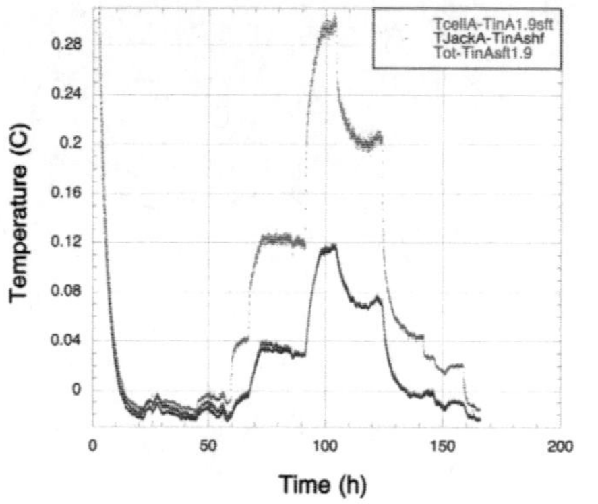


**Figure 5.** Tcell-Tin, Tjack-Tin and Tout-Tin for different input powers versus time where Tin is time delayed by 1.9 hours.

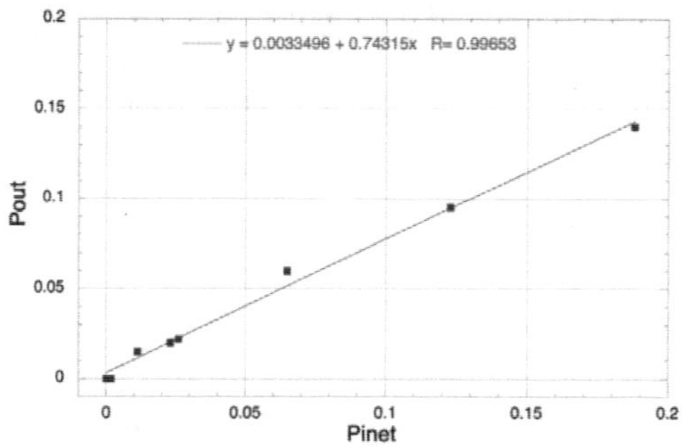


**Figure 6.** Mass flow calorimeter efficiency measured to be 74.3%.

and Tout lags Tcell in time by 7.2 minutes after a step change in input power. The calorimeter time constant is 4 hours so the several minutes time lags do not need to be included in a time correction. Figure 6 displays the measure of the efficiency of the calorimeter of 74%.

## 4. Excess Heat Detection

Figure 7 displays the first 560 hours of electrolysis for Tout-Tin (raw data) and Tout-Tin (Tin shifted by 1.9 hours) and the input power, Pinet, where the thermoneutral potential correction to the input power was employed. Notice shifting Tin 1.9 hours produces less temperature variation ($\pm 0.02$ Raw data vs $\pm 0.01$ shifted Tin) and the daily temperature swings are much reduced.

Several observations require comment. First, the excess power appears early at 4 hours after electrolysis is turned on and plateaus at 20 mW excess up to hour 65. During the first 48 hours, the nanoparticles were being slowly deposited and being slowly reacted with H and D. At 65 hours, excess power increases again to new level of 90 mW excess at $\sim$160 hours and remains at this level up to hour 560. Daily variations of Tout - Tin are not completely removed due to influence of Tamb on Tin.

Figure 8 displays the raw data for the entire 1500 hour run, uncorrected for the calorimeter efficiency. The system equilibrates for the first 18 hours prior to turning on the electrolysis at 25 mA under current control. After $\sim$160 hours,

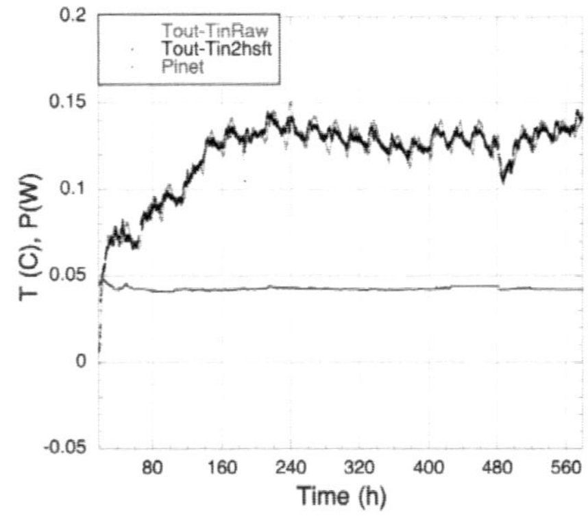


**Figure 7.** Tout-Tin (raw data, red) and Tout-Tin (Tin shifted by 1.9 hours, black) and input power, Pinet (green), for first 560 hours of electrolysis. The plotted values of delta T are proportional to the amount of power being produced. For a mass flow calorimeter, the deltaT is directly measured as power in Watts, uncorrected for the calorimeter efficiency.

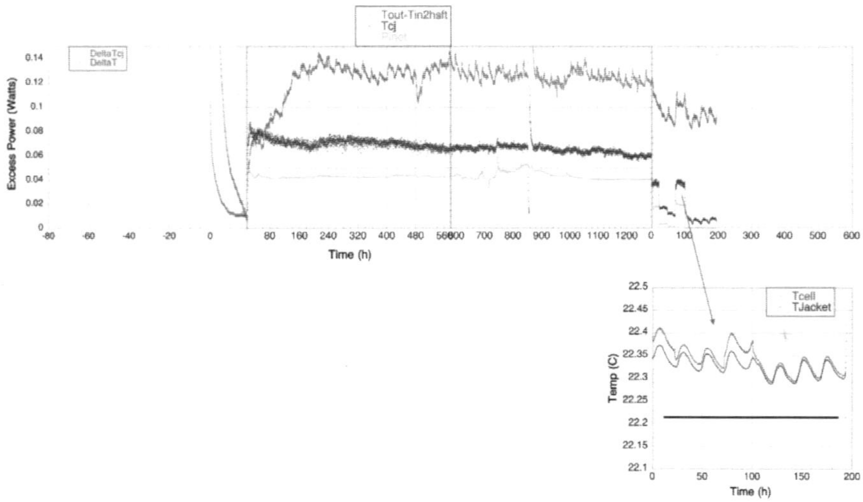


**Figure 8.** Raw temperature and power data, uncorrected for the calorimeter efficiency, for entire 1500 hour run. First 18 hours - Tout (red), Tin (blue). 18-1500 hours, power out as Tout-Tin (red), Tcell-Tjack (blue), power in (green). Arrow points to raw data for Tcell and Tjack for last 200 hours. Black horizontal line is T of Tcell and Tjack before electrolysis and after cathode removed from cell.

the excess remains (red line) at $\sim$90 mW until the power is reduced at 1300 hours. The total energy released from this raw data was $0.09 \times 1400 \times 3600 = 453{,}000$ Joules over 58 days. The blue line is Tcell – Tjack and the green line is the input power. Note that the daily variations are completely removed from the Tcell-Tjack data. This is due to the fact that Tcell and Tjack are both influenced by Tcab and there is no influence from Tamb. A plot of Tcell-Tout (not shown) produces an identical curve since Tjack, and Tout are highly correlated as we have already established.

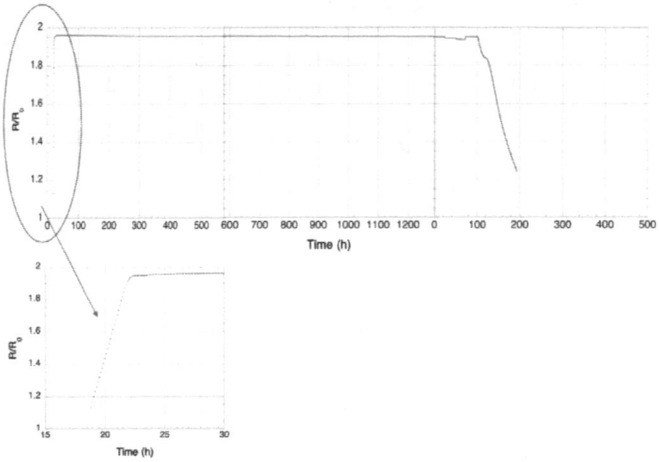


**Figure 9.** The resistance ratio R/R0 versus time. Scale change after 1300 hours.

There is a temperature excursion at 860 hours caused by loss of power to the constant temperature bath as confirmed by the timing of the temperature excursions reported by the sensors that showed Tin leading the excursions in time.

The last 200 hours of the run is especially interesting. During the last 200 hours the input power was manipulated where it was reduced in steps, increased and then set to 0 as shown in Fig. 8. The Tcell-Tjack (blue curve) faithfully reports the power changes (green curve) since the power is sensed directly by Tcell and there are no daily temperature excursions to mask the changes. On the other hand, Tout-Tin shows the general features of the power changes but does not drop to zero and keeps reporting the ~90 mW excess power. The arrow points to the raw data for Tcell-Tjack where the daily variations are clearly visible but note that both remain elevated and do not drop to 0 after power is removed. Therefore, Tcell, Tjack and Tout all report elevated temperature for the last 100 hours when the input power was 0. The horizontal black line is the temperature of Tcell and Tjack before electrolysis. Both sensors returned to this value after the cathode was removed and the cell was reassembled. The 90 mW of excess heat for 1400 hours and 100 hours of heat after death were reported by all T sensors.

Figure 9 shows the resistance ratio R/R0 for the entire 1500 hour run. With application of charging current the ratio rose to 1.95 in 4 hours and remained there until the current was manipulated during the last 200 hours. For H/Pd, the literature maximum value for the ratio is 1.8 and for D/Pd it is 2.0 [9]. Since the value of 1.95 is not possible for H insertion, one might surmise the Pd loaded with D even in the mixed electrolyte. The presence of the Pd NP on the surface complicates the interpretation of the resistance data. Consequently, the D/Pd ratio of the deposited Pd is unknown and the amount of D and H in the Pd cathodes is also uncertain. The ratio only slightly responds to halving the input power at 1300 hours, reloads when full power is reapplied, then slowly deloads over 25 hours when the power is shut off before more rapidly deloading for the next 75 hours. At the end of the run the ratio of 1.25 corresponds to D/Pd = 0.25 remaining in the cathode. Ed Storms has for years stated that the AHE can appear in Pd cathodes for loading as low as 0.2 [10].

Finally, we show the raw 90 mW excess power data for the entire 1500 hour run in Fig. 10. As already stated, the excess appears within 4 hours of the beginning of electrolysis, reaches a plateau up to 60 hours, climbs to a new plateau after ~160 hours and remains at ~90 mW for the remainder of the run, including the last 100 hours with the

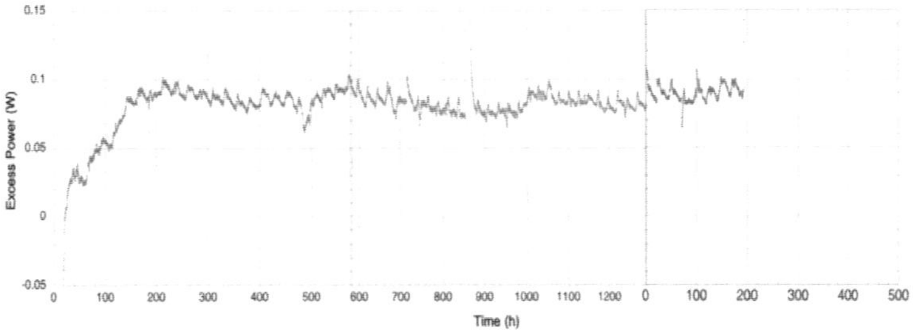


**Figure 10.** Excess power versus time. Scale change at 1300 hours,

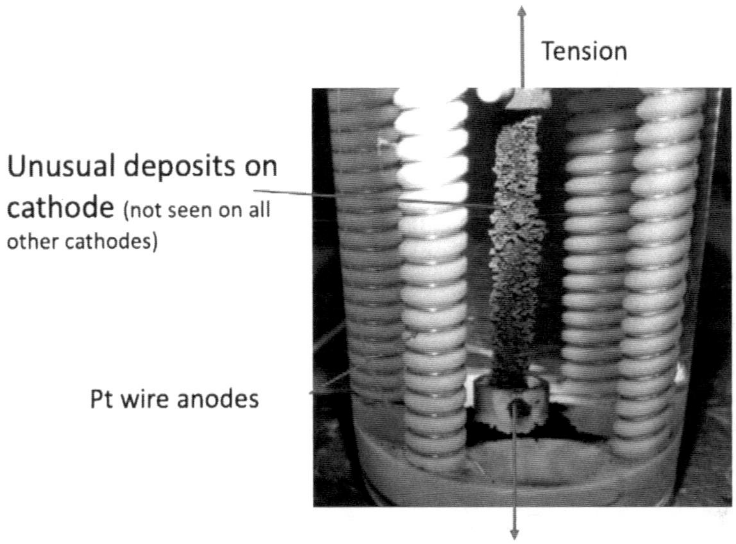


**Figure 11.** Cathode/anode support structure with stretched cathode still immersed in the electrolyte after removal of cell from the calorimeter.

input power shut down. Correcting the raw total energy released energy of 450,000 J by the calorimeter efficiency yields 610,000 J, the average power becomes ~120 mW and the average coefficient of performance (COP) was 2.6.

Figure 11 is a photograph of the cathode in its holder in the electrolyte just after removal of the cell from the calorimeter. The deposits on the cathode were unexpected and unusual, not normally seen after experimental runs.

## 5. Discussion

The excess heat detection is substantiated by evidence that calorimeter was working as expected:

- Tcell greatly exceeds Tjack, Tout, as it should for heat generated only at cathode
- Tjack, Tout lag Tcell in time by about 6 and 7 minutes, respectively, also very reasonable
- Tout, Tjack, track each other quite precisely
- Tout, Tcell, Tjack all consistently report excess heat

- Daily room temperature variations substantially accounted for by subtracting time adjusted Tin
- Tcell, Tjack, Tout settled down to pre-run values after cathode removed and cell reassembled
- Mass flow calorimeter provides method to accurately overcome daily temperature variability
- Noise floor $\pm 10$ mW.

Somewhat unusual behavior of this cell in our experience of over 1000 electrochemical cell runs using Pd foils are:

- Clear 120 mW excess power for 1400 hours or 58 days (COP = 2.6)
- Total energy liberated is 610,000 Joules
- Heat after death for 100 hours until run deliberately stopped
- Unusual deposits on cathode post run are not usually seen
- Excess heat appears within hours of the onset of electrolysis.

Non-standard parameters used to run a Fleischmann-Pons type cell that may have contributed to the heat production are:

- Stretched cathode
- 9 nm nanoparticles in electrolyte
- 0.1M LiOD in $H_2O$ electrolyte.

For the 58 days of heat reported in detail here, it could be the case that liberated H or D is recombined with O in the cell by an inadvertent or subtle recombination pathway and this caused the excess heat. As a check on this, 7 runs were conducted in thermoelectric and isoperibolic calorimeters, 3 with $D_2O$ and 4 with $H_2O$ electrolyte and 50 nm Pd NP. The cathode was Pd for 5 runs and Pt for 2 runs lasting from 6 to 14 days. The electrolyte volume was measured before and after each run. If there was no recombination, then a calculation of the electrolyte loss expected from Faradays law would predict the volume loss of electrolyte. The measured and calculated values for the electrolyte were equal within an average of 0.8%, thus proving that recombination in the cell was not the origin of the measured excess heat.

The run reported here was by far our best result. Many combinations of Pt or Pd cathodes, DD, HH, DH, HD ($1^{st}$ is 0.1 M LiOH or LiOD, $2^{nd}$ is $H_2O$ or $D_2O$) electrolyte. Sub-9 nm nanoparticles were also run, but not with a stretched cathode. Some produced <60 mW heat from a few hours to several days with COP of $\sim$0.2 to 0.5, but about half produced no heat. Those that produced heat were shut down to try new material combinations since the effect was small. HH never produce heat. Heat was observed in mass flow, differential, and $^4$He tight mass flow calorimeters, in open cells with no recombiner catalyst or closed cells with recombiner catalysts. In one case, an HH cell produced no heat for 3 days, heat appeared after 4 hours when LiOD was added to the electrolyte. These runs were unremarkable due to the small amount of excess heat which was not far above the sensitivity of the calorimeters and would not have been published by themselves. They are included here to make it clear that one should not expect a COP of 2 in every run.

Prior to using 9 nm NP's, SKINR conducted many runs with 50 nm Pd NP. Some of these runs produced small amounts of excess heat but the defining characteristics were a general inconsistency where no 2 runs were alike and often the heat would appear for a few days and then drop to 0. Other material combinations that never produced heat were:

- 3 nm Au NP, magnetically polarized
- 50 nm Fe coated Pd NP
- 50 nm Ni coated Pd NP.

## 6. Conclusions

We report 58 days of 120 milliwatts anomalous excess heat in an electrochemical cell which integrates to 610,000 Joules of liberated energy for a COP of 2.6, including 4 days of anomalous heat after electrolysis power was shut off.

The Pd cathode was placed in tension and 9 nm Pd nanoparticles were deposited on the cathode in situ at the onset of the experimental run. The electrolyte was 0.1M LiOD in $H_2O$, D/Pd ratio ~0.7 during run until last 200 hours, electrolysis current 25 mA for most of the run (Pinet = 0.048 W), and the Pd cathode loaded to R/R0 = 1.95 in 4 hours presumably with a mixture of H and D. Non-standard parameters used to run this Fleischmann-Pons type cell that may have contributed to the heat production are:

- Pd foil cathode placed in tenson at 1/3 the yield strength
- 9 nm nanoparticles in electrolyte
- 0.1M LiOD in $H_2O$ electrolyte

Evidently NP size matters since over 100 runs with 50 nm Pd np never produced consistent heat but 9 nm NP did produce some heat in about 50% of runs. The mixed LiOD in $H_2O$ electrolyte coupled with the Pd NP on a Pd foil creates an unknown situation as to how H and D may be partitioned between the NP and the bulk Pd foil. Any attempt to assign the partition would be speculation so we offer no conclusion this detail. These new features are no guarantee that excess heat will be observed, but they may increase the probability of observation of small amounts of excess heat.

**Acknowledgement**

This work was supported in full by Mr. Sidney Kimmel. We thank the reviewer for helpful comments and suggestions.

**References**

[1] M. Fleischman, S. Pons, M.W. Anderson, L.-J. Li, M. Hawkins, J. Electroanal. Chem. **287** (1990) 293.

[2] Y. Arata, Y.C. Zhang, Proc. Jpn. Acad. **70B** (1994) 106.

[3] Akira Kitamura, Akito Takahashi, Koh Takahashi, Reiko Seto, Takeshi Hatano, Yasuhiro Iwamura, Takehiko Itoh, Jirohta Kasagi, Masanori Nakamura, Masanobu Uchimura, Hidekazu Takahashi, Shunsuke Sumitomo, Tatsumi Hioki, Tomoyoshi Motohiro, Yuichi Furuyama, Masahiro Kishida, Hideki Matsune, "Excess heat evolution from nanocomposite samples under exposure to hydrogen isotope gases", J. of Hydrogen Energy **43** (2018) 16187.

[4] G. Levi, E. Foschi, B. H€oistad, R. Pettersson, L. Tegner, H. Essen, "Observation of abundant heat production from a reactor device and of isotopic changes in the fuel", http://www.sifferkoll.se/sifferkoll/wp-content/uploads/2014/10/LuganoReportSubmit.pdf

[5] Parkhomov AG., Int. J. Unity Sci. **6** (2014) 57; ibid. **7** (2015) 68; ibid. **8** (2015) 34.

[6] S. Jiang, "New result on anomalous heat production in hydrogen loaded metals at high-temperature", (2015). http://ja.scribd.com/doc/267085905/

[7] Dardik, I., Banover, H., El-Boher, A., Gazit, D., Golbreich, E., Greenspan, E., et al. "Intensification of low energy nuclear reactions using Superwave Excitation", $10^{th}$ International Conference on Cold Fusion", Cambridge MA, 24-29 August, (2003) World Scientific Publishing Co., p. 61.

[8] Kattesh V. Katti, Orchideh Azizi, Sagar Gupta, Kavita K. Katti, Arie El-Boher, Robert Duncan, Graham Hubler, "EGCG Stabilized Pd nanoparticles, method for making and electrochemical cell", US Patent 20170009366.

[9] McKubre, M. C., F. Tanzella, "Using resistivity to measure H/Pd and D/Pd loading: Method and significance", Condensed Matter Nuclear Science, ICCF-12, Yokohama, Japan, World Scientific, (2005) p. 392.

[10] Edmund Storms, "Anomalous energy produced by PdD", J. Condensed Matter Nucl. Sci. **20** (2016) 81.

J. Condensed Matter Nucl. Sci. 36 (2022) 48–55

Research Article

# Abnormal Absorption of Hydrogen in Nickel at Ambient Temperature With Associated Emission of Neutrons

Ubaldo Mastromatteo*

*A.R.G.A.L. Via S. Stefano, 27/B – 20010 Bareggio (MI), Italy*

**Abstract**

While it is known that nickel at a temperature of a few hundred degrees if in hydrogen can slowly absorb a certain amount of this gas, there is no evidence that this can occur at room temperature and at pressures below 1 bar. On the contrary, by conducting studies and experiments on LENR anomalies in the ARGAL laboratory in Bareggio, Italy, it has been experimentally verified several times that nickel in the form of wire, thin ribbon, foam, if properly covered with a thin layer of palladium, can absorb hydrogen in considerable quantities even at room temperature. Very often the neutron monitoring always active in the lab and close to the active reactor showed a significant rise in the background counts with two distinct peaks in the distribution of n / h (neutrons per hour) as will be better described in the following experiment description.


*Keywords:* Nickel, palladium on nickel, neutrons, neutron peaks, room temperature

## 1. Introduction

The ARGAL laboratory in Bareggio is equipped with several reactors to make experiments on LENR anomalies. The configuration used in the experiments that we are going to describe is depicted in the Figure 1 below.

Reactor 1 consists of a steel chamber with a volume of 290 cubic centimeters inside which is a glass tube wrapped with a resistor that serves as a heater, the material to be tested can be housed in the tube. A Pt100 in contact with the material allows us to follow the trend of its temperature on a PC. We also have a thin film Palladium resistance that allows to check if we have hydrogen or deuterium inside the reactor.

In the figure 3 it is represented the scheme for detecting the pressure inside the reactor. Pressure is the most important parameter for controlling the absorption of hydrogen by the sample [1], [2]. For this purpose we have a piezoelectric sensor that measures the pressure and an interface that displays its value and is connected to the computer that controls the experimental parameters.

Figure 4 is a screen shot showing the most important parameters that allow to follow the progress of the experiment. In particular we have three windows which display the parameters over time, in the top left we have two parameters: the

*Corresponding author: ubaldo.mastromatteo@libero.it

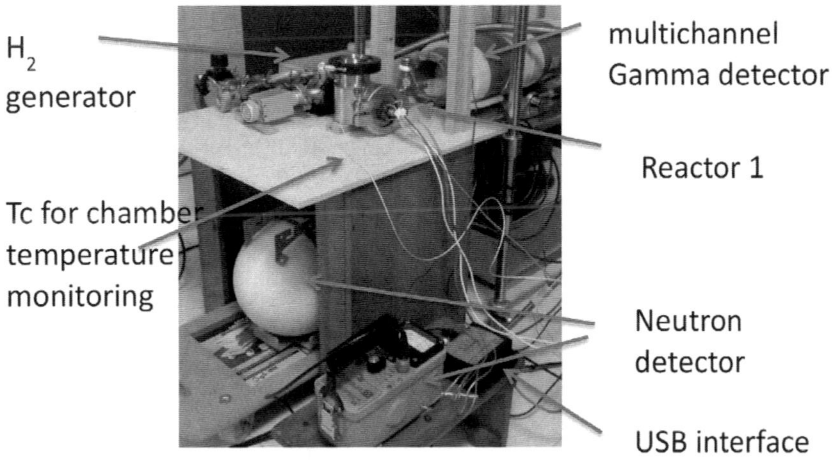


**Figure 1.** Experiment setup.

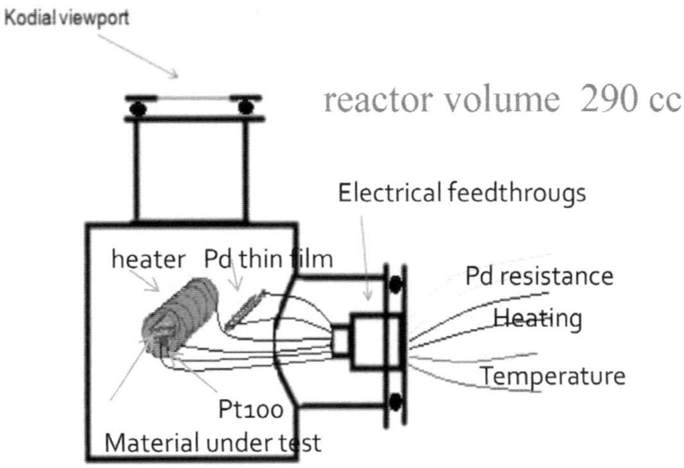


**Figure 2.** Reactor R1 sketch.

power sent to the heater and the temperature of the sample, in the top right window we have the ambient temperature and the reactor temperature, then in the bottom right window we have two important parameters which are the value of the thin film Palladium resistance which allows to know if we have the correct gas inside the reactor and then the pressure trend inside the reactor. The pressure is the most important parameter for checking absorption.

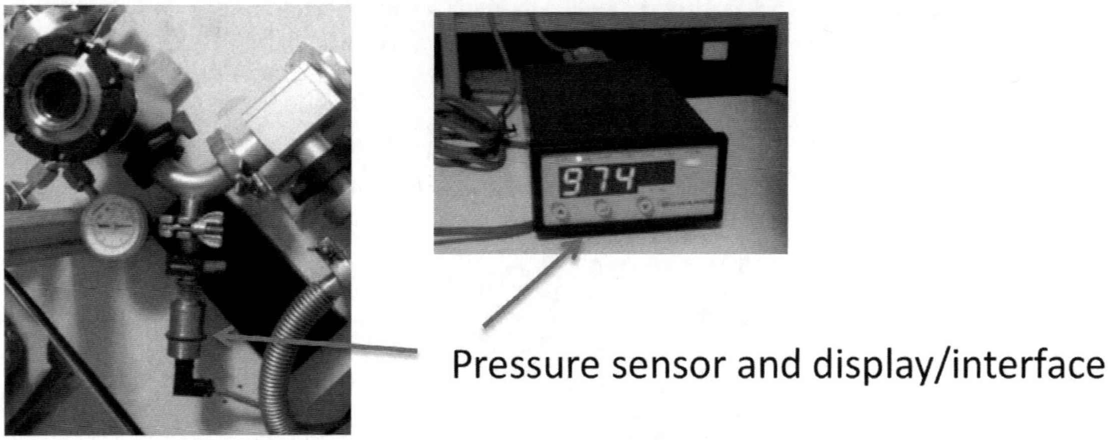


**Figure 3.** Setup for the pressure measurements.

## 2. The Material

The sample under test was a nickel strip with a deposit of about 200 nm of Palladium on both sides by electroplating. The deposit was made in the ARGAL laboratory with a solution of Palladium chloride and lithium chloride by placing the nickel strip on the cathode of a small elctroplating cell. To prepare the sample for deposit I used a mechanical action with a diamond file followed by cleaning with pure ethyl alcohol.

The deposit was calibrated for a thickness of about 200 nm and the aspect was similar to those reported in codeposition experiments

Weight of Ni sample 0.63 gr. The sample was folded several times to be able to introduce it into the glass tube on which a heating resistor is wound. Heating was used in the degassing phase of the material under vacuum before introducing hydrogen. Then the absorption experiment continued at room temperature, which remained fairly stable around 22 degrees centigrade.

## 3. Absorption Test Results

The graph in the figure 6 shows a clear absorption of the $H_2$ introduced into the reactor at a pressure of 315 mbar. In about 2 hours and 30 minutes the pressure dropped to 147 mbar.

The black curve illustrates this trend.

Taking into account that the absorption of $H_2$ by the small amount of palladium deposited on the nickel would have been able to make the pressure drop by about 3 mbar, it is evident that the reduction in pressure is mainly due to $H_2$ absorption by the nickel. The red curve indicates the increase of a thin film palladium resistance always into the reactor as presence of $H_2$ or $D_2$ monitor.

The graph in the figure 7 shows a clear absorption of the $H_2$ by a different sample in the same reactor R1.

Here the black curve illustrates this trend, and the red curve indicates the increase of the thin film palladium resistance that is in the reactor as a monitor of the presence of $H_2$ or $D_2$ in it.

The volume in liters of the absorbed gas is obtained multiplying the ratio between the delta pressure and the standard pressure at ambient temperature by the volume of the reactor.

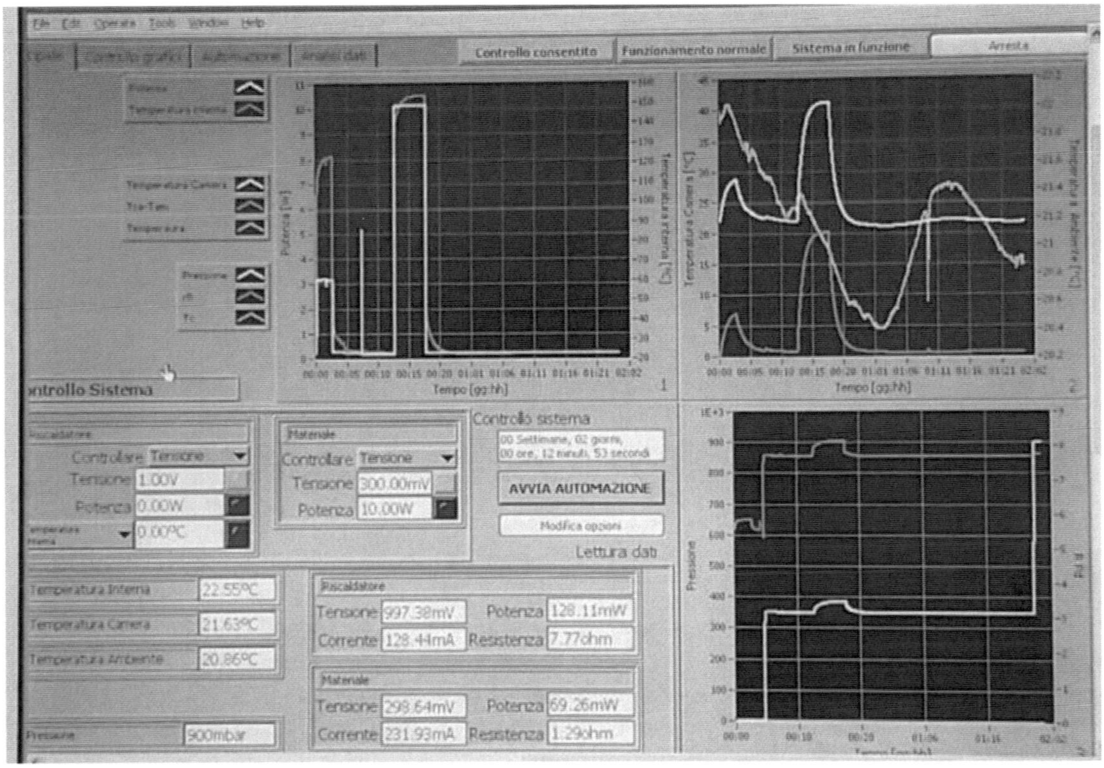


**Figure 4.** Screen shot of the LabView program controlling the reactor parameters.

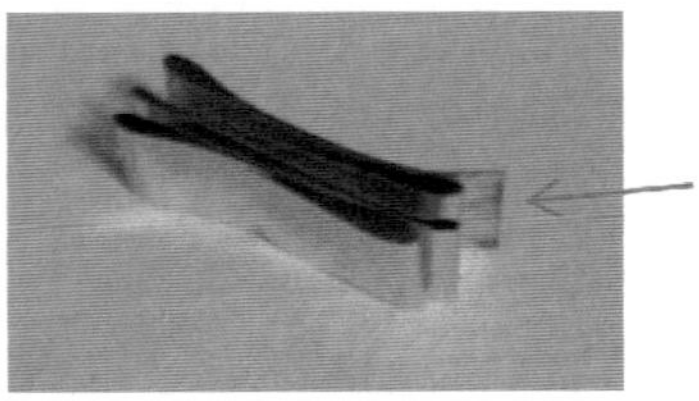


**Figure 5.** The sample folded to be adapted to the heater support.

Then the ratio between this volume and the standard volume of one mole of gas gives the $H_2$ absorbed moles. Dividing the H moles by the nickel moles gives the <H/Ni> ratio.

After the first absorption a hydrogen refill brought the pressure to 0.91 bar and after the second absorption to 0.981 bar. Here below the description of the adopted procedure.

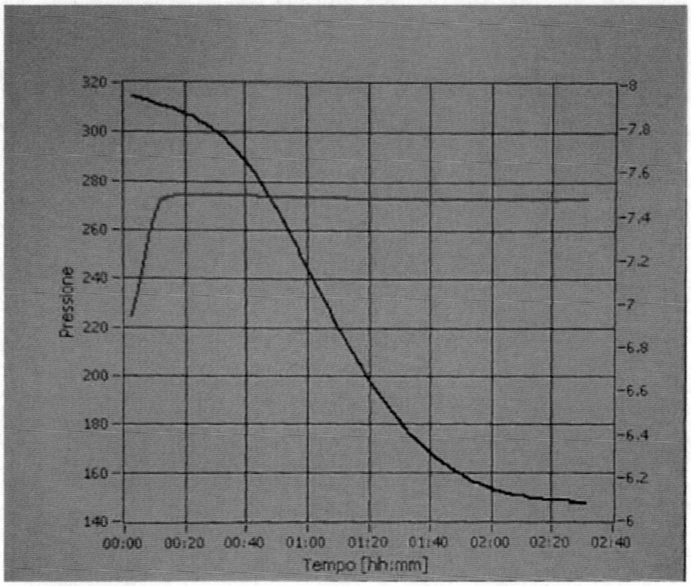


**Figure 6.** Pressure behavior when the reactor is filled with Hydrogen.

**Table 1.** Numerical data for absorbed Hydrogen moles calculation

Delta press. 1 (0,315 bar to 0,147 bar); Delta press. 2 (0,91 bar to 0,836 bar);
Delta press. 3 (0,981 to 0,968 bar);

$$(((0,315 - 0,147)/1,013) * 0,29)/24,22 + (((0,91 - 0,836)/1,013) * 0,29)/24,22 +$$
$$(((0,981 - 0,969)/1,013) * 0,29)/24,22$$

| First absorption | Second absorption | Third absorption |
|---|---|---|

**Table 2.** From absorbed moles to Hydrogen to Nickel atoms ratio

| H2 absorbed moles | H moles | Ni moles | H/Ni ratio |
|---|---|---|---|
| 0.0030 | 0.0060 | 0.0107 | 0.56 |

For the calculation of the absorbed moles we used the following procedure for each absorption phase: the difference between the initial and final pressure ($\Delta P$) is divided by the pressure value (here we indicate with Ps) at which one mole of gas occupies 24.22 liters at 22 degrees centigrade (we denote this volume by Vs), then we multiply the result by the reactor volume (we denote this volume by Vr) and divide the obtained value, which represents the part of the reactor volume absorbed, by the volume of one mole of gas and we will have as a result the moles of $H_2$ absorbed. By repeating the procedure for each absorption phase and adding together we will obtain the total amount of moles absorbed. The summary formula is:

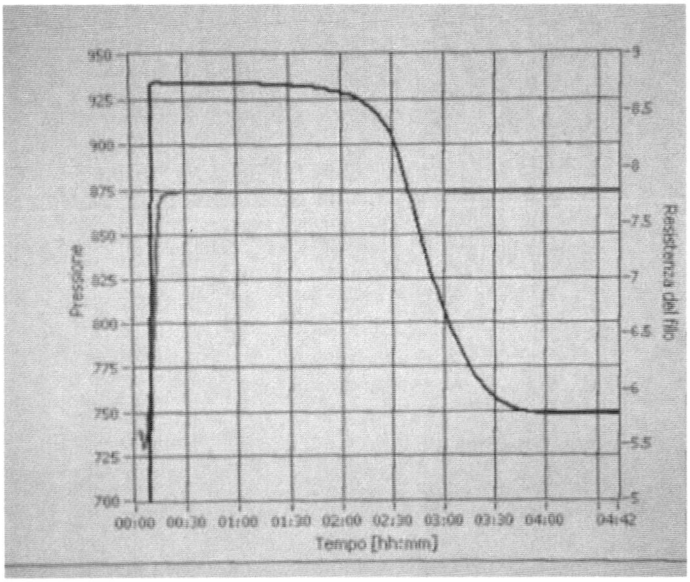


**Figure 7.** For another sample in a previous test the absorption started 2 hours after filling Hydrogen into the reactor.

## 4. Neutrons Measurements

The ARGAL laboratory in Bareggio is continuously monitored nearby the reactors by a Helium3 neutron detector (Ludlum) and with a Multichannel detector (Ludlum) with a 3" sodium iodide scintillator for gamma radiation [3].

In particular, for neutrons monitoring, is used the Ludlum 12-4 neutron meter detector. This detector is interfaced to a PC that uses a LabView program to record the events reported by the detector. The acquisition program records the data in two different modes, one in the form of events (counts) as a function of time (Figure 8), indicating the number of counts every minute (left axis) and every hour (right axis); then in a different screen (Figure 9) the normalized histogram for counts every hour. The program performs an acquisition for 8000 minutes, followed by a reset and the beginning of a new acquisition. The data of each complete acquisition are stored in a remote hard disk for any subsequent analysis.

The graph of figure 8 shows how it is changed the pattern of neutrons counts after the reactor was filled with Hydrogen.

The duration of the monitoring was 65 hours and up to the 33rd hour indicated by the yellow line the reactor had been kept under vacuum.

In the figure 8 the neutrons per minute are plotted as isolated dots while the neutrons per hour are plotted with dots united by the white line and it is clearly seen that at a certain point when hydrogen was introduced into the reactor the number of neutrons every hour is significantly increased as will be better shown in the histogram of figure 9.

During the absorption test, the neutron/hour counts showed an increase of about 80%.This situation has also been verified several times with samples having similar structure. So, in the neutron/hour histogram in figure 9 above on the left there are two distinct data populations, while for a reactor without hydrogen the neutron/hour histogram, in figure 9 on the right, get from a blank experiment, basically overlaps the background.

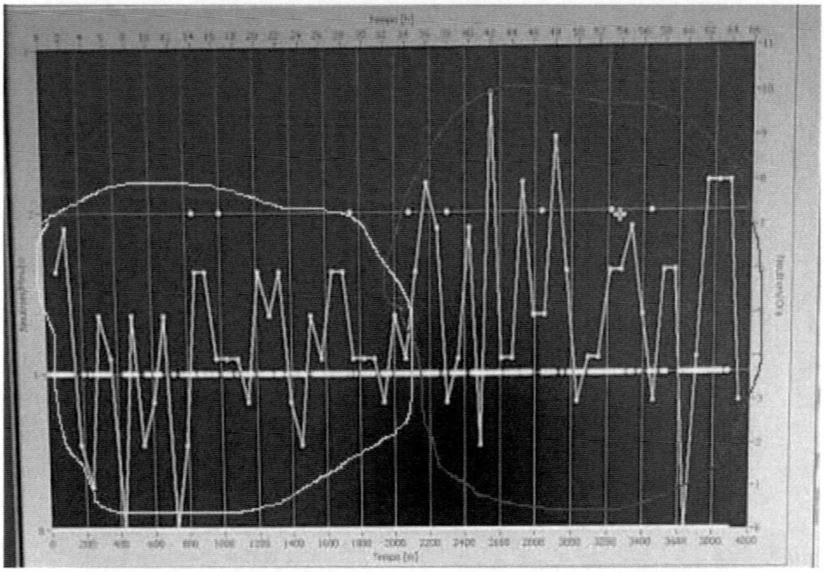

**Figure 8.** neutrons per minute and neutrons per hour trend. Into the yellow line the counts when the reactor was maintained in vacuum; into the red line when the reactor was filled with Hydrogen.

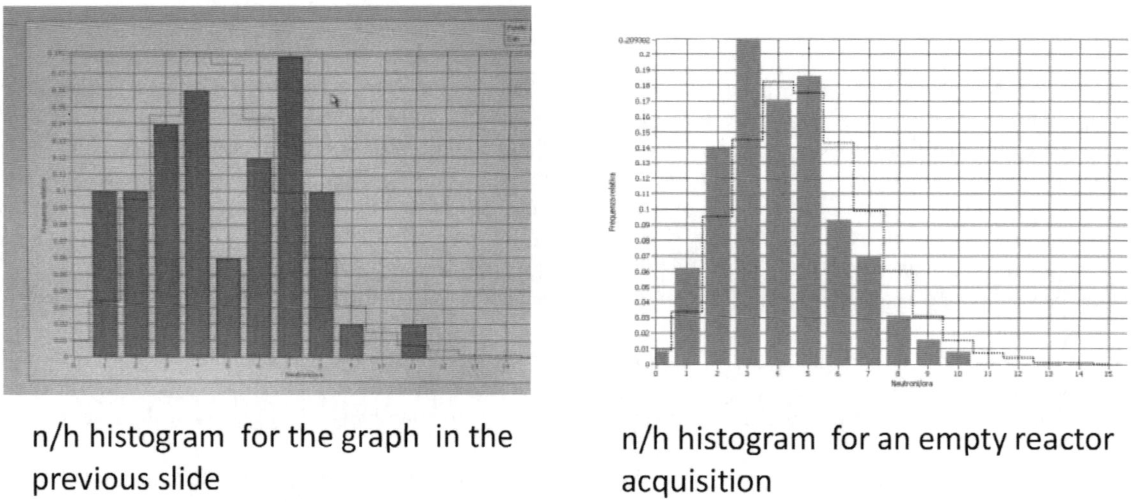


**Figure 9.** Neutron histogram during the experiment (left) and a typical background neutron histogram (right).

The production of neutrons showed peaks during the absorption phase, but a neutron emission in similar tests was seen to continue as long as hydrogen or deuterium remain in the reactor, as shown in the paper related to the iccf22 proceedings [4].

## 5. Final Remarks

In previous experiments the ability of nickel samples on which a thin layer of palladium was electrodeposited to absorb hydrogen even at room temperature in considerable quantities was verified several times, using nickel foam and nickel thin wires samples.

Once more the unexpected ability to absorb hydrogen at room temperature of such kind of structure, was verified on a different nickel sample: in particular a strip of 12 $cm^2$ of surface and 0.15 mm thickness.

In the specific, the pressure decrease at the end of three successive phases of absorption has been about 254 mbar (72 cc in volume of $H_2$ for a reactor volume of 290 cc ), such as to bring the ratio between hydrogen and nickel atoms to a value around 0.6, despite the fact that the deposition of the palladium was not particularly uniform.

Neutron emission was associated to hydrogen absorption and was clearly detectable not only during the absorption phase, which lasted several hours, but also the entire time the sample remained in hydrogen atmosphere after the absorption phase.

## References

[1] S. Focardi, V. Gabbani, V. Montalbano, F. Piantelli, S, Veronesi, "On the Ni-H System", SIF Conference Proceedings, Vol. 64, pp. 35–47, 1997.

[2] Campari, E. G., et al. (2000). Ni-H systems. 8th International Conference on Cold Fusion, Lerici (La Spezia), Italy, Italian Physical Society, Bologna, Italy.

[3] Campari, E. G., et al. (2005). Nuclear reactions in Ni-H systems. 6th International Workshop on Anomalies in Hydrogen/Deuterium Loaded Metals. Siena, Italy, ttp://www.iscmns.org/siena05/program.htm.

[4] Mastromatteo, U. (2020). Nuclear Signature in LENR Gas Loading Experiments, J. Condensed Matter Nucl. Sci. 32 (2020) 1–10

J. Condensed Matter Nucl. Sci. 36 (2022) 56–88

Research Article

# Input Stimuli and Output Signals for LENR Experiments

David J. Nagel*

*The George Washington University*

**Abstract**

One purpose of this paper is to draw attention to particular past LENR experiments and suggest some attractive future LENR experiments. Hence, we consider both past and prospective stimuli into and signals out of LENR experiments. Another goal is to provide a brief compilation of experimental data that can be used to test LENR theories. Rigorous challenges to theoretical results are needed to advance the understanding and practical applications of LENR.


*Keywords:* LENR, Experiments, Input stimuli, Output signals, LENR theory tests

## 1. Introduction

There are two primary motivations for attention to results from over three decades of both input stimuli and output signals from LENR experiments. The first is to identify empirical factors that have produced or resulted in unusual data that ought to be considered for future LENR experiments. That is a matter of looking backward in preparation for looking forward. The second motivation is to accumulate a diverse data set against which to test various theoretical ideas, if they are developed adequately for such tests. Many concepts about how LENR occur, that is, the mechanism that leads to LENR, are poorly developed. They are only ideas, without accompanying equations and numerical evaluations based on them. However, some concepts for LENR mechanisms have been reduced to equations that were evaluated to obtain numbers for comparison with data. The various types of parametric variations and measurements discussed in this paper serve to challenge the concepts, equations and numbers from diverse LENR theories.

LENR experiments usually consist of three parts, the central experiment, and the means to excite and monitor the experiment. These are shown in Figure 1. LENR experiments have been highly varied, usually to explore various ideas and sometimes to test theories. Many past experiments deserve review and additional consideration. We provide a short review of loading methods in the next section. The third section reviews the ranges of various quanta that are either input to or measured from LENR experiments. Their ranges and characteristics determine their uses. Then, Section 4 presents and discusses a framework for considering stimuli and signals in LENR experiments. This paper primarily focuses on the diverse inputs and outputs from LENR experiments. It essentially has major two parts, one

*Corresponding author:

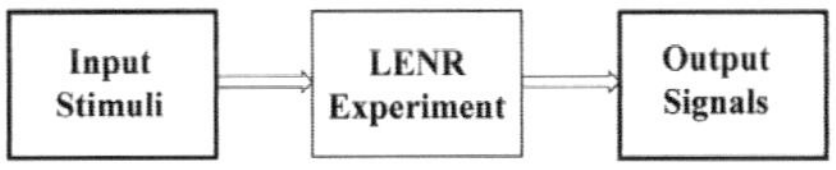


**Figure 1.** The three parts of almost all LENR experiments.

on input stimuli (Section 5) and the other on output signals (Section 6). Section 7 deals with quantities that have served as both inputs and outputs. Those three sections all look backward and forward. Section 8 discusses some factors that have not yet been used in LENR experiences, so it looks ahead. The conclusion highlights four topics. One involves parametric LENR experiments, which can be of great value scientifically and maybe practically. The second is on the testing of LENR theories. There is great need for such testing of the many concepts for how LENR actually happen. The third topic is the requirement for more analyses of LENR data. There have been too few serious analytical studies of data from LENR experiments. Data commonly contain additional useful information which can be extracted. Finally, we have a short discussion of engineering design of LENR experiments. That is another topic, which has not received adequate attention in the field.

## 2. Loading Methods and Materials

The core requirement for the production of LENR is to bring together either protons or deuterons on or in materials, usually metals or alloys. The methods for doing that are commonly called "loading", since they involve loading the hydrogen isotopes onto or into materials. That is, LENR experiments almost always involve two matter inputs, the materials and the sources of the protons or deuterons. As will be clear below, this paper is on inputs other than those needed to satisfy the basic loading requirement for production of LENR. So, in this section, we provide a short review of loading methods and materials as background for the rest of the paper.

Many ways have been used to perform loading over the decades of research on LENR. However, there have been three primary approaches, which are summarized graphically in Figure 2. Each of the three approaches has variations. Aqueous electrochemical loading is the original method of Fleischmann and Pons [1]. It has been done using both light and heavy water from the earliest days of the field. In most cases, the metallic cathode material was placed into an electrochemical cell at the start of an experiment. However, in many experiments, the palladium or other metal has been simultaneously deposited with deuterium from heavy water during the experiment. That method, called co-deposition, was introduced to the study of LENR by Szpak [2]. Molten salt electrolytes have also been tried with success. Electrolytic loading involves relatively low powers, and is a good tool for scientific research on

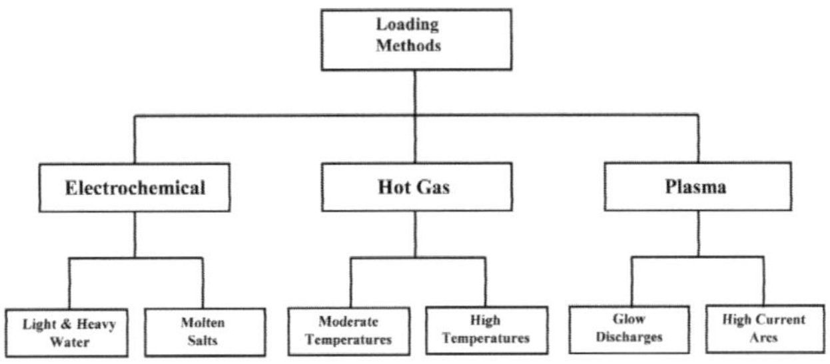


**Figure 2.** The three primary methods for loading protons or deuterons onto and into materials.

LENR. However, without pressurization of experiments [3], it is limited to the boiling point of water, a temperature too low for efficient production of electrical power.

Hot gas loading has involved higher input powers, which are sufficient to produce moderate (about 300 deg C) and high (up to and even beyond 1200 deg C) temperatures. High temperatures can also be generated in lower-power glow discharge plasmas and higher-power plasma arc experiments. Both hot gas and plasma methods of loading involve temperatures sufficient for efficient electricity generation, so they are the two most promising loading methods for commercial LENR generators.

As noted, loading involves two types of matter, (a) the hydrogen isotopes to be loaded and (b) the materials being loaded. Both determine the dynamics and results of loading processes. The paragraphs above in this section dealt with the sources of protons or deuterons. Logically, we would now turn to details on the materials part of loading. There have been many studies of materials before, during and after LENR experiments. They have dealt with both the composition and structures of materials, that is, what atoms are present and how they are arranged. Elemental materials, alloys and compounds have been considered. The majority of LENR experiments have involved palladium or nickel of various purities, and also alloys and compounds including them. Many LENR studies have focused on processes involving such materials before and during experiments.

The occurrence of LENR depends on the composition and structure of the materials in an experiment. Those characteristics depend both on the starting materials and on changes that occur during an experiment due to loading. It is known that success in producing LENR depends on the starting materials, especially for palladium. Variations in composition or structure from batch-to-batch, and even from sample-to-sample within one batch of materials, often makes a difference in the results obtained. Loading can lead to two beneficial changes in materials, (a) the production of regions conducive to the occurrence of LENR and (b) population of those regions with adequate concentrations of reactants. A recent example of changes in materials during loading is Storms' measurements of variations in the enthalpy of formation of palladium hydride as a function of H/Pd ratios for six loading and deloading cycles [4]. A more detailed consideration of LENR materials here would not contribute significantly to the main thrusts of this paper. Hence, we will not review materials issues further, due to a focus on input stimuli and output signals for LENR experiments. Such stimuli into materials and signals out of materials have different ranges, which are important, so they are considered next.

## 3. Ranges of Photons and Particles in Matter

What quanta can be put into or gotten out of LENR experiments depends on three primary factors, the particular quanta (photons, neutrons, electrons and particles), their energy (from small fractions of one electron volt to over millions of electron volts), and the materials through which they pass (both their chemistry, and their density ranging from high vacuum to dense condensed matter). The type of quanta and the energy of the various quanta are especially important in determining what can be done in LENR experiments. A thorough examination of all of these factors would be long, and is not needed in this paper. Rather, we will cite some of the factors and dependencies most relevant to understanding stimuli into and signals out of LENR experiments. It is not necessary to consider unusual particles like muons. A useful diagram relating energy and momentum transfers for various quanta and techniques is available in a review on Mossbauer spectroscopy with nano-eV resolution [5].

Photons have been widely used for LENR research, and there have been thousands of electrochemical LENR experiments involving water-based electrolytes. Figure 3 [6] shows the attenuation lengths of electromagnetic radiation in water from the UV region through the high frequency part of the radio-frequency region. High inverse length (1/m) values mean that the radiation with certain wavelengths will be able to penetrate into or escape from only small thicknesses of water. Since LENR electrochemical experiments have dimensions in the range of 1 to 100 cm (0.01 to 1 m), cells will transmit only photon radiation in range from 0.2 to about 1 micron for the wavelengths shown in

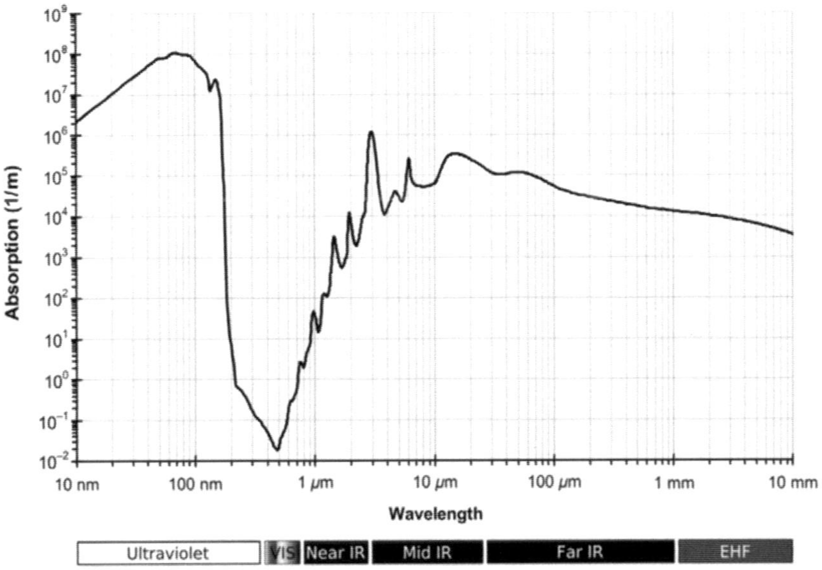


**Figure 3.** Inverse absorption length (1/meters) of electromagnetic radiation in water from the UV through the Visible (VIS)and Infrared (IR) to Extremely High Frequency (EHF) spectral regions. The peak of the solar spectrum is near 500 nm, very near the minimum in the absorption of water.

Figure 3. Very high energy gamma rays and hard x-rays, and very long wavelength radio waves, will also penetrate LENR cells.

Electromagnetic radiation can be made to access or escape most chambers for hot gas or plasma loading LENR experiments. That can be accomplished by using thin and low-absorption materials for most spectral regions, except for the ultraviolet range from about 10 to 100 nanometers. Internal antennas and waveguides are needed for input or detection of long wavelength photons in experimental chambers for hot gas and most plasma experiments.

Neutron energies fall into about ten classes in one accounting [7] and six classes in another [8]. They have ranges in different materials that can vary widely and rapidly with energy. Energetic (MeV) neutrons have long ranges in most materials, so they can be put into LENR experiments, but might be little absorbed within such experiments. They will also escape readily even from metallic reaction chambers. We will note below that few LENR experiments involved input neutrons, and most LENR experiments do not emit significant neutron intensities.

Electron beams are rarely put into or measured from LENR experiments. Their ranges in matter for energies up to a significant fraction of 1 MeV are relatively small. That is also true for ions, from protons to alpha particles (helium nuclei) to heavier particles, even for multi-MeV energies. Hence, it is necessary to use vacuum techniques for input or measurement of electrons or ions to or from LENR experiments. That means that only low-pressure hot gas or plasma LENR experiments are amenable to the use or measurement of electrons or ions. A very useful capability for obtaining data on electron, proton or alpha particle ranges in a wide variety of materials over many orders of magnitude for their energy is on the internet [9]. The available graphs show that the ranges for charged particles in materials vary smoothly with energy, unlike those for photons in and below the x-ray range or for neutrons.

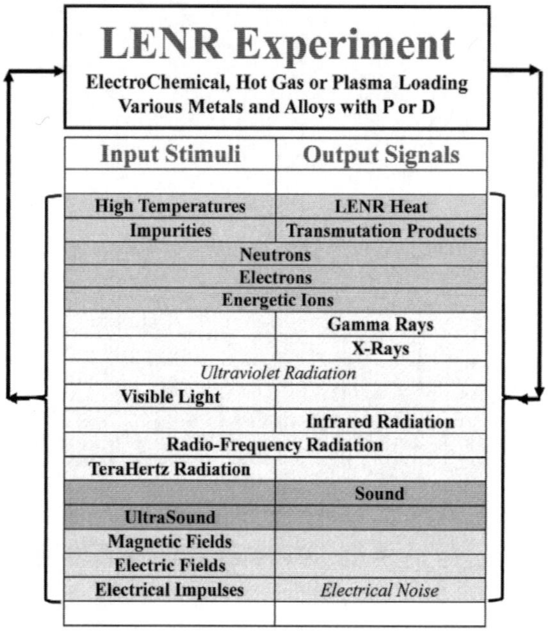


**Figure 4.** Schematic relationship between varied LENR experiments, the diverse input stimuli that have been applied to them (left column) and the many output measurements from them (right column). Blanks in the columns are possibilities that have not yet been applied or measured in actual LENR experiments. Ultraviolet Radiation has yet to be explored by LENR scientists, and Electrical Noise has not been measured from LENR experiments.

## 4. Inputs and Outputs

Experimental papers on LENR now numbers in the thousands. They raise many questions. Of all the published inputs and outputs, which might be most valuable to emphasize in future experiments? What other stimuli or signals might be used to further the understanding and commercialization of LENR? Considering such questions requires having some overview of what has already been done and found in LENR experiments. Figure 4 gives a way to organize much of the laboratory work on LENR. The LENR experiments might involve any type of loading, and any materials with either protons (P) or deuterons (D). It is seen that there are many cases in which particular entity can serve as either an input stimulus or output signal. The various stimuli and signals fall into classes. At the top, those that involve energy and matter are shown in pink. The orange grouping deals with particles, including neutrons, electrons and ions. Several regions of the electromagnetic spectrum, shown in yellow, have played input or output roles in past LENR experiments. Sound has been measured from one LENR experiment, and ultrasound put into a few experiments, as shown in green. Fields have played roles in many LENR experiments, as shown in the blue area in the bottom of Figure 4. The blanks in the diagram, as well as Ultraviolet Radiation and Electrical Noise, indicate cases where no LENR experiments are known. They are worth examining as new opportunities.

With this systematic way to classify the inputs and outputs for LENR experiments, the next section deals with most of the inputs. Section 6 involves many of the outputs. Factors that have or can serve as both inputs and outputs are the subject of Section 7. Section 8 considers the blanks in Figure 4. The concluding section provides additional comments on the various inputs and outputs, plus some discussion of parametric experiments, testing LENR theories, opportunities for more thorough data analyses, and the use of the tools of Systems Engineering.

## 5. Input Stimuli

This section will briefly review eight means of inputting stimuli into LENR experiments. The amount of work on each of them varies widely. However, many of the methods and results are worth consideration for future experiments.

High Temperatures. The question of the effects of temperature on LENR experiments arose early in the field for two reasons. First, it is a basic consideration for any reactions taking place on or in materials. Second, some alternative approaches to the aqueous electrochemical method of Fleischmann and Pons required high temperatures. One early example of the latter was the use of electrolysis experiments in which the electrolyte was a molten salt at 250 deg C by Liaw and his colleagues in 1991 [10]. They reported significant, albeit irreproducible power gains of 1.2 to 1.4 (after subtraction of heater power). The power gain at any specified time is the ratio of the output thermal power to the input electrical power. They report producing as much as 25.4 watts at about 460° C. Anomalous helium was also reported in a later paper by the group [11]. Another example was the use of a glow discharge by Karabut in 1992 [12], who got power gains of as much as 5. Lonchampt and others [13], in a 1996 reproduction of the Fleischmann and Pons electrochemical experiments, showed that higher temperatures favor excess power. They stated "Our results concerning the relative excess heat (percentage of excess heat to enthalpy input) can be summarized as follows: -below 70°C, between 0 and 5%, -between 70°C and 99°C, about 10%, and -at boiling, up to 150% especially in the final phase which appears as the best condition to get a large amount of excess heat." In 1998, Mengoli and colleagues [14] varied the temperature of aqueous LENR electrochemical experiments to near the boiling point. They reported "After prolonged electrolysis the loaded electrodes were found to continue heat generation in open circuit conditions." and "The major achievement of this work is to have devised a temperature range in which the generation of excess power is a totally reproducible phenomenon." In short, the benefits of using elevated temperatures in LENR experiments were well established within the first decade after the 1989 Fleischmann and Pons announcement.

Scientists at Tsinghua University have compiled useful summaries of the temperature dependence of six diverse LENR experiments [15]. Their results are summarized in Figure 5. All the plots show that (a) high temperatures favor production of LENR, and (b) the data fit an exponential variation with an activation energy (the Arrhenius equation) [16]. While clear and agreeable on those two points, the data raise two questions: (1) why do LENR obey such an activation equation, and (2) what determines the values and varieties of the activation energies, which range from 0.14 eV to 0.95 eV? These questions challenge theoreticians for explanations of the relatively simple and well-established behavior. Besides such clear scientific importance, the issue of the temperature dependence of LENR has strong practical implications for potential future commercial power generators. The temperatures at which LENR generators will operate will determine the materials for their construction, and influence their longevity.

Very recently, Storms [17] reported temperature variations experiments for both electrochemical and hot gas loading of palladium and nickel with deuterons. His data also exhibited behaviors similar to those shown in Figure 5, that is, Arrhenius variations with activation energies. He found two regimes with different temperature dependence in experiments with both Ni and Pd, some with electrolytic loading and others with hot gas loading. The behavior at lower temperatures was ascribed to the filling of active sites, in which LENR can occur, with deuterons. The high temperature behavior was thought to be controlled by the diffusive resupply of deuterons to the active sites to make up for deuterons consumed during nuclear reactions.

Impurities. In the early days of the semiconductor industry, sodium impurities led to irreproducible behavior of silicon-based integrated circuits, and control of sodium remains a challenge during modern chip production [18]. Impurities could also play a role in LENR experiments in a few different ways [19]. If impurities are a cause of LENR, they can improve the rates of LENR, or result in lower LENR rates, as they are consumed or move within

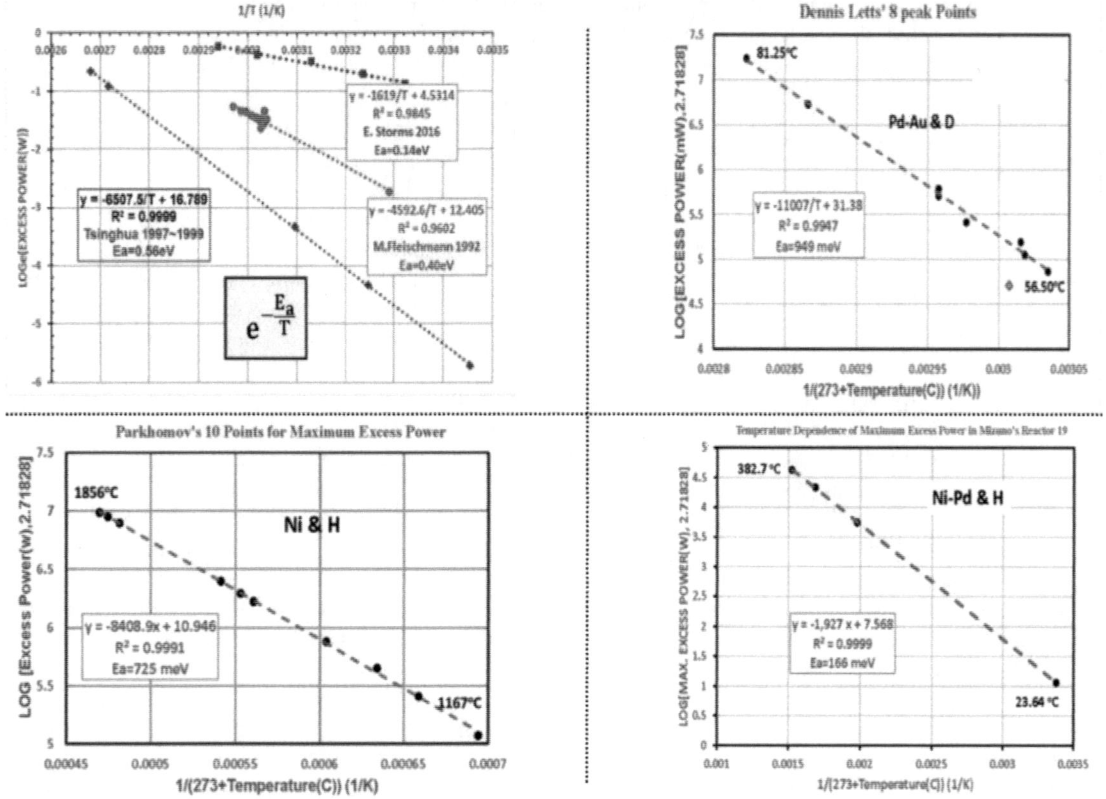


**Figure 5.** Four plots with data from six experiments, all having the log of LENR power vertically vs the inverse of the absolute temperature horizontally. All plots obey the Arrhenius activation equation, the main factor of which is in the yellow box.

materials. If materials are a result of the occurrence of LENR, they can similarly increase or decrease LENR activity. Impurities might create or destroy needed conditions, for example, the attainment of very high ratios of D to Pd in some locations within a lattice. Segregation of impurities to grain boundaries could impede the motions of reactants (protons or deuterons) along grain boundaries, which normally have high diffusion coefficients. That is, impurities might alter the rates for diffusion of hydrogen isotopes to LENR active sites. And, they might even change the number of such sites. Impurities could conceivably promote or retard reaction rates without participating in the reaction, much as do catalysts or inhibitors for chemical reactions. And, impurities could actually participate in the reactions. The numbers involved in this last possibility are especially interesting. If 1 part-per-billion of an impurity in a typical LENR cathode undergoes a nuclear reaction which yields 1 MeV each second, a power of about 1 W is produced. One watt is a nominal excess power from many LENR experiments that produced excess heat. If 10 or 24 MeV resulted per reaction, also normal sorts of nuclear reaction energies, correspondingly less material or a lower reaction rate would be required, or a higher excess power would result for the same rate. It should be noted that the large positive charges on most impurity nuclei argue against their being participants in LENR.

The important point is that few LENR experiments have been analyzed for impurities to the parts-per-billion level in the various phases within the experiments before, during or after runs. Some experiments seeking to quantify the appearance of transmutation products after LENR reactions have included such analytical work. Two famous examples are the studies of reaction products across the periodic table by Miley [20] and by Mizuno [21]. Although very different experiments, their results strongly support each other. Data from both Miley and Mizuno show peak LENR transmutation rates at the same values of the atomic mass across the entire periodic table. In these experiments, the impurities that were produced were apparently the result of LENR. It is not known if the increasing concentrations of such LENR products (a) influenced the rates of further LENR or, (b) if there was some feedback, whether it was positive or negative. Impurity levels in materials generally vary on the parts-per-billion level from batch to batch and, even, from sample to sample within a batch or single piece of material, or from centimeter to centimeter along a wire or rod of material. It is possible that, if there are unknown processes at work in LENR experiments that involve impurities, the reproducibility and controllability riddles will be explained on the basis of inadequate knowledge and control of impurity levels before and during experiments.

Given the potential roles of impurities in determining the outcomes of LENR experiments, and the dearth of quantitative studies of impurities, it would be reasonable to prepare materials for LENR experiments with a range of types and concentrations of impurities. That would be challenging, given both the wide variety of possible impurities, and the many combinations and concentrations that could be relevant. However, such parametric experiments could be useful both scientifically and practically. They might increase the probability of producing LENR. Also, they could have a major impact on the longevity of materials in LENR generators. The best strategy for such experiments with impurities is uncertain. Possibly, it would be good to start with the impurities normally found in palladium or nickel, since those metals are often used to produce LENR. A common approach to screening for impurity effects is to introduce only a few impurities, say six or less, into each sample. Then, if any sample with multiple impurities shows interesting behavior, the impurities in it can be put separately into further samples. That approach works if multiple impurities in one sample do not interact with or influence each other.

Low levels of materials dissolved in electrolytes have played a role in some electrochemical LENR experiments. In these cases, the impurities might be restricted to the surfaces of the cathode materials. Examples of such work were done by the groups in the Sydney Kimmel Institute for Nuclear Renaissance [22] and the company Coolescence [23].

Visible Light. Letts and Cravens presented a paper in 2003, which stated they had been exploring the effects of stimulation of LENR experiments by visible laser radiation for over a decade [24]. Figure 6 is one of the graphics in that paper. It shows that shining a 30 mW laser onto a gold-coated palladium cathode in an electrochemical LENR experiment within a calorimeter caused a 500 mW increase in LENR power over the course of about 15 hours. Removal of the light then caused a drop in the LENR power. The fraction of the laser power actually absorbed was not reported. The increase in LENR power was stated to be 5 to 30 times the laser input power. The paper reported on the polarization dependence of the 257 nm laser light, but did not involve variation of the wavelength. The Letts-Cravens paper has been cited about four dozen times due to both replication experiments and the overall importance of the work.

Another early study of the beneficial effects of laser light on LENR output was reported by Swartz and his colleagues [25]. Their abstract included these statements: "There is a positive photo-thermoelectric response for optically-irradiated [670 nm laser, 3.5 milliwatts] spiral-wound, electrically-polarized palladium cathodes in very low electrical conductivity heavy water. An incremental photoinduced excess heat of $\sim$89 milliwatts results from a $\sim$3 milliwatt incident optical beam, but only in the presence of a functioning active loaded cathode." Later, Violante and his team showed that application of laser light to electrochemical experiments led to an improvement in their reproducibility [26].

The use of incident visible light to stimulate and control the generation of LENR power calls for much more research. The variation of LENR production with the wavelength of the incident light is an open and important question. Similarly, how LENR varies with incident and absorbed intensities, angles of incidence and surface characteristics

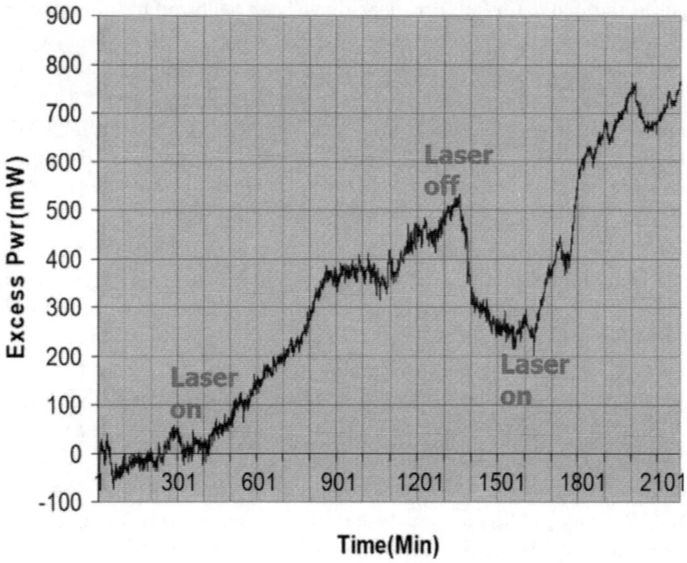


**Figure 6.** Time history of LENR power showing responses to laser light shined onto the cathode by Letts and Cravens.

should be determined. It would not be difficult to make a setup where light emitting diodes (LEDs) surround a cell, either on the outside or even in the electrolyte, to provide uniform and intense illumination of a cathode. Variations in LENR output with intensity could be studied with such an arrangement. The size of the area affected by focused laser light is unknown, and also calls for further experiments. Does the light affect only material that is illuminated, or is a larger area outside of the illuminated region induced to produce LENR?

Terahertz Radiation. Electromagnetic radiation of a very different wavelength was used to stimulate LENR in another experiment by Hagelstein, Letts and Cravens [27]. Beams from two lasers were beat against each other to produce terahertz radiation, which was again put onto the cathode in an electrochemical experiment in a calorimeter. The main graphic from that work is in Figure 7. It shows that the LENR power varies strongly with the frequency of the THz radiation. The abstract of the paper reads in part: "The cell responded to three difference frequencies in the THz range at 8.2 THz, at 15.1 THz, and at 20.8 THz. The first two of these frequencies can be associated with optical phonon frequencies of PdD with zero velocity." Those data were taken as evidence that phonons are important in the production of LENR. That is, it is possible that THz optical absorption produces phonons, and phonons contribute to the production of LENR. It should be noted that Vysotskii and his colleagues offered alternative interpretations of the peaks in Figure 6 in terms of coherent correlated states [28].

A later paper by Hagelstein and Letts [29] dealt with the temperature dependence of the experiment that produced Figure 7. Interestingly, it concluded "A picture for the temperature dependence is described in terms of the elimination of $^4$He which blocks active sites when the excess power is high." The effects of reaction products on further LENR reactions have received some discussion in the LENR literature. Helium can become trapped as gas bubbles in grain boundaries as result of its low solubility [30]. However, that might not affect further LENR production. Wang and Arata addressed the problem of He piling up in nano-particles [31]. The nano-scale particles are expensive, so it would be economically beneficial to have a way to resuscitate them for further use by processing them to remove LENR products. The presence of the He might reduce the rates for H and D diffusion within the materials, and hence cut

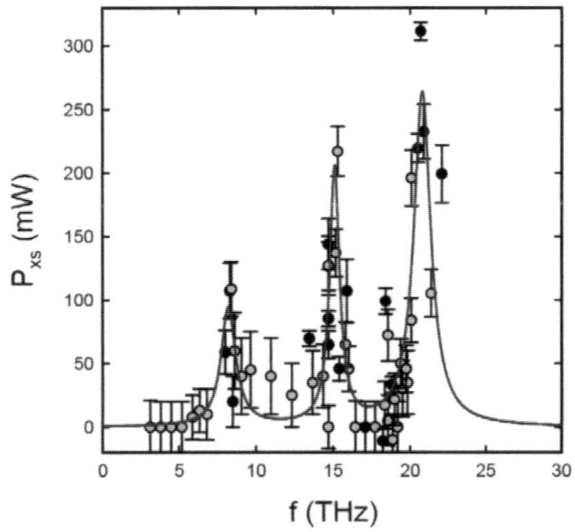


**Figure 7.** LENR power as a function of THz frequency put onto the palladium cathode in an electrolytic experiment.

down on the availability of the fuels for the LENR. Storms found that the diffusion of deuterium to reactions regions was the rate limiting step in LENR production.

The Thz experiment described above apparently has not been repeated in another laboratory. It seems like another good candidate for additional attention. The same variables noted above for visible light (polarization, geometry and intensity) are germane to LENR experiments with THz radiation. It seems particularly attractive to see if the effect observed with palladium, shown in Figure 7, could be seen also with other material systems, especially nickel.

Ultrasound. Several LENR experiments have involved ultrasound, which is defined as sound with frequencies greater than 20 kHz. One ultrasound experiment reportedly produced conventional deuteron-deuteron "hot" fusion within liquid media. That topic has been very controversial [32]. It is commonly called bubble-fusion or sonofusion. Such experiments probably have nothing to do mechanistically with LENR. The reported nuclear reactions do not occur on or in a solid medium, but rather in very small and very transient hot plasmas formed inside of collapsing bubbles within liquids.

Ultrasound has been used in two ways in actual LENR experiments. In one case, it led to the deuterium loading of solids and production of excess heat. Stringham used ultrasound for about two decades to stimulate LENR on and in thin foils of various metals. He reported producing nuclear reactions using both kHz and MHz excitations, with 40 W of excess power reported for one experiment [33]. In the second case, ultrasound was used as an adjunct to electrochemical LENR experiments. Energetic Technologies employed ultrasonic irradiation in addition to "superwave" electrical excitation. They reported a very high energy gain of 26 in one well-known experiment, which is discussed later in this section. That group also published an extensive paper on the application of ultrasound to the cathodes in electrochemical LENR experiments [34].

The situation for ultrasound inputs to LENR experiments is quite like that for putting electromagnetic radiation into such experiments. There is a similar large number of variables, again including frequency, intensity, and how the inputs do or do not vary with time. Here again, thorough experiments to address the effects and possible values of various sonic inputs would require (a) a major systematic research program and (b) a reproducible LENR experiment, so that the effects due to input parameter changes could be separated from other measured variations.

Magnetic Fields. The intersection of magnetic fields and LENR experiments is asymmetric. There have been many more experiments in which such fields have been applied as input to LENR experiments than situations where magnetic sensors have been used to monitor output fields. So, we first consider the application of magnetic fields to LENR experiments. The following is a sampling of papers about magnetic field effects on LENR, not a thorough review.

Letts reported results in 2011 from an experiment done in 2002. He showed that the orientation of a DC magnetic field from a permanent magnet directly influenced the production of excess power [35]. He used a 500 Gauss (0.05 T) magnet external to an electrochemical LENR cell which contained palladium co-deposited with deuterium onto a planar copper substrate. The variation of the LENR power with magnetic field orientation is shown in Figure 8. Moving the magnetic field direction from parallel to orthogonal to the cathode foil increased the excess power by a factor of about three. Then, returning the magnet to the initial position led to a decrease in the rate of LENR power production. That is, the LENR power was highest when the magnetic field was parallel to the electric field between the electrodes.

Swartz also studied the effects of magnetic fields on LENR (LANR) in electrochemical cells [36]. The cathode was a coil of palladium. He found that the field increased the electrolyte resistance, which limited unwanted gas evolution and improved loading of deuterons into the Pd cathode. That led to a higher rate of LENR occurrence. He wrote, also in 2011, "An applied magnetic field ~0.3 T increases the LANR solution's electrical resistance ~10 –17% with a time constant in minutes. The incremental resistance increase from an applied H-field is greatest with the applied H-field perpendicular to the driving electrical field (E-field) intensity."

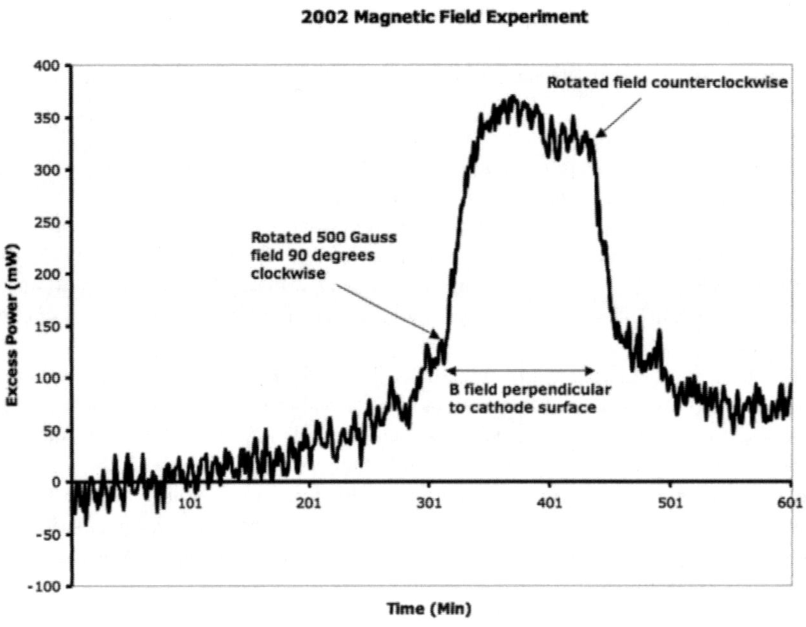


**Figure 8.** Dependence of the LENR power on the orientation of the magnetic field from a permanent magnet relative to the plane of a palladium cathode foil.

Letts and Swartz reported opposite orientation effects for a magnetic field on LENR rates, both in electrochemical systems. Whether or not that variation is due to the different geometries for the cathodes in their experiments is an unanswered question. Given the observed variations and the differences in the experiments, simulation of both experiments with software like COMSOL ought to be valuable [37].

There is a very basic question about the role of magnetic fields in electrochemical LENR experiments. It is possible that the fields will directly influence the rates of LENR, as indicated in an equation produced by Letts [38]. However, it is also possible that magnetic fields will alter the electrochemistry, and hence the rates of loading or deloading of the cathodes, as Swartz reported. There is a substantial non-LENR literature about magnetic field effects on electrochemical processes [39]. It appears that no one has yet digested that non-LENR literature in an attempt to determine if magnetic fields change the electrochemical dynamics in LENR experiments in ways that might indirectly affect LENR rates. Determination of whether the effects on electrochemical LENR experiments by magnetic fields are direct (only physical) or indirect (first chemical, and then physical) should be a near term goal.

In 2014, Swartz and others reported magnetic field effects on the rates of LENR in an experiment without electrochemistry [40]. They studied steady and pulsed magnetic field effects on a device called the NANOR$^{TM}$ which contains $ZrO_2$–PdD nanostructured components. DC and pulsed magnetic fields from three types of magnets were used. The pulsed fields were about 1.5 T with a 0.1 ms rise time, with 1 to 5000 pulses applied to the device. Energy gains were increased over a factor of four by use of DC fields. The pulsed fields produced large effects on the LENR output. To quote the authors, "The peak power gain was ranged from 22 to up to $\sim$80 times input electrical power or more beyond the control, as determined by calorimetry." It was found that the magnetic field effects could be both immediate (synchronous) and long lasting (metachronous). In both cases, the variation of LENR power with input power to the NANOR$^{TM}$ was different from the case with no magnetic fields. A 2016 paper by Swartz provided additional data and analyses of the response of his devices to magnetic fields, including discussion of the effects of magnetic domains within the materials of his devices [41].

As noted at the beginning of this section, there are some, but very few instances of magnetic fields being measured from LENR experiments. In fact, they are more ancillary measurements, than detection of magnetic fields generated during such experiments. In one instance, a magnetic antenna was used to pick up electromagnetic radiation [42]. In another, the magnetic field within a particle detector was measured [43]. It is clear that the potential emission of magnetic fields from LENR experiments is an unexplored territory.

Future LENR experiments involving magnetic fields might be designed to determine whether the effects seen in some past LENR experiments were direct or indirect. They might include variations in the field strengths and orientations, as well as temporal variations in those parameters. In all cases, the strength of the magnetic field at the cathode should be measured. Given the magnetic field effects already observed in various LENR experiments, it seems especially desirable to measure the effects of parametric variations of magnetic field parameters for loading of both solid and co-deposited materials. That ought to be especially true for time variations of magnetic field intensity, which have not been thoroughly explored. Such experiments can be conducted without having a reproducible LENR experiment, since various diagnostic methods and calorimetry can be used to determine the effects of diverse magnetic field characteristics.

Electric Fields. There are two types of electric fields applied to LENR experiments, those within experiments that accompany the applied voltages and those that are applied externally to the setup. Electric fields are closely related to the voltages applied to LENR experiments, since the value of an electric field is the gradient of the voltage. That is, electric fields can be expressed in Volts per Meter. Since Volts give the Energy per Charge, an electric field is an expression of the rate at which the energy of a charged particle changes spatially. Charged particles in an electric field will move to minimize their energy, much like a ball rolling downhill.

Internal electric fields due to applied voltages fall into two classes for electrochemical LENR experiments. The first is the field due to the voltage difference between the cathode and anode through the electrolyte, which is used in all experiments. The second, used in only some cases, is a voltage drop along an elongated cathode. The latter leads to an effect called electromigration. That term describes the motions of deuterons or protons within the cathode, which causes two beneficial effects for LENR production. One is an increase in the concentration of reactants in the positive end of the cathode wire. The other is a flux of reactants along the wire, which induces the interactions that are often prerequisite to reactions.

Electromigration has been used in several LENR experiments, including many by the Celani Group [44]. In 1996, the remarkable results of "about fifty" LENR electrochemical experiments with electromigration were reported in two papers by Preparata and his colleagues. The first dealt with an initial experimental campaign [45]. The second reviewed those results and provided data from the second campaign [46]. That paper provided many details of the experiments. The cathode was a wire of Pd 50 $\mu$m in diameter. It was configured to have a current flowing along its length of 250 cm, as well as between it and the anode. The experiment was run for 6,000 seconds, with the maximum input power of about 87 W and maximum excess power of over 83 W. Normalizing the excess power to the volume of the cathode gave surprising power densities in the range of 50 to 100 kW/cm$^3$. These values exceed those of nuclear fuel rods in fission reactors. They required fast heat loss from the thin wire to the electrolyte to avoid melting of the wire.

External electric fields have been applied to LENR experiments, in addition to the normal internal applied voltages in many electrochemical and gas loading LENR experiments. For gas loading, the fields (voltages) generally have been used to produce plasmas that can initiate and even control power production. That is, their effects are necessary, but not sufficient for LENR to occur. That is not the case for the use of fields in electrochemical LENR experiments. Then, the applied external fields can directly influence the ability to produce LENR.

Researchers at the SPAWAR Systems Center did co-deposition experiments in which static electric fields were applied to the regions of the cathode from outside of the cell [47]. The morphology of the co-deposited Pd-D materials varied greatly with the strength of the fields. Figure 9 provides one illustration of the influence of electric fields on (a) the structure of the materials produced in co-deposition experiments and (b) the results of the experiment.

While electric fields have been shown to greatly affect materials in co-deposition experiments, there seems to be no data on their influence on LENR rates. If it is shown in future LENR electrochemical experiments that there are electric field effects on LENR, then there will arise a basic question: do the fields directly affect LENR rates, or do they influence the dynamics of the electrochemistry which, in turn, determines such rates.

There are many sensors for electrical fields [48]. However, there has been little or no use of such sensors inside of LENR experiments. That is, electric fields have not been measured as either input or output signals within LENR experiments. Recent advances in electric field sensors based on fiber optics might enable measurements of fields within electrochemical LENR experiments. They offer spatial resolution of the scale of 10 micrometers [49], and might be used to determine internal fields even within aqueous electrolytes.

Overall, the situation for prospective experiments with electric fields applied to LENR experiments is very similar to that for the use of magnetic fields. Whatever their motivations, future LENR experiments with applied electric fields might include variations in the field strengths and orientations, as well as temporal variations in those parameters. Given the electric field effects already observed in co-deposition experiments, it seems highly desirable to measure the effects of parametric variations of electric field parameters on atomic and particle structure of materials that are deposited *in situ*. That ought to be especially true for time variations of electric field intensity, which have not been explored. Such experiments can be conducted without having a reproducible LENR experiment, since diagnostic methods like x-ray diffraction and scanning electron microscopy can be used to determine the effects of diverse electric fields.

<u>Electrical Impulses.</u> Disequilibrium, especially the presence of fluxes of deuterons in palladium cathodes, has been discussed for LENR experiments since the earliest days of the field [50]. We now review another type of disequilibrium, specifically the application of impulses to LENR experiments. It is not known yet if impulses act only to change the

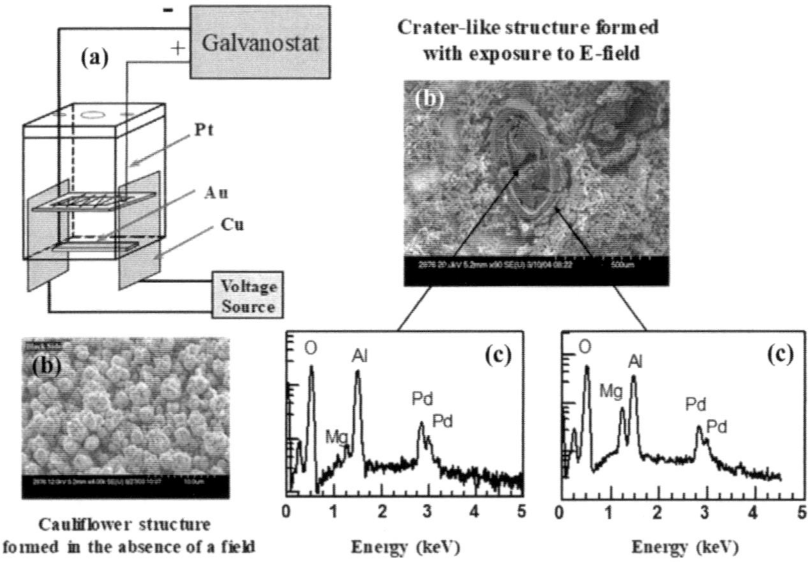


**Figure 9.** The experimental arrangement (a) for a co-deposition experiment with an applied electric field, (b) micrographs showing the different morphologies of the material without and with the field present, and (c) x-ray spectra showing the appearance of Mg and Al during the experiment at a crater.

chemical conditions within electrochemical LENR experiments, or if they act directly on the nuclear reactions. Two key experimental plots will illustrate the apparent importance of impulses for electrochemical LENR experiments.

The first data set is the well-known graphic from Energetics Technologies, first presented at ICCF-11 in 2004, and shown in Figure 10 [51]. The data in that figure showed instantaneous power gains as high as 35 and an integrated energy gain of 26. There are five times when the *input* power sharply decreased for a brief period. They are labelled A through E in Figure 10. At those times, the input power went to zero (once, namely D) or even apparently negative (A, B, C and E):

- A was at the start of the experiment, when there was an immediate, but small excess power.
- B marks two negative excursions, when the output power shot up from about 2 W to 18 W.
- C marks a second large jump in thermal power, from 20 to 29 W, after an apparently negative input change.
- D marks a time when the input power went to zero, which was followed closely by an increase of output power from about 13 W to 34 W.
- E notes another apparently negative excursion of the input power near the end of the experiment, which did not result in an increase in the output power.

In addition to the large increases in output LENR power following closely the small decrements in the input power, there are other correlations between input and output power. They are indicated by the green lines in Figure 10. In general, there are significant increases in output power due to most of the transient very small increases in the input power. The rise times in the output power are limited by the time constant of the calorimeter.

Overall, the three sharp decreases in input power in Figure 10 led to three positive excursions in output power. The meaning of the apparently negative input power is interesting. During those brief periods, the experiment appeared to

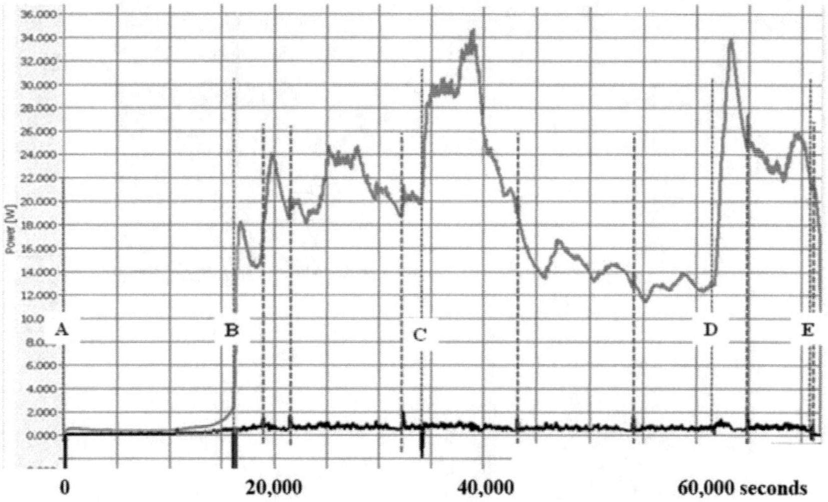


**Figure 10.** Time variation of the electrical power input (black) and thermal power output (red) for an experiment conducted by Energetics Technologies in Israel. Times when the input power dropped precipitously are indicated by the letters.

be acting as a battery, supplying rather than consuming power. It is not clear if the instrumentation in use to record the input electrical power was capable of detecting and properly measuring such behavior. However, it is evident that negative excursions in the recorded input electrical power resulted in positive increases in the output thermal power. Some of the half dozen brief positive excursions in input power did not produce significant changes in the output power. That might indicate that fluxes are a necessary, but not sufficient condition for the production of LENR.

A second strong LENR heat generation result, which involved small changes in input power with large effects, was obtained at the Naval Research Laboratory, as shown in Figure 11 [52]. The letters indicate the times when there were discontinuous increases in the input power. In most cases, save for times E, F and J, increases in the output power occurred at the times of, or soon after those increases in input power. Discontinuous electrical power inputs generally had beneficial effects on the LENR output power. Figures 10 and 11 both show that very small changes in input power led to much larger changes in output LENR power. However, in one case (Figure 10), sudden *decreases* in input power triggered large increases in output power. In the second case (Figure 11), it was sharp small *increases* in input power that triggered larger increases in the output power.

The natural question of why both positive and negative excursions in input power should lead to increases in output LENR power might be easy to answer. Much laboratory evidence, and some LENR theories, indicate that a flux of deuterons is desirable to produce LENR. However, there is no indication if an inward or outward flux is better. Both produce deuteron motions that can lead to higher LENR production rates.

The data in the past two figures have interesting and maybe important implications. It is possible that using either or both increases or decreases in input power, with any waveform, could provide a means to vary the output power of a LENR experiment, prototype or product that uses a LENR cell of any type. This suggests that experiments be performed in which the input electrical power is willfully changed suddenly. Figure 12 shows one possibility. It is also possible to input pulses by the use of coils around the leads to the electrochemical cell. There are many choices to be made in such experiments in addition to picking either an increase or decrease in the power. The magnitude and time duration of the changes have to be chosen. The frequency with which changes are made is another variable. By

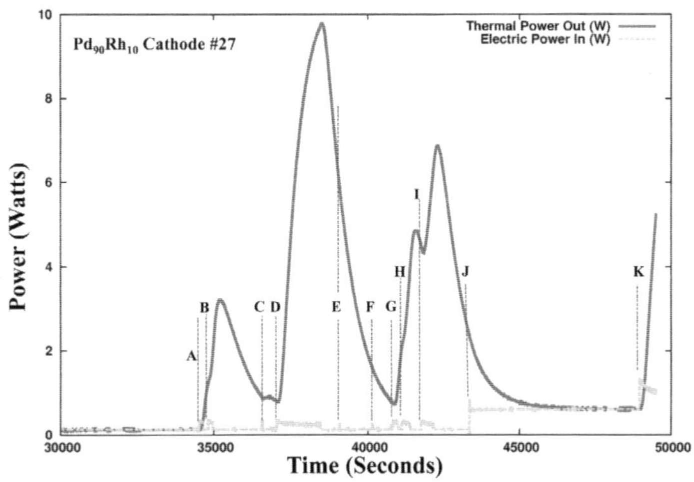


**Figure 11.** Time variation of the electrical power input (green) and thermal power output (red) for an experiment conducted by the Naval Research Laboratory in Washington DC. The letters indicate times when the input electrical power increased suddenly.

themselves, those four variables of direction, magnitude, duration and frequency, present a large experimental matrix. Execution of experiments in such a matrix would be challenging, due to the large number of possibilities, even if one had a reliable means of producing LENR at will. However, the current inability to reproducibly and controllably cause LENR to happen in an experiment makes it doubtful if such parametric experiments can be done successfully now. It must be noted that input pulses are being used by Brillouin Energy Corporation in their LENR research and development [53]. They believe that the pulses directly affect nuclear reactions.

While waiting for reproducible LENR experiments to be developed, it is possible to consider the effects of sudden changes in the input power, either or both due to changes in electrolysis or the use of heater pulses. Step functions [54], and also square waves [55], contain many high frequency components. That is, the Fourier Transforms of such input waveforms contain many frequencies, some of which might be useful in producing LENR. That might be the case if there are one or more resonances that might be excited, which will improve LENR rates. It is possible that the use of "superwaves" in some productive LENR experiments produced desirable excitations [56].

## 6. Output Signals

This section deals with six various signals, which have been measured from LENR experiments. As with the input stimuli considered in the last section, the amount of work on the diverse signals varies greatly. The difference is that input stimuli can be controlled, but it is rarely possible to completely or even significantly control what comes out of a particular LENR experiment. It is interesting that a list of 11 different types of output observations appeared as early as 1990 in a paper by Fleischman [57]. He listed separately steady and burst observations.

Heat. Thermal energy is probably the most measured variable from LENR experiments. One reason might be the fact that it was measured in the original Fleischmann and Pons experiments. The level of LENR energy is often so high that potential chemical reactions are excluded as the origin, leaving only nuclear reactions as the likely source. Another, and probably more compelling reason for interest in heat measurements, is the possibility that commercial LENR generators will be developed to produce at least heat and maybe also electricity.

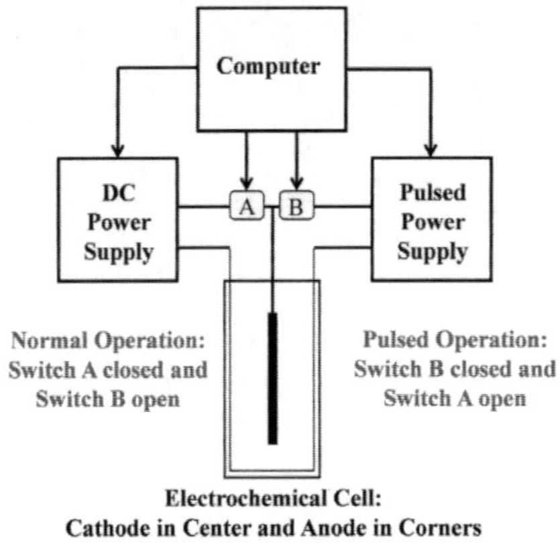


**Figure 12.** Circuit for putting impulses into electrochemical LENR experiments.

There have been thousands of LENR experiments in which heat production was measured by thermometry or calorimetry. Searching the lenr.org library [58] with the word "heat" produces a list of over 1100 papers and reports. Since this paper is focused on the less common inputs to and outputs from LENR experiments, we will not consider heat measurements further here. Thermal power measurements were part of the discussion in the last section of the effects of impulses on LENR experiments.

Transmutation Products. The situation for LENR reaction (transmutation) products is similar to that for heat measurements. There has been great interest in products from LENR, since they provide much information on what happened in an experiment. Another search of the lenr.org library with the word "transmutation" produced over 400 hits. As with thermal measurements, there is much further work on LENR products that could be done. One motivation is the fact that some really unusual transmutation products have been reported. The papers by Miley [20] and Mizuno [21], already noted above, both show the same peaks in LENR rates as a function of atomic mass. The reasons for the existence of those peaks remains to be explained, which calls for further measurements of transmutations, as well as additional theoretical considerations. Similarly, Iwamura reported transmutations that apparently involved as many as six deuterons [59]. There is no theoretical explanation for those results. And, there are several reports of isotope anomalies after LENR experiment, which are not understood [60]. The variety of transmutation results raise the question about whether or not there is only one mechanism for LENR. That is possible, but there might be multiple mechanisms by which to produce LENR. This is one of the more fundamental open questions in the field.

Gamma Rays. Interest in measuring gamma rays from nuclei in LENR experiments immediately followed the 23 March 1989 Press Conference by Fleischmann and Pons for two reasons. One motivation was the known gamma ray emission from ordinary deuteron fusion in one out of ten million events. Although relatively rare, the gamma emission from deuteron fusion is readily measured. The other reason was reports of gamma ray measurements by Fleischmann and Pons, as documented in their 1989 initial paper [61], which was submitted ten days prior to their Press Conference, and in another paper the same year [62]. Searching the lenr.org library with the word "gamma" produces about 230 hits. There are reports of gamma ray spectra from various LENR experiments, but they are relatively rare. One early example is from a glow discharge experiment [63]. The spectrum contained 25 gamma ray lines from 51 to 335 keV, nine of which were identified with isotopes of Sr, Rh and Pd. It is noted that gamma ray emission can originate from radioactive decay, as well as from the decay of nuclear excitations. The low levels of radioactive products from LENR experiments are challenging scientifically and potentially important commercially.

It is now known that LENR experiments do not generally emit significant gamma ray intensities. That is true both during and after experiments. LENR experiments do not usually produce significant residual radioactivity. That is good, of course, since there is no radioactive waste problem for LENR, as there is with existing fission and anticipated fusion power systems.

X-Rays. There are two classical sources of x-rays. One is line radiation from electrons filling vacancies in atomic orbitals and the other is continuous radiation called Bremsstrahlung from deceleration of energetic electrons by nuclear fields. The situation for x-ray emission from LENR experiments is a lot like that for gamma ray emission from such experiments. There are about 200 papers in the lenr.org library that contain the word "x-ray". However, many of them have to do with using x-ray diffraction to probe the lattice parameters of materials in LENR experiments, rather than measurements of x-ray emission during or after experiments.

Similar to the case for gamma rays, some examples of x-ray emission and spectra can be found in the LENR literature. One of the early cases, Iwamura and his colleagues [64] measured x-rays from an electrochemical LENR experiment. They observed a clear peak in the x-ray emission near 75 keV during the run. Spectrochemical analysis of the palladium cathode after the run showed the presence of lead. The strongest K$\alpha$ x-ray emission line of lead is at 75 keV.

Much lower energy x-rays were measured by Storms and Scanlan in an experiment with a glow discharge in deuterium gas [65]. The conclusion of that study contained the following statements "Anomalous radiation is produced when a glow discharge is produced in deuterium gas. This discharge consists of low-energy X-radiation and occasional energetic particles. The energy of the X-radiation increases as voltage applied to the discharged is increased, with an energy equal to applied voltage at 600 V. Intensity increases with applied voltage and with increase in O/D ratio. The source does not appear to be K-radiation from oxygen or nitrogen. The intensity of X-radiation is at least 1000 times greater than the particle flux. Insertion of Mylar absorbers suggests the particles are deuterons with energy peaks having various values between 0.5 and 3 MeV. The energy is sensitive to voltage applied to the cell and to gas composition."

Infrared Radiation. There are only a few LENR papers where infrared (IR) emission was reported. The first was taken from a cell in which palladium and deuterium were co-deposited on a substrate near the wall of an electrochemical cell [66]. One of the images from a video [67] of that experiment is in Figure 13. The red area from the cathode is warmer than the surroundings. Within that region, there are small white spots that are even warmer. The video showed

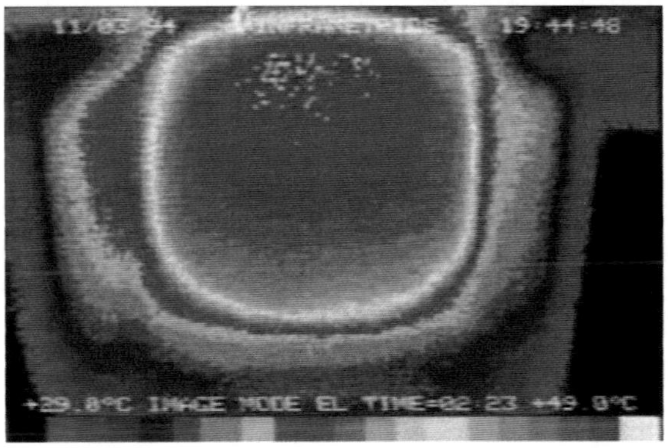

**Figure 13.** Infrared image of the cathode in an electrochemical LENR co-deposition experiment.

that the hotter regions occurred at different locations and lasted for only short times. The behavior is consistent with fast heating due to the release of LENR energy in small regions, which then cool quickly. Estimates of the temperature increases based on the particle sizes in the co-deposited materials are consistent with nuclear levels of energy: release of 1 MeV in a cube of Pd 100 nm on a side gives a temperature (T) rise of $\Delta T = 380$ K using $(3\ k\ \Delta T)/2$ as the increase in vibrational energy, or $\Delta T = 55$ K using the specific heat for Pd = 26 J/K mole [68]. Additional experiments of this type should be very valuable, especially if the infrared imaging were augmented with time-resolved infrared spectroscopy.

In 2008, Swartz reported the measurement of non-thermal IR radiation from different LENR experiments [69]. To quote from that paper, "Non-thermal near-IR electromagnetic radiation is emitted from LANR active electrodes when excess heat is observed at their optimal operating point. The appearance of excess energy in several types of high impedance and co-deposition Phusor®-LANR (lattice assisted nuclear reaction) devices [Pd high impedance, Pd co-deposition, and possibly Ni with light water] was simultaneously detected from loaded, active nickel and palladium devices, by *in situ* monitoring."

The only other LENR experiment involving IR measurements was reported in 2019 [70]. It involved multiple layers of hydrogenated copper–lithium–boron alloy electrodes separated by a liquid dielectric containing fine graphite particles in a capacitor configuration. Infrared imaging was used to view the edges of the stacks of electrodes through an IR transmissive window in the cell.

Infrared imagers, which can be attached to smart phones, are now available for modest costs [71]. That fact, and the likelihood that IR imaging of hot gas and plasma loading experiments would be valuable, indicates that their use might be relatively productive in some future LENR experiments.

Audible Sound. The range of audible sounds has frequencies from 20 Hz to 20 kHz. The emission of acoustic radiation was measured in only one known LENR experiment. Szpak and his colleagues co-deposited Pd and deuterium onto a piezoelectric disk [72]. Such a disk is a reciprocal transducer for sound and electrical signals. Deformations due to sound produces voltages, and application of voltages causes deformation in piezoelectric materials. In the LENR experiment, deflection of the disk produced voltage signals indicative of the emission of sound in the experiment. That experiment, which is relatively easy to perform, has not been replicated.

Unlike heat, reaction products and photons, there is likely to be sound from all LENR experiments. That may be most certain for electrochemical experiments due to the bubbles that form on the electrodes. It might also be true of both hot gas and plasma experiments. Sounds can arise from both chemical and nuclear processes. If sound can be attributed to episodes of LENR heat release, [73] the intensity and frequency of acoustic emission will provide data on a time scale much faster than that offered by calorimeters. Microphones are small and cheap, and the circuits for capturing sound signals are simple. Listening for sounds ought to be done in many more LENR experiments.

## 7. Inputs and Outputs

We now turn to four entities that have served as both inputs and outputs for LENR experiments.

Neutrons. Since ordinary deuterium fusion results in the production of neutrons, there were many early attempts to measure neutrons from LENR experiments. They produced varied results. There have been very few cases where neutrons were used as input to LENR experiments. Motivations for such experiments include the fact that neutrons can lead to nuclear reactions that result in activation of some nuclei and fission of other nuclei.

One paper [74] that described neutron input to a LENR experiment stated "…..metallic deuterated Pd samples have been irradiated with partly moderated Am/Be neutrons and the resulting neutron intensity has been measured by the FERMI apparatus….. the resulting rate corresponds to several outgoing neutrons for every neutron impinging on the Pd-D sample……. The underlying process can be interpreted as D-D fusion in a Pd-D lattice perturbed by neutrons. The excess, predominantly due to thermal incident neutrons, demonstrates that the Palladium lattice strongly increases the probability for D-D fusion even almost at rest."

The literature on neutron measurements from LENR experiments is quite extensive in both the number of papers and the sophistication of the experiments, which were described. In the next paragraph, we give one example from that literature to illustrate the complex efforts made to measure very low levels of neutrons from LENR experiments. The challenge was born of the fact that low numbers of neutrons come out of most LENR experiments, much smaller than would occur if ordinary deuterium fusion were happening. Hence, it was necessary to achieve low noise and low background performance of the neutron detection systems, including going underground to avoid cosmic ray neutrons.

Part of the abstract from the report [75] on a major program by scientists from the Los Alamos National Laboratory, which searched for low levels of neutrons from LENR experiments, reads "A wide variety of neutron detector systems have been used at various research facilities to search for anomalous neutron emission from deuterated metals. Some of these detector systems are summarized here together with possible sources of spurious signals from electronic noise. To improve our detection sensitivity, we have increased the shielding in our counting laboratory, changed to low-background $^3$He tubes, and set up additional detector systems in deep underground counting stations." That paper reported on the results of exposing titanium to various pressures and temperatures of deuterium gas.

It must be noted that a few non-electrochemical LENR experiments did provide substantial neutron emission rates. Early in the field, an Italian group subjected titanium samples to rapid temperature changes, and observed neutron emission [76]. The results of similar experiments were reported 25 years later, again with stronger neutron measurements [77]. Reports such as those raise a question: do different types of LENR experiments cause major variations in both rates and branching ratios?

Another question involving neutron measurements from LENR experiments concerns the origin of the neutrons. Are they the result of LENR, however stimulated, or do they come from deuterium fusion reactions caused by energetic processes that arise during stress-induced cracking of materials? The latter process is called fractofusion. Charge separation during cracking can produce potentials high enough to accelerate charged particles, like deuterons, to energies sufficient to cause nuclear reactions. Attention to fractofusion began soon after the Fleischmann-Pons announcement [78]. Reports of fractofusion are in the 2015 book by Carpenteri [79]. Determining the source of the neutrons measured in various LENR and related experiments is another of the old challenges in the field.

Finally, we note that, like issues with magnetic fields and other LENR variations, there remains the question of what influences and determines neutron production from LENR experiments. Is it a direct effect, or the result of some prior, possibly chemical effect, which is active?

Electrons. The use of electron beams in LENR experiments has been rare, with only a few studies.

Lipson and his colleagues did experiments in which an electron beam with a current of 100-300 nA and energy of keV was used to irradiate samples with high concentrations of deuterons [80]. The targets were Pd/PdO:$D_x$ and $TiD_x$. Passive plastic CR-39 detectors covered with absorbing foils were used to detect and measure the types and energies of emitted particles. It was found that protons with energies near 3 MeV and alpha particles with energies in the range of 11 to 20 MeV were emitted only when the electron beam was active. Otherwise, the targets did not emit the protons or alpha particles.

Reports of electrons emitted by LENR experiments have also been infrequent. In a series of studies with a glow discharge plasma in a deuterium gas atmosphere, Storms and Scanlan reported electrons, as well as heavier particles and x-rays. The conclusion of their first paper [81] reads "The observations show that energetic electrons make up part of this emission and energetic particles that might be protons and/or alpha particles add to the radiation, depending on the chemical composition of the cathode and gas. The type of radiation, its energy, and the rate are all sensitive to the presence of certain elements in the environment." Later, the same authors tried to deflect the emissions with a magnetic, which was not possible. Hence, they concluded that they were measuring photons, not electrons [64]. The certain measurement of electrons emitted from LENR experiments remains another challenge.

Energetic Ions. The situation for ions into and out of LENR experiments is unusual. There are many instances of such stimuli and measurements. However, they involve very different experiments. The relatively short ranges of energetic ions in condensed matter mean that it is generally not possible to put them into electrochemical LENR experiments. Injection of ions into hot gas and plasma experiments is possible, depending on the pressures used, but that does not seem to have been done during any experimental study of LENR. The experiments, where beams of energetic ions have been used, involve impacts on solid and, in a few cases, liquid surfaces within vacuum chambers.

Papers on the impact of ions with keV-scale energies have been presented at LENR conferences since early in the field [82]. They showed that the deuterium fusion cross sections are enhanced above expected levels as the beam energy is decreased. It was thought, and might still be possible, that the enhancement, which is called anomalous screening, is due to a mechanism similar to what is active in LENR at ordinary temperatures. However, there is a difference between the beam energies (keV) and room temperature (1/40 eV) of over a factor of ten thousand. It remains to be determined if the physics of fusion cross section enhancement at low ion beam energies is related to the physics of LENR.

There are some theoretical papers on the relationship of anomalous screening and LENR. In 2009, the abstract of a paper by Czerski read "The enhanced electron screening effect observed in accelerator experiments for the $^2$H(d,p)$^3$H and $^2$H(d,n)$^3$He reaction in deuterized metallic targets may be a breakthrough in understanding the phenomenon of cold fusion. The dielectric function theory enables an extrapolation of experimental cross sections determined at higher energies down to room temperature, leading to an enhancement of the fusion reaction rates by a factor of $10^{40}$ compared to the value predicted for the deuterium molecule. An additional enhancement can be obtained due to a 0+ resonance which should exist in the compound nucleus $^4$He very close to the D-D reaction threshold. Combination of both processes offers a simple explanation of high penetration probability through the Coulomb barrier and the reaction branching ratio preferring the $^4$He channel in heavy-water electrolysis experiments.". Two recent theoretical papers by Kálmán and Keszthelyi note the potential connection between anomalous screening and LENR. In one, they wrote "The 3rd-order heavy particle assisted and the 4th-order electron assisted resonance-like double nuclear processes may partly be responsible for the so-called anomalous screening effect observed in low energy accelerator physics investigating astrophysical factors of nuclear reactions of low atomic numbers [83]." In the second, they stated "Consequently the anomalous screening phenomenon may be connected to the processes discussed here." [84] They were referring to LENR processes.

Given that ordinary deuterium fusion produces energetic protons, tritons and helium-3 ions, as well as neutrons, much early effort was devoted to measuring ions from LENR reactions. However, as noted, ions with even MeV energies have very short ranges in condensed matter, as in electrolytic LENR experiments. So, the experiments were challenging. A paper from 1998 [85], after many searches for ion emission, concluded "We searched for nuclear products in electrolytic cells in which were observed excess heat, in gas phase of Pd + D systems and in D + single crystal. The detectors were run for long period to measure background and check stability. The signals from the detectors were analyzed by the Pulse Height Analyzer and recorded in the personal computer as spectra. Having done that, we could surely identify the signals due to noise or true events. The results of experiments reported here contradict other groups' claims of finding various sorts of nuclear products. We could only obtain the upper limit of existing nuclear products in our systems." In short, measurements of ions from LENR experiments are similar to searches for neutrons in that usually only low-level emission is found.

Radio-Frequency Radiation. Frequencies in the radio-frequency range have been both applied to and measured from LENR experiments [86]. Figure 14 is a graphical summary of the specific frequency ranges used by various investigators. It includes notes about whether or not excess heat was obtained in the experiments. It can be seen that there is no clear relationship between the measurement of radio frequencies from the LENR experiments and the appearance of excess heat. However, application of such frequencies appeared to be beneficial in some cases.

Radio-frequency excitation was applied to electrochemical experiments by Bockris and his team early in the field [87]. Figure 15 shows the setup and results. A coil was wrapped around the cell and excited by an RF gen-

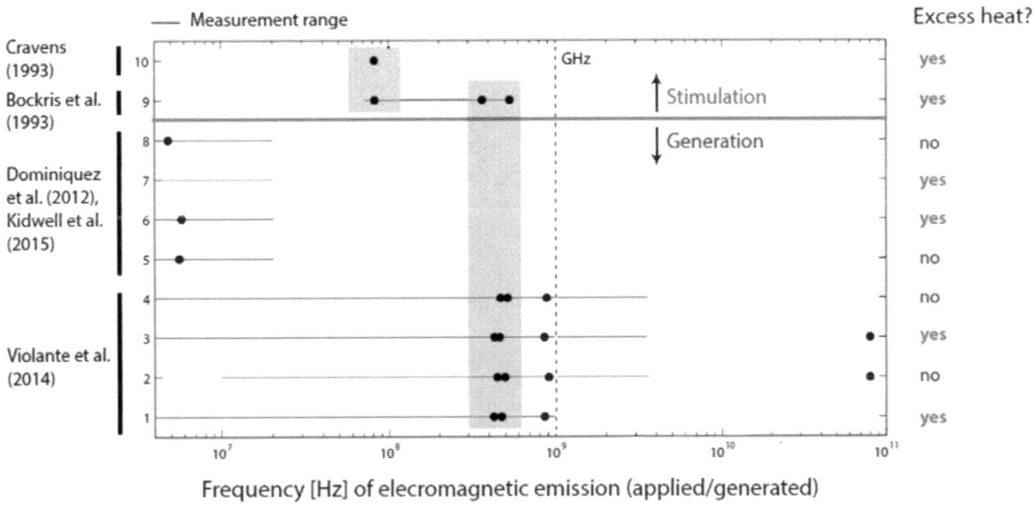


**Figure 14.** The ranges of radio frequencies that have been input to (above green line) or measured from (below green line) LENR electrochemical experiments.

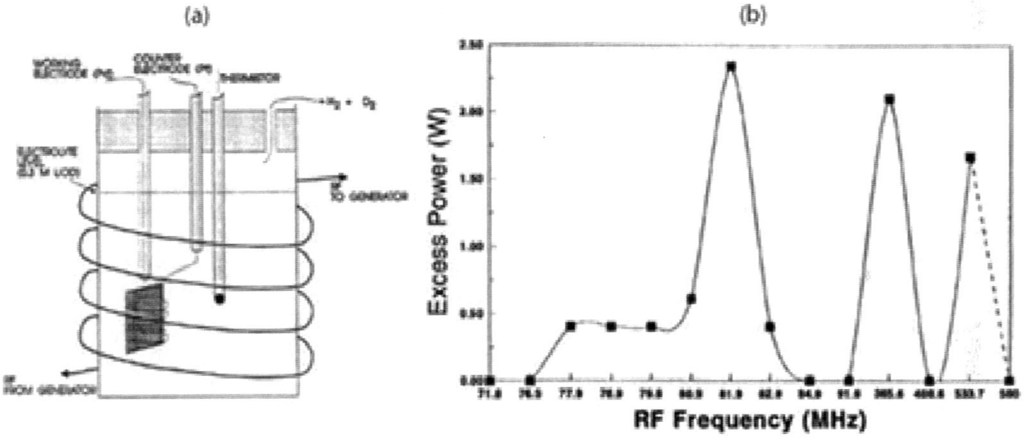


**Figure 15.** Left: The experimental setup in which a coil encircled the LENR electrochemical cell. Right: Variations in the excess power as the excitation frequency was tuned.

erator. Peaks in the excess power were seen at frequencies of 81.9 and 365.6 MHz. Radio-frequency signals have also been detected within and from LENR experiments. Frequencies in the range from 0.45 to 4.82 MHz were observed during the production of excess power at the Naval Research Laboratory [88]. RF emissions in the 400–800 MHz and 79–81 MHz ranges were seen by scientists at ENEA [89].

A surprising recent example of RF emission was published by Swartz and discussed in a pair of papers [90]. He found radio-frequencies were produced by two types of LENR experiments, a dry pre-loaded NANOR$^{TM}$ and an aqueous system. When the NANOR$^{TM}$ was driven above a voltage threshold, but below the point of avalanche operation,

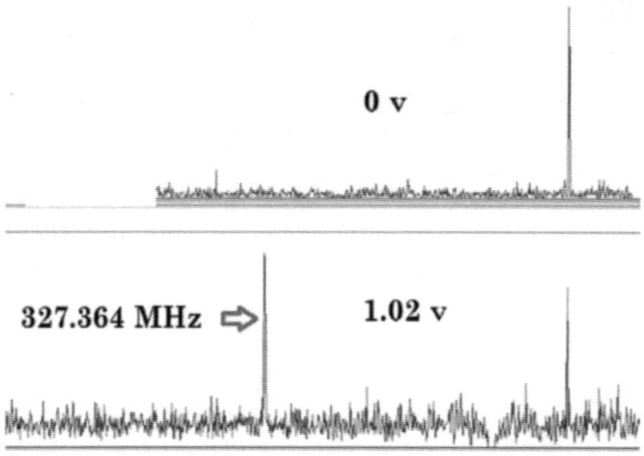


**Figure 16.** RF emission from a NANOR$^{TM}$ for two applied voltages. Frequency increases to the right.

maser action at a specific frequency was measured. Figure 16 gives the spectra measured with the NANOR$^{TM}$ off and in the maser mode, the latter showing a line at 327.364 MHz. The other line in the spectrum is of galactic origin, and serves as a fiducial. Swartz interpreted his observations as indicating that deuterons in vacancies in the Face-Centered Cubic palladium lattice are the source of LENR [91].

## 8. Unexplored Possibilities

The blanks in Figure 4 might exist for two reasons. One is the simple fact that no one has thought that experimenting with the missing input or output possibilities would be worthwhile. Another is the perception that such experiments are either not worthwhile or too difficult. Whatever the reason, we consider each of those blanks in this section.

Input Ionizing Radiation. We already noted that low energy (∼keV) ion beams have been used to produce enhanced deuterium fusion cross sections. We now turn to three other types of ionizing radiation that do not appear to have been input to LENR experiments. They are all electromagnetic, namely gamma rays, x-rays and ultraviolet (UV) photons. The reason for the lack in incident gamma ray experiments seems clear. There seems to be no particular scientific motivation. Further, the production, collimation and aiming of gamma rays are challenging, and relatively few gamma rays might be stopped within a LENR experiment (depending on the gamma ray energy). By contrast, it is relatively easy to produce beams of x-rays and UV radiation, if there were reasons to use them for excitations or other effects in LENR experiments. That is most true in synchrotron radiation facilities, but also applies to laboratory x-ray and UV sources. X-rays of appropriate wavelengths can get into the cathodes of electrolytic experiments and be absorbed by them. Although their absorption would produce heating and electronic excitations, it is unclear how those effects might contribute to the occurrence of LENR. X-rays would be little absorbed in hot gas or plasma experiments. X-rays have been used as probes of LENR experiments, as will be noted in the concluding section.

There are no papers in the lenr.org library on the emission of ultraviolet radiation from LENR experiments. That is understandable for the case of aqueous electrochemical experiments. UV radiation is heavily absorbed in water. Figure 3 shows that UV radiation in and near the visible region can escape from water-containing cells of the size of those used in LENR experiments. However, for wavelengths below about 180 nm, the escape depth is on the order of micrometers or smaller. Then, it is not possible to measure UV emission from aqueous cells.

The absorption of UV radiation by hot gases or plasmas, such as those used in some LENR experiments, is much less than the absorption in aqueous cells due to the lower density of the atmospheres in those cases. So, it ought to be possible to measure the UV emission from hot gas and plasma loading LENR experiments. Such data could be useful, especially for plasmas, which commonly contain many emission lines in the UV regions. The wavelengths and intensities of those lines can be used to obtain information on the temperatures and densities of plasmas, as well as whether or not their electrons and ions are in equilibrium [92]. UV might be absorbed in some hot gas and plasma LENR experiments. However, here again, there does not seem to be any motivation for using input UV radiation.

Output Visible Light. The motivation for measuring output visible light from LENR experiments seems clear and strong. Visible spectra are easy to measure with modern instruments, and can provide a great deal of information on the excitations and ionizations of materials in experiments. However, that seems not to have been done for electrochemical LENR experiments. When one looks for papers on light emission from electrochemical experiments, the topic of LECs (Light-emitting Electrochemical Cells) appears [93]. They are solid-state devices, which seem to have little in common with the wet electrochemical cells in many LENR experiments. It is possible that electrochemical experiments, in general, and LENR electrolytic experiments, in particular, do not emit easily-measured intensities of visible light. However, both hot gas and plasma LENR experiments ought to emit significant and useful visible light. It seems that there should be papers on spectra from such experiments.

Input Infrared Radiation. The situation for input non-ionizing radiation might be similar to that for input ionizing radiation, as just detailed. What is the motivation for shining IR radiation into LENR experiments? Doing that could produce localized heating. However, no other reason for shining IR radiation into any type of LENR experiment is known to this author.

Input Sound. It is easy to put sound of any frequency into almost any type of LENR experiment. We noted the input of ultrasound in Section 5. There are many sources and transducers for sonic frequencies. Consider the billions of audio systems of any kind in use in the world. The apparent lack of work on audible insonification of LENR experiments might again be due to the absence of motivations for such research. The question of what would sound do in an LENR experiment was unanswered until recently. A new theoretical paper [94] dealt with the effects of vibrations on the occurrence of LENR. It concluded: "So, the main consequences for LENR caused by the presence of classical vibration (coherent phonon field) are the following: (a) If a vibration is switched on then nuclear reactions may occur even at very low energy (in the $\varepsilon \to 0$ case), (b) If a vibration has a high enough amplitude then de-localization of nuclear reactants happens with high probability, (c) Nuclear reactions may follow delocalization, (d) The rate of nuclear reactions has strongly nonlinear dependence on the amplitude of vibration, and (e) Especially during electrolysis, accidental mechanical vibrations caused by e.g. bubble formation can be connected to accidental processes (bursts, explosions and crater formations)." That work indicates that it could be useful to input sound to LENR experiments at some yet-unknown frequencies and amplitudes. As with some of the other input effects, it is unlikely that the sound would directly affect the nuclear reactions. Rather, sound (like ultrasound) might influence prior effects that lead to conditions that might change rates of LENR.

Output Ultrasound. Running water is a rich source of ultrasound. It is possible that LENR experiments also generate ultrasound, although that does not seem to have been measured yet. The search of emitted ultrasound from LENR experiments might, like many other earlier exploratory LENR measurements, produce something of interest. The sudden release of nuclear-level energies in a small region of space and time by LENR could produce high frequency vibrations, as well as electrical noise. If LENR do generate ultrasound, it should be possible to detect the occurrence of LENR at rates well below the lower limits of calorimeters by measuring ultrasound, similar to the following discussion of electrical noise measurements.

Electrical Noise. Noise is a feature of all experiments, including those that seek to produce LENR. There are multiple kinds of noise in LENR experiments, including both the input stimuli and the output signals. The noise on the thermal power measured in an experiment limits the ability to measure small powers with calorimeters. Swartz

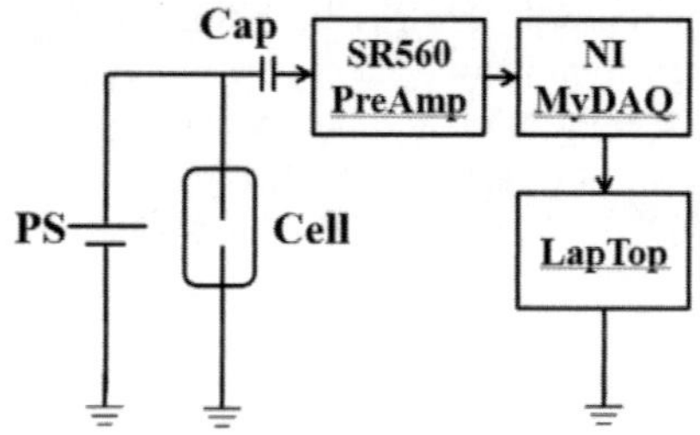


**Figure 17.** The electrochemical LENR circuit on the left, consisting of the Power Supply (PS) and the cell. Electrical noise is coupled out of that circuit with a Capacitor (Cap), amplified by the Stanford Research 560 unit and digitized with the National Instruments MyDAQ box.

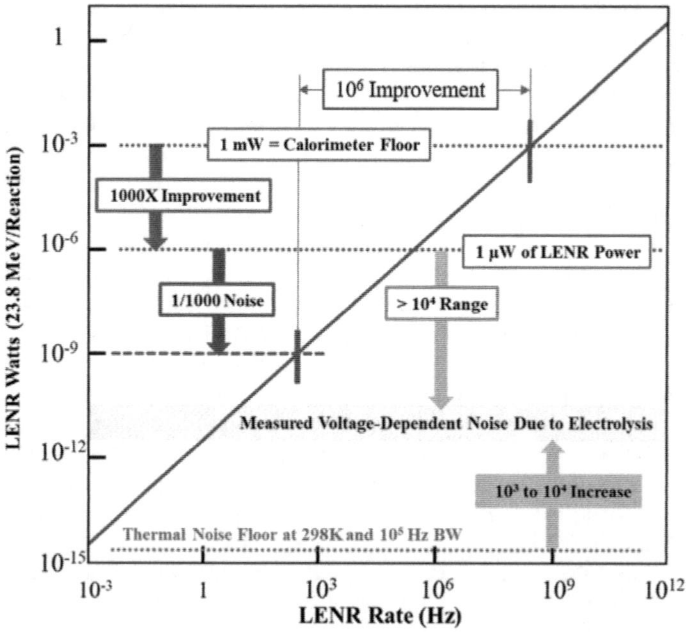


**Figure 18.** Plot of LENR power as a function of the rate at which LENR occur. See text for details.

discussed that type of noise in a 1997 paper [95]. There is also noise on all electrical currents within a LENR experiment, including the current that flows through electrochemical cells. Electrical noise is widely used to understand the behavior of electronic devices [96]. It could also provide a new way to detect LENR at low rates, that is, at rates below those needed for calorimetric detection of LENR heat. Such a method is being developed at the George Washington University [97]. It might offer faster determination and lower limits for detection of the occurrence LENR. The method is based on quantitative measurements of electrical noise on the cell voltage. It depends on what fraction of the energy produced by LENR goes into electrical noise.

The idea and status of using electrical noise as a sentinel for LENR can be described with reference to Figures 17 and 18. The circuit for noise measurements is easily mated by a capacitor with the circuit for most electrochemical cells, quite independent of the design of the cell and any calorimeter in use. A Stanford Research low-noise preamplifier and a small analog-to-digital converter from National Instruments are used. Noise data are sampled at a rate of 200 kSamples/sec for 100 seconds. Then, performing a Fourier Transformation on that time series gives the spectrum of the electrical noise from 1 Hz to 100 kHz. The spectrum consists of 1/frequency (Flicker) noise up to a few Hz and thermal (White) noise for most of the range to 100 kHz. With no current in the electrochemical circuit, the noise floor is set by the temperature and the input impedance of the pre-amplifier.

The red diagonal line in Figure 18 is a plot of the LENR power (assuming 23.8 MeV per reaction) as a function of the rate at which LENR occur. The horizontal blue line at the bottom is the noise floor computed from the equation for Thermal Noise: $V^2 = (4 \text{ k x T x R x BW})$, where V is the noise-induced voltage, k is Boltzmann's constant, T is the absolute temperature, R is the input resistance, and BW is the bandwidth [98]. Since electrical power P = I x V = $V^2/R$, one has Noise Power/Hz = 4 kT. We measured the increase in White Noise from an electrochemical cell with both electrodes made of platinum. For voltages on that cell of 2, 3, 4 and 5 V, the noise increased by a factor between $10^3$ and $10^4$ above the level for zero voltage. The yellow band in Figure 3 gives the Noise Power when the cell is powered. Turning attention to the top of Figure 3, the 1 mW floor for a calorimeter is indicated, along with the power level of 1 $\mu$W. If only 1/1000 of a LENR power of 1 $\mu$W went into electrical noise, that noise would still be one to two orders of magnitude higher than the electrical noise in the powered electrochemical circuit. If the LENR power were higher than 1 $\mu$W, or a larger fraction of the LENR power went into electrical noise, this noise technique might make it possible to detect an even smaller fraction of the LENR power and rate. As indicated in Figure 18, an improvement in the ability to detect the occurrence of LENR might be as much as a factor on one million. That increase in capability could have a large impact on LENR research, especially for the screening of materials for LENR activity.

## 9. Conclusion

This paper has dealt with inputs to LENR experiments that are meant to influence their outcomes, and with signals, which show at least partially what happened during such experiments. Such stimuli and measurements do not have to be simultaneous. Here we make passing note of techniques used to study the conditions and dynamics of some LENR experiments that have involved simultaneous inputs and outputs. Such probe measurements are not intended to control or even influence the experiments. There are a few examples of probe measurements in the literature on LENR. X-rays have been used to determine the structure [99] and composition [100] of LENR experiments. Similarly, Raman spectroscopy (input and output laser radiation) of active LENR experiments has been demonstrated [101]. There is no evidence that such probe measurements have determined, or even influenced the course of the interrogated LENR experiments.

Input Stimuli. It is possible that the roles of incident radiations or fields in LENR experiments could be determined theoretically, at least in some cases. Work by Hagelstein [102] and other theoreticians [103] involves phonons within cathode materials. Such theoretical research might explain the role of incident electromagnetic fields. However, it is also possible that resolution of the questions regarding the effects of various radiations and fields on LENR will be decided experimentally. That could take a relatively long time due to the large number and complexity of the potential experiments, even if a reliable means to produce LENR were already in hand.

Output Signals. There is clear opportunity and need for the use of various additional measurement techniques in the experimental study of LENR. Modern laboratory tools are readily available. One example involves compact and high-resolution instruments for measurement of spectra in the UV, Visible and IR regions. Such spectrometers are relatively easy to integrate with most LENR experiments. They can produce many wavelengths and intensities of line spectra for comparison with each other and other measurements. If any two measurements are made, their comparison or correlation provides additional information on an experiment.

Both Inputs and Outputs. Some of the excitory and emitted entities discussed above involve oscillations. That is true of electromagnetic and sonic waves, of course. There are also oscillations within materials, such as the cathodes in electrochemical experiments. Phonons are a prime example. The presence of these oscillations, each with some frequency or range of frequencies, raises two types of questions. First, what role might they play in the basic mechanism(s) that produce LENR? Second, are there interactions between any of these vibrations? The second question leads to consideration of resonances and resonant interactions in LENR experiments. The natural frequencies of atoms, molecules and materials that are a part of LENR experiments require consideration, in addition to applied and emitted oscillatory entities. A discussion of potential resonances and their interactions has been published [104].

One can ask what are the best combinations of inputs and outputs for LENR experiments in terms of (a) scientific learning and (b) potential commercialization. Eventually, it should be possible to answer both of those questions on the basis of experimental results and maybe also theoretical considerations. It seems likely that the best input stimuli will be those for a given loading method that have led to the most certain, intense and controllable production of LENR. Similarly, the best measurements ought to be those that provide the most useful information on what happened during an experiment. Given the current experimental and theoretical uncertainties regarding LENR, it is unlikely that any consensus on the best input stimuli and output signals will develop very soon.

Parametric Experiments. There have been many systematic measurements in LENR experiments, in which some variable, such as temperature, is changed during a run or between runs. We have noted such experiments in a few instances earlier in this paper. They have multiple advantages. In the absence of well-developed and widely-accepted theory on the potential outcomes of LENR experiments, parametric measurements indicate which variables are important. They are invaluable in showing the importance of any variable in the outcome of the experiments. Further, they challenge theoreticians to take the variables of importance into account. Parametric experiments can now be performed with modern instrumentation, such as that available at Synchrotron Radiation and Neutron Scattering Facilities.

Testing LENR Theories. The potential inputs and measurements discussed in this paper challenge LENR experimentalists. Past and possible experiments also challenge those who have developed theories about the LENR mechanism(s). LENR theories start with a concept of the mechanism that causes nuclear reactions to happen at low energies. They next need to be developed further by deriving the equations based on the concept, and performing calculations using those equations. Only then are numbers available for comparison with past or future experimental outcomes. The development of theories about LENR varies widely from many with only concepts and nothing more, to some that have the entire range of requirements and features. For those few LENR theories that have been developed to the point of explicit equations and numbers computed from the equations, data from parametric experiments serves to test the validity of the theories. Both the input stimuli and the output signals discussed in this paper provide tests of any LENR theories. Just as LENR theories vary greatly in their development, there is great variety in the thoroughness with which the better developed theories have been compared with experimental results.

Thorough Data Analyses. There have been two chronic shortcomings in the literature on LENR. One involves reports that fail to include relevant information, which is needed to (a) fully understand what was done and found, or (b) replicate the reported experiments and their results. A recent example was published by Staker, in which cyclic loading and deloading of palladium cathodes prior to experiments was cited as beneficial, without specific information on the protocols (currents and times) used for those operations [105]. The other problem includes reports in which data are presented without clearly possible additional data analyses. A common example of the second difficulty involves time histories of measurements without subsequent Fourier Transforms, which reveal the spectral (frequency) content of the data. A search of the lenr.org library with the phrase "Fourier Transform" does not produce any papers. Another example of missed analytical opportunities is the failure to perform correlations between simultaneous measurements to see if they might be related in some fashion. As the cliché states, correlation does not prove causation. However, correlation is additional information that might be useful for empirical or theoretical insights. Only ten papers in the lenr.org library deal with correlations. One of the more notable is an experiment in which neutron emission was found to time correlate with excess heat production [106]. However, that is an observation, not the result of a sophisticated data analysis. In another example, a mathematical correlation was detailed between the time of day and unexpected (and still unexplained) variations in the loading ratio, temperature and pressure in transmutation experiments [107]. That paper does exemplify the type of data analyses that are possible and desirable after future LENR experiments.

Systems Engineering. Past and current LENR experiments, and future LENR generators, are systems. They consist of the inputs, core and outputs discussed in relation to Figure 1. It is worth noting that there have been a few examples where the tools of Systems Engineering were applied to LENR. In 2003, Bass and McKubre [108] developed a Cause (C) and Effect (E) Protocol for determination of the appearance of LENR in an experiment. They wrote "Instead of the usual eyeballing of a (C,E) plot and assigning of a figure of merit regarding linearity in a subjective manner, the experimenter may now plan to evaluate a sequence of similar experiments in a completely objective manner by computing the significance ratio derived above." Then, in 2008, Bass and Swartz [109] applied the methods of Empirical System Identification (ESID) to LENR. Systems Identification was also employed for the study of the calorimetry used in a recent Google-funded program [110]. It is expected that Systems Identification methods will be applied to both future LENR experiments and also the development of commercial LENR generators. Books on the methodology are available [111].

This paper has not been exhaustive, citing all papers on each of the topics, which were discussed. Further, it is likely that this review has missed several papers that should have been included in the discussion above. However, it is hoped that the provided organization and example papers serve to stimulate additional considerations of and experiments on LENR, as well as more thorough comparisons of theoretical results with experimental data.


## Acknowledgements

The author appreciates encouragement from Graham Hubler, and discussions with Robert Duncan, Trevor Dardik and Shelby Lacouture. Xing Zhong Li and Zhan M. Dong from Tsinghua University generously provided preprints of the temperature dependence data. Edmund Storms sent his papers with Scanlan on radiation detection. Mitchell Swartz and Dennis Letts provided papers and comments on magnetic field effects. Peter Kálmán and Michael McKubre provided useful comments on a draft of this paper. The searchable Library in Jed Rothwell's website lenr.org is a remarkably useful resource. A reviewer provided comments that improved the paper. It is a pleasure to acknowledge all this valuable assistance.


## References


[1] M. Fleischmann and S. Pons, "Electrochemically Induced Nuclear Fusion of Deuterium", J. Electroanalytical Chemistry, Vol. 261, pp. 261-267 (1989) and Errata in Vol. 263.

[2] S. Szpak, P.A. Mosier-Boss and J.J. Smith, "Reliable Procedure for the Initiation of the Fleischmann-Pons Effect", in Proc. of the Second Annual Conference on Cold Fusion, "The Science of Cold Fusion", pp.87-92, Societa Italiana di Fisica, (1991)

[3] M. Fleischmann and S. Pons, "Calorimetry of the Pd-$D_2O$ System: from Simplicity via Complications to Simplicity", in Proc. of Third International Conference on Cold Fusion, "Frontiers of Cold Fusion", p. 47, Universal Academy Press, Inc., Tokyo, Japan (1992)

[4] E. Storms, "The Enthalpy of Formation of PdH as a Function of H/Pd Atom Ratio", J. Condensed Matter Nuclear Science, vol. 29, pp. 275–285 (2019)

[5] https://www.intechopen.com/chapters/68850
[6] https://commons.wikimedia.org/wiki/File:Absorption_spectrum_of_liquid_water.png
[7] https://en.wikipedia.org/wiki/Neutron_temperature
[8] https://three.jsc.nasa.gov/articles/Heilbronn_Neutron_Supplement.pdf
[9] https://www.nist.gov/pml/stopping-power-range-tables-electrons-protons-and-helium-ions
[10] B.Y. Liaw et al., "Elevated-temperature excess heat production in a Pd + D system", J. Electroanalytical Chemistry, vol. 319, p. 161 (1991)
[11] B. Y. Liaw et al., "Helium analysis of palladium electrodes after molten salt electrolysis.", Fusion Technology vol. 23, pp. 92-97 (1993)
[12] A.B. Karabut et al., "Nuclear product ratio for glow discharge in deuterium", Physics Letters A, vol.170, p. 265 (1992)
[13] G. Lonchampt et al., "Reproduction of Fleischmann and Pons experiments", Proceedings of the Sixth International Conference on Cold Fusion: Progress in New Hydrogen Energy. New Energy and Industrial Technology Development Organization and The Institute of Applied Energy. Vol. I, pp. 113-120 (1996)
[14] G. Mengoli et al., "Calorimetry close to the boiling temperature of the D2O/Pd electrolytic system", Journal of Electroanalytical Chemistry, vol. 444, PP. 155-167 (1998)
[15] Z.M. Dong et al., "Temperature Dependence of Excess Power in Both Electrolysis and Gas-loading Experiments", J. Condensed Matter Nuclear Science, vol. 29, pp. 85-94 (2019) and X. Z Li et al., "Resonant Surface Capture Model", J. Condensed Matter Nuclear Science, vol. 29, pp 440-452 (2019) and S. X. Zheng et al., "Temperature Dependence of Maximum Excess Power in 3 New Experiments (Letts, Parkhomov, & Mizuno)", presented at ICCF-22 (2019)
[16] https://en.wikipedia.org/wiki/Arrhenius_equation
[17] E. Storms, "The Nature of the D+D Fusion Reaction in Palladium and Nickel", at ICCF-23 (2021)
[18] I. Consant et al., "Deposition and removal of sodium contamination on silicon wafers", Semiconductor Science and Technology, vol. 15(1), pp. 61-66 (2000) DOI: 10.1088/0268-1242/15/1/311 and https://www.researchgate.net/publication/241485164_Deposition_and_removal_of_sodium_contamination_on_silicon_wafers
[19] D. J. Nagel, "The Status of 'Cold Fusion'", Radiation Physics and Chemistry, vol. 51, pp. 653-668 (1998)
[20] G. H. Miley, "Possible evidence of anomalous energy effects in H/D-loaded solids – Low energy nuclear reactions (LENR), Journal of New Energy, Vol 2(3-4), pp. 6-13 (1997) and G. H. Miley and J.A. Patterson, "Nuclear transmutations in Thin-Film Nickel Coatings Undergoing Electrolysis", Journal of New Energy, Vol 1 (3): pp. 5-13 (1996) and G. H. Miley "On the Reaction Product and Heat Correlation for LENRs", Proc. of $8^{th}$ International Conference on Cold Fusion, F. Scaramuzzi (Editor), Lerici (La Spezia), Italy, pp. 419-424 (2000).
[21] T. Mizuno et al., "Isotopic Changes of the Reaction Products Induced by Cathodic Electrolysis in Pd", Journal of New Energy, vol. 1(3): pp. 31-45 (1996) and T. Mizuno, "Isotopic Changes of Elements Caused by Various Conditions of Electrolysis", 237th ACS National Meeting, Salt Lake City, UT (2009).
[22] O. Azizi et al., "Effect of Cathode Pretreatment and Chemical Additives on H/D Absorption into Palladium via Electrochemical Permeation", J. Condensed Matter Nuclear Science, vol 19, pp. 1–9 (2016)
[23] D. Knies et al., "A Method to Control Palladium Crystallographic Texture and Surface Morphology", presented at ICCF-19 (2015). https://www.lenr-canr.org/acrobat/KniesDamethodtoc.pdf
[24] D. Letts and D. Cravens, "Laser Stimulation of Deuterated Palladium: Past and Present", P. L. Hagelstein and S. R. Chubb (Editors), Proc of the $10^{th}$ Conference on Condensed Matter Nuclear Science, World Scientific Publishing Company (2005) ISBN 10: 9812565647
[25] M. R. Swartz et al., "Photoinduced Excess Heat from Laser-Irradiated Electrically-Polarized Palladium Cathodes in D2O", P. L. Hagelstein and S. R. Chubb (Editors), Proc of the $10^{th}$ Conference on Condensed Matter Nuclear Science, World Scientific Publishing Company (2005) ISBN 10: 9812565647
[26] V. Violante et al., "Progress in Excess Power Experiments with Electrochemical Loading of Deuterium in Palladium," Condensed Matter Nuclear Science, Proc. ICCF-12, A. Takahashi, K.-I. Ota and Y. Iwamura (Editors), World Scientific, 55-64 (2006)
[27] P. L. Hagelstein, D. Letts, and D. Cravens. "Terahertz difference frequency response of PdD in two-laser experiments." J. Condensed Matter Nuclear Science, vol. 3, pp. 59-76 (2010)

[28] V. I. Vysotskii et al., "Application of Correlated States of Interacting Particles in Nonstationary and Periodical Modulated LENR Systems", J. Condensed Matter Nuclear Science, vol. 13, pp. 624–636 (2014)

[29] P. L. Hagelstein and D. Letts, "Temperature Dependence of Excess Power in Two-laser Experiments", J. Condensed Matter Nuclear Science, vol. 13, pp. 165-176 (2014)

[30] P. Jung (Editor) "Diffusion and Clustering of Helium in Noble Metals. Fundamental Aspects of Inert Gasses in Solids", Plenum Press, NY (1991) and G. C. Abell et al., "Helium release from aged palladium tritide." Physical Review B: Materials Physics, vol. 41, p. 1220 (1990) and J. Xia et al., "A comparative study of helium atom diffusion via an interstitial mechanism in nickel and palladium." Physica Status Solidi B, vol. 243B, pp. 579-583 (2006)

[31] X. F. Wang and Y. Arata, Y. 2014. "The Importance of the Removal of Helium from Nano-Pd Particles after Solid Fusion," J. Condensed Matter Nuclear Science, vol. 13 pp. 13-18 (2014)

[32] http://en.wikipedia.org/wiki/Rusi_Taleyarkhan

[33] Stringham, R. 2006. "Low Mass 1.6 MHz Sonofusion Reactor," Condensed Matter Nuclear Science, Proc. ICCF-11, World Scientific, 238-252.

[34] I. Dardik et al., "Ultrasonically-excited electrolysis Experiments at Energetics Technologies" in Proc. of ICCF-14, D. J. Nagel and M. E. Melich (Editors), pp. 101-118 (2008)

[35] S. R. Chubb and D. Letts, "Magnetic Field Triggering of Excess Power in Deuterated Palladium," Infinite Energy, Issue 95, pp. 40-54 (2011)

[36] M. Swartz, "Impact of an Applied Magnetic Field on a High Impedance Dual Anode LANR Device" J. Condensed Matter Nuclear Science, Vol. 4, pp. 93–105 (2011) and Infinite Energy, Issue 98, pp. 18-25 (2011)

[37] https://www.comsol.com/

[38] D. Letts, "A Method to Calculate Excess Power", Infinite Energy, Issue 112, pp. 63-66 (2013)

[39] For example, T. Z. Fahidy, "The Effect of Magnetic Fields on Electrochemical Processes", in Modern Aspects of Electrochemistry, B. E. Conway, J. O'M. Bockris and R. E. White (Editors), pp 333-354, Springer (2002) and F. A. Garcés-Pineda et al., "Direct magnetic enhancement of electrocatalytic water oxidation in alkaline media", Nature Energy, vol. 4, pp. 519–525 (2019). https://doi.org/10.1038/s41560-019-0404-4

[40] M. R. Swartz et al., "Amplification and Restoration of Energy Gain Using Fractionated Magnetic Fields on ZrO2–PdD Nanostructured Components", J. Condensed Matter Nuclear Science, vol. 15, pp. 1–15 (2014)

[41] M. R. Swartz, "Oscillating Excess Power Gain and Magnetic Domains in NANOR$^{R}$-type CF/LANR Components", J. Condensed Matter Nuclear Science, vol. 23, pp. 1–12 (2016)

[42] D. Afonichev, "High-Frequency Radiation and Tritium Channel" in Proc. of Tenth International Conference on Cold Fusion, P. L. Hagelstein and S. R. Chubb (Editors), World Scientific, pp. 353-359 (2003)

[43] R. Cantwell and M. McConnell, "Partial Replication of Storms/Scanlan Glow Discharge Radiation", in Proc. of ICCF-14, Proc. of the International Conference on Condensed Matter Nuclear Science, D. J. Nagel and M. E. Melich (Editors), pp 288-298 (2008)

[44] F. Celani et al., "Deuterium overloading of palladium wires by means of high-power microsecond pulsed electrolysis and electromigration: suggestions of a "phase transition" and related excess heat." Physcs Letters. A vol. 214, p. 1-13 (1996)

[45] G. Preparata, M. Scorletti, and M. Verpelli, "Isoperibolic calorimetry on modified Fleischmann-Pons cells", J. Electroanalytical Chemistry, vol. 411. p. 9 (1996)

[46] G. Preparata, "Everything You Always Wanted to know About Cold Fusion Calorimetry?" in "Progress in New Hydrogen Energy", Proc. of the Sixth International Conference on Cold Fusion, pp. 136-143 (1996).

[47] S. Szpak et al., "The Effect of an External Electric Field on Surface Morphology of Co-Deposited Pd/D Films," J. ElectroAnanytical Chemistry, vol. 580, pp. 284-290 (2005)

[48] J. V. Wilmsdorff et al., "An experimental overview on electric field sensing", Ambient Intelligence and Humanized Computing, vol. 10, pp. 813–824 (2019). https://doi.org/10.1007/s12652-018-0877-1

[49] V. Calero et al., "An ultra wide band-high spatial resolution-compact electric field sensor based on Lab-on-Fiber technology", Science Reports, vol. 9, p. 8058 (2019). https://doi.org/10.1038/s41598-019-44644-y

[50] A. DeNinno et al., "Evidence of Emission of Neutrons from a Titanium-Deuterium System.", Europhysics Letters, vol. 9 (3), pp. 221-224 (1989)

[51] I. Dardik et al., "Excess Heat in Electrolysis Experiments at Energetics Technologies", Proc. of the 11th Int'l. Conference on Condensed Matter Nuclear Science, J.-P. Biberian (Editor), World Scientific, pp. 84-101 (2006).

[52] D. Dominguez et al., "Evidence for Excess Energy in Fleischmann–Pons-Type Electrochemical Experiments", J. Condensed Matter Nuclear Science, vol. 14, pp. 15–28 (2014)

[53] https://brillouinenergy.com/science-technology

[54] M. Fleischmann et al., "Calorimetry of the palladium-deuterium-heavy water system", J. Electroanalytical Chemistry, vol. 287, p. 293 (1990)

[55] A. Takahashi et al., "Neutron Spectra and Controllability by PdD/electrolysis Cell with Low-High Current Pulse Operation", in Proc. of Second Annual Conference on Cold Fusion, "The Science of Cold Fusion", pp. 93-7 Societa Italiana di Fisica, Bologna, Italy (1991)

[56] I. Dardik et al., "Ultrasonically-excited electrolysis Experiments at Energetics Technologies" in ICCF-14 International Conference on Condensed Matter Nuclear Science. (2008)

[57] M. Fleischmann, "An Overview of Cold Fusion Phenomena" in The First Annual Conference on Cold Fusion, pp. 344-50, National Cold Fusion Institute (1990)

[58] https://lenr-canr.org/wordpress/?page_id=1081

[59] Y. Iwamura et al., "Observation of Nuclear Transmutation Reactions induced by $D_2$ Gas Permeation through Pd Complexes", in Proc. of the Eleventh International Conference on Condensed Matter Nuclear Science, J.-P. Biberian (Editor), World Scientific, pp. 339-350 (2004)

[60] See, for example, Y. Iwamura et al., "Observation of Low Energy Nuclear Reactions Induced By $D_2$ Gas Permeation Through Pd Complexes", Proc. of ICCF-9 (2002)

[61] M. Fleischmann, S. Pons and M. Hawkins, "Electrochemically induced nuclear fusion of deuterium", J. Electroanalytical Chemistry, vol. 261, p. 301 (1989) and errata in Vol. 263

[62] M. Fleischmann et al., "Measurements of gamma-rays from cold fusion". Nature, vol. 339, p. 667 (1989)

[63] A. B. Karabut, Y.R. Kucherov and I.B. Savvatimova, "Nuclear product ratio for glow discharge in deuterium.", Physics Letters A, vol. 170, p. 265 (1992)

[64] Y. Iwamura et al., "Characteristic X-ray and Neutron Emissions from Electrochemically Deuterated Palladium", Proc. of ICCF-5, pp. 197-200 (1995)

[65] E. Storms and B. Scanlan, "Detection of Radiation Emitted from LENR" in D. J. Nagel and M. E. Melich (Editors), Proc of 14th Int'l. Conference on Condensed Matter Nuclear Science, pp. 263-287 (2008)

[66] S. Szpak, P.A. Mosier-Boss, and M. Miles, "Calorimetry of the Pd+D codeposition", Fusion Technology, vol 36, pp. 234-241 (1999)

[67] https://www.youtube.com/watch?v=OUVmOQXBS68

[68] D. J. Nagel, "Low Energy Nuclear Reactions: Experiment and Theory", presented at the American Nuclear Society Winter Meeting (2005)

[69] M. Swartz, G. Verner and A. Weinberg, "Non-thermal near-IR emission from high impedance and codeposition LANR devices", Proc. the 14th Int. Conf. on Condensed Matter Nuclear Science and the 14th Int. Conf. on Cold Fusion (ICCF-14), 10–15 August 2008, Washington, D.C., ISBN: 978–0–578-06694-3, pp. 343-361 (2010)

[70] W. McCarthy, "Light Hydrogen LENR in Copper Alloys", J. Condensed Matter Nuclear Science, vol. 29, p. 191-201 (2019)

[71] https://www.flir.com/products/flir-one-pro/

[72] Mosier-Boss, P. et al., "Review of Twenty Years of LENR Research Using Pd/D Co-deposition," J. Condensed Matter Nuclear Science, 4, 173-187 (2011).

[73] H. Kozima, W.-S. Zhang and J. Dash, "Precision measurement of excess energy in electrolytic system Pd/D/H2SO4 and inverse power distribution of energy pulses vs. excess energy", Proc. of the 13th Int. Conf. on Condensed Matter Nuclear Science, MATI Moscow, pp. 348–358 (2008)

[74] B. Stella et al., "Evidence for Stimulated Emission of Neutrons in Deuterated Palladium" in H. Ikegami (Editor). Proc. of the Third International Conference on Cold Fusion, "Frontiers of Cold Fusion". Universal Academy Press, Inc., pp. 437-440 (1992)

[75] H. O. Menlove et al, "Low-background Measurements of Neutron Emission from Ti Metal in Pressurized Deuterium Gas", in Proc. of the Second Annual Conference on Cold Fusion, "The Science of Cold Fusion", Societa Italiana di Fisica, Italy." (1991)

[76] A. De Ninno et al., "Emission of neutrons as a consequence of titanium-deuterium interaction", Nuovo Cimento Soc. Ital. Fis. A, vol. 101, p. 841 (1989)

[77] M. A. Prelas and E. Lukosi, "Neutron Emission from Cryogenically Cooled Metals Under Thermal Shock", J. Condensed Matter Nuclear Science, vol.13, pp. 455-463 (2014)

[78] F. Mayer et al., "Nuclear fusion from crack-generated particle acceleration", J. Fusion Energy, vol. 9(3), p. 269 (1990) and G. Preparata, "A New Look at Solid-State Fractures, Particle Emission and 'Cold' Nuclear Fusion." Nuovo Cimento A, vol. 104, p. 1259 (1991)

[79] A. Carpenteri et al. (Editors), "Acoustic, Electromagnetic, Neutron Emissions from Fracture and Earthquakes", Springer (2015)

[80] A. G. Lipson et al., "Charged Particle Emissions Upon Electron Beam Excitation of Deuterium Subsystem in the Pd and Ti- Deuteride Targets" in ICCF-14 International Conference on Condensed Matter Nuclear Science, pp. 220-249 (2008) and A. G. Lipson et al., "Charged Particle Emissions and Surface Morphology of Pd/PdO:Dx and TiDx Targets Under Electron Beam Excitation" in 15th International Conference on Condensed Matter Nuclear Science, pp. 187-196 (2009)

[81] E. Storms and B. Scanlan, "Radiation Produced by Glow Discharge in Deuterium" in 8th International Workshop on Anomalies in Hydrogen / Deuterium Loaded Metals, pp. 297-305 (2007) http://www.iscmns.org/catania07/index.htm

[82] J. Kasagi, Y. Honda and K. Fang, "Screening energy for low energy nuclear reactions in condensed matter" in J.-P. Biberian (Editor), "Cold Fusion: Advances in Condensed Matter Nuclear Science, Elsevier pp. 167-190 (2020)

[83] P. Kálmán and T. Keszthelyi, "Resonance-like nuclear processes in solids: 3rd and 4th order processes", https://arxiv.org/abs/1303.1262v2 (2013) and (2017) "Forbidden nuclear reactions", Physical Review C, vol. 99, p. 054620 (2019)

[84] P. Kálmán and T. Keszthelyi, "Forbidden nuclear reactions", Physical Review C, vol. 99, p. 054620 (2019)

[85] T. Aoki et al., "Search for nuclear products of the D + D nuclear fusion", Int. J. of the Society for Materials Engineering Resources, vol. 6(1), p. 22 (1998)

[86] F. Scholkmann et al., "Electromagnetic Emission in the kHz to GHz Range Associated with Heat Production During Electrochemical Loading of Deuterium into Palladium: A Summary and Analysis of Results Obtained by Different Research Groups", J. Condensed Matter Nuclear Science, Vol. 19, pp. 325-335 (2016)

[87] J. O'M. Bockris et al., "Triggering of Heat and Sub-Surface Changes in Pd-D System," Transactions of Fusion Technology, vol. 26, pp. 267-290 (1994)

[88] D. Dominguez et al. "Anomalous Results in Fleischmann-Pons Type Electrochemical Experiments," ICCF-17, Daejeon, Korea (August 2012) and D. A. Kidwell et al., "Observation of radio frequency emissions from electrochemical loading experiments", Current Science, vol. 108, pp. 578-581 (2015)

[89] V. Violante et al. 2014. "RF Detection and Anomalous Heat Production During Electrochemical Loading of Deuterium in Palladium," Energia, Ambiente e Innovaziene, Vol. 2-3, 63-77.

[90] M. R. Swartz, "Active LANR Systems Emit a 327.37 MHz Maser Line", J. Condensed Matter Nuclear Science, vol. 33 pp. 81-110 (2020) and "Pulsatile Superhyperfine Lines at 327.37 MHz Herald LANR Activity and Possible Mass–Energy Transfer", J. Condensed Matter Nuclear Science, vol. 33, pp. 111-125 (2020)

[91] M. R. Swartz, "FCC Vacancies in $ZrO_2PdD$ are the Active LANR Site" J. Condensed Matter Nuclear Science, vol. 33 pp. 126-144 (2020)

[92] D. J. Nagel et al., "X-Ray Emission from Laser-Produced Plasmas", Physical Review Letters, vol. 33, pp. 743-46 (1974)

[93] S. B. Meier et al., "Light-emitting electrochemical cells: recent progress and future prospects", Materials Today, vol. 17 (5), pp. 217-223 (2014)

[94] P. Kálmán and T. Keszthelyi, "Quantum mechanics of nuclear reactions in $PdD_\xi$ systems", submitted to Physical Review C (2021)

[95] M. R. Swartz, "NOISE MEASUREMENT IN COLD FUSION SYSTEMS", J. of New Energy, Vol 2 (2) pp. 56-61 (1997)

[96] L. K. J. Vandamme, "Noise as a Diagnostic Tool for Quality and Reliability of Electronic Devices", IEEE TRANSACTIONS ON ELECTRON DEVICES, vol. 41, pp. 2176-2187 (1994)

[97] A. Mehrabian and D. J. Nagel, unpublished

[98] J. Johnson, "Thermal Agitation of Electricity in Conductors", Physical Review, vol. 32, pp. 97–109 (1928) and H. Nyquist, "Thermal Agitation of Electric Charge in Conductors", Physical Review, vol. 32, pp. 110–113 (1928)

[99] D. L. Knies et al., "In-situ synchrotron energy-dispersive x-ray diffraction study of thin Pd foils with Pd:D and Pd:H concentrations up to 1:1", J. Applied Physics, vol. 112, p. 083510 (2012)

[100] Y. Iwamura et al., "Observation of Surface Distribution of Products by X-Ray Fluorescence Spectrometry During $D_2$ Gas Permeation Through Pd Complexes", Proc. of the 12th International Conference on Condensed Matter Nuclear Science, pp. 178-187 (2005)

[101] M. R. Swartz, "Increase of an Anti-Stokes Peak at the Cathode of an Electrically Driven, Active Aqueous Nickel/$H_2O$/Pt System", J. Condensed Matter Nuclear Science, vol. 27, pp. 22–29 (2018)

[102] P. L. Hagelstein, "Calculation of the Boosted Spin–orbit Contribution to the Phonon–Nuclear Coupling Matrix Element for $^{181}$Ta", J. Condensed Matter Nuclear Science, vol. 29, pp. 392–400 (2018)

[103] S. D. Seccombe, "Phonon-Assisted Nuclear Fusion-II", Presented at ICCF-23 (2021)

[104] D. J. Nagel, "Questions About Lattice Enabled Nuclear Reactions: Experiments, Theories and Computations", Infinite Energy, Issue 119, pp. 17-36 (2015)

[105] M. R. Staker, "Coupled Calorimetry and Resistivity Measurements, in Conjunction with an Emended and More Complete Phase Diagram of the Palladium–Isotopic Hydrogen System", J. Condensed Matter Nuclear Science, vol. 29, pp. 129–168 (2019)

[106] H. Ogawa et al., "CORRELATION OF EXCESS HEAT GENERATION AND NEUTRON EMISSION IN Pd-LiOD ELECTROLYSIS" in Proc. of the $6^{th}$ International Conference on Cold Fusion, pp. 448-454 (1996)

[107] F. Scholkmann et al., "Statistical Analysis of Unexpected Daily Variations in an Electrochemical Transmutation Experiment", J. Condensed Matter Nuclear Science, vol. 8, pp. 37-48 (2012)

[108] R. W. Bass and M.C.H. McKubre. "Generalized cause and effect demonstration protocol" in Proc, of Tenth International Conference on Cold Fusion. P. L. Hagelstein and S. R. Chubb (Editors), World Scientific, pp. 1005-8 (2003)

[109] R. W. Bass and M.R. Swartz. "Empirical System Identification (ESID) and Optimal Control of Lattice-Assisted Nuclear Reactors" in Proc. of ICCF-14, the International Conference on Condensed Matter Nuclear Science, D. J. Nagel and M. E. Melich (Editors), pp 497-502 (2008)

[110] B. P. MacLeod et al., "Calorimetry under non-ideal conditions using system identification", Journal of Thermal Analysis and Calorimetry vol. 138, pp, 3139–3157 (2019) and https://doi.org/10.1007/s10973-019-08271-z.

[111] L. Ljung, "System Identification: Theory for the User" (2nd Edition), Pearson (1998) and K. J. Keesman, "System Identification: An Introduction", Springer (2011)

J. Condensed Matter Nucl. Sci. 36 (2022) 89–96

Research Article

# Excess Heat From Cold Fusion Activates Nitinol

Mitchell R. Swartz*

*JET Energy, Inc. Wellesley, MA USA © September 28, 2021*

*23rd Int Conf on Condensed Matter Nuclear Science (Xiamen University, June 9-11, 2021)*

**Abstract**

Nitinol was investigated as a possible sensor of CF/LANR (Lattice Assisted Nuclear Reactions) activity. Several Nitinol devices were built to examine the responses of Nitinol to both ordinary and to CF/LANR-induced excess heat. A NANOR®-type component delivered a peak power output of ~12.17 Watts when driven with input power of 3.94 Watts. The metric of qualitatively derived RMS induced force was more useful than length measurement for both prosaic applied heat and excess heat. The Nitinol had derived RMS induced force-power response of about ~170-250 microNewtons/Watt. The Nitinol LANR/CF Detectors were equally sensitive to both ordinary heat and to CF/LANR excess heat, and the sensitivity was only qualitative.



*Keywords:* Nitinol, LANR activity, active calorimetry, accelerometer, gyroscope, time-of-flight sensor, NANOR®-type component, preloaded LANR material, $ZrO_2$-PdD, Shape Memory Alloy

## 1. Introduction

Nitinol [1,2,3,10] is a Shape Memory Alloy [SMA, also known as Flexinol and Muscle Wire]. The field of CF/LANR needs improved means to detect and semi-quantitatively measure the desired XSH-producing driven state. Therefore, what was investigated here was whether Nitinol can be a possible sensor of CF/LANR activity. Two questions were paramount: 'Will A Shape Memory Alloy [Nitinol] Respond to the Heat and Excess Heat from an ACTIVE LANR system?' and 'Can A Shape Memory Alloy [Nitinol] Distinguish an ACTIVE LANR system distinguishing CF XSH from ordinary heat?'

## 2. Background - Successful LANR has a Single Desired Active Driven State

Cold fusion is real and achievable, and active electrically-driven systems and components are obtainable as needed for experiments, making heat and electricity, and open demonstrations. The active driven state produces excess heat (XSH) and de novo 4Helium; and is distinguishable by unique antiStokes spectra [6], [7] and unique high-Q Deuterium-line

*Corresponding author:

© 2022 ICCF. All rights reserved. ISSN 2227-3123

radiofrequency emissions at 327.37 MHz [8.9]. However, the proverbial "bad news" is that, although not usually recognized, there are two (2) driven states but ONLY ONE (1) state/mode is DESIRABLE ["Active"] making the sought for excess heat [XSH]. The other is inactive. This is true both for LANR aqueous systems and dry preloaded components [4], [5]. The existence of these two states were first suggested theoretically by the Quasi-1-dimensional model of isotope loading [which led to codeposition, OOP control, etc. (ICCF-4)]. This was later confirmed experimentally by NANOR® components driven into avalanche mode without XSH (ICCF-19), by coherent Raman spectroscopy of electrically driven CF/LANR components [6, ICCF-20], by the important single run for an aqueous CF/LANR system (shown at ICCF-21), and by RF Deuterium-Line Emissions [8, ICCF-22].

### 2.1. Background – Nitinol Exhibits Shape Memory

Nitinol behaves differently from normal viscoelastic materials, which after being squeezed or pulled apart, relax to a different shape from the final unstressed condition [called "General elastic recovery", 10]. In contrast, after SMA undergoes a first deformation at a high temperature [austenite phase] and is then cooled, thereafter, when squeezed or pulled apart in a second deformation, the SMA will recover the original, first deformation shape when heated to any temperature above the critical "transformation temperature". This is 'Shape Memory', and a SMA can repeatedly recover the first shape - even from severely deformed shapes. Nitinol's properties of shape memory, dramatic kink resistance, and superelastic properties have enabled new inventions from bridge and building dampers, to HVAC actuators, to medical devices [10].

Nitinol is used here as sensor to detect CF/LANR activity based on its complex unique metallurgy and its response to heat. This unusual change in shape from heat, the "magic" of Nitinol, arises from its ordered structure, unlike homogeneously mixed intermetallic alloys where the metallic components are randomly arranged throughout the volume. With active Nitinol, the atoms are arranged precisely in rows. SMA alloys of nickel and titanium have near equal atomic percentages, and are named according to the weight percent of nickel (e.g. Nitinol 55 is 55% Ni). The NOL in nitinol stands for Naval Ordinance Laboratory, where the memory properties were investigated by William J. Buehler and Fredrick Wang.

### Acquisition of the Nitinol Shape ["Shape Memory"]

The high-temperature austenitic phase, where shaping occurs, is $\sim$500 °C (932 °F). In this "parent phase", the SMA is "shape set". Thereafter, the martensite phase ("daughter phase") is produced by cooling. The Shape memory results from the crystallographic reversibility of the many martensitic phase transformations of martensite's crystal structure. Beginning as monoclinic, it has the unique ability to undergo limited deformation ("strain") by twinning without breaking atomic bonds and without diffusion. Energy-dispersive and X-ray analyses reveal that the base nitinol material is TiNi in the B2 phase, and confirm the presence of the $Ti_2Ni$ intermetallic inclusions. The microdiffraction patterns reveal a polycrystalline structure. SEM images of Nitinol wire reveal nanofibers [diameter 30 –70 nm] with a longitudinal fiber size of several microns, all aligned parallel to the wire axis. Astoundingly, the martensitic state will revert back to austenite phase instantaneously, regardless of whatever deformation it was in. The original shape is recovered by a solitary "one path" transition back to the austenitic phase and shape; and 35,000 to 100,000 psi (689 MPa) become available to make this happen.

## 3. Experimental

Several Nitinol devices were built and examined for their responses to ordinary and CF/LANR-induced excess heat. Nitinol wire is springy, and not very difficult to bend. Soldering is very difficult, so crimping was used. For this work,

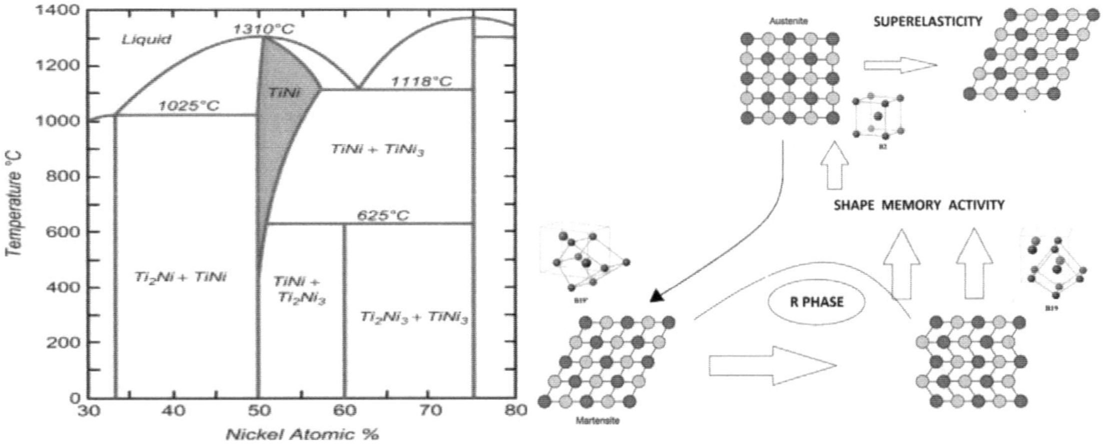


**Figure 1.** The Ni Ti Phase Space of Nitinol as a Function of Nickel/Titanium Ratio and Temperature. Attention is directed to the fact that that the desired region of shape memory activity (SMA) is only in the blue area. The "magic" of Nitinol arises from itfs ordered structure, unlike homogeneously mixed intermetallic alloys where the metallic components are randomly arranged throughout the volume. With active Nitinol, the atoms are arranged precisely in rows.

the high temperature Nitinol shape was a line. The LANR XSH was obtained from a dry preloaded ZrO2-PdNiD NANOR®-type LANR Component used is NANOR #7-24 which has a ZrO2-PdD core, and which delivered a peak power output of ∼12.17 Watts when driven with input power of 3.94 Watts. Controlled heating was by electrically driving an ohmic control alternating with an active NANOR®-type component (#N7-24) with a ZrO2-PdD core [4], [5]. The metrics measured were heat induced force, temperature, and changes in the Nitinol wire's length as determined by using in situ 3D accelerometers and gyroscopes, thermocouples, and a time-of-flight (TOF) optical reflection measurement along the nitinol itself, during the heating.

## Instrumented, Tethered CF or Ohmic Control Driven Nitinol 3D-Printed Springs

Types 2 and 3 Instrumentation was attached to a pillared single-turn spring. The PLA 3D-printed half spring had the nitinol set across the ends. The system was 3D printed as a single piece with the pillar and platform, with two 3D-printed holders to restrain the nitinol. A restoring force was included consisting of one to three elastic bands. In the Type 4 Instrumentation, the system was turned sideways and removed from the PLA half-turn spring entirely. As before, at the center of the system was a 3D-printed clip holding the ohmic control (98 ohms), the NANOR®-type component, a thermocouple, and covering all that is the thermally conductive, electrically insulating, boron nitride (the blue soft material that looks like clay) (Figure 2).

This central core consists of a "sandwich" of an ordinary ohmic resistor as a thermic control (for ordinary heat) and a dry preloaded NANOR®-type LANR component for fast, active CF/LANR excess heat [XSH]. Also present is sensitive temperature monitoring instrumentation and the Nitinol. Some restoring force on the spring was found to be needed, and provided by an elastic band added around the 3D-printed PLA semicircle.

The NANOR®-type component was NANOR® #7-24 with a $ZrO_2$-PdD core. It had an incremental power gain of ∼300% at peak for an electrical input power of 3.94 Watts, generating a thermal power output of ∼12.17 Watts.

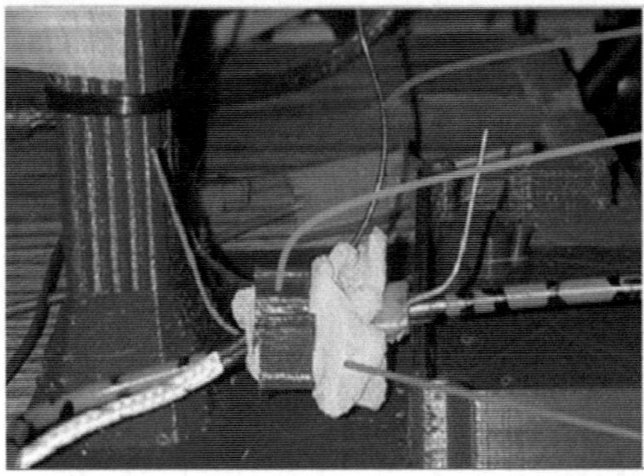

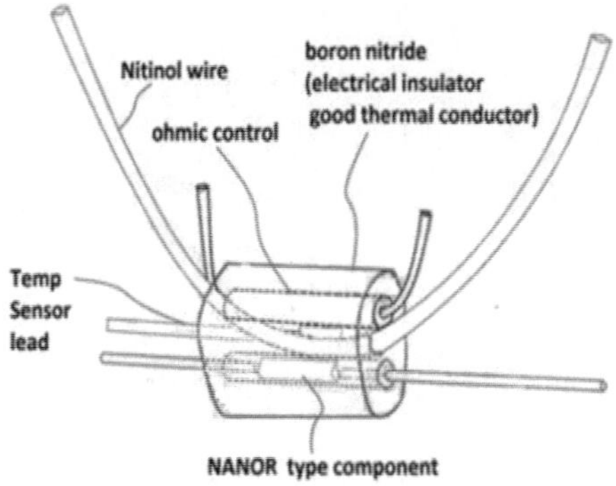


**Figure 2.** Experimental Core to test Response of Nitinol to Applied Heat and Excess Heat.

### 3.1. Metrics Examined

**Temperature** - Specialized thermocouples were used having very fast response times, small probe size, and low thermal mass. Two MAX6675 performed cold-junction compensation for the type-K thermocouples.

**Spring diameter** - Distance was determined by time-of-flight ["TOF"] optical measurement using VL6180X Time-of-Flight Distance Sensor. An integration period of 50 ms yields an incident light resolution of 0.32 lux/count and a range reading in mm.

**Forces wrought** - Qualitatively determined by accelerometers followed by vectorial RMS addition, and multiplication by the effective mass.

**Electric Power Drive** - Precision computer controlled power supply (Tekpower TP3005P) or Arduino driving a MOS-FET using PWM. Power delivered sequentially to the ohmic control and the NANOR®-type component. The applied electric power dissipation makes heat over time which makes the nitinol expand, increasing both the force and distance between ends.

**Calorimetry** - For good, calibrated measurements, the electrically-driven active components are always timely compared to an ohmic control. The components are wrapped in boron nitride [BN] and a larger thermal mass, along with a thermocouple.

**Planned Paradigm** - Confirmation of the algorithm, checking for linearity and time-invariance, and calibrating the output of the component under test, is made by reproducing the heat generated from a square wave sent to an ohmic control. As always, time integration of the power, in and out, is necessary to prove the analytic system is correct, and to rule out phase changes, and the like.

## 4. Results – Proof of CF/LANR Activity in NANOR®-type component Nanor 7-24

Figure 3 is a control demonstrating reproducible CF/LANR activity in NANOR®-type component Nanor 7-24 from electrical drive. The graph shows the calorimetry of electrical power delivered to an ohmic control bracketing an active Nanor®-type component. Power and energy are shown [Input power (blue lines), Output power (red lines), Electrical energy (blue, red, dashed lines)]. It is also a graph showing the calorimetry of electrical power delivered to this ohmic control (a carbon composition resistor). The thermal control and the active Nanor®-type component are in the Nitinol Arch4. Shown is the input electrical power (blue lines) to the ohmic control, the NANOR®-type component Nanor 7-24, and then the ohmic control again, as a function of time. The output thermal power are red lines.

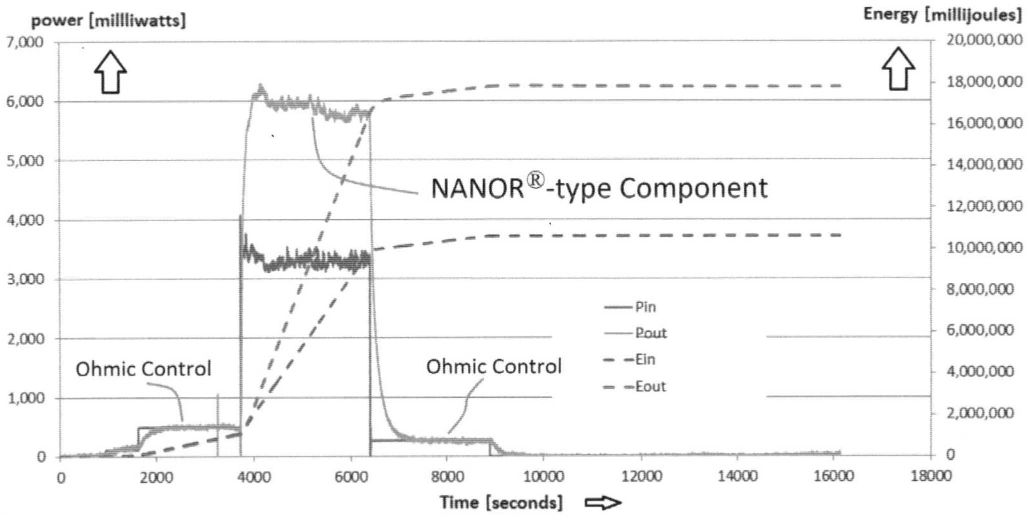


**Figure 3.** Test of Ohmic Control and LANR component for Emission of heat and excess heat.

It can be seen that the controls are connected and the NANOR®-type CF component is active. Excess power and excess energy are easily seen, but only for the LANR component. There is excellent LANR activity.

## 4.1. Results – Response of Nitinol to Input Power Delivered to the Ohmic Control

Figure 4 shows the thermometry of the ohmic control alone against the Nitinol. The peak force wrought was about 224.24 $(+/-26.1)$ microNewtons/Watt. This figure shows the thermometry of the ohmic control alone against the Nitinol. Shown is the input electrical power (red line), and the responses: including temperature (pink line) and the qualitative derived induced RMS force ("milliNewtons").

Note the short delay of the Nitinol response. It can be seen that the input electric power-induced temperature changes lag as the component and surroundings absorb heat according to their heat capacity, before the temperature is sufficient to induce the mechanical change anticipated.

## 4.2. Results - Response of Nitinol to Ohmic Control & Nanor®-Type Component 7-24

Figures 5 and 6 are two coupled graphs of same experiment which show the Calorimetry of, and then the Nitinol response to, and active electrically driven CF/LANR NANOR®-Type component.

Figure 5 is the Calorimetry. The graph shows the input electrical power delivered to an ohmic control and then NANOR®-type component, and the resulting output thermal power from both. The input electrical power (blue lines) to the ohmic control and CF/LANR component are shown as a function of time. Also shown is the output thermal power (red lines) from each. Finally, also shown is the integrated energy in and out from each, which is a requisite to check the calorimetry.

Figure 6 shows a graph presenting the Nitinol response wrought first from the heat delivered from the electrically-driven ohmic control and then the response to the heat from the electrically-driven NANOR®-type component 7–24. The data shows the thermometry (left) and Nitinol response (right) wrought by the heat during a single run using the

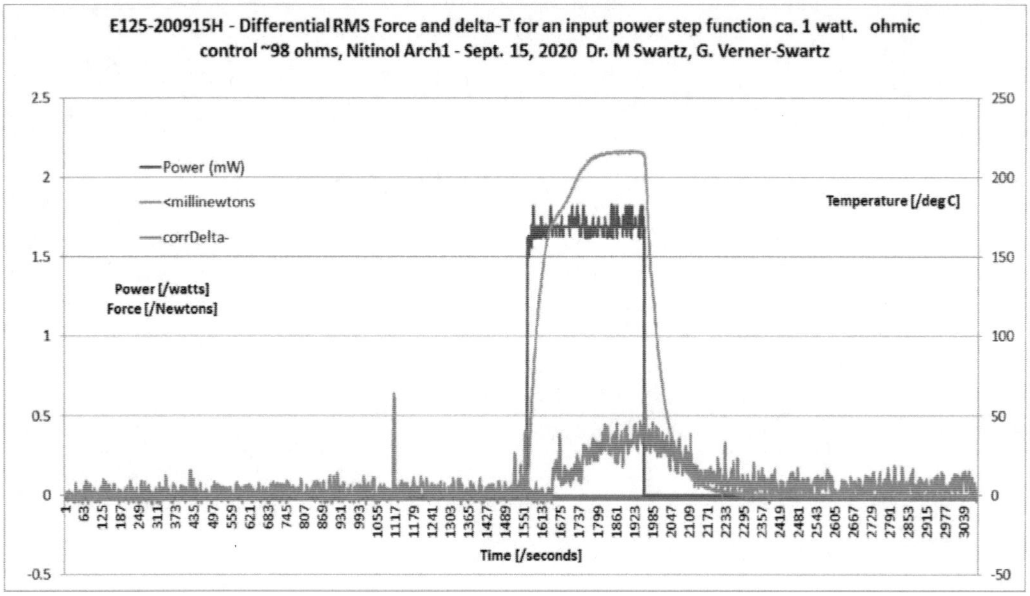


**Figure 4.** Impact of Ohmic Control on Nitinol - Step Response, Note Delay of Nitinol Response.

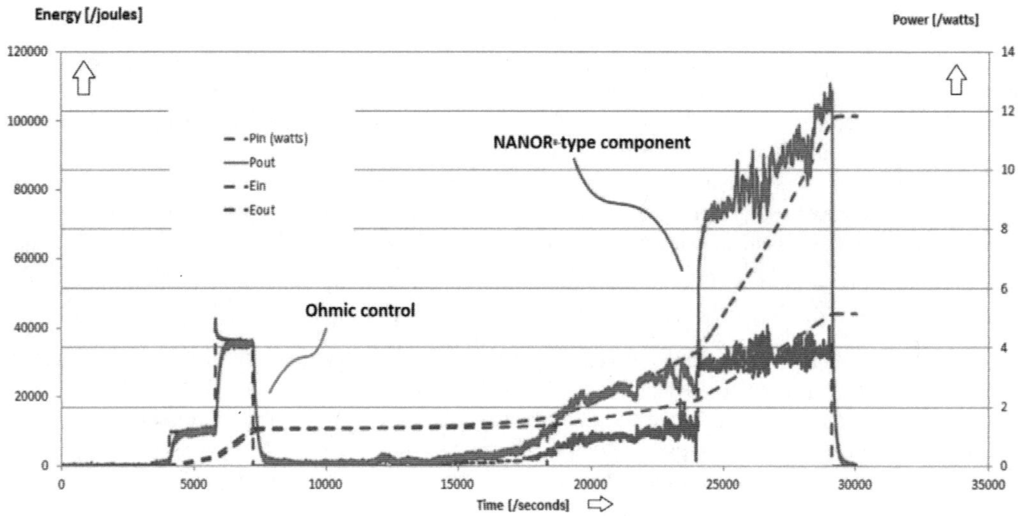


**Figure 5.** Calorimetry of an Ohmic Control & Nanor®-type component 7-24 next to Nitinol.

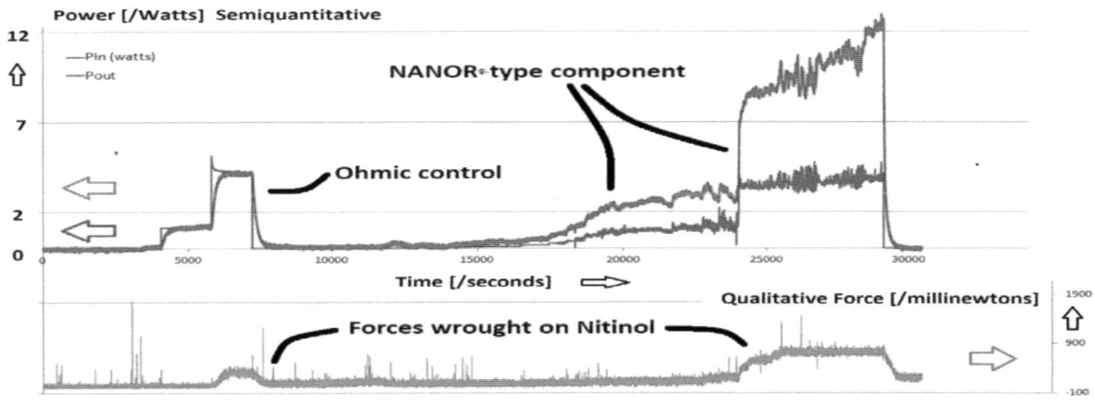


**Figure 6.** Response of Nitinol to Heat from Ohmic Control & Nanor®-type component 7-24.

ohmic resistor as a control. The input electrical power delivered to each (blue lines), and the resulting output thermal power (red lines) from each are shown, along with the kinematic responses from the Nitinol [as an RMS force].

There is excellent LANR activity. Excess power and excess energy are easily seen, but only for the LANR component. The ohmic control has an incremental power gain of 1.00 (100% or less of the input electrical power), The NANOR®-type component at the first lower power input level, has an incremental power gain of 239%. At the end, the incremental power gain is about ~300% [For an electrical input of 3.94 W there is a thermal power output of circa 12.17 Watts. At the end of the run the excess energy produced is ~51, 147.6 Joules.

## 5. Conclusion and Summary of Results

Several Nitinol devices were built to examine the responses of Nitinol to both ordinary and to CF/LANR-induced excess heat. As a control for heat, a 98 ohm resistor was used. For excess heat, a NANOR®-type component delivered the XSH. It had a peak power output of ~12.17 Watts when driven with input power of 3.94 Watts.

Ordinary heat and true CF/LANR excess heat are not distinguished; as the Nitinol Detectors were equally sensitive to both ordinary heat and to CF/LANR Excess Heat.

Although the biggest problem is that Nitinol CF/LANR sensors are non-selective for ordinary heat versus true excess heat from active CF/LANR devices/components, the second biggest problem is that Nitinol is non-linear. This is in part because it is associated with a preliminary phase change, and its transitions have hysteresis. For whatever combination of reasons, there do occur irregular movements in Nitinol which must be considered, then controlled/minimized/avoided to make this type of measurement/monitoring device useful even for qualitative measurements. Then, the Nitinol preliminary phase change must be addressed, as it requires a precise heat absorption before the Nitinol will work as expected. Without this, the Nitinol will not change shape immediately and accurately as, and when, expected.

It has been discovered that the metric of qualitatively derived RMS induced force was more useful than length measurement for both prosaic applied heat and excess heat. The Nitinol had a derived RMS induced Force-Power response of about ~170-250 microNewtons/Watt.

Thus, Nitinol can be used as a qualitative heat sensor in this field BUT it is not selectively specific for LANR/CF-induced excess heat, although it can detect the excess heat. Furthermore, Nitinol sensors are not as sensitive, and are not as reliable, and also are more difficult to use, than other sensors described in the cold fusion literature. They are probably limited in the future to qualitative corroboration [10]. By contrast, full calorimetry (with ohmic controls, time integration and checks of the algorithm and calorimeter), and coherent antiStokes (CMORE) spectroscopy [6], [7], and Deuterium Line radiofrequency (RF) spectroscopy near 327 MHz [8], [9] remain the sole, superior methods to selectively, specifically, and semi-quantitatively detect ACTIVE states/modes in driven CF/LANR/LENR systems and components.

**Acknowledgements**

The author gratefully acknowledges, and thanks, David Nagel, Gayle Verner, and Isidor Straus for their very helpful comments, editorial assistance, and additional material. This effort has been supported by JET Energy Inc., Nanortech Inc., and Anthropocene Institute.

## References

[1] K. Ōtsuka & C.M. Wayman, Eds., *Shape Memory Materials*, Camb. Univ. Press, 1998, ISBN 0-521-44487-X, 2013.
[2] H. R. Chen, ed., Shape Memory Alloys, Nova Science Publishers, Inc., 2010, ISBN 978-1-60741-789-7.
[3] Pentagon admits testing wreckage from UFO crashes. https://www.the-sun.com/news/2321251/pentagon-admits-testing-wreckage-ufo-crashes/
[4] Swartz M. R., Verner G., Tolleson, J, Energy Gain From Preloaded $ZrO_2$–PdNi–D Nanostructured CF/LANR Quantum Electronic Components, JCMNS 13, 528, (2014).
[5] Swartz M. R., Hagelstein P.I., Demonstration of Energy Gain from a Preloaded $ZrO_2$–PdD Nanostructured CF/LANR Quantum Electronic Device at MIT, JCMNS 13, 516 (2014).
[6] Swartz, M. R, Peter L. Hagelstein, Increased PdD anti-Stokes Peaks are Correlated with Excess Heat Mode, JCMNS 24, 130-145 (2017).
[7] Swartz, M. R, Increase of an Anti-Stokes Peak at the Cathode of an Electrically-Driven, Active Aqueous Nickel/H2O/Pt System, JCMNS, 279, 22 (2018).
[8] Swartz, M. R, Active LANR Systems Emit a 327.37 MHz Maser Line, Proc.ICCF-22, JCMNS, 33, 80-110 (2020).
[9] Swartz, M. R, FCC Vacancies in $ZrO_2$PdD RF are the Active LANR Site, Proc. ICCF-22, J. Cond.Matter Nucl. Sci., Volume 33, pages 126-144 (2020)
[10] Swartz, M. R, Nitinol Can Detect CF/LANR Excess Heat Activity, J. Cond. Matter Nucl. Sci. to be published.

J. Condensed Matter Nucl. Sci. 36 (2022) 97–108

Research Article

# Some Novel Analytical Techniques Applied to LENR Active Materials

Francis Tanzella*

*Energy Research Center LLC, San Carlos, CA 94070, USA*

Robert Godes and Robert George

*Brillouin Energy Corporation, Berkeley, CA 94705 USA*

Anis Rahman

*Applied Research & Photonics, Inc., Harrisburg PA 17111 USA*

**Abstract**

We explore the results from the analyses of several low energy nuclear reactions (LENR)-active tubes run by Brillouin Energy (BEC) using several different techniques. The most novel analytical techniques are terahertz (THz) imaging and THz spectroscopy. This imaging technique gives insight into the lattice spacing of crystalline and micro-crystalline metals and ceramics complimentary to that from X-ray diffraction (XRD) measurements. XRD of powder removed from this sample showed the normal lattice spacing for pure Ni powder. However, THz imaging showed lattice dilation in the Ni coating. We discuss the potential source of this dilation as well as its possible importance to LENR. We will also discuss the results of THz spectroscopy performed on materials from this and similar tubes This report discusses metallurgical and physical measurements, such as optical microscopy, density and porosity measurements, terahertz (THz) imaging and spectroscopy and scanning electron microscopy (SEM). We also cover chemical and elemental measurements such as energy dispersive X-ray (EDX) inductively coupled plasma-optical emission spectroscopy (ICP-OES), ICP-mass spectroscopy (ICP-MS) and He mass spectroscopy. With the latter method we present an intriguing result that shows tritium production in a unique Pd electrolysis experiment. EDX, ICP-OES, and ICP-MS can help the researcher understand the role of minor and major elemental impurities in these materials. If present, ICP-MS may also detect isotopic anomalies in these materials post experiment. The ultimate point of this exercise should be to help the LENR field reproduce the materials that are best suited for use in LENR experiments such as that presented as a worked example of designing and manufacturing a Pd rod from measurements like these.



*Keywords:* LENR, Metallurgy, Elemental analysis, Chemical analysis, Palladium

*Corresponding author: consulting@tanzella.name

© 2022 ICCF. All rights reserved. ISSN 2227-3123

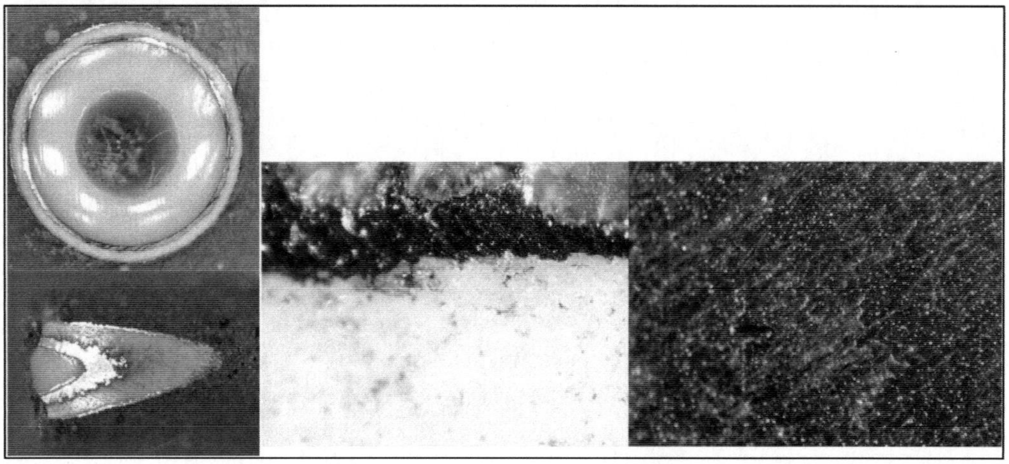

**Figure 1.** SEM micrographs of one of Brillouin's tubes.

## 1. Introduction

Since there has been a significant amount of discussion recently about finding the best material for low energy nuclear reactions (LENR), this paper discusses existing methods to determine the properties of the relevant materials used and to take the opportunity to introduce new techniques.

This report discusses metallurgical and physical measurements, such as optical microscopy, density and porosity measurements, terahertz (THz) imaging and spectroscopy and scanning electron microscopy (SEM) in the first half of the paper. The second half will cover chemical and elemental measurements such as energy dispersive X-ray (EDX) inductively coupled plasma-optical emission spectroscopy (ICP-OES), ICP-mass spectroscopy (ICP-MS) and He mass spectroscopy. Then we will end with a rather old worked example of designing and manufacturing a Pd rod from measurements like these.

## 2. Metallurgical and Physical Analyses

### 2.1. Optical Microscopy

The old-fashioned optical microscope can tell you quite a bit about your materials.

Figure 1 shows micrographs from two pieces of Brillouin Energy Corporation's three-layer coated tubes. These have been cut in a diamond saw, mounted in an epoxy casting resin, and sequentially polished from coarse grit sandpaper down to 1 μm diamond paste on a buffing cloth. The upper left photo shows the cross-section of the tube after mounting and polishing. The white inner ring is the tubular ceramic substrate below the first coating of copper, followed by a plasma-sprayed ceramic coating and one of nickel. The distance from the inside the copper to outside of the nickel is approximately 400 μm. The lower left photo shows the tube mounted lengthwise with a metal shim holding it $\sim 4^{\circ}$ from horizontal so the polishing exaggerates the depth radially. In the middle micrograph you can see the outer two layers of dielectric and Ni, and the epoxy, with the voids evident. The middle micrograph shows the voids in the ceramic coating. The micrograph on the right shows the Cu from the oblique cut sample exposing its less than 1 μm grains.

**Figure 2.** SEM micrographs of electrolysed Pd cathode.

In Figure 2 are two optical micrographs of a professionally cut, mounted and polished 3mm Pd cathode after a lithium deuteroxide electrolysis experiment from SRI International. The Pd grains and grain boundaries can be seen in both micrographs. The middle micrograph is about 130μm wide while the higher magnification one on the right is 52 μm wide. The color saturation and exposure has been adjusted the for better contrast in these two micrographs. After electrolysis the grains are about 5μm, probably oriented as received (drawn). Normally a researcher would make note of the grain size and orientation before use via non-destructive microscopy, for later comparison.

### 2.2. Density, Porosity and Stress/Strain Measurement

At SRI we regularly used instruments [1] regularly to characterize solids and powders. Although these instruments are normally used for catalyst and catalyst support materials, they can reveal important information about pure metals or metal/ceramic composites. Understanding the physical voids and internal surface area of cathodes and gas-exposed solids is important as recently noted. For pure materials, the measured density reveals the percent density of the actual material compared to that 100% dense. Knowing the non-molecular absorption or adsorption on the surface or in internal voids, allows one to better calculate how much of the hydrogen isotopes are in the interstitial sites. Various mechanical properties can be determined using instruments from Instron® [2].

### 2.3. X-Ray Diffraction

Another important physical measurement tool is X-ray diffraction. A powder or solid sample is held in a monochromatic X-ray beam while an X-ray detector is swung through an arc of 90° or more. The diffracted beam intensity is plotted against the angle in 2-theta. Generally, the resultant peaks are compared to a very large data base of known diffraction patterns. Impurities much below 1% are generally missed by this technique. The three major peaks shown in the diffractogram [3] in Figure 3 are from the face centered cubic (fcc) lattice of Ni. The minor peaks are from the alumina powder impurity, since this Ni was scraped from a Ni coating on top of alumina. The structure of unknown materials created during the LENR process may be determined using a process called the Reitveld refinement.

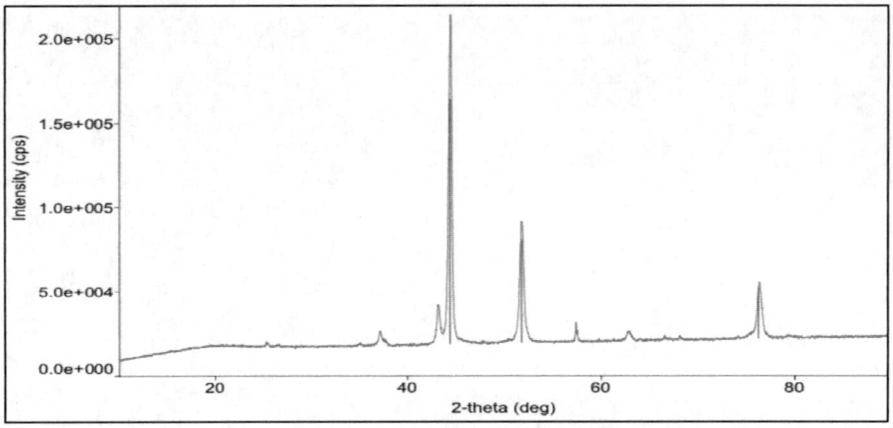


**Figure 3.** X-ray diffractogram of NI powder from one of Brillouin's tubes.

## 2.4. Terhertz Imaging and Spectroscopy

Recently, we have been practicing the THz imaging technology practiced Anis Rahman at Applied Research and Photonics. Rahman has published [4] calibrations that have shown that this technology yields the same measurement results seen via SEM and TEM on 14nm metal lines on Si wafers. By plotting the grayscale depth vs horizontal position across an image, the top being a black line and the bottom the white space peaks are seen. The FWHM of each peak, used to calculate the line width, averages to $15 +/- 1$ nm compared to the nominal 14nm lines. It is important to note that this technique overcomes the Abbe diffraction limit of traditional microscopic methods.

From a tube constructed similarly to that shown in earlier optical micrographs, the outer Ni coating was analyzed using THz imaging. The sample in its holder is shown in the left photograph. The image in the middle shows the lines seen from a 100nm x 100nm measurement [5]. Applying the analysis shown from the calibration, the FWHM measurement from the grayscale depth plot on the right gives a significantly greater lattice constant, as coated, than found by scraping some of the powder off and running X-ray powder diffraction. The plasma spraying process seems to have yielded Ni with a dilated lattice constant known as lattice expansion. This level of expansion would yield a unit cell volume approximately twice that of pure Ni. Representations of stationery and rotating cube images can also be produced by this technique.

Four similar samples were measured using these imaging and spectroscopic techniques [6]. THz spectroscopic measurements were collected from 0.1 to 32 THz. The THz laser spectrometer setup is shown in Figure 5. The absorbance spectrum from 0.1 to 10THz is also shown on the bottom right graph. Several peaks are noted in the range of those calculated for Ni phonon excitations. These are in a similar range to those reported by Dennis Letts [7] in his dual-laser beat-frequency experiment on Pd.

Figure 6 shows close ups of the same data from Figure 5 presented in wavenumber mode. The absorbance peaks are consistent with those reported in the literature for NiO. We predict that the Ni was oxidized after leaving the plasma spray gun and before cooling on the tube. Although the metal travels through a reducing atmosphere plasma at the spray gun, it traverses room air on its way to the target tube. There is still some work yet to be done in understanding these materials, before and after exposure to hydrogen in the reactors. Although these results agree with some published results, there are still unidentified peaks that should be able to tell us about the chemistry and physics of these coatings.

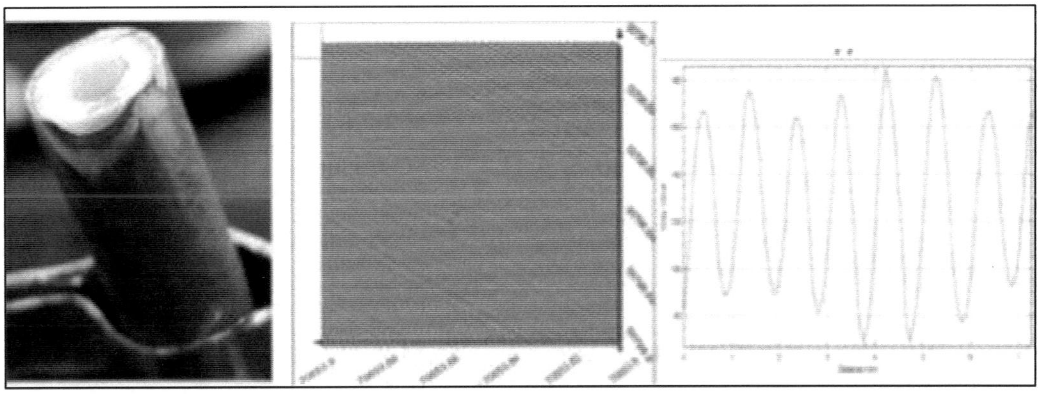

**Figure 4.** Sample and results from lattice-scale THz imaging measurement.

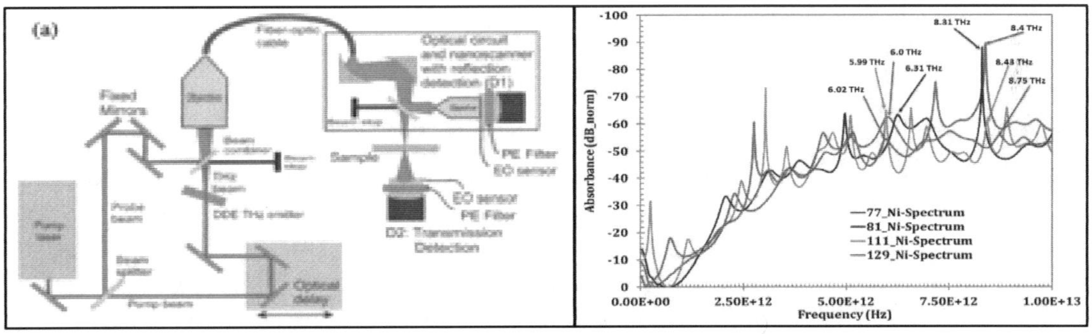


**Figure 5.** THz spectroscopic setup spectrum of Brillouin tube sample.

## 2.5. Scanning Electron Microscopy

Figure 7 displays three SEM micrographs taken by Alan Goldwater of MagicSound Labs, on Brillouin tubes similar to those shown earlier. The conducting materials show up as white and the insulators show up darker. The legend bar on the left two micrographs of the cross-section and oblique angle cut samples is 2mm. In this mounting the vertical direction is expanded by a factor of 10 in the oblique cut sample compared to that of the cross section. The legend bar on the lower right is 7μm, allowing for much finer measurements than in the optical photographs. Those green dots on the right micrograph are for EDX elemental composition measurements.

Figure 8 shows a micrograph of a Pd foil cathode of much higher magnification - 5,000x post experiment. This is from SRI's reproduction of Energetic Technologies SuperWave® electrolysis experiments [8]. Taken at the Naval Research Lab features of less than 1μm are visible. This sort of SEM can tell you quite a bit about a sample before and after exposure in a reactor.

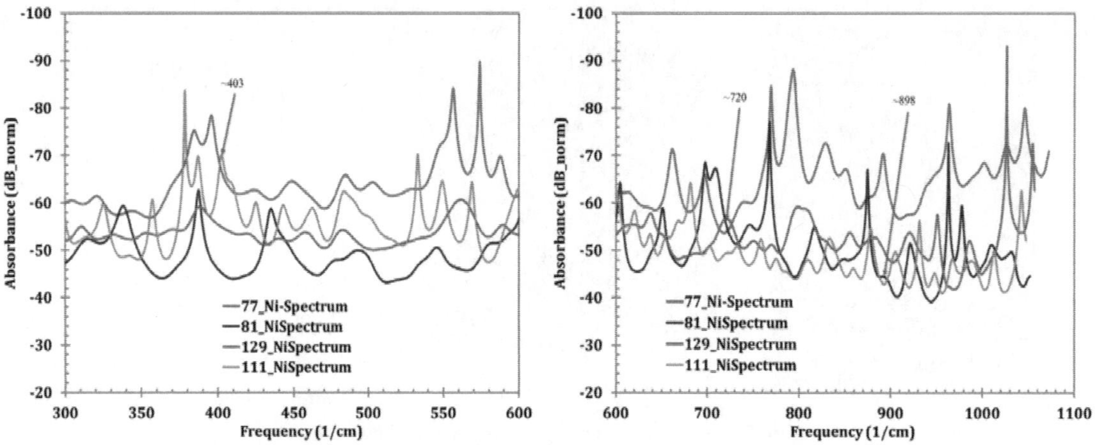


**Figure 6.** Wavenumber representation of THz spectrum form Brillouin tube sample.

**Figure 7.** Scanning electron micrographs of pieces of a Brillouin coated tube.

## 3. Chemical and Elemental Analyses

### 3.1. Energy Dispersive X-Ray Analysis

Energy dispersive X-ray analysis uses the known X-ray energies of individual elements when exposed to an X-ray beam in the SEM's vacuum chamber. Figure 9 displays an EDX map of the same oblique cut sample shown in Figure 7 helps elucidate the sample's composition. These elemental maps taken by Alan Goldwater are a great visual aid in understanding what's going on in multi-layer coatings. The aluminum dominates both the sprayed and extruded $Al_2O_3$ map in the lower left map. However, the 5 images on the right show the Al in orange and the oxygen in red separately, which verifies the composition. The carbon map in yellow is from the epoxy mounting resin. The thin blue map is of the outer Ni coating. The inner Cu coating is shown in green.

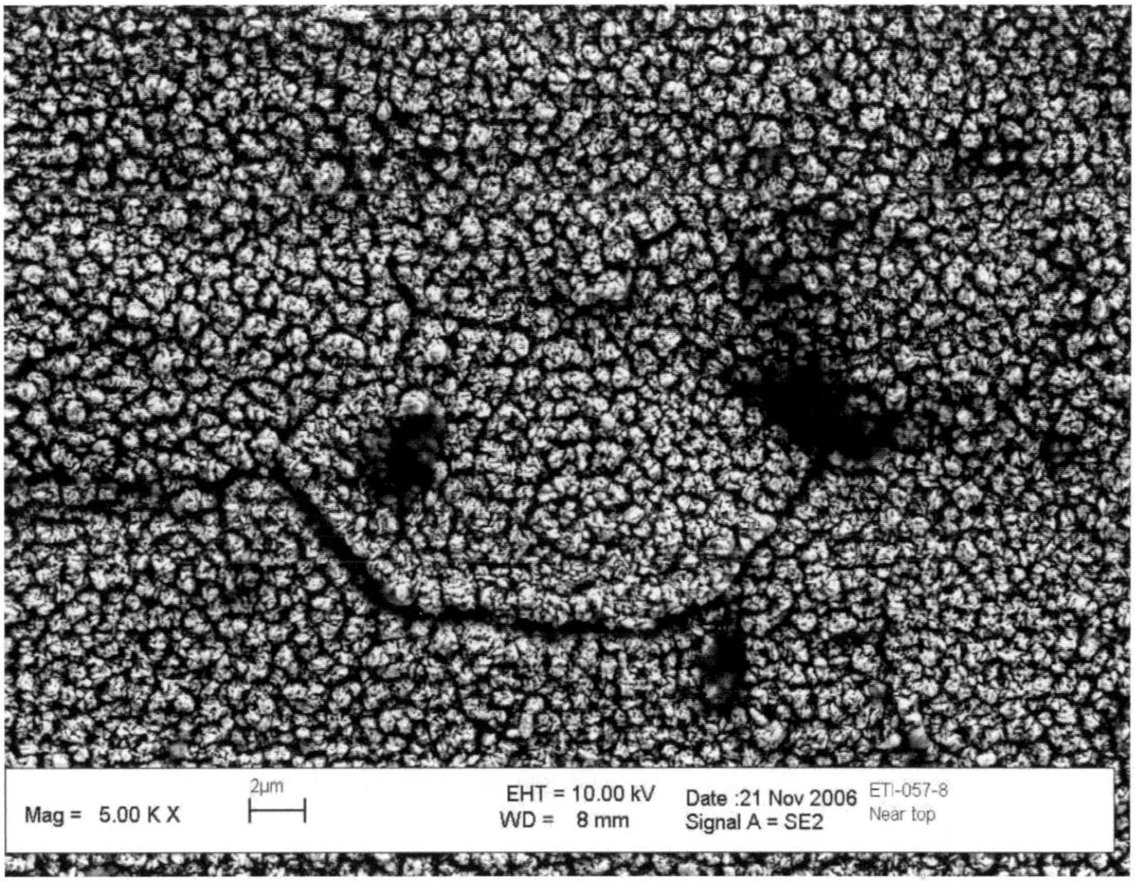


**Figure 8.** High magnification SEM image of electrolyzed Pd foil. (Courtesy of the Naval Research Laboratory).

Another way to view this multi-layer coating is from the cross-section sample. On the bottom left of Figure 10 is a thin section of a SEM image of this sample covering 682µm horizontal. The line designates where the EDX spectrometer has taken elemental measurements along the image, which is a novel way to view the elemental constituents across a multi-layered coating. In the graph in the upper left, which aligns with the SEM image, the red line designates the Ni composition, the blue line the aluminum in the dielectric, and the green line the Cu coating. The voids in the image line up with the reduced Al strength in the graph. There is obviously some sodium impurity (yellow line) in the Cu. The right image shows the elemental analyses of different points (green dots) in the SEM micrograph taken in the middle of the alumina coating. Although concentrations below 1 % are suspect, notice the 1.22 % sodium in the 234-2089 sample. It is interesting to note that the Al:O ratio is relatively constant. The carbon concentration is from machine background or from epoxy that moved into the voids during polishing.

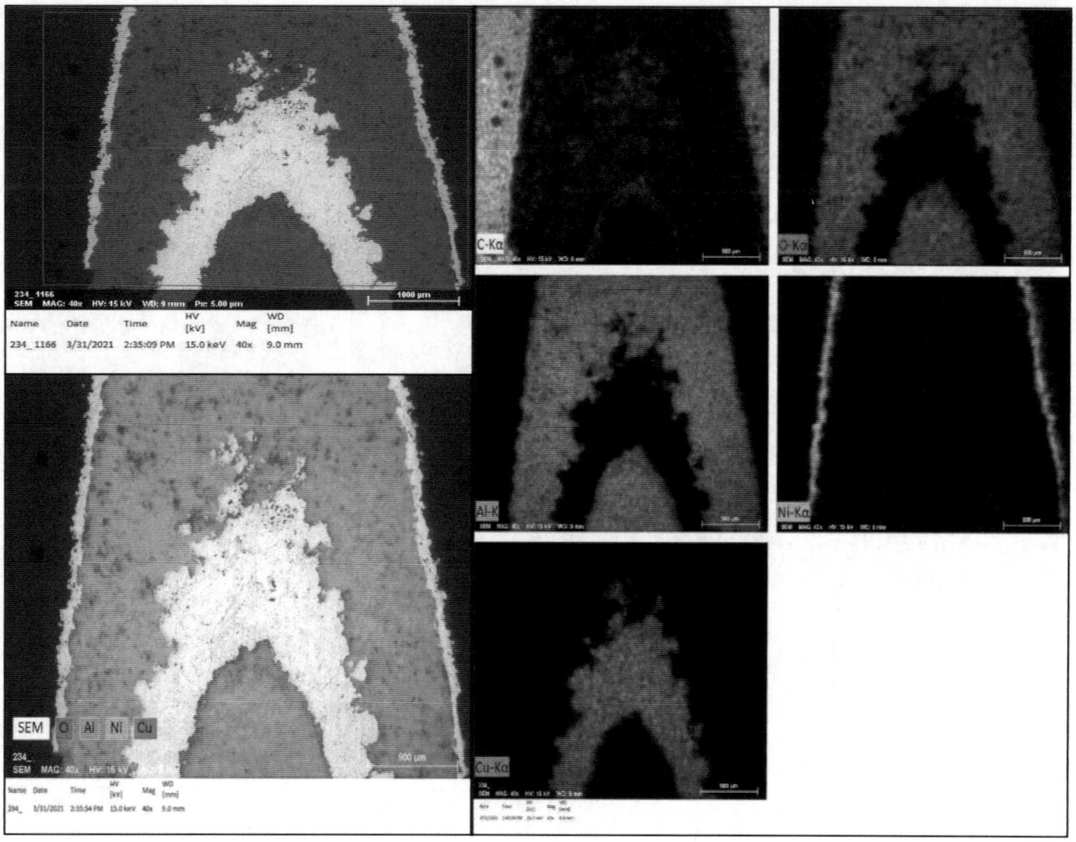


**Figure 9.** EDX elemental map of oblique-cut Brillouin tube sample.

### 3.2. X-Ray Fluorescence

A tool similar to EDX is X-ray fluorescence. It uses the same principle measuring the X-ray energies of materials exposed to incident X-rays to identify the elements present in solids, powders, and even liquids. As with XRD and EDX the results are usually compared to known library peaks, even though the X-ray peaks can be calculated from first principles. Impurities below the 1% level are generally not seen. The same powder scraped from the Brillouin coated rod and measured using XRD in Figure 3 was measured using an XRF instrument at SRI [9] and is shown in Figure 11. The middle-energy plot on this slide shows the nickel K-alpha and K-beta peaks, along with a minor amount of iron. Other elements called out in the plot are not detectable. The lower energy plot, not shown, also showed the aluminum from the alumina impurity. Unlike EDX, no vacuum is necessary for XRF measurements since it is performed in room air.

### 3.3. Inductively-Coupled Plasma Optical Emission Spectroscopy

ICP-optical emission spectroscopy is a destructive way to determine the major, minor, and impurity elements in a material. Because it obtains a spectrum from the emission of many elements, several elements can be analyzed for

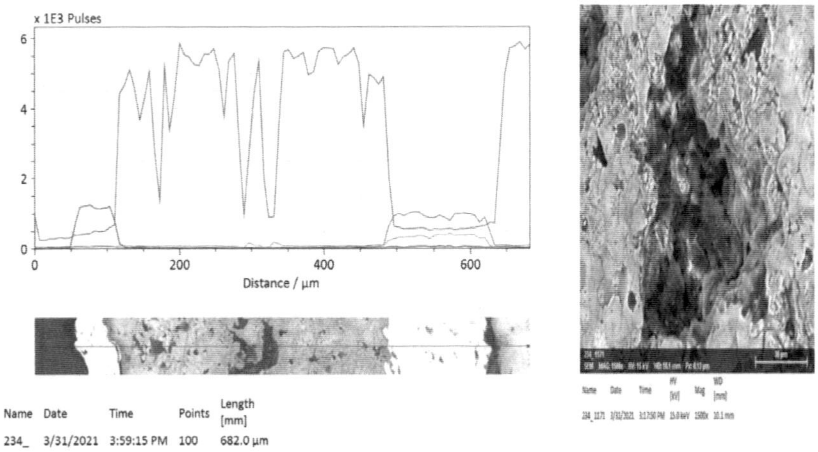


Atomic concentration [%]

| Spectrum | C | O | F | Na | Al | Si |
|---|---|---|---|---|---|---|
| 234_2084 | 29.02 | 50.36 | 0.61 | | 20.02 | |
| 234_2085 | 42.85 | 40.04 | 0.50 | | 16.60 | |
| 234_2086 | 50.00 | 38.06 | 0.53 | | 11.41 | |
| 234_2087 | 57.67 | 30.06 | | 0.56 | 11.72 | |
| 234_2088 | 57.85 | 31.86 | | 0.53 | 9.75 | |
| 234_2089 | 77.10 | 18.78 | 0.38 | 1.22 | 2.12 | 0.40 |
| 234_2090 | 69.82 | 24.80 | | 0.90 | 4.48 | |
| **Mean** | **54.90** | **33.42** | **0.51** | **0.80** | **10.87** | **0.40** |
| **Sigma** | **16.18** | **10.44** | **0.09** | **0.32** | **6.27** | **0.00** |
| **SigmaMean** | **6.12** | **3.95** | **0.04** | **0.12** | **2.37** | **0.00** |

**Figure 10.** EDX line-scan of a cross-section of a Brillouin tube sample.

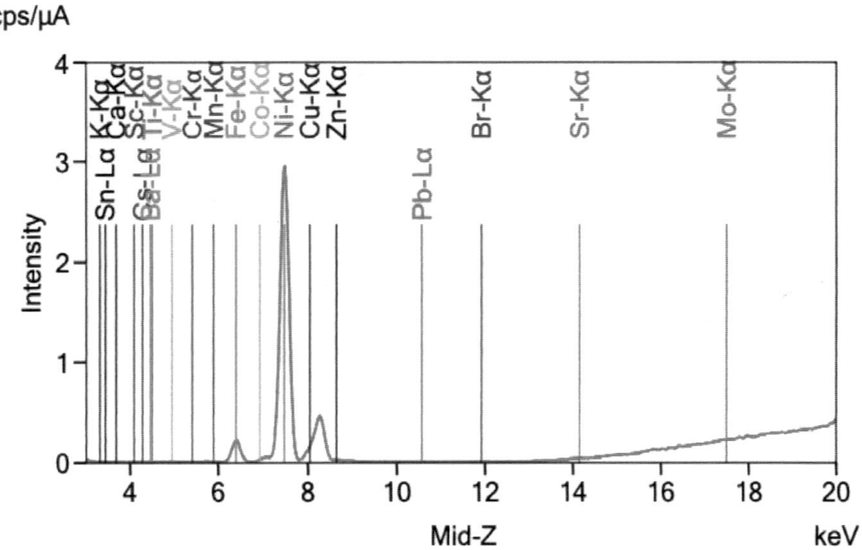


**Figure 11.** X-Ray fluorescence spectrum of nickel powder from Brillouin coated tube.

simultaneously, as opposed in the older atomic absorption method. Concentrations in the part per trillion level are common from ICP-OES analyses. The numbers in Table 1 have been normalized to the Ni major component. Results are usually reported as ppm or pptr. The analyses presented on Tables 1 and 2 are from Cerium Labs® in Austin Texas.

### 3.4. Inductively-Coupled Plasma Mass Spectroscopy

ICP-MS can tell you some of the same information that you get from ICP-OES, but may also tell you about isotopic changes in your material. Note the results in the bottom line of Table 2. The best way to assure reliable isotopic analysis is to use good isotopic standards. The example shown here has not been verified but is shown as a possible example analysis. In this case, the analyst felt that their mass resolution was not good enough to comfortably differentiate between Ni-64 and a possible Zn-64 impurity.

### 3.5. He Mass Spectroscopy

This measurement from SRI was first presented at ICCF11 [10]. He MS is another use of destructive metals analysis using mass spectrometry. This Arata double-structured cathode [11] consisted of a thick walled Pd bottle filled with Pd black and electrolyzed in $D_2O$ for 3 months. Brian Clarke and Brian Oliver analyzed [12] the He-3 as an off-gas from the heated Pd black and Pd wall from this double-structured cathode, shown in Figure 12. Slicing up the wall as illustrated in the cartoon, they created a depth profile of He-3 from the inner wall to the outer wall of the cathodes from both experiments. The results from the $D_2O$ experiment showed that the tritium produced in the inside diffused out achieving the expected concentration gradient as it decayed to He-3 along the way. A fair amount of tritium was also measured in the electrolyte from the $D_2O$ experiment. No tritium or excess He-3 was measured in the $H_2O$ experiment.

## 4. Palladium Rod Manufacture from these Type of Measurements

IMRA-Japan measured 21 different elements from a Pd rod that yielded impressive results. They also performed many of the other physical and chemical measurements listed in this paper. This was one of 42 samples subjected to this type of analyses. With the right capability, you can reproduce a desired material that my no longer be available. The details are not important but using these chemical analyses in addition to other materials analysis, IMRA Japan was

**Table 1.** ICP-OES analyses of samples from Brillouin's coated tubes.

| Analyte Name | Al | Ba | Ca | Co | Cu | Na | Si | Sn | Zn |
|---|---|---|---|---|---|---|---|---|---|
| 1N | 1.545 | 1.933 | 0.307 | 1.005 | 0.752 | N/D | 0.381 | 0.362 | 4.127 |
| 2N | N/D | N/D | 1631.5 | N/D | 115.507 | 11815.39 | 5212.185 | 483.268 | N/D |
| 3N | 0.691 | 1.797 | 0.873 | 0.946 | 0.147 | N/D | 0.243 | N/D | 4.425 |
| 4N | 0.113 | 0.017 | N/D | 0.009 | 0.121 | 0.0008 | 0.169 | 0.0007 | 0.0431 |

**Table 2.** ICP-MS isotopic results from a Brillouin Ni powder sample.

| Analyte | Nominal Mass | Accurate Mass | Nominal Abundance | Ni Tube 75 | Percent Diff |
|---|---|---|---|---|---|
| Ni | 58 | 57.93535 | 68.27 | 65.74 | 3.77% |
| Ni | 60 | 59.93079 | 26.10 | 25.17 | 3.62% |
| Ni | 61 | 60.93106 | 1.13 | 1.21 | 6.70% |
| Ni | 62 | 61.92835 | 3.59 | 3.89 | 8.06% |
| Ni | 64 | 63.92797 | 0.91 | 3.99 | 125.65% |

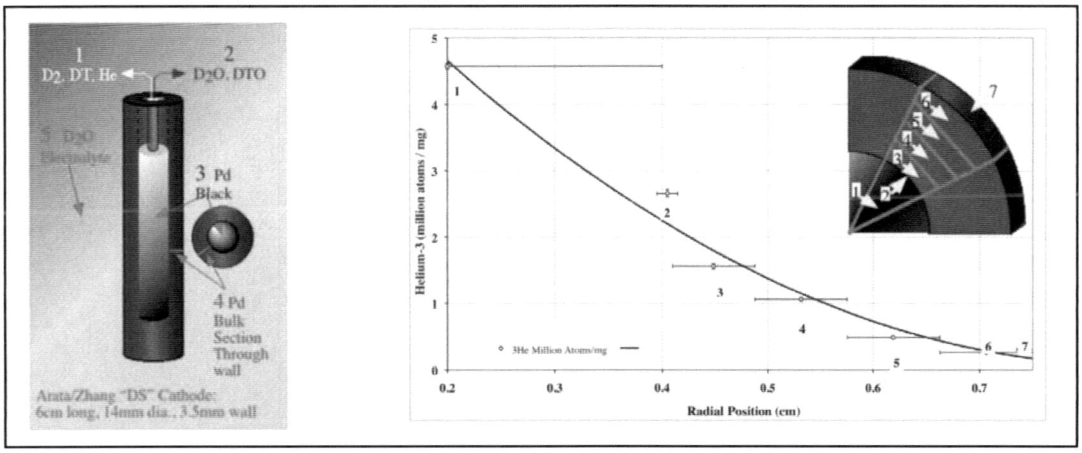


**Figure 12.** He-3 mass spec measurements from an electrolyzed double-structured cathode.

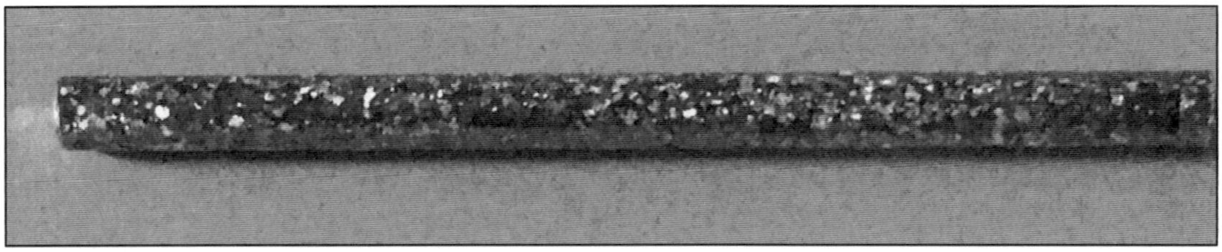

**Figure 13.** Palladium rod reproduced from extensive analyses of a good cathode.

able to reproduce one of these materials, shown in Figure 13. Many other Pd rods were produced for the NHE lab in Shin-Sapporo using these analyses.

## 5. Summary and Conclusions

The authors hope that the readers appreciate and the value of, and perform, some of these physical and chemical analysis techniques in documenting their better LENR materials.

Optical and electron microscopy and THz imaging can tell you about the metallurgy of metals and metal-ceramic composites. Density and surface area measurements can help you manufacture metals or powders that are LENR-active. EDX, ICP-OES, and ICP-MS can help understand the role of minor and major elemental impurities in these materials. If present, ICP-MS may also detect isotopic anomalies in these materials post experiment. In addition, you can produce pretty pictures to use in presentations to potential funders.

The ultimate point of this exercise may be to help the LENR field reproduce the materials that are best suited for use in LENR experiments such as that shown in Figure 13, or to create and verify the properties of a material of one's own design.

## 6. Acknowledgements

The authors would like to thank Alan Goldwater for the use of his images and results performed for Brillouin. We would like to acknowledge IMRA Japan for its analysis and Pd rod manufacture done for the NHE lab in Sapporo. The authors would also like to thank the Anthropocene Institute for its financial support.

## References

[1] Micromeritics Instrument Corporation, https://www.micromeritics.com.

[2] Instron, https://www.instron.com/en/.

[3] Rigaku MiniFlex X-ray Diffraction Spectrometer, https://www.rigaku.com/products/xrd/miniflex

[4] A. Rahman and A. K. Rahman, IEEE Transactions on Semiconductor Manufacturing, **32**(1), 7-13, (2019)

[5] A. Rahman *et al.*, Novel Research in Sciences, **2**(4), (2019), NRS.2019.2.000545. DOI.10.31031/NRS.2019.2.000545

[6] A. Rahman and D. Tomalia., Spectroscopy, **36**(4), 20-27 (2019).

[7] P. Hagelstein, D. Letts and D. Cravens, "Terahertz difference frequency response of PdD in two-laser experiments", Journal of Condensed Matter Nuclear Science, 3 (2010) 59.

[8] E. Castagna, E.; M. Sansovini, S. Lecci, A. Rufoloni, F. Sarto, V. Violante, D. Knies, K. S. Grabowski, G. K. Hubler, M. C. H. McKubre, F. L. Tanzella, "Metallurgical characterization of Pd electrodes employed in calorimetric experiments under electrochemical deuterium loading", Proceedings of 14th International Conference on Condensed Matter Nuclear Science, Washington, DC, (2008) 44.

[9] Rigaku NEX QC+ QuantEZ X-ray Fluorescence Spectrometer, https://www.rigaku.com/products/edxrf/quantez

[10] P. L. Hagelstein, M. C. H. McKubre, D. J. Nagel, T. A. Chubb, and R. Hekman. In New Physical Effects in Metal Deuterides, Eleventh International Conference on Condensed Matter Nuclear Science, Marseille, France, Marseille, France, 2004.

[11] Y. Arata and Y. C. Zhang, "Anomalous difference between reaction energies generated within D2O-cell and H2O-cell". Jpn. J. Appl. Phys. Part 2, 37 (1998) L1274.

[12] B. W. Clarke, B. M., Oliver; M. C. H. McKubre, F. L. Tanzella, and P. Tripodi, "Search for 3He and 4He in Arata-Style Palladium Cathodes II: Evidence for Tritium Production", Fusion Sci. & Technol. 40 (2001) 152.

J. Condensed Matter Nucl. Sci. 36 (2022) 109–114

Research Article

# LENR Transmutation of Stable Sr and K Isotopes in Activated Microbiological Syntrophic Anaerobic Association

Alla Kornilova and Sergey Gaydamaka

*AV Systems Inc. (54, Harpsichord TPKE, Stamford, CT, USA)*

Vladimir Vysotskii*

*Taras Shevchenko National University of Kyiv, Ukraine*

**Abstract**

In our previous works [1]–[5], an effective method for accelerated deactivation of the reactor $Cs^{137}$ isotope during nuclear reaction of transmutation $Cs^{137} + p = Ba^{138}$ in growing microbiological cultures was presented and discussed. In the present study we have attempted to solve the problem of accelerated transmutation of another very dangerous radioactive $Sr^{90}$ isotope. This paper presents the results of investigation of the nuclear reaction $Sr^{88} + p = Y^{89}$ a transmutation to a stable $Sr^{88}$ analog of radioactive $Sr^{90}$ isotope. The research was carried out on the basis of optimal anaerobic syntrophic associations under the stimulated long-distant action of weak undamped thermal waves.



*Keywords:* Nuclear reaction in biological systems, Biotransmutation of isotopes, Coherent correlated states, Utilization of radionuclides, Microbiological syntrophic association

## 1. INTRODUCTION

It is well known that among the large number of long-lived reactor radioactive isotopes, there are two that are most dangerous to humans - $Cs^{137}$ and $Sr^{90}$. These chemical elements (Cs and Sr) are direct biochemical analogs of vital elements K and Ca and are very efficiently absorbed by living organisms in the event the organisms lack these vital elements.

In our previous studies, methods of accelerated deactivation of $Cs^{137}$ isotope were presented, carried out using a transmutation reaction $Cs^{137} + p = Ba^{138}$ stimulated by the growth of microbiological syntrophic associations [1]–[5]. During these studies, we have developed and optimized methods to accelerate this reaction, which is equivalent to reducing the lifetime of this isotope from 30 years to several (2 - 3) weeks. These experiments with a real radioactive

*Corresponding author: vivysotskii@gmail.com

$Cs^{137}$ isotope were first carefully studied based on the $Cs^{133} + p = Ba^{134}$ reaction with a stable analogue $Cs^{133}$ of this radioactive $Cs^{137}$ isotope.

We are currently implementing a similar research technique to create a technology for the deactivation of another dangerous isotope, $Sr^{90}$, in the reaction $Sr^{88} + p = Y^{89}$ of transmutation of a stable $Sr^{88}$ analogue of this isotope. In the course of the experiment, new methods of optimization of such reactions were investigated – the impact of shock-cavitation waves on the syntrophic association in order to increase the viability in life-threatening conditions of this association.

## 2. MATERIALS AND METHODS

As the object of our research, we chose an anaerobic syntrophic association grown on waste from the food and light industries. The optimum temperature for the life of the microbial community used for the experiments was 34 - 36°C.

One of the main goals of the experiments was the search for methods to optimize the transmutation reaction with the participation of heavy chemical elements and isotopes. In this series of experiments, we investigated the distant effect of weak undamped temperature waves acting for a relatively short time on a given microbiological association. The possibility of the existence and features of the propagation of such undamped temperature waves directly follows from the refined equations of thermodynamics, if we take into account the time $\tau$ of local thermal relaxation (time of thermalization) at any point of the field of the thermal (temperature) wave. Such waves can propagate over a long distance without dissipation, and their frequency is determined by the formula $\omega = (n + 1/2)\pi/\tau, n = 0, 1, 2 \ldots$. These waves were predicted, observed, and studied in detail (both theoretically and experimentally) in our works [6]–[9].

The action of such waves leads to the formation of weak shock-acoustic waves in the volume, which contains the used microbiological association.

Earlier, we carried out similar detailed studies on the stimulation of nuclear (dd)-fusion in remote Ti + D targets due to the action of the same waves [9]. In these older experiments, we very effectively stimulated this fusion reaction through the formation of coherent correlated states in the target [10]–[16].

Such quantum states correspond to a coherent quantum superposition of the eigenfunctions of a particle in a potential well (in particular, to a coherent superposition of proton vibrations modes in an interatomic potential well).

These states are formed by, for example, a fast pulse deformation of a potential well, or when it is periodically modulated at certain frequencies [10]–[16]. The main feature of such states is the possibility of the formation of very large fluctuations in the momentum and energy of the particle, which leads to a very sharp increase in the probability of the tunneling effect and, accordingly, in the probability of a nuclear reaction at low (e.g., room) temperature. These states and the associated process of stimulation of nuclear processes have been studied in detail in works [10]–[16].

In the current experiment with $Sr^{88}$ isotope (natural abundance 82.6%) we also wanted to test the effectiveness of such undamped temperature waves on a distant microbiological association before the start of a long process of $Sr^{88} + p = Y^{89}$ nuclear transmutation. The synthrophic association biomass concentrate with a volume of 30 ml was left under the action of these waves at a distance of 20 cm from the wave's source for 30 minutes. These waves were generated at the exit surface of the cavitation unit with an internal water jet. Experiments carried out by us many times have shown that the fundamental frequency of such continuous waves depends on the state of the air (temperature, pressure, and humidity) and lies in the range of 75-85 MHz. There were several possible reasons for such an additional action (both nuclear-physical and purely microbiological) and the final results confirmed our expectations.

A photo of the irradiation process is shown in Fig. 1.

In control experiments, there was no such preliminary irradiation of microbiological association.

Both experiments (the main one with preliminary irradiation of the microbiological association and the control one without such irradiation) were carried out simultaneously on spatially separated installations.

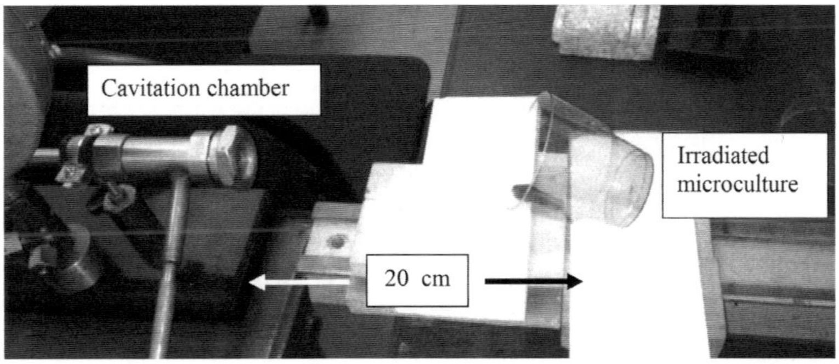


**Figure 1.** Photo of the system for remote irradiation of microculture before the start of the nuclear transmutation process.

A wet biomass centrifugate of the syntrophic association weighing ~1.41 g was placed in a culture medium with a volume of 0.075 L, with a strontium content of about 50 mg/L ($Sr(NO_3)_2$ salt). In addition to the high content of heavy metal, the absence of biogenic elements, including sources of potassium, phosphorus, and calcium, was a vital factor.

As a control experiment, a concentrate of syntrophic association biomass without wave action (~1.48 g) was used.

Glucose was used to ensure the physiological growth of microcultures.

Glucose was introduced at the rate of 0.1 - 0.8 g/L per day, taking into account the assessment by the photometric method of its consumption for the growth of microcultures. To carry out this periodic current concentration of glucose analysis from the composition of the working nutrient medium, microsamples of the medium with a volume of 25 microliters were taken. Such tests were carried out 5 times per day. The liquid taken for such analyzes was taken into account when calculating the content of elements before and after incubation.

After the end of the incubation, the biomass of the syntrophic association was separated by centrifugation under the same conditions.

## 3. RESULTS AND DISCUSSION

The weight of the wet sediment of the biomass was 2.4455 g. The content of Sr, Ca, Y and K was analyzed by the ICP-MS method under initial conditions and after incubation in all components of the working volume including supernatant (liquid phase that remains after insoluble substances are precipitated during centrifugation).

Table 1 shows the results of the analysis of the content of these elements in the initial biomass, the original nutrient medium, supernatant and biomass sediment after incubation.

The measurement error was $\pm$ 4% (or less) for each of the presented results.

A typical series of experiments lasted 21 days under anaerobic conditions.

The results of the analysis of the elemental composition of a closed cell with a microbiological association and all components of the culture medium at the beginning and after the end of the optimal experiment with preliminary irradiation are presented in Table 2.

The main result of the experiment is a significant decrease in the concentration of strontium, as well as an increase in the concentration of yttrium isotope in the optimal experimental closed volume.

In addition, a significant increase in the concentration of calcium was found as a result of the process $K^{39} + p = Ca^{40}$, as well as a decrease in the concentration of potassium.

The lack of a complete balance in the change in the concentrations of pairs of elements (strontium-yttrium) and (potassium-calcium) in Table 1 is most likely caused by other additional transmutation reactions that we have not yet discovered.

**Table 1.** The results of the analysis of the total content of Sr, Y, K and Ca in all components of the experimental system.

| Sample name | Content mg/L solution or mg/g wet biomass | | | |
|---|---|---|---|---|
| | Sr | Y | Ca | K |
| Initial solution | 56.32 | < 0.005 | 0.3 | 264.25 |
| Initial biomass | 0.013 | 0.000266 | 0.748 | 0.433 |
| Supernatant | 19.4 | <0.005 | 1.133 | 205.00 |
| Biomass sediment | 1.043 | 0.000206 | 0.462 | 1.25 |

**Table 2.** Content of elements at the beginning and at the end of the optimized microbiological LENR experiment.

| **Element name** | **Total mass at the beginning of the experiment, mg** | **Total mass at the end of the experiment, mg** | **Increase or decrease, %** |
|---|---|---|---|
| Sr | 4.24 | 4.01 | ↓5.4 |
| Y | 0.00037 | 0.00050 | ↑35.1 |
| K | 20.4 | 18.4 | ↓9.8 |
| Ca | 1.08 | 1.21 | ↑12.0 |

In the control experiment, practically no glucose was consumed. This result confirms the very low metabolic rate in such a microculture. It was found that the control biomass lost weight from 1.4801 to 1.3991 grams, which indicates that there was no growth of the untreated biomass wave due to the lack of necessary microelements (e.g., P and Ca). It should be noted that in the control experiment the balance for the same elements at the beginning and at the end of incubation converged and did not exceed the statistical error.

It should be noted that in nuclear reactions stimulated by the process of formation of coherent correlated states, the channels of nuclear reaction, the daughter products of which are radioactive nuclei, are completely impossible and forbidden by the strict laws of quantum mechanics [11], [14]. For this reason, the additional reaction $Sr^{86} + p = Y^{87}$ with participation of another $Sr^{86}$ isotope (natural abundance 9.86%) which leads to the formation of radioactive $Y^{87}$ isotope, is fundamentally impossible.

## 4. CONCLUSIONS

The fact of the vitality of the syntrophic association and its effect on the Sr content as a result of treatment with shock-cavitation waves in such life-threatening conditions as a lack of macro- and microelements was revealed. It was shown that in the optimized experiment, effective processes of synchronized changes in strontium and yttrium, as well as potassium and calcium take place.

The biophysical reason for this transmutation is most likely due to the fact that the synthesized isotope of yttrium is a direct biochemical analogue of calcium (for yttrium, this radius is $R_Y^{(2+)} \approx 0.99A$ and for calcium $R_{Ca}^{(2+)} \approx 0.99 \ldots 1.011A$). The same reason justifies the process of another observed transmutation (potassium to calcium).

Using the same considerations and estimates, it is easy to make sure that with an optimal experimental setup, the direct use of radioactive strontium by a growing microculture (its ionic radius $R_{Sr}^{(2+)} \approx 1.12A$ is not as close to calcium $R_{Ca}^{(2+)}$ as yttrium $R_Y^{(2+)}$) can be much less effective than its "processing" into yttrium with subsequent absorption.

Based on our preliminary estimates, we tried to reduce the initial (impurity) calcium content in all components of the system. It is very difficult to completely eliminate calcium, since initially calcium was contained in the medium in a bound form during the introduction of biomass.

As is well known, calcium is one of the essential trace elements and such transmutation processes are very efficiently supported by growing microcultures.

We discussed this process in detail in our early works [1]–[4]. It consists in the fact that the fluctuationally formed nucleus (that is, formed during the transmutation reaction) of the required isotope, together with its electron shell, very quickly (practically instantaneously) integrates into the structure of the growing microculture, which fixes the process of transmutation.

Obviously, with a more complete "purification" of the system from impurity calcium, as well as eliminating the prerequisites for an alternative pathway of calcium formation in the reaction $K^{39} + p = Ca^{40}$ (which will require minimizing the initial potassium concentration), the conversion efficiency of strontium (including the radioactive $Sr^{90}$ isotope) will be significantly higher. We are planning such experiments in the near future.

The nuclear-physical aspects of this transmutation, in our opinion, are associated with the formation of coherent correlated states [6]–[12].

In our opinion, these reactions were stimulated by the same processes of optimization of nuclear reactions at low energies due to the formation of coherent correlated states in growing microbiological associations [2], [3], [4], as in the case of $Cs^{137}$ isotope transmutation.

Pre-irradiation of the syntrophic association with shock-cavitation waves optimized the structure of the multicomponent syntrophic association and launched the process of an initial nuclear transformation, which then proceeded for a long time in the absence of such stimulation.

## References


[1] V.I.Vysotskii, A.A.Kornilova. Nuclear transmutation of stable and radioactive isotopes in biological systems. Pentagon Press, India, 2010, 187 p.

[2] V.I.Vysotskii, A.A.Kornilova. Transmutation of stable isotopes and deactivation of radioactive waste in growing biological systems. Annals of Nuclear Energy, **62**, (2013) 626–633.

[3] V.I.Vysotskii, A.A.Kornilova Microbial Transmutation of Cs-137 and LENR in growing biological systems. Current Science, **108(4)**, (2015), 636–640.

[4] Vysotskii V.I., Kornilova A.A. Effective LENR and transmutation of stable and radioactive isotopes in growing biological systems, Cold Fusion. Advances in Condensed Matter Nuclear Science, Ed. by Jean-Paul Biberian, Elsevier, 2020. CHAPTER 12, p. 205–232.

[5] Kornilova A.A., Vysotskii V.I., Gaydamaka S.N., Gladchenko M.A.. Nuclear reactions in living nature: the possibility of biological processing and deactivation of liquid radioactive waste. Physical and Mathematical Modeling of Earth and Environment Processes. Springer Proceedings in Earth and Environmental Sciences. Springer, book series (SPEES), 2019, p 213–230.

[6] V.I.Vysotskii, A.O.Vasilenko, V.B.Vassilenko. Generation and propagation of undamped temperature waves under pulse action on a target surface. Journal of Surface Investigation: X-ray, Synchrotron and Neutron Techniques, **8(2),** (2014) 367–373.

[7] V.I.Vysotskii, A.A.Kornilova, A.O.Vasilenko. Observation and investigation of X-ray and thermal effects at cavitation. Current Science, **108(4)**, (2015) 608–613.

[8] V.I.Vysotskii, A.A.Kornilova, A.O.Vasilenko, T.B.Krit, M.V.Vysotskyy. The prediction, observation and study of long-distant undamped thermal waves generated in pulse radiative processe. Nuclear Instruments and Methods in Physics Research B, **402,** (2017) 251–255.

[9] V.I.Vysotskii, M.V.Vysotskyy, A.A.Kornilova, T.B.Krit, S.N.Gaydamaka, P.L.Hagelstein. Distant behind-screen action of undamped temperature waves. Journal Condensed Matter Nucl. Science, **33**, (2020) 296–304.

[10] Vysotskii V.I., Vysotskyy M.V. Coherent correlated states and low-energy nuclear reactions in non-stationary systems. European Phys. Journal A, **49**, issue 8, (2013) 99.

[11] V.I.Vysotskii, M.V.Vysotskyy. Universal mechanism of LENR in physical and biological systems on the base of coherent correlated states of interacting particles. Cold Fusion. Advances in Condensed Matter Nuclear Science, Ed. by Jean-Paul Biberian, Elsevier, 2020. CHAPTER 17, p. 333–370.

[12] V.I.Vysotskii, M.V.Vysotskyy. Formation of correlated states and tunneling for a low energy and controlled pulsed action on particles, Journal of Experimental and Theor. Phys., **125(2),** (2017) 195–209.

[13] V.I.Vysotskii, M.V.Vysotskyy, S.Bartalucci. Features of the formation of correlated coherent states and nuclear fusion induced by the interaction of slow particles with crystals and free molecules, Journal of Experimental and Theor. Phys. **127,** (2018) 479–490.

[14] V.I.Vysotskii, M.V.Vysotskyy. Features of the propagation, evolution, and remote collapse of a correlated wave packet. Journal of Surface Investigation: X-ray, Synchrotron and Neutron Techniques, **13(6),** (2019) 1116–1121.

[15] V.I.Vysotskii, M.V.Vysotskyy, S.Bartalucci. Using the method of coherent correlated states for production of nuclear interaction of slow particles with crystals and molecules. Journal Condensed Matter Nucl. Science, **29,** (2019) 358–367.

[16] S.Bartalucci, V.I.Vysotskii, M.V.Vysotskyy. Correlated states and nuclear reactions: An experimental test with low energy beams. Phys.Rev.Accel. and Beams, 2019, v.22, #5, 054503.

[17] V.I.Vysotskii, M.V.Vysotskyy. Features of correlated states and a mechanism of self-similar selection of nuclear reaction channels involving low-energy charged particles. Journal of Experimental and Theor. Phys. 2019, v.128, p. 856–864.

J. Condensed Matter Nucl. Sci. 36 (2022) 115–129

Research Article

# LENR Solution of the Cosmological Lithium Problem

V.I. Vysotskii* and M.V. Vysotskyy

*Taras Shevchenko National University of Kyiv, Kyiv, 01601, Ukraine*

Sergio Bartalucci

*INFN Laboratori Nazionali di Frascati, Frascati, 00044 Italy*

**Abstract**

This paper considers a mechanism for solving the "lithium problem" in cosmology: a very significant difference in the observed (measured) concentration of lithium isotopes in the Universe in comparison with the results of a detailed analysis of the process of primary nucleosynthesis in the Big Bang model. It is shown that such isotope anomalies (a great decrease of the $Li^7$ concentration and a 500-fold increase of the $Li^6$ concentration) can be associated with a very significant increase of the probability of $Li^6(d, He^4)He^4$ and $Li^7(p, He^4)He^4$ reactions involving protons and deuterium in the near-surface region of stars with temperature $T \leq 100\,eV$. The process of optimization of nuclear fusion reactions is associated with the formation of coherent correlated states of protons and deuterons, which leads to a short-term generation of very large fluctuations in the momentum and energy of these particles, and occurs under the action of shock waves and impulse changes of the magnetic field of the star. These effects are identical to the processes that correspond to the standard LENR concepts and proceed at even lower energy.


*Keywords:* Big Bang nucleosynthesis, LENR in cosmology, Coherent correlated states, Lithium problem

## 1. INTRODUCTION

The basis of modern cosmology is the Big Bang theory. Its reliability is based on three fundamental results: a) redshift of radiation from distant stars due to the expansion of the Universe; b) the presence of microwave background radiation, the spatial structure of which is associated with the dynamics of the expansion of the Universe; c) the theory of primary nucleosynthesis (Big Bang nucleosynthesis, BBN), which lasted in the range from 13.8 sec to 38 minutes after the Big Bang and led to the formation of light nuclei $H^2$, $He^3$, $He^4$, $Li^6$ and $Li^7$ in the volume of an expanding plasma with a temperature that has changed during this time from $4.5 \cdot 10^9 K$ (390 keV) to $3.5 \cdot 10^9 K$ (30 keV).

The quantitative characteristics of the BBN theory were obtained from a very detailed analysis of the evolutionary processes of different isotopes, taking into account their mutual transformations during the time after the Big Bang (e.g. [1]). From this analysis, reasonable estimates were obtained for the ratio of the concentrations of the main light

*Corresponding author: vivysotskii@gmail.com

isotopes ($n_X$) and hydrogen ($n_p$), calculated both on the basis of the model of primary nucleosynthesis (BBN) at $t = t_0$, and obtained during the process of making astronomical measurements (short designation - am) at the present time $t = t_{am}$

$$(\tilde{n}_d)_{am} \equiv \tilde{n}_d(t_{am}) = n_d(t_{am})/n_p \approx 1.4 \cdot 10^{-5},\ (\tilde{n}_{He^3})_{am} \equiv \tilde{n}_{He^3}(t_{am}) \approx 3.5 \cdot 10^{-6},$$
$$(\tilde{n}_{He^4})_{am} \equiv \tilde{n}_{He^4}(t_{am}) \approx 0.06;$$
$$(\tilde{n}_d)_{BBN} \equiv \tilde{n}_d(t_0) \approx 1.3 \cdot 10^{-5},\ (\tilde{n}_{He^3})_{BBN} \equiv \tilde{n}_{He^3}(t_0) \approx 3.1 \cdot 10^{-6},$$
$$(\tilde{n}_{He^4})_{BBN} \equiv \tilde{n}_{He^4}(t_0) \approx 0.06. \quad (1)$$

These data demonstrate very good agreement between the calculation results and the experimental data.

A fundamentally different situation occurs in the analysis of lithium isotopes [1], [2], [4]. These isotopes were formed in the reactions of the initial nucleosynthesis [1], [5]–[8]

$$He^4 + H^3 \to Li^7 + \gamma,$$
$$He^4 + He^3 \to Be^7 + \gamma + 1.59MeV \to Be^7 + \beta^- \to Li^7 + \gamma,$$
$$He^4 + d \to Li^6 + \gamma. \quad (2)$$

The currently recorded relative concentration of the nuclei of the $Li^7$ isotope turns out to be 3 times less of the calculated initial value:

$$(\tilde{n}_{Li^7})_{BBN} \equiv \tilde{n}_{Li^7}(t_0) = n_{Li^7}(t_0)/n_p \approx 6.68 \cdot 10^{-11}$$
$$(\tilde{n}_{Li^7})_{am} \equiv \tilde{n}_{Li^7}(t_{am}) = n_{Li^7}(t_{am})/n_p = K_{Li^7}\tilde{n}_{Li^7}(t_0) \approx 2.2 \cdot 10^{-11}, K_{Li^7} \approx 1/3, \quad (3)$$

which, according to theoretical estimations, should remain approximately the same at the present time $t_{am}$. This decrease of the $Li^7$ isotope, which currently corresponds to 92.41% of all lithium, leads to a decrease of the total amount of lithium in the Universe.

In contrast, the recorded relative concentration of the nuclei of the $Li^6$ isotope (which now corresponds to 7.59% of the total lithium) turns out to be $K_{Li^6} \approx 500$ times greater than the calculated by BBN theory initial value:

$$(\tilde{n}_{Li^6})_{BBN} \equiv \tilde{n}_{Li^6}(t_0) = n_{Li^6}(t_0)/n_p \approx 1.8 \cdot 10^{-15},$$
$$(\tilde{n}_{Li^6})_{am} \equiv \tilde{n}_{Li^6}(t_{am}) = n_{Li^6}(t_{am})/n_p = K_{Li^6}\tilde{n}_{Li^6}(t_0) \approx 0.9 \cdot 10^{-12}. \quad (4)$$

These fundamental inconsistencies are called the "lithium paradox" in astrophysics, and their presence requires either an explanation or a revision of the entire BBN theory.

## 2. ANALYSIS OF THE POSSIBLE EVOLUTION OF LITHIUM ISOTOPES BASED ON THE THEORY OF "STANDARD" NUCLEAR PROCESSES

### 2.1. Abnormal Features of the Evolution of $Li^6$ Isotope in the Universe

Let us first consider the dynamics of changes of the concentration of lithium isotopes after the initial nucleosynthesis based on standard thermonuclear fusion.

The process of changing the concentration of the $Li^6$ isotope nuclei at a constant concentration of hydrogen, deuterium, and helium is described by the balance equation

$$\frac{dn_{Li^6}}{dt} = n_{He^4}n_d{<}\sigma_{He^4+d=Li^6}v{>} - n_{Li^6}\{n_p[{<}\sigma_{Li^6+p=He^3+He^4}v{>} + {<}\sigma_{Li^6+p=Be^7}v{>}] + n_d{<}\sigma_{Li^6+d=2He^4}v{>}\}, \quad (5)$$

**Table 1.** Parameters of the rates of thermonuclear reactions.

| Reaction | $Li^6(p,He^3)\alpha$ | $Li^6(p,\gamma)Be^7$ | $Li^6(d,\alpha)\alpha$ | $\alpha(d,\gamma)Li^6$ | $Li^7(p,\alpha)\alpha$ | $H^3(\alpha,\gamma)Li^7$ |
|---|---|---|---|---|---|---|
| A | $5.8{\cdot}10^{-14}$ | $2{\cdot}10^{-18}$ | $1.3{\cdot}10^{-13}$ | $9{\cdot}10^{-23}$ | $1.2{\cdot}10^{-15}$ | $1.4{\cdot}10^{-18}$ |
| B | 8.415 | 8.415 | 10.14 | 6.993 | 8.473 | 8.08 |
| C | 3.659 | 3.659 | 4.409 | 3.04 | 3.684 | 3.513 |

which is determined by a combination of two opposite processes: an increase of this concentration due to the basic reaction of $Li^6$ isotope synthesis

$$He^4 + d = Li^6 + \gamma + Q(1.474\,MeV) \tag{6}$$

and its decrease in three similar reactions

$$\begin{aligned} &Li^6 + d = 2He^4 + Q(22.373\,MeV),\\ &Li^6 + p = He^4 + He^3 + Q(4.019\,MeV),\\ &Li^6 + p = Be^7 + \gamma + Q(5.606\,MeV). \end{aligned} \tag{7}$$

The solution of this equation has the form

$$\begin{aligned} \tilde{n}_{Li^6}(t) &= \tilde{n}_{He^4}\tilde{n}_d < \sigma_{He^4+d}v > \tau_{Li^6} n_p \left(1 - e^{-t/\tau_{Li^6}}\right) + \tilde{n}_{Li^6}(t_0)e^{-t/\tau_{Li^6}}\\ &\equiv \tilde{n}_{Li^6}(\infty)\left(1 - e^{-t/\tau_{Li^6}}\right) + \tilde{n}_{Li^6}(t_0)e^{-t/\tau_{Li^6}}\\ &= \tilde{n}_{Li^6}(\infty) + \left[\tilde{n}_{Li^6}(t_0) - \tilde{n}_{Li^6}(\infty)\right]e^{-t/\tau_{Li^6}},\\ \tau_{Li^6} &= 1/n_p\left\{\left(<\sigma_{Li^6+p=He^3+He^4}v> + <\sigma_{Li^6+p=Be^7}v>\right) + \tilde{n}_d<\sigma_{Li^6+d=2He^4}v>\right\}. \end{aligned} \tag{8}$$

In this solution, the first term characterizes the synthesis process, which determines the increase of the concentration of the nuclei of the $Li^6$ isotope, and the second determines its decrease due to the participation of these nuclei in the synthesis of other elements. These oppositely directed processes proceed with the same characteristic relaxation time $\tau_{Li^6}$.

From the general solution (8) follows the expression for the limiting (at $t \gg \tau_{Li^6}$, i.e., formally at $t \to \infty$) relative concentration of $Li^6$

$$\tilde{n}_{Li^6}(\infty) = \frac{\tilde{n}_{He^4}\tilde{n}_d < \sigma_{He^4+d}v >}{\{< \sigma_{Li^6+p=He^3+He^4}v > + < \sigma_{Li^6+p=Be^7}v >\} + \tilde{n}_d < \sigma_{Li^6+d=2He^4}v >}, \tag{9}$$

which is the amplitude of the first term.

From the same general solution (8), one can determine the time

$$t_{K,Li^6} = \frac{\ln\left\{\left[\tilde{n}_{Li^6}(\infty) - \tilde{n}_{Li^6}(t_0)\right] / \left[\tilde{n}_{Li^6}(\infty) - \tilde{n}_{Li^6}(t_{am})\right]\right\}}{n_p\left(< \sigma_{Li^6+p=He^3+He^4}v > + < \sigma_{Li^6+p=Be^7}v >\right) + n_d < \sigma_{Li^6+d=2He^4}v >} \tag{10}$$

during which the relative concentration of the $Li^6$ isotope increases in $K_{Li^6} \approx 500$ times with respect to the initial calculated value $\tilde{n}_{Li^6}(t_0)$ to the current experimentally measured value $\tilde{n}_{Li^6}(t_{am}) = K_{Li^6}\tilde{n}_{Li^6}(t_0)$.

The parameters of the rates of the corresponding thermonuclear reactions [10]

$$< \sigma v >= (A/T_9^{2/3})e^{-B/\sqrt[3]{T_9}} \equiv (A/T_9^{2/3})10^{-C/\sqrt[3]{T_9}} \tag{11}$$

are presented in the Table (here $T_9$ is the temperature in units of $10^9$ K).

In these and subsequent formulas the reaction rate is measured in [$cm^3$/sec].

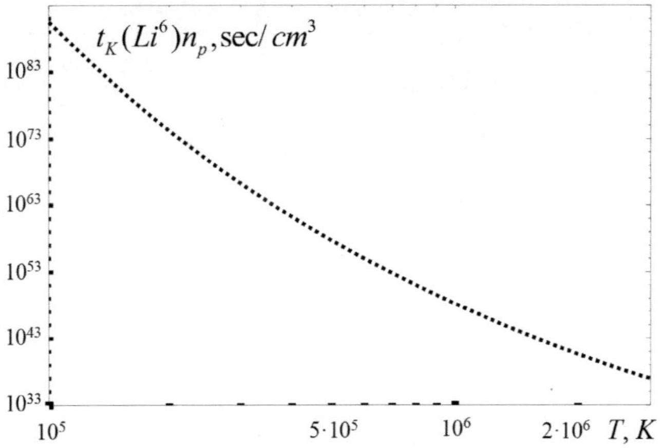


**Figure 1.** Dependence of the duration $t_{K,Li^6}$ of the process of increasing the concentration of the $Li^6$ isotope by K = 500 times on temperature in "standard" nuclear reactions (6) and (7).

Using these parameters, it is easy to verify that the maximum possible (at $t \to \infty$) relative concentration of $Li^6$ isotope (9) exceeds both the initial (calculated) relative concentration $\tilde{n}_{Li^6}(t_0) \approx 3.5 \cdot 10^{-16+0.426/\sqrt[3]{T_9}}$ of this isotope and the observed (registered) value $\tilde{n}_{Li^6}(t_{am}) = K_{Li^6}\tilde{n}_{Li^6}(t_0) \approx 0.9 \cdot 10^{-12}$ only if the temperature does not exceed $T_9 \leq 0.0012$ $(T \leq 1.2 \cdot 10^6 K,\ kT \leq 100\,eV)$.

From this condition it is obvious that such processes can occur only in the near-surface and relatively cool layers of stars, where the probability of thermonuclear reactions is very small.

Figure 1 shows the dependence (10) of the time $t_{K,Li^6}$ of $Li^6$ isotope formation on temperature and proton concentration.

From these data, it follows that even at a very high (for the surface layers of a star) proton concentration $n_p$ = $10^{23}$ $cm^{-3}$, the duration of the increase $Li^6$ isotope concentration at a temperature $T \leq 1.2 \cdot 10^6 K$ is many orders of magnitude longer than the duration of the existence of the Universe, which proves the impossibility of such a process if we consider it based on the standard approach! This is one side of the cosmological lithium paradox.

## 2.2. Paradoxes of $Li^7$ Isotope Evolution in the Universe

The basic relations characterizing the change of the concentration of the nuclei of the $Li^7$ isotope can be found in a similar way.

After the completion of the stage of the primary nucleosynthesis of the $Li^7$ isotope (including $He^4 + H^3 = Li^7$ reaction with the participation of unstable tritium), the peculiarities of the change of its concentration are determined by two oppositely directed processes: an increase due to the reactions of $Be^7$ isotope synthesis and subsequent electron capture

$$He^4 + He^3 \to Be^7 + 1.59MeV \to Be^{\,7} + \beta^-(100\%) \underset{53.2day}{\longrightarrow} Li^7 + \gamma$$

and a decrease due to the $Li^7 + p = 2He^4$ reaction, which is described by the balance equation

$$\frac{dn_{Li^7}}{dt} = n_{He^4}n_{He^3}{<}\sigma_{He^4+He^3=Be^7+\beta^-=Li^7}v{>} - n_{Li^7}n_p{<}\sigma_{Li^7+p=2He^4}v{>} \qquad (12)$$

with the corresponding solution

$$\tilde{n}_{Li^7}(t \geq t_0) = \frac{\tilde{n}_{He^4}\tilde{n}_{He^3} < \sigma_{He^4+He^3=Be^7+\beta^-=Li^7} v >}{< \sigma_{Li^7+p=2He^4} v >} \{1 - \exp[-n_p < \sigma_{Li^7+p=2He^4} v > t]\}$$
$$+ \tilde{n}_{Li^7}(t_0) \exp\left[-n_p < \sigma_{Li^7+p=2He^4} v > t\right]$$
$$\equiv \tilde{n}_{He^4}\tilde{n}_{He^3} < \sigma_{He^4+He^3=Be^7+\beta^-=Li^7} v > \tau_{Li^7} n_p (1 - e^{-t/\tau_{Li^7}}) + \tilde{n}_{Li^7}(t_0) e^{-t/\tau_{Li^7}} \quad (13)$$

and the time of relaxation

$$\tau_{Li^7} = 1/n_p < \sigma_{Li^7+p=2He^4} v >. \quad (14)$$

From formula (14) it follows that the limiting (at $t \gg \tau_{Li^7}$) relative concentration of the $Li^7$ isotope should be equal to

$$\tilde{n}_{Li^7}(\infty) = \frac{\tilde{n}_{He^4}\tilde{n}_{He^3} <\sigma_{He^4+He^3=Be^7+\beta^-=Li^7} v>}{<\sigma_{Li^7+p=2He^4} v>} \quad (15)$$

The final expression for the current relative concentration of this isotope takes the form

$$\tilde{n}_{Li^7}(t \geq t_0) = \tilde{n}_{Li^7}(\infty)(1 - e^{-t/\tau_{Li^7}}) + \tilde{n}_{Li^7}(t_0) e^{-t/\tau_{Li^7}}$$
$$= 2.3 \cdot 10^{-10+0.172/\sqrt[3]{T_9}}(1 - e^{-t/\tau_{Li^7}}) + 6.3 \cdot 10^{-11} e^{-t/\tau_{Li^7}} \quad (16)$$

From solution (17), one can determine the time

$$t_{K,Li^7} = \frac{1}{n_p < \sigma_{Li^7+p=2He^4} v >} \ln \left\{ \frac{\tilde{n}_{Li^7}(t_0) - \tilde{n}_{Li^7}(\infty)}{\tilde{n}_{Li^7}(t_{am}) - \tilde{n}_{Li^7}(\infty)} \right\} \quad (17)$$

during which the relative concentration of the $Li^7$ isotope changes (decreases in 3 times, $K_{Li^7} \approx 1/3$) to the current measured value $\tilde{n}_{Li^7}(t_{am}) \equiv \tilde{n}_{Li^7}(t_{K\approx 1/3, Li^7}) = K_{Li^7}\tilde{n}_{Li^7}(t_0)$.

From (16) it follows that the calculated limiting concentration of the $Li^7$ isotope $\tilde{n}_{Li^7}(\infty) \approx 2.3 \cdot 10^{-10+0.171/\sqrt[3]{T_9}}$ is much higher than both the initial $\tilde{n}_{Li^7}(t_0) \approx 6.68 \cdot 10^{-11}$ and current measured $\tilde{n}_{Li^7}(t_{am}) \approx 2.2 \cdot 10^{-11}$ concentrations of this isotope.

This directly contradicts the experimental data, which indicate a decrease in this concentration. This is another side of the cosmological paradox of lithium.

Based on these results, it is obvious that it is impossible to solve the "lithium paradox" on the basis of standard thermonuclear fusion models, and explain the very large difference between the BBN theory and the results of experimental measurements of the concentrations of $Li^6$ and $Li^7$ isotopes.

## 3. FORMATION OF COHERENT CORRELATED STATES AND OPTIMIZATION OF REACTIONS INVOLVING LITHIUM ISOTOPES TAKING INTO ACCOUNT DYNAMIC PROCESSES IN STAR VOLUME

### 3.1. Features of Dynamic Processes in the Near-Surface Layers of Stars

All traditional methods for analyzing nuclear transformations in stars (including the analysis of the evolution of lithium) are based on the standard concept of an active medium as an isolated plasma, and do not take into account other significant factors that strongly influence nuclear reactions, such as:

a) the action of shock waves generated in the volume, on the surface and in the atmosphere of stars;

b) the action of a nonstationary magnetic field, which is formed in a moving inhomogeneous plasma.

Let us consider these processes using the example of the Sun.

In the center of the Sun there is a *core*, within which the main (standard) thermonuclear reactions take place. In the range of 25% - 70% of the Sun's radius $R_0$, there is a *zone of radiative transfer*, within which the density and temperature change in the intervals $n_p \approx 10^{25}$ - $10^{23} cm^{-3}$ and $T \approx (7 - 2) \cdot 10^6\, K$. This zone rotates as a whole and is like a very compressed metal. Above the surface of this zone, there is a thin ($\sim 4\%\ R_0$) transitional layer (*tachocline*).

Above the tachocline (in the interval (74% - 100%) $R_0$) there is a *convective zone*, where there is a continuous transfer of heated matter and thermal energy from the inner region of the Sun to its surface due to directed convection. Typical temperature and density (hydrogen concentration) vary within this zone from $T \approx 10^{-3} T_9 = 10^6\, K$, $n_p \approx 10^{23}\, cm^{-3}$ to $T \approx 6 \cdot 10^{-6}$ - $10^{-5} T_9 = 6 \cdot 10^3$ - $10^4\, K$, $n_p \approx 10^{17}\, cm^{-3}$ on the border with the atmosphere (photosphere) of the Sun.

The special significance of this zone is that it directly connects two very important processes: 1) the possibility of carrying out of nuclear reactions in the region of high density and temperature in the lower part (near the bottom) and 2) the possibility of photometric remote registration of the products of these reactions carried out due to the processes of thermally stimulated diffusion near the upper boundary (in the photosphere), where the registration and analysis of the considered isotopic anomalies is possible with the help of astronomical observations. The movement of matter in the convective zone corresponds to circulation in the form of hexagonal cells, ordered along the surface. In this case, the substance rises along the axis of each cell, and falls at its periphery.

The rate of transfer of matter in the volume of cells increases from 50 - 100 m/sec at the border with the tachocline to 1 - 2 km/sec at the border with the outer surface (photosphere). The size of these cells decreases by ten to a hundred times as the ions move to the surface of the star. Upon reaching the surface, these cells form a structure in the form of a two-dimensional grid of granules. Such circulation provides a fairly fast vertical exchange of ions and mixing of matter within the entire convective zone in a time $\Delta t \approx 10^7$sec for stars like the Sun.

A very significant change in the structure and properties of the plasma occurs in the volume of the tachocline. This is due to the fact that the rotation of the convective zone is similar to the rotation of a liquid - the circumpolar regions rotate much slower than the radiative transfer zone, and the equatorial region rotates much faster. As a result, the tachocline layer is under a very strong shear stress, which should lead to the formation of shock waves. It also leads to fast and intense fluctuations of the magnetic field due to the fact that the lines of force of the electromagnetic field are drawn out by plasma flows, increasing their strength.

The convective zone exists in all types of stars: in cold red stars its thickness reaches half the radius, and in hot stars it is less than that of the Sun.

The region of the star adjacent to the tachocline seems to be the most attractive for explaining the existing anomalies of lithium isotopes. A significant drawback of the generally accepted scenario of the evolution of isotopes is the neglect of the influence of fundamental quantum effects, which greatly increase the probability of nuclear reactions. They can occur, for example, during the action on interacting nuclei of repeated magnetic field pulses or shock waves, connected with global dynamical processes in the bulk of stars.

In this paper, the effect of shock waves is considered, and the features of the effect of a variable magnetic field will be considered in another paper.

Such shock waves are reliably recorded on the Sun (for example, [11]), and their amplitude decreases as they approach the edge of the solar disk. It confirms that such shock waves are global in nature and their formation is connected with very powerful explosive processes in the deep layers of stars.

The energy released during such pulse processes can reach $6 \cdot 10^{25}$ J, which exceeds the explosion energy of hundreds of millions of the most powerful atomic bombs. This very large energy release is associated with intense magnetohydrodynamic processes in the plasma caused by impulse switching (reconnection) of field lines of a strong magnetic field frozen into the plasma, the magnitude of which reaches 3000 Oe. Very strong electric currents are associated with these non-stationary pulsed magnetic phenomena. The total effect of these processes leads to the fact that the primary energy of the magnetic field is converted into the energy of fast particles and pulsed heating of the plasma, which causes both the generation of electromagnetic radiation of all ranges (from radio waves to hard X-rays) and fast protons, as well as the formation of very powerful shock waves. These shock waves propagate both in the volume of the star in the region including the tachocline and, as a result, in its photosphere.

The compression parameter of an optically dense (non-ionized or partially ionized) medium at the front of a kinematic adiabatic shock wave

$$\frac{n_{p1}}{n_{p0}} = \frac{(\gamma+1)p_1 + (\gamma-1)p_0}{(\gamma+1)p_0 + (\gamma-1)p_1} \tag{18}$$

is determined by the shock adiabat $\gamma$ [11]–[14].

In the case of a very intense shock wave (at a large pressure difference $p_2 \gg p_1$), this parameter corresponds to the value

$$g_{sw} = n_{p1}/n_{p0} = n_{e1}/n_{e0} = (\gamma+1)/(\gamma-1). \tag{19}$$

For a nonrelativistic plasma $\gamma = 5/3$, and the maximum compression ratio is $g_{sw} = 4$. In such a medium, the maximum change in the proton oscillation frequency at the shock front corresponds to the value $\omega_1/\omega_0 = \sqrt{g_{sw}} = 2$. These parameters correspond to the most convective zone of the star, except for the lowest region adjacent to the tachocline.

A fundamentally different situation occurs in the zone of radiative transfer, within which the medium of the star is completely ionized, and the transfer of energy is carried out mainly due to radiation. In this region, the motion of the shock wave is very close to the isothermal non-adiabatic process, for which the adiabatic parameter $\gamma \approx 1$ and the parameter of compression of the medium at the shock front is determined by a fundamentally different relation [12]–[14]

$$g_{sw} = n_{p1}/n_{p0} = M_s^2. \tag{20}$$

Here $M_s = D/v_s$ is the Mach number, $D \equiv v_{sw}$– the speed of the shock wave, $v_s \approx \sqrt{kT/m_p}$ is the isothermal speed of sound in a given medium.

The speed of a strong shock wave for the inner regions of the star, close to the tachocline (for this region $n_{p0} = n_{e0} \approx 10^{23}\mathrm{cm}^{-3}$, $T \approx (1 - 2) \cdot 10^6\ K$), can reach $v_{sw} \approx (3 - 5) \cdot 10^3$km/s [12], [13]. With such parameters, the Mach number in this region, taking into account the isothermal speed of sound $v_s \approx 100$km/s, corresponds to $M_s \approx 30 - 50$.

While analyzing these parameters, it was taken into account that the main components of the convective zone and the zone of radiative transfer are protons (their share in the total concentration is 94%) and electrons. Let us consider one of the most effective mechanisms of the influence of such global processes on nuclear reactions.

In the region of any star adjacent to the tachocline, there is a fully ionized hot plasma. The oscillation frequency of heavy ions (in particular, protons with mass $m_i = m_p$) in such a plasma is equal to

$$\omega_{i0} = \sqrt{4\pi n_{e0} e^2/m_i}. \tag{21}$$

For protons in the considered spatial region with $n_{e0} \approx n_{p0} \approx 10^{23}\, cm^{-3}$ this frequency is equal to $\omega_{p0} \approx 5 \cdot 10^{14}\, Hz$, and for deuterons $\omega_{d0} \approx 3.5 \cdot 10^{14}\, Hz$.

In a dense plasma medium, the "thickness" of the shock front depends on the mean free path $<l> = 1/\sigma_{ii} n_p$ of heavy ions (protons) and is determined by the standard expression

$$\delta_{sw} = N<l>, <l> \approx (kT)^2/4\pi n_e e^4 \Lambda, N \approx 5 - 10, \tag{22}$$

here

$$\Lambda = \begin{cases} \ln(r_D kT/e^2) = (1/2)\ln\{(kT)^3/4\pi n_e e^6\}, & e^2/\hbar v_{sw} \gg 1, \\ \ln(\sqrt{m_p kT} r_D/\hbar) = (1/2)\ln\{m_p (kT)^2/4\pi n_e e^2\hbar^2\}, & e^2/\hbar v_{sw} \ll 1 \end{cases}$$

- Coulomb logarithm, $v_{sw}$- the shock wave velocity.

If we take into account that within the shock front, the concentration of heavy ions changes sharply from $n_{p0}$ to $n_{p1}$, then the corresponding expression for the mean free path in an isothermal plasma can be obtained using the replacement $n_p \to < n_p > \approx (n_{p0} + n_{p1})/2$ and has the form

$$<l> \approx (kT)^2/2\pi n_{p1} e^4 < \Lambda >, \ < \Lambda > = \Lambda(n_{p0}) + \ln\sqrt{n_{p0}/n_{p1}}. \tag{23}$$

In these relations, the indices 0 and 1 correspond to the parameters of the medium ahead of the shock front and directly behind the front.

Using these parameters, the maximum change in the frequency of oscillation of protons at the front of a shock wave with a thickness of $\delta_{sw} \approx 4 \cdot 10^{-8}$cm is equal to $\omega_1/\omega_0 = \sqrt{g_{sw}} \approx 30$ - $50$. This change takes place within a very short time interval $\tau_{sw} \approx \delta_{sw}/v_{sw} \approx 10^{-16}$sec (in normalized form $\omega_0\tau_{sw} \approx 0.04$) synchronously with the change of the density of the medium and with acceptable accuracy can be described by the formula

$$\omega(t) = \omega_0\{e^{-t/\tau_{sw}} + \sqrt{g_{sw}}(1 - e^{-t/\tau_{sw}})\} = \omega_0\{e^{-t/\tau_{sw}} + M_s(1 - e^{-t/\tau_{sw}})\}. \tag{24}$$

### 3.2. Formation of Coherent Correlated States and a Special Mode of Lithium Isotopes Evolution in the Near-Surface Layers of Stars

The oscillatory state of the particle in such a dynamic environment corresponds to the nonstationary (compressing) harmonic oscillator. The presence of shock modulation of the parameters of such an oscillator leads to the formation of coherent correlated states (CCS) [15]–[29] of this particle, which, in turn, leads to the generation of very large and relatively long fluctuations of its momentum and kinetic energy.

It was shown in [17]–[29] that one of the most effective methods for a very significant increase of the probability of nuclear reactions involving charged particles due to an increase of the transparency of the potential barrier is associated with the use of such states that characterize the dynamic parameters of a particle in a superposition state in a potential well. In a concentrated form, the features of CCS are demonstrated by the Schrödinger-Robertson uncertainty relation [15], [16]

$$\begin{gathered} \sigma_A\sigma_B \geq G^2 | < [\widehat{A}\widehat{B}] > |^2/4, \\ G = 1/\sqrt{1-r^2}, r = \sigma_{AB}/\sqrt{\sigma_A\sigma_B}, \\ \sigma_{AB} = \left(< \widehat{A}\widehat{B} + \widehat{B}\widehat{A} >\right)/2 - <A><B>, \end{gathered} \tag{25}$$

which corresponds to these states.

Here $r$ and G are, respectively, the correlation coefficient and the coefficient of correlation efficiency, which determine the degree of mutual coupling of dynamic variables $A$and $B$ in a coherent superposition state described by the wave function $\Psi(q)$. These parameters are specified in intervals $0 \leq |r| < 1$, $1 \leq G < \infty$.

In cases $A = q$, $B = p_q$ and $A = E$, $B = t$, relations (25) take the form of modified Heisenberg uncertainty relations

$$\delta q \delta p_q \geq G_{pq}\hbar/2 \equiv \hbar^*/2, \ \delta E \delta t \geq G_{Et}\hbar/2 \equiv \hbar^*/2. \tag{26}$$

It was shown in [17], [18] that the values of the correlation efficiency coefficients $G_{pq}$and $G_{Et}$ for different pairs of variables ($p, q$ and $E, t$) are close to each other for each specific system.

At the absence of correlation $r \to 0$, $G \to 1$, $\hbar^* \to \hbar$ and relations (25) take the form of the "usual" Heisenberg uncertainty relations for the mean-square fluctuations of specific variables (coordinate and momentum or energy and time) and the dispersion of these values

$$\delta q \delta p_q \geq \hbar/2,\ \delta E \delta t \geq \hbar/2,$$
$$\delta\sigma_q \delta\sigma_p \geq \hbar^2/4,\ \delta\sigma_E \delta\sigma_t \geq \hbar^2/4. \tag{27}$$

One more important clarification has to be made. The "standard" Heisenberg relations (27) correspond, in particular, to such a situation when the particle, prior to the beginning of the CCS formation process, was in the ground (lowest energy) quantum state. However, in the case when the particle before the beginning of this process was at the excited N-th level in such an oscillator, then the Heisenberg uncertainty relation has a significantly different form [19]

$$\delta\sigma_q \delta\sigma_p \geq (2N+1)\hbar^2/4. \tag{28}$$

Accordingly, when a coherent correlated state is formed on the basis of such an excited system, the Schrödinger-Robertson uncertainty relation takes the form [19]

$$\delta\sigma_q \delta\sigma_p \geq G_{pq}^2 (2N+1)\hbar^2/4. \tag{29}$$

If we substitute the expression for the dispersion of the momentum

$$\delta\sigma_p \geq G_{pq}^2 (2N+1)\hbar^2/4\delta\sigma_q \tag{30}$$

into the formula for the fluctuations of kinetic energy corresponding to this momentum

$$\delta E = (\delta p)^2/2m_i \equiv \delta\sigma_p/2m_i, \tag{31}$$

we obtain a dependence

$$\delta E_{N,G>1} \geq G_{pq}^2 (2N+1)\hbar^2/8m_i\delta\sigma_p \equiv G_{pq}^2 \delta E_{N,G=1}, \tag{32}$$

that shows that the formation of the CCS on the basis of an already excited state (in particular, a state with a specific temperature) increases the energy fluctuation by $G_{pq}^2$ times with respect to $\delta E_{N,G=1}$, i.e. that is, in fact, in relation to this temperature. It can be seen from this formula that the presence of a CCS is equivalent to an increase of temperature (and so an increase of the average kinetic energy) in $G^2$ times during the time of action of the fluctuation of the kinetic energy of a particle.

The presence of a correlation in the Schrödinger-Robertson uncertainty relation with an acceptable accuracy can be taken into account by substitution $\hbar \to \hbar^* \equiv \hbar/\sqrt{1-r^2} \equiv G\hbar$ in the corresponding expressions for the probability of the tunnel effect. It is very important that in the limiting case $|r| \to 1$ the transparency coefficient of any potential barrier

$$P_{r\neq 0} \approx \exp\left\{-\frac{2\sqrt{1-r^2}}{\hbar}\int_R^{R+L(E)} \sqrt{2\mu\{V(q)-E\}dq}\right\} = (P_{r=0})^{\sqrt{1-r^2}} \equiv \sqrt[G]{P_{r=0}}, \tag{33}$$

increases to the limiting value $P_{|r|\to 1} \to 1$ at any (including very low) particle energy.

A similar dependence on the correlation coefficient corresponds to the averaged thermonuclear reaction rate at a specific temperature

$$< \sigma(v)v >= \sqrt{\frac{8/\pi\mu}{(kT)^3}} \int_0^\infty \sigma(E)Ee^{-E/kT}dE. \tag{34}$$

If we use the standard expressions for the reaction cross section taking into account the $S$-factor

$$\sigma(E)=\frac{S}{E}\exp\left(-\frac{\pi e^2 Z_1 Z_2}{\hbar^*}\sqrt{\frac{2\mu}{E}}\right)=\frac{S}{E}\exp\left(-\frac{\pi e^2 Z_1 Z_2}{\hbar}\sqrt{\frac{2\mu}{G^2 E}}\right) \tag{35}$$

then the final expression for the rate of thermonuclear interaction takes the form

$$<\sigma(v)v> \approx 1.3\cdot 10^{-16}\frac{S[keV\cdot b]}{(G^2T_9)^{2/3}}\left(\frac{Z_1Z_2}{\mu_A}\right)^{1/3}\exp\left\{-\frac{4.246(Z_1^2Z_2^2\mu_A)^{1/3}}{(G^2T_9)^{1/3}}\right\}\text{cm}^3/\text{s}, \tag{36}$$

which differs from the "standard" expression for the value $<\sigma(v)v>$ by the presence of the correlation efficiency coefficient G.

Here $Z_1, Z_2$ are the charges of the interacting particles participating in the reaction, $\mu = m_1m_2/(m_1+m_2)$ is the reduced mass of these particles in grams, $\mu_A = A_1A_2/(A_1+A_2)$ is dimensionless reduced mass of these particles,$A_1, A_2$ are their mass numbers.

It should be noted that this effect of a very significant (by many orders of magnitude) increase of the transparency coefficient and significant increase of the reaction $Li^7+p=2He^4$ probability for protons with a relatively low energy $E\approx 500$eV during the automatic formation of CCS was found in the experiments that were analyzed in [20], [21]. In these experiments, the formation of CCS was carried out during the channeled motion of moderately low energy protons in the periodic field of the atomic lattice of lithium.

The physical reason for a very significant increase of the transparency of the barrier at $|r|\to 1$ is the constructive interference of mutually correlated eigenwave functions of the energy and momentum of a particle in a nonstationary potential well. It leads to the formation of giant fluctuations of momentum and kinetic energy, the amplitude of which is many orders of magnitude greater than both the total energy and average kinetic energy of the same particle.

The simplest way to form a CCS is associated with non-stationary modulation of the parabolic potential well, which corresponds to the location of the particle under consideration in a non-stationary harmonic oscillator with an initial frequency $\omega(0)=\omega_0$.

In [22]–[29], different methods of forming a CCS with a close to maximum correlation coefficient $|r|\to 1$ were considered. These methods were connected with both an increase and decrease of the size of such a potential well and periodic modulation of its parameters. An alternative mechanism for the formation of CCS can also be connected, for example, with the action on charged particles of a pulsed or alternating magnetic field, the influence of which on the state of the particle is similar to the influence of the parabolic potential.

In works [17]–[29] it was shown that the process of formation of the CCS is described by the non-stationary Schrödinger equation with a non-stationary parabolic potential with a specific law of frequency variation $\omega(t)$. In the particular case of an isolated quantum system, the main characteristics of this process (the correlation coefficient and the correlation efficiency coefficient) can be determined on the basis of the relation

$$r = Re\left\{\varepsilon^*\frac{d\varepsilon}{dt}\right\}\Big/\left|\varepsilon^*\frac{d\varepsilon}{dt}\right|, \tag{37}$$

in which $\varepsilon(t)=q(t)/q_0$ is the dimensionless (normalized to $q_0=\sqrt{\hbar/m_i\omega_0}$) complex coordinate of the particle, which is the solution of the equation of motion of the classical oscillator

$$\frac{d^2\varepsilon}{dt^2}+\omega^2(t)\varepsilon=0,\ \ \varepsilon(0)=1,\ \ \left.\frac{d\varepsilon}{dt}\right|_0=i. \tag{38}$$

In all the relations under consideration, the parameters are normalized using the same frequency $\omega_p\equiv\omega_{p0}\approx 5\cdot 10^{14}\,Hz$, which corresponds to the state of protons ahead of the shock front. The effect of dephasing collisions on the formation of the CCS in real plasma is taken into account below.

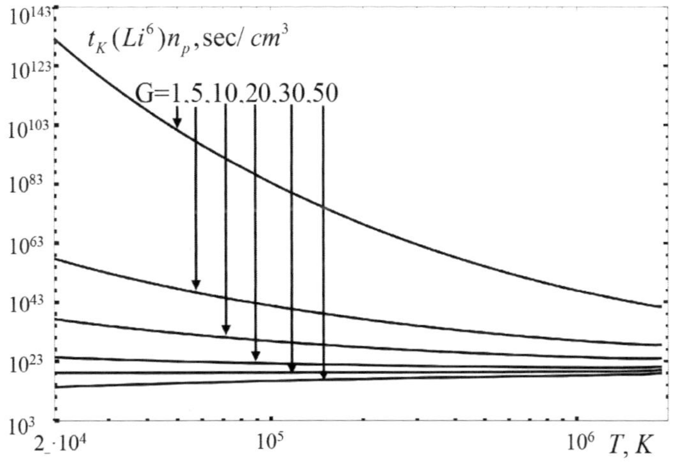


**Figure 2.** Dependence of the duration $t_{K,Li^6}$ of the process of increasing the concentration of the $Li^6$ isotope by K = 500 times on temperature at different values of the coefficient of correlation efficiency. The value G = 1 corresponds to the "standard" uncorrelated state.

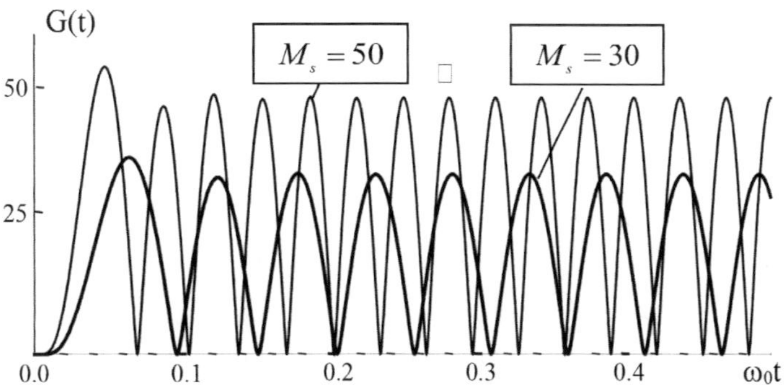


**Figure 3.** Change of the correlation efficiency coefficient G at the front of an intense shock wave in the volume of the star for reactions leading to a decrease in the concentration of the $Li^7$ isotope. The dynamics of the formation of CCS for the reaction $Li^6 + d = 2He^4$ corresponds to the change of the normalization parameter of the time scale $\omega_0 t \to \sqrt{m_p/m_d}\omega_0 t \approx 0.707\omega_0 t$.

Figure 2 shows the results of calculation of the durations $t_K(Li^6)$ of formation of the experimentally measured concentrations of $Li^6$ isotope at different values of the coefficient of correlation efficiency G.

These data show a very rapid decrease in time $t_{K,Li^6}$ in the presence of a correlated state.Figure 3 shows the results of calculating the correlation efficiency coefficient $G = 1/\sqrt{1-r^2}$ for the reaction $Li^7 + p = 2He^4$, obtained using equations (24), (37), (38) and taking into account described above parameters of the part of the star volume spatially adjacent to the tachocline (in particular, the duration of the leading front of the shock wave $\omega_0 \tau_{sw} \approx 0.04$) for two values of the Mach number ($M_s$ = 50 and $M_s$ = 30) characterizing the parameters of shock waves in the volume of the star.

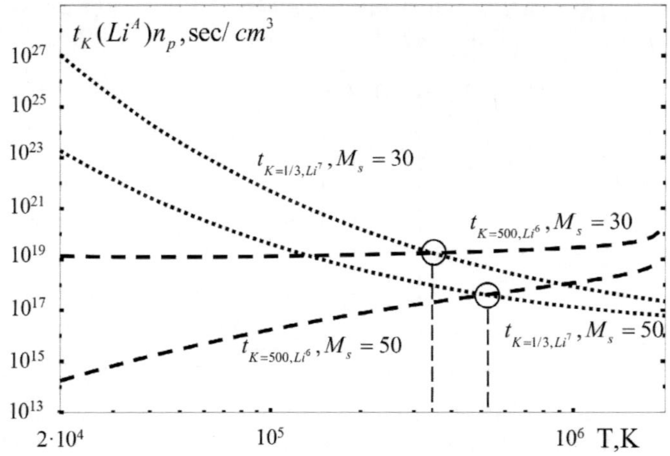


**Figure 4.** The durations $t_{K,Li^A}$ of the processes of formation of the experimentally observed concentrations of Li$^6$ and Li$^7$ isotopes depending on temperature at the presence of a coherent correlated state, the parameters of which are determined by the Mach number of the shock wave M$_s$.

The same data correspond to another very fast reaction $Li^6 + d = 2He^4$ taking into account the replacement of the parameters of the time axis $\omega_0 t \to \sqrt{m_p/m_d}\omega_0 t \approx 0.707\omega_0 t$ in this figure.

From the analysis of these dependences, it can be seen that the value $G(t)$ periodically changes in the interval $0 \leq G \leq G_{\max}$ and at each period is characterized by almost the same amplitude $G_{\max}$. The value $G \geq 0.9G_{\max}$ corresponds to approximately 25% of each period. On the basis of such a regularity, it is possible to evaluate the effectiveness of the influence of the process of the formation of the CCS on nuclear reactions using a simple interpolation: we can replace the real value $G(t)$ by the value $G = 0.9G_{\max}$, taking into account that this state is attained during $\eta \approx 0.25$ of each period.

Figure 4 shows the results of joint calculation of the durations $t_K(Li^A)$ of formation of the experimentally observed concentrations of Li$^6$ and Li$^7$ isotopes in the case of such interpolation for two different parameters ($M_s = G_{\max} =$ 30, 50) of the shock wave. The calculation corresponds to equations (10) and (17) for the times $t_{K,Li^6}$ and $t_{K,Li^7}$, modified taking into account the replacement $T \to G^2T$ only in the formulas for the $Li^6 + d = 2He^4$ and $Li^7 + p = 2He^4$ reaction parameters and taking into account the fact that, according to the above interpolation, these values $t_{K,Li^A}$ should be increased in $1/\eta \approx 4$ times.

It was theoretically shown in [17], [18], [20] (and was experimentally confirmed in [21], [30]) that among the reactions (2), (6), and (7) corresponding to the transformation of these isotopes, only for these two ultrafast reactions ($Li^6(d, He^4)He^4$ and $Li^7(p, He^4)He^4$) the effective formation of CCS is possible. In other reactions, also leading to the transformation of Li$^6$ and Li$^7$ isotopes, the temperature *T* remains unchanged.

This circumstance (the presence of two different temperatures, one of which is real T, and the second $T \to G^2T$ formally reflects a very short and very sharp increase in the kinetic energy of particles) provides a solution that is in good agreement with the results of astrophysical measurements. The short duration of such energy fluctuations means they are too short to generate the other, slower reactions.

The presence of such features leads either directly to an increase in the rate of utilization reaction for Li$^7$ isotope or to a decrease in the relaxation time $\tau$, which also leads to an acceleration of the final synthesis (for Li$^6$ isotope).

It can be seen from these results that for each value of $M_s$ an important result takes place: an approximate coincidence of the values of the normalized durations

$$t_{K=1/3,Li^7} = t_{K=500,Li^6} \approx (10^{19} - 10^{18})/\eta(n_p[cm^{-3}])\text{sec} \tag{39}$$

of significantly different processes of formation of those current values of the concentrations of both lithium isotopes that correspond to the data from astronomical observations and correspond to close values of the temperature of the medium.

These features are easy to explain based on the analysis of specific relationships. The dynamics of the change in the population of the $Li^7$ isotope is described by the balance equation (12) and the corresponding solution (13). The main mechanism for the decrease in the concentration of this isotope is associated with the acceleration of the $p + Li^7$ loss reaction, which sharply increases due to the use of the correlated state of protons. For the Li-6 isotope, the main channel for increasing its concentration is the reaction $He^4 + d = Li^6$, and the dynamics of the change in this concentration is described by Eq. (5) with solution (8). It is important to note that although the probability of a loss reaction $Li^6 + d = 2He^4$ in this process increases in a correlated state, a decrease in the relaxation time $\tau_{Li^6}$ turns out to be more significant, which allows a very fast (500 times) increase in the concentration of $Li^6$ in the direction of the limiting value $\tilde{n}_{Li^6}(\infty)$.

For the final interpretation of these results, it is necessary to determine the duration of the existence of the formed CCS and to estimate the dephasing effect of other particles (local environment of the considered nonstationary oscillator) on the formation of the CCS and, as a result, on the values of $r$ and $G$.

It is obvious that the most significant effect on the efficiency of the formation of the CCS in a particular particle is caused by the surrounding protons, which make up the vast majority (94%) in the convective zone and the zone of radiative transfer in the volume of the star.

This influence can be estimated on the basis of the condition that the oscillator can be taken as an isolated system, the state of which is described by the given above equations, during a time limited by the interval

$$\delta t \approx <l>/v_s \approx 4 \cdot 10^{-15}\ \text{sec} \tag{40}$$

from the beginning of the compression process to the first collision with the nearest proton, which corresponds to the normalized time $\omega_0 \delta t \approx 2$. Here $v_s \approx \sqrt{kT/m_p}$ is the isothermal speed of sound (and the speed of protons) in a given medium.

During such collisions, there is a significant violation of the optimal phase relations between different eigenfunctions of the particle in the superposition state of the harmonic oscillator, which sharply decreases the correlation coefficient. It can be seen that the duration of this interval is 40 times longer than the duration $\tau_{sw} \approx \delta_{sw}/v_{sw} \approx 10^{-16}$ sec (or $\omega_0 \tau_{sw} \approx 0.05$) of the passage of the shock front. Thus, the main part of the existence of the CCS corresponds to the compressed state of the medium immediately behind the shock front. For this state, in the considered region of the star, the proton density is equal to (20)

$$n_{p1} = n_{p0} M_s^2 \approx (1 - 2.5) \cdot 10^{26} cm^{-3}, \tag{41}$$

and the total time of active nuclear transformation with the participation of CCS, which is needed to achieve the experimentally measured concentrations of $Li^6$ and $Li^7$ isotopes, is equal to

$$t_{K=1/3,Li^7} = t_{K=500,Li^6} \approx \left(10^{-7} - 10^{-8}\right)\ \text{sec}. \tag{42}$$

If we take into account that the duration of the existence of the correlated state corresponds to the value $\delta t \approx 4 \cdot 10^{-15}$ sec, it means that

$$N_{SW} = t_{K,Li^A}/\delta t \approx 2 \cdot \left(10^7 - 10^6\right) \tag{43}$$

powerful shock waves are needed to generate the observed change in the concentrations of both isotopes in a given volume. It should also be taken into account that the specific composition of the substance within the region adjacent to the tachocline is constantly renewed due to the transfer of the part the hot ions to the surface, and transfer of the colder ones in the opposite direction. From this fact it follows that $N_{SW}$ has to be multiplied by the number of cycles of such an exchange (about K≈100 cycles, based on the parameters of the exchange process and the relationship between the sizes of tachocline and the convective zone), during which all the substance entering the volume of the convective zone will be involved in the process of generating the nuclear reactions under consideration.

For stars larger than the Sun, this value K increases 5–10 times.

As a result, we find that the total number of powerful shock waves in the tachocline zone required to generate the reactions for all nuclei in the entire outer part of the star volume over the entire lifetime of old stars should be equal to

$$N_{SW}^{total} = KN_{SW} \approx 10^{10} - 10^{9}. \tag{44}$$

It follows from these estimates that for the implementation of such transformations of lithium isotopes in typical stars with a lifetime of several billion years, 1–10 powerful shock waves in the tachocline region are required per year. These estimates are in good agreement with the data of astrophysical observations (e.g. [11], [12]).

This cumulative nature of the formation of lithium isotopic concentrations is in good agreement with the data from astronomical observations, which confirm that these anomalies (with respect to the initial values of the parameters of primary nucleosynthesis) increase with the age of stars and are most clearly expressed in old stars.

## 4. CONCLUSION

The mechanism considered in this paper – transformation of lithium isotopes in the volume of stars in the framework of the astrophysical "lithium problem" – is based on the well-tested principles of quantum mechanics and nuclear physics, and does not require the involvement of fundamentally new and radical models of cosmology. The process of automatic formation of coherent correlated states of protons and deuterons, which is generated at the shock wave front, and which leads to the generation of very intense fluctuations of the momentum and kinetic energy of these particles, ensures the corresponding fusion reactions will occur, along with the gradual evolution of the concentrations of both lithium isotopes, synchronized with the shock waves on the stars. This mechanism of stimulation of nuclear reactions is similar to the processes observed in controlled LENR experiments at moderately low energies with the participation of protons and lithium nuclei [17], [20], [21].

An important significant aspect of such processes is the simultaneous satisfaction of several basic conditions: a synchronized decrease in the concentration of the $Li^7$ isotope by 3 times and an increase in the concentration of the $Li^6$ isotope by 500 times, which is fully consistent with the data from astronomical observations; the implementation of such a multidirectional change of lithium concentrations for the period from the time of the end of the primary nucleosynthesis to the present; agreement of these processes with data that show that such a change of the concentration of lithium isotopes agrees best with old stars.

Another factor in the stimulation of such processes is the action of a variable magnetic field in the volume of stars, which will be considered in another work.

It is important that the process of the CCS formation affects only the conversion of lithium isotopes, and does not change the conversion of other light isotopes ($H^2$, $He^3$, $He^4$). This is confirmed by the coincidence of the BBN theory with the data from astrophysical observations.

## References

[1] R. H. Cyburt, B. D. Fields, K. A. Olive, T.-H. Yeh, Rev. Mod. Phys. **88**, 015004 (2016); arXiv:1505.01076 [astro-ph.CO].

[2] L. Sbordone, P. Bonifacio, E. Caffau et al., Astron.Astrophys. (2010), 522, article A26; arXiv:1003.4510 [astro-ph.GA].

[3] S. Q. Hou, J.J. He, A. Parikh, D. Kahl, C. A. Bertulani, T. Kajino, G. J. Mathews, G. Zhao. The Astrophysical Journal, **834** (2017):165 (5pp).

[4] M. Asplund, D. L. Lambert, P. E. Nissen, F. Primas, V. V. Smithl. Astrophysics J., **644**, (2006) 229–259.

[5] Schramm D. N., Wagoner R. V. Element Production in the Early Universe. Ann. Rev. Nucl. Sci. **27**,(1977) 37–74.

[6] H. Reeves. Rev. Mod. Phys. **66**, (1994) 193.

[7] C.R. Brune, R.W. Kavanagh, C. Rolfs. Phys. Rev. C. **50**, (1994) 2205.

[8] C. Angulo et al. Nuclear Physics A. **656,** (1990) 3–183.

[9] Andreas Goudelis, Maxim Pospelov, Josef Pradler. arXiv:1510.08858v2 [hep-ph] 24 May 2016.

[10] S. B. Dubovichenko, A. V. Dzhazairov-Kakhramanov and N. A. Burkova, arXiv:1706.05245v2 [nucl-th].

[11] M. I. Fisenko. Journal of Physical Science and Application, **7 (1)**, (2017) 35–38.

[12] I.A. Klimishin. Shock waves in the shells of stars. - M., 1984, 216 p. (In Russian).

[13] T.A. Weaver. Astrophys. Jour., **32,** (1976) 233.

[14] Ya.B. Zeldovich, Yu.P. Raizer. Physics of shock waves and high-temperature hydrodynamic phenomena. "Science", M., 1966.

[15] E. Schrödinger, Ber. Kgl. Akad. Wiss. **24**, (1930) 296.

[16] H. P. Robertson, Phys. Rev. A, **35**, (1930) 667.

[17] V.I.Vysotskii,, M.V.Vysotskyy. Jour. of Experimental and Theoretical Physics, **128**(6), (2019) 856–864.

[18] V.I. Vysotskii, M.V. Vysotskyy. Universal mechanism of LENR in physical and biological systems on the base of coherent correlated states of interacting particles. Cold Fusion. Advances in Condensed Matter Nuclear Science, Edited by Jean-Paul Biberian. Elsevier, 2020. CHAPTER 17, pp. 333–370.

[19] V.V. Dodonov, A.B. Klimov, V.I. Manko. "Physical e?ects in correlated quantum states", in: Squeezed and Correlated States of Quantum Systems, Nova Science, Commack, New York, **295** (1993) 61–107.

[20] V.I. Vysotskii, M.V. Vysotskyy., S. Bartalucci. Jour. of Experimental and Theoretical Physics, **127**(3), (2018) 479–490.

[21] S. Bartalucci, V.I. Vysotskii, M.V. Vysotsky. Phys. Rev. Accel. and Beams. **22** (5), (2019) 054503.

[22] V.I. Vysotskii, S.V. Adamenko. Technical Physics, **55**(5) (2010) 613.

[23] V.I. Vysotskii, M.V. Vysotskyy. European Phys. Jour. A**49**, issue 8: 99, (2013) 1–12.

[24] V.I. Vysotskii, S.V. Adamenko, M.V. Vysotskyy. Annals of Nuclear energy, **62**, (2013) 618–625.

[25] V.I. Vysotskii, M.V. Vysotskyy. Current Science, **108 (4),** (2015) 524.

[26] V.I. Vysotskii, M.V. Vysotskyy. Jour. of Experimental and Theoretical Physics, **152**(8) (2017) 234.

[27] V.I. Vysotskii, A.A. Kornilova, M.V. Vysotskyy. Jour. of Experimental and Theoretical Physics, **131**(4), (2020) 566–571.

[28] V.V. Dodonov, A.V. Dodonov, J. Russ. Laser Res. **35,** (2014) 39.

[29] A.V. Dodonov, V.V. Dodonov, Phys. Lett. A, **35,** (2014) 1071.

[30] S. Lipinski, H. Lipinski, Hydrogen-Lithium fusion device, Patent WO 2014/189799 A9 (22.05.2013).

J. Condensed Matter Nucl. Sci. 36 (2022) 130–136

Research Article

# A Search for Correlated Quantum States in Nuclear Reactions: First Exciting Results From an Experimental Test

Sergio Bartalucci*
*INFN Laboratori Nazionali di Frascati, Frascati, 00044 Italy*

V. I. Vysotskii and M. V. Vysotskyy
*Taras Shevchenko National University of Kyiv, Kyiv, 01601 Ukraine*

**Abstract**

The first experimental test of the Correlated-Coherent quantum States (CCS) model is described in this paper, showing its potentialities in the explanation of anomalous effects in nuclear physics and astrophysics. Some very interesting events, which cannot be ascribed to background, have been observed. The occurrence of nuclear reactions at very low energy is a clear indication of a strong enhancement of Coulomb barrier transmissivity. Some technical issues, which are related to this difficult experiment are discussed, and possible suggestions for the next activity on this topic are also presented.


*Keywords:* Correlated quantum states, effective Planck constant, proton-lithium reaction at sub-keV energy, monocrystal target, vapor target

## 1. Introduction

The well-established theoretical concept of Correlated-Coherent quantum States (CCS) [1], [2], which was envisaged from the early years of quantum mechanics, may provide a clue to several open questions in nuclear astrophysics, such as the primordial lithium problem [3] and the enhancement of the cross section and other anomalous effects which have been observed in some nuclear reactions [4] in metallic environments [5]. It may also offer a new perspective in energy-production-related nuclear physics, which would open in turn an entirely new technological field of applications.

The mechanism of formation of CCS in lithium, either in crystalline form lattice or in gaseous phase containing both single *Li* atoms and free $Li_2$ nanoclusters, when bombarded by a proton beam, was described in a previous paper [6]. On these premises a preliminary test was proposed [7], by focusing on the well-known $^6Li(p,\alpha)^3He$ and $^7Li(p,\alpha)^4He$ reactions which, however, have to be investigated at very low proton beam energy (E $<$ 0.7 keV) in a special, unprecedented, experimental environment.

*Corresponding author: Sergio.Bartalucci@lnf.infn.it

**Figure 1.** The experimental setup.

In this paper, the first encouraging results from the test are reported, which are suggestive of the occurrence of nuclear reactions in an energy region corresponding to a CCS based model predictions.

## 2. Experimental Setup

It must be stressed that all the equipment required by this test was put at disposal of our research team by an external laboratory (external with respect to our home institutions), and it has not been possible for the authors of this paper to get careful control of the different phases of test preparation. It has also not been possible for lack of time.

A picture of the experimental setup is shown in Fig. 1. A solid target made of lithium polycrystalline foil (area of approximately 1 $cm^2$, thickness 0.6 mm, purity 99.99%) is irradiated by a proton beam from an ECR ion source. The foil was delivered by the manufacturer in an unspecified gas atmosphere and manipulated in air for insertion in the loadlock chamber. Preliminary surface cleaning was made by using an abrasive pen.

The ion source is a TPIS Plasma Source [9], which is also used to irradiate the target with an Ar ion beam for cleaning purposes. Typical irradiation time is about 20 mins for Ar sputtering.

The Li target is mounted on an Aluminum support (sample holder, SH), isolated from the UHV chamber, with electrical contacts to measure the total impinging ion current $I_{SH}$ and apply a biasing voltage.

The UHV chamber is equipped with a Silicon PIPS (Passivated Implanted Planar Silicon) detector [11], which is offline calibrated for detection of alpha particles in the energy range of interest for the experiment by means of a $^{241}$Am source. The detector is located about 15 mm from the SH. No online calibration was foreseen.

No calibration data were provided by the laboratory staff, unfortunately, but we have received assurance about the correctness of the operation.

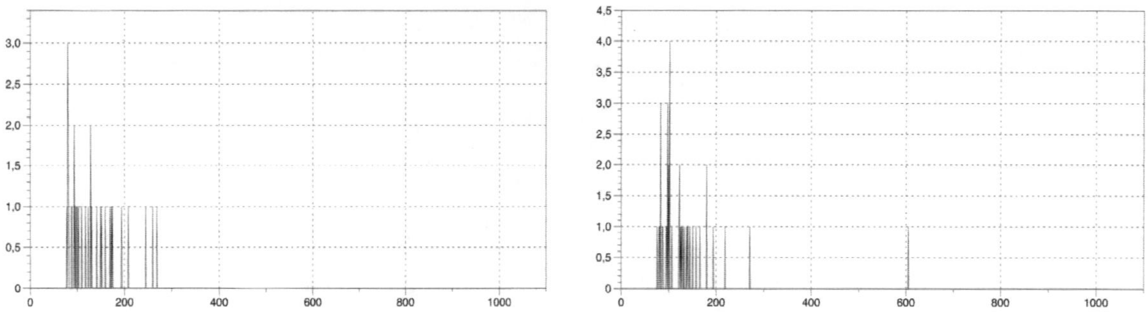


**Figure 2.** Typical observed spectrum during a 30 mins. Run with ion source off (a) and on (b).

A sample manipulator is used to position and rotate the SH with respect to the ion beam. The beam exit port of the TPIS source is located about 30 mm from the SH. Current flow as measured by a Faraday cup is typically about 150 $\mu$A/cm$^2$ for a current on the SH of 600 $\mu$A.

The peak energy definition of the ion beam is dictated by the intrinsic plasma energy, which is around 20 - 25 eV and by a multiple grid assembly, where the bias voltage $V_g$ at the inner "plasma grid" is adjustable by an independent power supply in steps of 1 V in the range 0. – 2000 eV and the extraction grid is usually negatively biased at a fixed potential, $V_b = -200$ V. It must be emphasized that the beam peak energy is not influenced by the latter parameter, but only the energy spread and beam intensity (see Fig. 5 in Ref. [9]).

A better energy definition seems impossible with the present setup: it would require a higher energy beam, with a magnetic analyzer for momentum measurement and precise ion species separation, or the use of time-of-flight or Wien filter techniques, and a decelerating unit (electrostatic lens assembly) to decrease the energy down to the desired value, while providing adequate beam collimation.

Source absolute energy, energy spread, and energy stability are in general crucial issues for cross-section measurements in nuclear astrophysics experiments, and known narrow resonances (or threshold energies for neutron emission) are used as standards to calibrate other instruments, but at the very low energies of our test no such reference exists, to our knowledge.

At the maximum energy of the ion source ($< 2$ keV) background is mainly due to proton backscattering from target and SH. The entrance window of the PIPS detector is covered by an Au foil thick enough to stop the intense flux of elastically scattered protons.

No nuclear reaction is known at these low energies that may simulate the expected signals.

Nuclear reactions such as (p, n$\gamma$), (p, $\alpha\gamma$) which may give rise to spurious signals due to contaminants in the target are also excluded, so the only possible source of background are higher energy cosmic rays. A full characterization of background, and calibration monitoring of the PIPS detector as well, would only be possible by installing a radioactive source with a proper mechanical support inside the vacuum chamber, but this measure unfortunately was not considered from the beginning by the technical staff of the host laboratory.

A typical example of background yield during a 1 hour running time is reported in Fig. 2 with the ion source off (a), and with ion source on (b) with $V_g = 210$ V, $V_b = -200$ V, showing that the occurrence of counting in the higher part of the spectrum is of the order of 1 event/hour, and this is what has been observed in all runs, with one exception, as described below.

The noise generated by ECR source is difficult to estimate since the discriminator threshold was adjusted to cut off on the left-hand side of the pedestal and it was not possible to change it.

## 3. Experimental Procedure

The model [6] describing a method for the formation of CCS in a crystal lattice of pure lithium when traversed by slow protons is used to determine the experimental procedure. This analysis is based on the fact that the longitudinal motion of a proton in an interplanar crystal channel with potential barriers periodically located on the channel walls is equivalent to its localization in a nonstationary harmonic oscillator.

According to this method [2], [6], the maximum correlation coefficient $r(z)$, hence the maximum yield of reaction products should be reached at a proton longitudinal velocity $v \approx 2a_z \langle\omega\rangle$. This condition corresponds to the parametric resonance frequency $\omega_{res} = 2\langle\omega\rangle$. Here $a_z$ is the crystal period along the particle motion and $\langle\omega\rangle$ is the average frequency which modulates particle oscillations in the transverse potential, corresponding to the characteristic channeling frequency.

This choice should give rise to a very fast increase of the $r(z)$ coefficient within the $4^{th}$ period of Li lattice, up to a maximum theoretical value of $|1 - r(z)| = 10^{-9}$. This value of the correlation coefficient corresponds to very large fluctuations of proton transverse kinetic energy (many tens of keV) [2], [6]. The estimated range of optimal proton longitudinal kinetic energies $mv^2/2$ for such regime is 400–600 eV for a Li lattice.

According to the results of CCS analysis and the parametric oscillator theory [6], another much weaker and much narrower (1 - 2%) maximum is expected at the main unperturbed frequency $\omega_{main} = \omega_{res}/2 = v/a_z \approx \langle\omega\rangle$ of the particle's transverse motion.

Several series of energy scanning runs were made therefore in the range 180 - 640 V to fully cover the expectation area for optimum energies.

In some cases the experimental conditions, such as the angle between the beam axis and the normal to the SH, were changed from one run to the next.

A systematic scanning in $V_g$ range 300 - 480 V in steps of 20 V for a duration of 30 mins was made. In the first series 380 - 480 V the extractor grid voltage was set to 0, while in the other it was set to $-200$ V, according to TPIS manual prescriptions, which suggests that the voltage at the extractor grid should be always lower than $-200$ V, although, according to user's experience this plays no significant role in the operation of the ion source.

Two further scans, one at higher energy 500 - 640 V and the other at lower energy, 180 - 280 V were also done. In all these measurements the H gas flow was often adjusted to keep chamber pressure at values around $2 \cdot 10^{-2}$ mbar, resulting indeed between 1.4 and $3.2 \cdot 10^{-2}$ mbar.

The SH current, which in principle should be a monitor of the ion current at the target, was also quite unstable during these runs, oscillating between 9 and 30 $\mu$A. A visual inspection of SH after the end of the test shows that ion irradiation appears not well focused on the sample, but it is spread out on the whole surface of the SH. This might be avoided by using a concave grid, according to the reference paper [9]. This is important for Ar sputtering, too.

To reproduce as much as possible the experimental conditions under which a very high counting rate of alpha particles was observed by an independent team [8], a bipolar power supply is used to apply negative voltage $V_{bias}$ to the SH. This alleged extraordinary effect in fact was observed at a nominal $E_p = 225$ eV and $V_{bias} = -222.6$ V, but it is unclear from their unpublished work how the proton energy is determined, and if the given value is comprehensive of the plasma energy.

## 4. First Results

A total measurement time of more than 30 hours over 7 full days was devoted to the search for significant events. None was detected, with one exception: on the first run of day 19.11.2019 some counts were observed indeed (Fig. 3). Experimental parameters were $V_g = 210$ - 220 V, $V_{bias} = -200$ V, P$\sim$5.4.0$\cdot 10^{-3}$ mbar, impact angle $= 0°$, run duration 60 minutes. The proton energy on target at the time that energetic ions were observed is estimated to be centered around 440 eV.

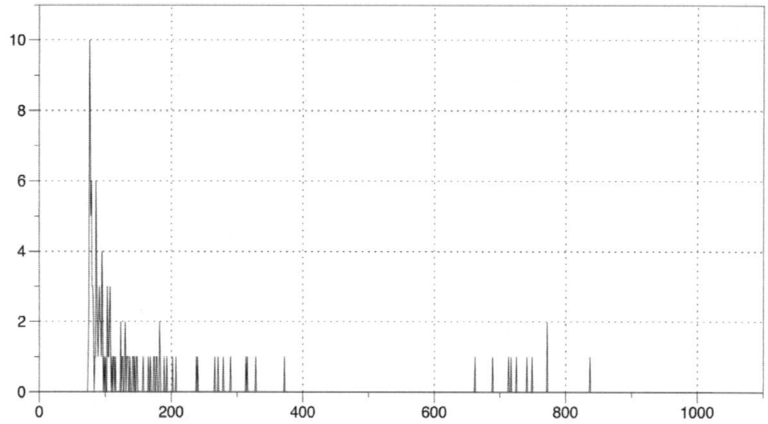


**Figure 3.** Measured spectrum in the run of 19.11.2019.

The energy attributable to such events is in the region around $\sim$ Q/2, according to the detector's calibration, where $\alpha$-particles from $^{7}Li(p,\alpha)^{4}He$ (Q = 17.35 MeV) are expected.

A significant counting at lower energy, separated from higher energy counting, was detected as well. These events occurred in a very short time interval, of the order of a few seconds. No precise time recording is possible with the present setup, unfortunately. No similar effects have been observed during any other run. Attempts to reproduce the results of this run by changing the nominal impact angle and repeatedly Ar cleaning of the target surface failed.

## 5. Target Characterization

Lithium target characterization is the crucial issue in this experiment. Metallic Li is highly reactive with nitrogen, carbon, oxygen and water vapour from the atmospheric air. This reactivity has to be taken into account for the production process as well as for the quality of the target when exposed to ambient air. A further complication in our case consists of the requirement of a pure monocrystalline Li target, unlike all other experiments on the same reactions, which have used LiF or metallic lithium alloys so far. A regular crystalline structure without defects is needed, although this might be necessary only for the very first layers, since the correlation coefficient is shown to grow rapidly for impinging protons at the optimum velocity. Another basic requirement is crystal orientation: optimum conditions are found for protons moving along the system of crystal planes in analogy with the physics of "classical" channeling of positively charged particles. On the other hand, unlike what occurs in the channeling model, the onset of CCS is not due to an average atomic potential along the crystal axis, but because of the presence of a periodic structure of the crystal potential along the direction of motion. The difference between these modes lies in the value of the energy of the longitudinal motion: the channeling condition and the potential averaging are fulfilled only at a relatively high energy E $>$ 30-50 keV, and the formation of the CCS in periodic crystal potential at a much lower optimal energy E $<$ 1 keV.

Fabrication of pure Li monocrystals with the Czochralski system is a rather complex procedure, thereby resulting in a very expensive product, if careful surface polishing is also taken into account [10]. Furthermore, absolute crystal orientation cannot be specified by the manufacturer, as it is defined as "random", although its accuracy is quite good (typically $< 0.1°$). This difficulty can be overcome nevertheless, e.g., by single crystal orientation measurement with the well-known method of X-ray diffraction goniometry.

Otherwise, a polycrystalline, rather cheap, Li foil can be used but quite clearly the condition for the onset of CCS would occur only if the proton beam could impinge on a crystal grain of the specimen that has the proper crystallographic orientation. The observation of CCS-related particle reactions thus would be a matter of chance entirely, as in the present test and in the previous experiment [8], unless a comprehensive and more sophisticated tool to spatially characterize grain geometry and orientation such as the Electron Backscatter Diffraction (EBSD) method is employed [11].

Owing to the mentioned issues in handling solid Li targets, it is important to foresee in future tests the study of CCS formation also in Li vapour targets, as theoretically described in a previous work [6], and experimentally observed, with a much higher alleged intensity than in a Li solid target [8].

## 6. Conclusions

What can be learnt from this very preliminary but also very encouraging test of the CCS model?

- Proton beam quality has to improve substantially: an ion irradiation facility with accurate energy definition, high current and beam spot size control down to energy of the order of 100 eV is required. An electrostatic decelerator would probably be necessary to fulfill the latter requirement.
- The target is perhaps the crucial issue in this context: the future experiment must be flexible enough to host both solid and gas Li targets, if a complete and accurate investigation of the CCS model as applied to the p-Li reaction is desired. Clearly, the use of a very thin oriented target made of lithium single crystal is the best solution.
- Much more time must be allowed for test preparation and run, because this is a search for rare events. The present test was carried out over a period of two-and-a-half weeks, which is clearly too short, especially if compared with the previous experiment which took 25 series of long duration tests performed at 4 different facilities and spanning over seven years [8].

Fortunately, such a facility might be soon available in Europe, after some upgrade.

## Acknowledgements

Dr. Stefan Facsko (Head of Ion Beam Center, Helmholtz- Zentrum Dresden-Rossendorf, HZDR) is warmly thanked for the preparation of the experimental setup which has been used in this test. Technical assistance by Mario Steinert (HZDR) during the data taking is gratefully acknowledged. Dr. Alessandra Fantoni (INFN) is thanked for useful discussions and support.

## References

[1] Dodonov V.V., et al., Generalized uncertainty relation and correlated coherent states, Phys. Lett. **A79**, (1980) 150, 10.1016/0375-9601(80)90231-5.

[2] V. I. Vysotskii et al., Coherent correlated states and low-energy nuclear reactions in non-stationary systems, Eur. Phys. J. **A 49**, 99 (2013), and references therein.

[3] Fields B.D., The primordial lithium problem, Annu. Rev. Nucl. Part. Sci. 2011, **61**:47-68, 10.1146/annurev-nucl-102010-130445; Bertulani C.A., et al., "Cosmological lithium problems", EPJ Web of Conferences, **184** 01002, (2018), 10.1051/epj-conf/201818401002.

[4] Raiola F. et al., Enhanced electron screening in d(d,p)t for deuterated Ta*, Eur. Phys. J. A **13**, 377–382 (2002) 10.1007/s10050-002-8766-5; Fiorentini G. et al., Fusion rate enhancement due to energy spread of colliding nuclei, Phys. Rev. **C67**, (2003) 014603, 10.1103/PhysRevC.67.014603; Kasagi J. et al., Strongly Enhanced Li + D Reaction in Pd Observed in Deuteron Bombardment on PdLix with Energies between 30 and 75 keV, Jou. Phys. Soc. of Japan **73**, No. 3, 608–612 (2004), 10.1143/JPSJ.71.2881.

[5] Czerski K. et al., Screening and resonance enhancements of the $^{2}H(d,p)^{3}H$ reaction yield in metallic environments, Eur. Phys. Lett. **113** (2016) 22001 and refs. therein, 10.1209/0295-5075/113/22001.

[6] Vysotskii V.I., Vysotskyy M.V.and Bartalucci S., Features of the Formation of Correlated Coherent States and Nuclear Fusion Induced by the Interaction of Slow Particles with Crystals and Free Molecules, J. Exp. Theor. Phys., **127**, (3), p. 479 (2018) and refs. therein, 10.1134/S1063776118080253.

[7] Bartalucci S., Vysotskii V.I. and Vysotskyy M.V., Correlated states and nuclear reactions: an experimental test with low energy beams, Phys. Rev. Acc. and Beams **22**, (2019) 054503, 10.1103/PhysRevAccelBeams.22.054503

[8] Lipinski S. and Lipinski H., Hydrogen-Lithium fusion device, Patent WO 2014/189799 A9 (22.05.2013).

[9] Anton R. et al., Design and performance of a versatile, cost-effective microwave electron cyclotron resonance plasma source for surface and thin film processing, Rev. Sci. Instr. **71**, no. 2 (2000), 1177, 10.1063/1.1150420.

[10] see e.g.: https://mateck.com/info/introduction-crystals-and-crystal-growth.html

[11] Quinn A. et al., Cell Reports Physical Science **1**, 100137 August 26, 2020 10.1016/j.xcrp.2020.100137

J. Condensed Matter Nucl. Sci. 36 (2022) 137–183

Research Article

# The Electromagnetic Considerations of the Nuclear Force—PART II: The Determination of the Lowest Energy Configurations for Nuclei

N. L. Bowen*
*Colorado Mountain College, Glenwood Springs, Colorado, USA*

**Abstract**

This paper explores how the electromagnetic energies of the quarks within the atomic nucleus affect the behavior of the Nuclear Force. By examining the electromagnetic energies and forces, many questions about nuclear behavior can be answered, and many insights into the atomic nucleus can be gained. Previous theoretical models for the Nuclear Force include only the Coulomb electric force of the protons, but with little or no consideration of the electromagnetic characteristics of the quarks. By incorporating the electromagnetic energies and forces into nuclear theory, this model has been able to achieve predictions of binding energy better than any previous model, doing so by using only one parameter instead of the five parameters used in the semi-empirical Weizsäcker formula of the Liquid Drop Model. The Electromagnetic Model unifies the Nuclear Force to the Electromagnetic Force. The Electromagnetic Model of the Nuclear Force includes the calculation of electromagnetics of the quarks, and by doing such, it is shown that the Nuclear Force is significantly influenced by the Electromagnetic Forces of the quarks. This paper, Part II of this series, illustrates the ground state configurations of the atomic nuclei, showing the basic segments of how the protons and neutrons cluster together, and how these segments bond to form larger atomic nuclei. A pattern emerges for the ground state configurations due to the uniformity of the electromagnetic laws. Diagrams are shown for this basic pattern, for both stable and radioactive atomic nuclei. By incorporating the electromagnetic energies and forces into nuclear theory, this model has been able to achieve not only excellent predictions of binding energy, but the ability to answer many other questions regarding the various behaviors of atomic nuclei.


*Keywords:* Electromagnetics, Nuclear Force, Quarks, Nuclear molecules, Clustering, Nuclear diagrams, Lowest energy state

## 1. Introduction

The Nuclear Force is the force which binds the protons and neutrons together in the atomic nucleus. The Electromagnetic Model of the Nuclear Force states that the Nuclear Force is a direct result of the Electromagnetic Forces of the quarks within the nucleons. The model further answers and clarifies how this Electromagnetic Force is able to achieve nuclear bonding. Historical objections and misunderstandings regarding the Electromagnetic Force within atomic nuclei are answered and clarified in Part I of this series on the Electromagnetic Considerations of the Nuclear

*Corresponding author: nbowen@coloradomtn.edu

Force [1]. Thus, it is recommended that the first paper in this series is reviewed and understood, prior to continuing with this second paper.

In this second paper, the ground state configurations of the atomic nuclei are determined and illustrated, to show the basic segments of the clustered protons and neutrons and the bonding of the segments. The lowest energy configurations of the atomic nuclei are determined by using the laws of electromagnetics as applied to the quarks. Because thermodynamics insists that the configuration of the atomic nuclei are in their lowest energy states, a determination of these lowest energy states is necessary before a better understanding of the subsequent nuclear behavior can be achieved. Unless a Nuclear Force model describes the lowest energy configuration of nuclear structure, then the results for that model are irrelevant, and such approaches are not realistic for the proper consideration of binding energy and other nuclear behaviors. For a genuine and credible understanding of any structured atomic nucleus, the lowest energy configuration must first be determined. This lowest energy configuration is strongly dependent upon the forces and energies, and the associated equations, within the atomic nucleus. Thus, any type of nuclear behavior cannot be properly determined unless there is a mathematical formulation of these forces and energies internal to the atomic nucleus to determine the lowest energy. In this paper the lowest energy configuration of the structured atomic nucleus is determined by using the electromagnetic equations applied to the electromagnetic forces and energies of the quarks.

This Electromagnetic Model of the Nuclear Force assumes the laws of electromagnetics are valid inside an atomic nucleus, and that these laws should not be disregarded. This model asserts that the electromagnetic properties of the quarks are significantly responsible for holding the nucleons together in an atomic nucleus. The thermodynamics of the electromagnetic energies of the quarks cause the atomic nuclei to fall into the lowest energy state and configuration. Therefore, to understand nuclear behavior of the atomic nuclei, the lowest energy state must be known. The lowest energy configurations of the atomic nuclei are responsible for giving the atomic nuclei specific behaviors—such as the unique shape of the binding energy curve, the large quadrupole moments, excited states, and particle decay.

When taken into full account, rather than being disregarded and ignored, the Electromagnetic Forces of the quarks can explain much about nuclear behavior. The role of electromagnetics within the atomic nucleus is a field that deserves further research and serious analysis by theoretical nuclear physicists. This paper serves as further on-going theoretical investigations into the electromagnetic considerations of the Nuclear Force.

In this paper, the lowest energy state configurations of the atomic nuclei are determined and illustrated, showing the basic segments of how the protons and neutrons cluster together to form the larger atomic nuclei. The lowest energy configurations of the atomic nuclei are determined by using the laws of electromagnetics as applied to the quarks.

### 1.1. Definition, Explanation, and Clarification of Terms

The "Nuclear Force" is the force that binds together the protons and neutrons in an atomic nucleus. Historically, the Nuclear Force was called the "Strong Nuclear Force", and it was considered to be one of the four forces of nature—along with the Gravitational Force, the Electromagnetic Force and the Weak Nuclear Force. Soon after the discovery of quarks, a force called the "Chromodynamic Force" was used to describe that force which holds the quarks together inside a proton or neutron. The Chromodynamic Force is considered to be much stronger than the Nuclear Force. In the 1970's, the term "Chromodynamic Force" was changed to the term "Strong Nuclear Force". The "Strong Nuclear Force" is now used to describe the behavior and interactions of all sub-particles contained inside of a proton or neutron, such as quarks, gluons, and other partons. The name of the force that holds the protons and neutrons together in an atomic nucleus was also changed to the "Nuclear Force", also known as the Nucleon-Nucleon Force . Unfortunately, this series of name changes has caused much confusion.

Even more confusing, one model of the Nuclear Force is called the "Residual Strong Force", also known as the "Residual Chromodynamic Force". For clarity to the reader in this paper, the terms "Strong Force" and "Strong Nuclear Force" will not be used, since they are confusing and ambiguous terms. Rather the term "Quantum Chromodynamic

Force" will be used to reference the force that holds together the quarks inside a proton or neutron. The term "Nuclear Force" will be used to describe the force that holds together the protons and neutrons in an atomic nucleus. The term "Residual Color Force" will be used to describe the one specific model of the Nuclear Force, which is also known as the "Residual Strong Force".

Linguistically, it is also unfortunate that the words "neutron", "nucleon", "nucleons", "nuclides", "nuclei", "nucleus", and "nuclear" are all so similar. This similarity makes it easy for a reader to misread them and misunderstand what is being communicated. To mitigate the confusion between these linguistic similarities, the term "protons and neutrons" will occasionally be used instead of "nucleons". Also, the terms "atomic nucleus" or "atomic nuclei" will be used instead of "nucleus", "nuclide", "nuclides" or "nuclei". (Other writers often use the word "isotopes" to mean "nuclides" in order to avoid this same confusion; however, since that is a misuse of the word "isotope", this will not be done for this paper.)

Another source of confusion is caused by the similarity of the terms "Chromodynamic Force", and the "Residual Chromodynamic Force". Thus, to prevent confusion, the terms "Residual Color Force", and "Quantum Chromodynamic Force" will be used to better distinguish these two different forces for the reader. The Quantum Chromodynamic Force is the force inside of the protons and neutrons, and it holds the quarks together in a proton or neutron. The Nuclear Force is outside of the protons and neutrons, and it holds the protons and neutrons together in an atomic nucleus. There are many models of the Nuclear Force, one of which is the Residual Color Force. It is related to the Quantum Chromodynamic Force, theorized to be caused by virtual pi-mesons going back and forth between the protons and neutrons. It is thought to be similar to the van der Waals force for molecules. The "Quantum Chromodynamic Force" is the force that holds the quarks together inside of a proton or neutron.

This paper is about the Nuclear Force, the force that holds the protons and neutrons together inside an atomic nucleus.

This paper is not intended to be an overview, an explanation, or a tutorial of the Quantum Chromodynamic Force or of the Weak Nuclear Force. For a detailed review and explanation of the Quantum Chromodynamic Force model, see [2], [3]. For a review and explanation of the Weak Nuclear Force model, see [4].

### 1.2. The Electromagnetic Model of the Nuclear Force and the Intent of this Research

The Electromagnetic Forces cause the atomic nucleus to fall into the lowest energy state and lowest energy configuration. It is asserted that this electromagnetic energy, as well as the specific lowest energy configurations of the atomic nucleus, are the attributes that dictate the nuclear behaviors—such as binding energy, large quadrupole moments, excited states, and particle decay.

Rather than disregarding the Electromagnetic Forces of the quarks, when taken into full account and understanding, the Electromagnetic Forces inside an atomic nucleus can explain much about nuclear behavior. By applying this knowledge and insight, a better understanding and explanation of the nuclear behaviors can be gained through this model.

The role of electromagnetics within the atomic nucleus is a research field that deserves further analysis and serious consideration by theoretical nuclear physicists. This paper serves to advance the theoretical investigations into the electromagnetic considerations of the Nuclear Force.

For detailed calculations of the electromagnetic internucleon quark-to-quark energies, please refer to Part I of this series of papers on the Electromagnetic Considerations of the Nuclear Force. These detailed calculations will not be repeated in this paper. Rather only a brief review of the electromagnetic energies will be given here.

The intent of this paper, as well as for the series of papers about the Electromagnetic Considerations of the Nuclear Force, is to explore how much of nuclear behavior can be explained by the Electromagnetic Force. In other words, the goal is to explore what fraction of the Nuclear Behavior can be explained by the Electromagnetic Force.

It is not the intent of this paper to explore the Residual Color Force. Other theoretical research groups have explored the theoretical Residual Color force, with only minimal success in duplicating experimental data, as is discussed in the appendix. To repeat this extremely difficult task of exploring the Residual Color Force is not the intent of this paper. Rather, the intent is to explore how much of the nuclear behavior can be explained by the Electromagnetic Force. That is the intended endeavor of this entire series of papers.

For that reason, the force between nucleons is assumed, from the onset, to be an internucleon quark-to-quark Electromagnetic Force. Using that assumption, the resulting predicted nuclear behavior is examined and compared with the actual experimental data. This is why the Residual Color Force Model—or any of the other numerous models of the Nuclear Force such as the Liquid Drop Model, the Shell Model, or the Effective Field Models—are not included in the calculations for the simulations in this paper. The intended endeavor of this series of papers about the Electromagnetic Considerations of the Nuclear Force is to learn what parts of nuclear behavior can be explained by the Electromagnetic Force.

### 1.3. A Clarification of Chromodynamic/Color Forces

A clarification should be made here. The topic of this paper is the Nuclear Force, that force which holds the protons and neutrons together in an atomic nucleus. The subject of this paper is the lowest energy configurations of atomic nuclei, as determined by the Electromagnetic Force. The information described in this paper is consistent with the current theories and models about the Quantum Chromodynamic Force, quark containment, the Standard Model, and the Weak Nuclear Force. Thus these topics are not reviewed in this paper; readers interested in these topics are referred to other scientific papers for more information on these ancillary topics.

A further clarification must be made regarding the Quantum Chromodynamic Force and the Residual Color Force. The Quantum Chromodynamic Force acts inside the protons and neutrons. It is believed, by current models, that the Quantum Chromodynamic Force also has a residual force, which is called the Residual Color Force in this paper. The Residual Color Force acts outside the protons and neutrons. The Quantum Chromodynamic Force and the Residual Color Force are related to each other, but they are not exactly the same force. The two forces are related because they are both associated with the color of the quarks; however, a distinction between these two forces should be made. The Quantum Chromodynamic Force is mediated by massless gluons, it is extremely strong, and it has a long range. The Residual Color Force is thought to be mediated by virtual pi-mesons, it is much weaker than the Quantum Chromodynamic Force, and it has a very short range. Although both forces are related to quark color, one force is inside the protons and neutrons and the other force is outside the protons and neutrons. Also, they have different strengths, ranges, and mediating particles.

### 1.4. A Brief Review of Quarks

In 1964, the existence of quarks was proposed independently by Gell-Mann and Zweig [5], [6], [7], changing the concept of the proton and neutron from homogeneously-charged particles to particles having electrical inhomogeneity. A proton is made up of three quarks, two up quarks and one down quark. A neutron is also made up of three quarks, two down quarks and one up quark. Up quarks have an electrical charge which is 2/3 of an elementary charge. Down quarks have a charge which is $-1/3$ of an elementary charge. The electrical charge and magnetic moments of a proton or neutron are confined to the quarks, rather than being homogeneously distributed.

Gell-Mann believed that quark charges could be localized, and Richard Feynman asserted that high energy experiments showed that quarks are indeed real particles. Feynman surmised the quarks have a distribution of position or momentum, like any other particle, and he correctly believed that the diffusion of quark momentum explained diffractive scattering. James Bjorken proposed that point-like partons would imply certain relations in deep inelastic scattering of electrons and protons, which were verified in experiments at SLAC in 1969. Thus, it is believed that the quarks have a distribution of position and momentum, due to uncertainty principles. However, when averaged over time, they can be modeled as point-like particles within those same distributions. This is an important concept to understand, and it is related to the Expectation Value of the position of the quarks, when the probability function of the quarks is analyzed and averaged over time. Stated again for clarity, there are quantum fluctuations, due to the uncertainty that is inherently involved; however, the position of the quarks can nonetheless be modeled as a point-like charge.

Quarks have both color and electric charge. (The word color does not mean an actual visual color, rather the word color is simply a quantum attribute of the quarks—either red, green, or blue.) The Quantum Chromodynamic Force controls the quarks and confines the quarks, and it is dependent upon the color of the quarks. Outside of the proton and neutron, only a slight residual of the Quantum Chromodynamic Force remains; however, when that slight residual force is outside of the nucleon, it is called the "Residual Color Force". (To avoid confusion, the residual force should be thought of, conceptually, as a different force—one that is related to, but different from, the Quantum Chromodynamic Force. This is similar to the way the van der Waals force is conceptually different from the Electromagnetic Force.) Also outside of the proton and neutron, and also related to the quarks, is the electromagnetic force. The Electromagnetic Force is a long range force and it does not abruptly "turn itself off" or neutralized itself at the edge of the proton or neutron. Conversely, the Quantum Chromodynamic Force does essentially neutralize itself at the edge of the proton or neutron. To reiterate, the Quantum Chromodynamic Force abruptly changes to almost zero at the edge of the nucleon, but the Electromagnetic Force continues outside of the nucleon, unaffected by the edge of the nucleon.

It is important for the reader to understand that the quarks have both a color and an electric charge. Because the quarks inside the nucleons are in the three colors of red, green, and blue, this causes the protons and neutrons to be color neutral. Thus, the Quantum Chromodynamic Force, which is due to the color of the quarks, behaves with an abrupt step-like function near the edge of the nucleon, dropping suddenly to almost zero. In contrast to that, the Electromagnetic Force, and its associated electromagnetic field, are unaffected by the edge of the nucleon. The Electromagnetic Forces of the quarks, therefore, continue to be felt by all of the other quarks in the atomic nucleus, and these Electromagnetic Forces have an influence on the other quarks within the entire atomic nucleus.

The Quantum Chromodynamic Force is inside of the nucleon, and it is what controls and contains the position of the quarks, inside of the proton and neutron. The electromagnetic charges of the quarks do not contain and control the quarks inside the proton or neutron. However, the electromagnetic charges of the quarks are felt outside of the proton and neutron, whereas the Quantum Chromodynamic Force is not. The Electromagnetic Force of the quarks extend beyond of the edge of the nucleon, essentially unaffected by it. The Electromagnetic Forces, thereby, can electromagnetically attract or repulse the other quarks in the atomic nuclei. That attractive or repulsive force depends on the polarity of the associated electric charges and magnetic dipoles of the quarks.

As a result, a quark that is inside of one nucleon, feels the electromagnetic influences of a quark that is inside of another nucleon, in a quark-to-quark internucleon force. Those two quarks, which are in two different nucleons, are contained within their two different respective nucleons by the Quantum Chromodynamic Force, which acts inside of the nucleon. However, these two same quarks are influenced by each other due to the electromagnetic forces outside of their respective nucleons. Furthermore, that electromagnetic influence can be either attractive or repulsive.

### 1.5. Current Concepts Regarding the Quantum Chromodynamic Force

The Quantum Chromodynamic Force is thought to be much stronger than the Nuclear Force, by several orders of magnitude. However, an actual definitive measurement of the strength of the Quantum Chromodynamic Force has not been made because particle physicists have been not able to separate the quarks from inside a nucleon. Presently there are numerous theoretical models about what is inside a nucleon, which are not only quite complex, but also conflicting.

Particle physicists believe that there might be several hundred non-valence quarks inside a nucleon. In this paper, only the three valence quarks are considered.

### 1.6. The Numerous Models for the Nuclear Force

Currently, there is no one theory of the Nuclear Force that can explain all of nuclear behavior [8], [9]. Rather, there are numerous models for the Nuclear Force, each one describing one or two specific aspects of nuclear behavior. Even with these numerous models, there are still many nuclear behaviors that remain unexplained. A brief review of the numerous other models of the Nuclear Force is given in Part I of this series on the Electromagnetic Considerations of the Nuclear Force, and will not be repeated here. Another excellent and comprehensive review of the numerous models of the Nuclear Force is given by Cook [10]. One of the many models of the Nuclear Force is the Residual Color Force Model.

### 1.7. The Residual Color Force Model

The Residual Color Force Model is based on the concept that the Nuclear Force is a residual of the Quantum Chromodynamic Force. The Residual Color Force Model is similar to the meson exchange model of 1935, in that the force is believed to be due to an exchange of virtual pi-mesons.

Of the numerous models of the Nuclear Force, the Residual Color Force Model is one of the few models that incorporates the concept of quarks, thereby making it consistent with the Standard Model. Unfortunately, there is no closed mathematical form for the Residual Color Force, and it is extremely difficult to calculate any type of nuclear behavior.

As described previously, the Quantum Chromodynamic Force is the force that holds the quarks together inside the proton or neutron, mediated by gluons. In contrast, the Residual Color Force occurs outside the protons and neutrons, mediated by virtual pi-mesons. The Residual Color Force Model theorizes that the Residual Color Force binds the protons and neutrons together via an internucleon quark-to-quark force based on the color of the quarks. In other words, in the Residual Color Force Model, the protons and neutrons are theorized to be held together by a bond that is formed between a quark in one nucleon binding to a different quark in another nucleon; this is what is meant by an internucleon quark-to-quark bond.

Because the Residual Color Force Model includes the concept of quarks, it is consistent with the Standard Model. However, due to its extreme mathematical complexity, the Residual Color Force Model cannot sufficiently explain nuclear behavior. For example, Residual Color Force Model cannot accurately predict binding energy or duplicate the binding energy curve. Also, it does not explain large quadrupole moments, excited states, radioactivity, or particle decay. Conceptually, the Residual Color Force Model, based on its internucleon quark-to-quark interaction, is an appealing theoretical concept to most particle physicists. However, as an applied theory, it has been relatively unsuccessful due its extreme difficulty and mathematical complexity.

The Residual Color Force Model is a very complicated mathematical problem, requiring brute computing power and using a discretized lattice of time and space. There are two difficulties with a derivation of Nuclear Forces from Quantum Chromodynamics. The first difficulty is that each proton or neutron consists of three quarks, turning even the simplest binding energy calculation, such as for deuteron, into a six-body problem. The second difficulty is that the Quantum Chromodynamic Force a very strong force compared to the Nuclear Force. This large difference in the strengths of these two forces makes it difficult for the iterative mathematical computer calculations to converge to a solution.

To solve the six-quark problem with brute computing power, the six-quark system is put into on a lattice of discrete points, with the three spatial dimensions and one temporal dimension, in a method known as lattice QCD. However, such calculations are computationally very expensive and cannot be used as a feasible nuclear physics tool. As a result, the Residual Color Force Model of the Nuclear Force has not been useful from a practical standpoint [11], [12], [13]. More details about the Residual Color Force Model are given in the appendix for the interested reader.

### 1.8. A Minor Change to the Residual Color Force Model

The Residual Color Force Model postulates that the Nuclear Force is a residual effect of the color forces between a quark in one proton or neutron with another quark in a different proton or neutron, and that the force is related to the color of the two quarks involved. The Residual Color Force Model is postulated to be due to an exchange of virtual pi-mesons between the quarks of the two different nucleons. However, one simple variation to this model makes a very interesting difference in the understanding of the forces involved within atomic nuclei.

If the Nuclear Force between the protons and neutrons is dependent upon the up and down flavor of the quarks, which then considers the positive and negative electric charge of quarks rather than only the color of the quarks, then this small change is able to answer many questions about nuclear behavior.

The concept here is that a negatively-charged down quark is electrically attracted to a positively-charged up quark, one that is in a different nucleon from itself. Similarly, a positively-charged up quark is attracted to down quark, one that is in a different nucleon. Up quarks and down quarks that are in two different nucleons are electrically attracted to each other. Up quarks are electrically repulsed by other up quarks in other nucleons, and down quarks are repulsed by other down quarks in other nucleons.

These attractive and repulsive Electromagnetic Forces are internucleon quark-to-quark forces that are outside of the protons and neutrons, where the term "internucleon" means it is between two nucleons. These electromagnetic internucleon quark-to-quark forces bind the protons and neutrons together in an atomic nucleus. An internucleon quark-to-quark force is not the Quantum Chromodynamic Force. (The Quantum Chromodynamic Force is the force that bonds, confines the quarks together inside a proton or neutron.) Rather, these electromagnetic internucleon quark-to-quark forces are a Nuclear Force; they are a force that binds the protons and neutrons together to form an atomic nucleus. The bonding force between the nucleons is, quite simply, the attraction of the up quark in one nucleon to the down quark in another nucleon. When a bond is made between these two different quarks in two different nucleons, the two different nucleons are, therefore, similarly bonded. This force bonding these two different nucleons together is the Nuclear Force; it is not the Quantum Chromodynamic Force. This force bonding these two different nucleons together is an electromagnetic internucleon quark-to-quark attractive force.

This simple change of concept, that the Nuclear Force is dependent upon the electromagnetic charge of the quarks rather than the color of the quarks, easily explains why a system of 6 protons and 6 neutrons is at a lower overall energy (and thus at a higher binding energy) than 5 protons and 7 neutrons. It is because 6 protons and 6 neutrons can form one bond for every pair of up-down quarks. There are 18 up quarks and 18 down quarks in the system of 6 protons and 6 neutrons. Thus there could be 18 up-to-down quarks bonds. However, for 5 protons and 7 neutrons, there are 19 down quarks and 17 up quarks, thus only 17 bonds could be formed. The nucleus with 5 protons and 7 neutrons would be at a higher overall energy (and thus at a lower binding energy) than the system of 6 protons and 6 neutrons. With this simple change, it easily can be understood why a nucleon can only bond to its nearest neighbors. Specifically, a nucleon can only bond to three other nucleons, because there are only three quarks in every nucleon with which to bond. The internucleon quarks involved in the bonding must be of opposite electromagnetic polarity.

This single concept immediately explains the asymmetry term of the Weizsäcker formula, because the greatest number of bonds occurs when there are equal numbers of up quarks and down quarks, which in turn means an equal number of protons and neutrons. Also, this new concept–that the Nuclear Force is dependent upon the electromagnetic charge rather than the color of the quarks–clarifies the "constant" term of the Weizsäcker formula. It is because each nucleon has a limited number of bonds it can form with other nucleons. As will be clarified later when looking at the representations of the lowest energy configurations, the "surface" term applies to the unbonded quarks on the ends of the chain-like configuration, as well as the two unbonded quarks that are in the star segment. The Coulomb energy term of the Weizsäcker formula is also easily explained as being related to the electrical energy of the net positive charges. Without going into a long and detailed explanation, the pairing term can cursorily be explained as being related to the fact that there are two protons and two neutrons for each alpha segment, and that the alpha segments have no unbonded quarks. Thus the highest binding energy occurs when there are even numbers of protons and neutrons, and the most number of alpha segments. Thus, for this one simple change—that the internucleon quark-to-quark bond is related to electromagnetics rather than color—the Electromagnetic Model can explain all of the five terms of the semi-empirical Weizsäcker formula. This is indeed why it is able to duplicate the binding curve so accurately.

Quick calculations have been made to test this hypothesis [14]. These simple calculations show that this concept of an electromagnetic internucleon up-to-down quark bond reproduces the binding energy curve surprisingly well, using only one variable—the strength of the bond. Such good replication, achieved with simple mathematics, strongly implies that this concept is correct. The quick calculation of this concept did not assume any particular type of force for the bond; it only assumed that a bond was formed between an up quark and a down quark. A more rigorous and detailed calculation follows when the Electromagnetic Force is used as the force between the up and down quarks, as was done in Part I of the Electromagnetic Considerations of the Nuclear Force [1]. The detailed calculations for the Electromagnetic Forces are found in Part I, and will not be repeated in this paper.

### 1.9. The Limitation of the Electromagnetic Force

Prior to the 1960's, the proton was incorrectly thought to be homogeneously charged and to have a radius of about 1.2 femto-meters. As a result, the strongest electrical energy between two such protons was thought to be $9.62 \times 10^{-14}$ joules.

The experimental energy required to free a single nucleon from an atomic nucleus is much larger than $9.62 \times 10^{-14}$ joules. For this reason, the Nuclear Force was believed to be much stronger than the Electromagnetic Force. As a result, this incorrect concept of a homogenously charged proton created an erroneous limitation of the Electromagnetic Force. Unfortunately, this incorrect concept is still often perpetuated.

If the minimum internucleon quark-to-quark distance (defined as the minimum distance of one quark in one nucleon to another quark in another nucleon) is 0.2111 femtometers, which is the parameter we use for the minimum internucleon quark-to-quark distance in our calculation, and if we include the magnetic force between the quarks into calculation, the value for the energy of an electromagnetic bond is $1.18014 \times 10^{-12}$ joules, which is 7.36584 MeV.

Mathematically, in the limit as the distance goes to zero, the electromagnetic energy goes to infinity. Hence, the electromagnetic energy can be extremely large if the quarks are close enough to each other. Since the charge of the nucleons resides within the quarks, a quark from one nucleon can bond electromagnetically with a quark in another nucleon, and that resultant force between two such quarks can be extremely large—large enough, indeed, to be the Nuclear Force.

The Electromagnetic Model of the Nuclear Force assumes the laws of electromagnetics are valid inside an atomic nucleus, and that these laws should not be disregarded. This model asserts that the electromagnetic properties of the quarks within the atomic nucleus are the forces that create the Nuclear Force. Thus, these Electromagnetic Forces are the actual Nuclear Forces that hold the protons and neutrons together in an atomic nucleus, a force that was previously thought to be a mysterious and complicated force.

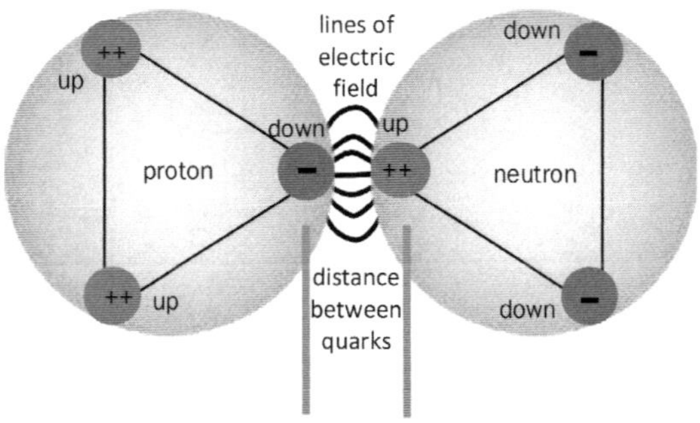


**Figure 1.** The electromagnetic bond between quarks.

## 2. The Electromagnetic Energies Inside the Atomic Nucleus

### 2.1. The Internucleon Quark-to-Quark Bond

The electric charges and magnetic dipole moments of the protons and neutrons are contained in the quarks, and this fact must be taken into consideration for any model of the Nuclear Force. The attractive Electromagnetic Force between an up quark and a down quark is strong enough to bond the nucleons together. No other force, different from the Electromagnetic Force, is needed to account for the strength of this bond. This electromagnetic bond is shown in Fig. 1, along with the minimum internucleon quark-to-quark distance.

The force binding the proton and neutron together need not be anything other than the Electromagnetic Forces between the internucleon quarks.

### 2.2. The Electric Energy

The electric energy [15], [16], [17] between two electrically charged particles is shown in Eq. (1):

$$U_{E12} = \frac{-q_1 q_2}{(4\pi\epsilon_0)(r_{12})} \tag{1}$$

where: $U_{E12}$ is the electric energy between particles 1 and 2, $r_{12}$ is the distance between particles 1 and 2, and $q_1$ and $q_2$ are the electric charges on particles 1 and 2, respectively.

For additional charges, the electrical energies are simply summed for every pair of charges, as shown in Eq. (2).

$$U_{electric\_total} = \sum_{i=1}^{n-1} \sum_{j=i+1}^{n} \frac{-q_i q_j}{(4\pi\epsilon_0)(r_{ij})} \tag{2}$$

This is the equation used to calculate the electric energy on a configuration of electric charges, in a three dimensional configuration.

### 2.3. The Magnetic Energy

The magnetic energy [18], [19], [20] between two magnets is more complicated than the electric energy because it has vector and position dependence. Given two magnets, with magnetic moments $\mu_1$ and $\mu_2$, the magnetic field of magnet$_1$ must first be determined at the position of the magnet$_2$, symbolized as $B_{12}$. This is shown in Eq. (3).

$$\overrightarrow{B_{12}} = \frac{\mu_0}{4\pi}\left(\frac{3\left(\overrightarrow{\mu_2}\cdot\overrightarrow{r_{12}}\right)\overrightarrow{r_{21}} - r_{21}^2\left(\overrightarrow{\mu_2}\right)}{r_{21}^5}\right) \tag{3}$$

where $r_{21}$ is the vector distance from magnet$_2$ to magnet$_1$, and similarly $r_{12}$ is vector distance from magnet$_1$ to magnet$_2$.

The resultant energy, $U_{magnetic12}$, is the negative dot-product of the vector of the magnetic moment $\mu_2$ with the vector of $B_{12}$, as shown in Eq. (4).

$$U_{magnetic\_12} = -\overrightarrow{\mu_2}\cdot\overrightarrow{B_{12}} \tag{4}$$

For a collection of magnets, the total magnetic energy is the double summation over all magnet pairs, as shown in Eq. (5).

$$U_{magnetic\_total} = \sum_{i=1}^{n-1}\sum_{j=i+1}^{n} -\overrightarrow{\mu_i}\cdot\overrightarrow{B_{ij}} \tag{5}$$

where $\overrightarrow{B_{ij}}$ is the vector magnetic field established by the i$^{th}$ magnet at the location of the j$^{th}$ magnet.

Combining equations (2), (3), and (5), the total electromagnetic energy of a distribution of charges and magnets is shown in Eq. (6):

$$U_{EM\ total} = \sum_{i=1}^{n-1}\sum_{j=i+1}^{n}\frac{q_i q_j}{\left(4\pi\epsilon_0\right)\left(r_{ij}\right)} + \sum_{i=1}^{n-1}\sum_{j=i+1}^{n}\frac{\mu_0}{4\pi}\left(\frac{3\left(\overrightarrow{\mu_j}\cdot\overrightarrow{r_{ji}}\right)\left(\overrightarrow{\mu_i}\cdot\overrightarrow{r_{ji}}\right) - r_{ji}^2\left(\overrightarrow{\mu_i}\cdot\overrightarrow{\mu_j}\right)}{r_{ji}^5}\right) \tag{6}$$

This is the equation used to calculate the total electromagnetic energy on a configuration of electric charges, in a three dimensional configuration.

Due to the vector properties of this energy, the lowest energy configuration for two magnets is a stacked bond, in which the magnetic dipole moments of the magnets are aligned in a stacked orientation with respect to each other, and the magnets are as close as physically possible. Another way for two magnets to bond is a side-by-side bond, with the magnetic diole moments anti-parallel, and the magnets oriented side-by-side, as a close as physically possible. A bond that is in between a stacked and a side-by-side bond is an angled bond, and this will give intermediate results for its energy. The various orientations are shown in Fig. 2.

From quantum field theory, it is known that quarks behave as point-like Dirac particles, each having their own inherent magnetic moment [21], [22].

The calculated binding energy is the difference of the total energies of the configuration and the total energies of the isolated constituent parts. Although binding energy is considered to be positive, this binding energy is subtracted from the energy of the isolated constituent parts, resulting in the bound configuration being at an overall lower energy than the isolated parts. Unfortunately, this is often a source of confusion for many people, including some particle physicists.

To reiterate, the *higher* the binding energy, $U_{\text{binding energy}}$, of a bound configuration, the *lower* the overall total energy (and mass) of the configuration. In other words, binding energy of a bound configuration does not contribute to the mass of the bound configuration. Rather, a positive binding energy *reduces* the mass of the configuration compared to the mass of the isolated constituent parts. This is shown in Eq. (7).

$$U_{total\ of\ configuration} = U_{total\ of\ isolated\ parts} - U_{binding\ energy} \tag{7}$$

For nuclear binding energy, in this calculation, the mass of the electrons is taken into consideration, as is the binding energy of the electron to the atom. Any difference in the energy due to angular momentum must also be taken into consideration in this equation. For a stable configuration, the bond configuration will be at a lower total mass and

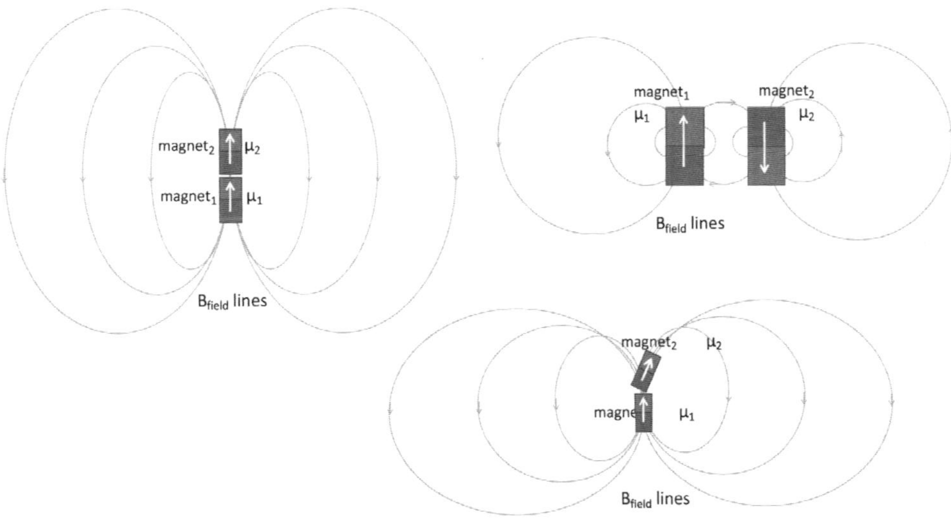


**Figure 2.** Three orientations for magnetic bonds, stacked, side-by-side, and angled.

lower total energy. This is true regardless of whether the energy is quantum energy or some other type of classical energy. Indeed, quantum energy is not different in this regard.

For the electromagnetic theory, it is assumed that no other energies, other than the electromagnetic energy and the angular momentum energy, contribute to the binding energy. Thus, the binding energy can be expressed as the difference in the electromagnetic energy, $\Delta U_{\mathrm{electromagnetic}}$, plus the difference in the angular momentum energy, $\Delta U_{\mathrm{angular\ momentum}}$, as shown in Eq. (8).

$$U_{binding\ energy} = \Delta U_{electromagetic} + \Delta U_{angular\ momentum} \tag{8}$$

The electromagnetic energy of the quarks inside the individual protons and neutrons is taken into account as part of the electromagnetic energy of the individual constituent parts.

### 2.4. Short-Range versus Long-Range Forces

Historically, since there was no indication of a mysterious Nuclear Force at long-range distances, it was previously thought that the Nuclear Force must be a short-range force. Due to the misconception that protons were homogeneously charged, it was also thought that the Electromagnetic Force could not be same as the Nuclear Force. Thus, the existence of a force, different from the Electromagnetic Force, was postulated, and it was assumed to be some sort of a mysterious short-range force. Given the understanding that the Nuclear Force is electromagnetic, the presumed requirement of a short-range force is no longer necessary.

### 2.5. Electromagnetic Energies and the Schrödinger Equation

The Schrödinger equation is a useful tool and has been successful in many areas of quantum physics. The Schrödinger equation is a second-order differential equation that can only be solved for a given potential energy, kinetic energy, and initial boundary conditions. Previous nuclear models have modeled the nuclear potential as one large energy well, using different formulas for the Hamiltonian to obtain the probability function of the atomic nuclide in question.

However, if the Schrödinger equation is to be used properly for a quark-based Nuclear Force, the correct solution should have a Hamiltonian that is related to each individual quark within the atomic nuclide. Also, the proper potential energy well must take into account an energy that can be both attractive and repulsive, as well as any vector properties of the force. However, even without polarity and vector dependence, the quark-based solutions using the Schrödinger equation have proven to be extremely difficult, even for an atomic nucleus as simple $^{2}$H, (as mentioned previously when discussing the Residual Color Force Model). Thus using the Schrödinger equation with a quark-based model has proven to be almost intractable for larger atomic nuclei.

### 2.6. The Concept of a Nuclear Bond Is Similar to the Atomic Bond

For both the Residual Color Force Model and the Electromagnetic Model of the Nuclear Force, there is an energy well between the two internucleon quarks, and this energy well forms an internucleon quark-to-quark bond. (Recall, a nucleon is a proton or a neutron, and an "internucleon quark-to-quark bond" is a bond between a quark in one nucleon and another quark in a different nucleon.) For the Residual Color Force Model, the energy well is hypothesized to be between two quarks of two different colors, one quark from one nucleon and the other quark from another nucleon. For the Electromagnetic Model, the energy well is between the positively-charged up quark from one nucleon and the negatively-charged down quark of another nucleon. The energy of the Residual Color Force Model between quarks has no closed mathematical form, making it difficult to calculate. This is in contrast to the energy of the Electromagnetic Model, which is very easy to calculate.

Regardless of which energy or which force is used—the Residual Color Force or the Electromagnetic Force—the Nuclear Force is an internucleon quark-to-quark bond, which lowers the overall energy of the atomic nucleus.

This aspect, the lowering of the overall energy when a bond is formed, is similar to the electronic bonds that are formed between the atoms of a molecule. In the electronic bonding of atoms, one atom cannot bond to an indefinite number of other atoms. Rather the number of times that one atom can bond to another atom is limited by the number of valence electrons for that particular atom. Similarly, one proton or neutron cannot bond an indefinite number of times to other nucleons. Similar to the electronic bonding of atoms to form a molecule, the number of bonds that one proton or neutron can make with another proton or neutron is limited by the number of valence quarks available for bonding. The number of valence quarks available for bonding is three.

Thus each nucleon, having only three valence quarks each, can only bond a maximum of three times to other nucleons. Also note that two quarks, one from each of the two different nucleons, are needed for one bond.

### 2.7. The Assumed Shape of the Protons and Neutrons

In order for the energy of the internucleon quark-to-quark bond to be calculated, the distance between the quarks must be determined. To make this determination, the general intrinsic shape of the proton must be known. However, the intrinsic shape, unfortunately, is not experimentally or theoretically known. As a result, some assumptions about the shape are necessary in order to proceed. For this paper, these assumptions will be as few as possible and as simple as possible.

Experimental and theoretical particle physicists do not know the intrinsic shape of the proton or neutron. There are many conflicting concepts about the nucleon shape, but the overall consensus is that protons and neutrons are not spherical. The most likely shape is an oblate ellipsoidal shape [23], [24], [25].

There is a quantum probability function associated with the quarks, depending on their Hamiltonian and potential energy wells. To the extent that the quarks vibrate at speeds in accordance with uncertainty principles, the Expectation

Value of their location is defined by this probability function. The vibrational speed of the quarks depends on variables in quantum physics that are unknown—variables such as the quark mass, the correlation of the movement, and the distortion from a spherical shape of the quarks' three-dimensional potential energy well. Another unknown is which uncertainty principle should be appropriately applied to the quarks. The Robertson-Schrödinger uncertainty principle [26], [27] should be applied if there is a strong correlation between the position and momentum of the quarks, and such correlation could very well exist. Other physicists even go so far as to suggest that uncertainty principles should not apply, since quarks are not isolated particles.

These same unknown variables in quantum physics are problematic with the Residual Color Force Model as well. To avoid this problem, the calculations for the Residual Color Force Model use an artificially inflated rest mass for the quarks, a rest mass that is overly large by several orders of magnitude. The calculations are done at several different values of these artificially large masses, followed by an extrapolation down to the estimated value for the smaller quark mass [28], [29], [30], [31]. This procedure is done to assist in the convergence of the computer simulations; however, this procedure also conveniently sidesteps the uncertainty principle problems. Thus, the complicated and difficult issue of the uncertainty of the quark position and momentum is essentially sidestepped for the calculations of the Residual Color Force Model.

Fortunately there is some justification for sidestepping this problem. As mentioned previously, the quark has a distribution in position and momentum, which averaged over time is similar to a point-like particle, wherein the location in three-dimensional space is the position of the Expectation Value of its probability function. Even though this probability function and the resulting Expectation Value depend on the exact shape of the potential energy well, the result is a point location in three-dimensional space. Thus, even with the various unknowns, quantum fluctuations, and uncertainties, the result is nevertheless similar to a point-like particle at a specific location, when averaged over time. In other words, regardless of its extreme vibrational velocity and regardless of its positional uncertainty, the quark's location, when averaged over time, can be understood as the Expectation Value of its positional probability function.

This concept correlates to the concept that the quark is a point source of electric charge and a point-like magnetic dipole moment. Confined inside a proton or neutron by the Quantum Chromodynamic Force and the gluon field, the quarks vibrate within a potential energy well, at very high speeds. At any given instant in time, the quark has an uncertainty in its momentum, position, and energy. Also, the uncertainty in the quark position can extend outside of the mathematically defined radius of the nucleon, which is exactly what gives the Quantum Chromodynamic Force a residual force. The assumption that the Residual Color Force is several orders of magnitude smaller than the Quantum Chromodynamic Force implies that the probability function of the quarks is very tightly contained in three-dimensional space of the nucleon. Thus, when averaged over time, the Expectation Value for the location of the quark looks like a point source of charge and magnetism.

To review, each quark has an associated electric field and magnet field. When averaged over time, that electric field can be modeled as the field from a point-like electric charge. Similarly, when averaged over time, the magnet field can be modeled as the field from a point-like magnetic dipole. In actuality, the quarks do indeed move, and they have an extended probability function in three-dimensional space due to their vibrations, movements, fluctuations, and uncertainty. Experimentally, it is known that the quarks are tightly confined by the energy well of the Quantum Chromodynamic Force. However, the shape of that energy well for the Quantum Chromodynamic Force is not currently known, nor can this energy well be represented in a closed mathematical form. However, it is known experimentally that the energy well of the Quantum Chromodynamic Force is very strongly and tightly confining. When averaged over time, the Expectation Value for the position of quarks will give the quarks a characteristic that is similar to a point-like particle of charge and magnetism. All of these concepts are similar to those currently implied within the calculations of the Residual Color Force Model.

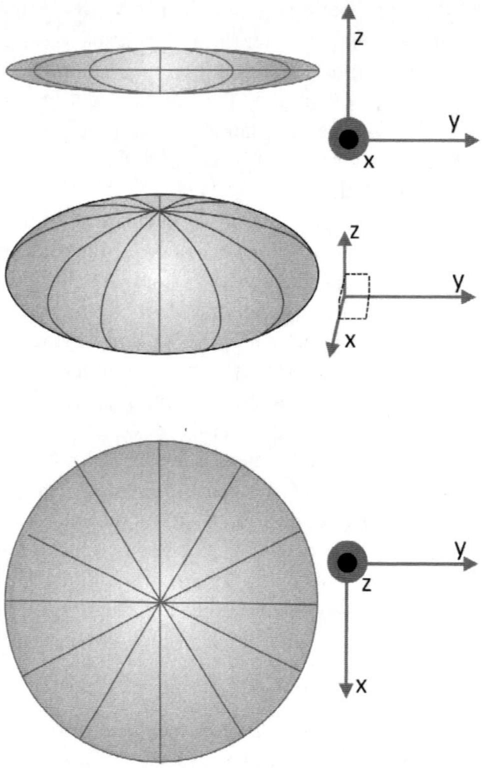


**Figure 3.** An oblate ellipsoidal proton, with three quarks in an equilateral triangle.

## 2.8. The Minimum Assumptions Possible

Consistent with what is observed experimentally, an oblate ellipsoid shape for the proton and neutron is assumed, with the quarks in an equilateral triangle inside that ellipsoid. This is shown in Fig. 3.

A different kind of triangle could be assumed, like an isosceles triangle or a scalene triangle; however, those are not a reflection of the simplest assumption, involving more variables for the shape. If a different sort of triangle were assumed, it would change some of the binding energy values of the smallest atomic nuclei, but it would have an overall minor effect on the other nuclear properties. Thus, a different type of triangle would not cause a significant alteration of the overall results, but it would unfortunately include more variables. Should it be determined, in the future, that the Expectation Value of the positional probability function of the quarks are arranged in a different type of triangle, it would cause only minor variations in the binding energy curve. In this paper, an equilateral triangle is used, to reduce the number and complexity of assumptions.

No other assumptions are made concerning the intrinsic shape of the protons and neutrons.

## 2.9. Nuclear Bonds and the Quantum Hard-core Repulsion

The Pauli Exclusion principle states that nucleons cannot overlap in their physical dimensions; this is also known as the hard-core repulsion. For this reason, a pair of double-bonded nucleons, defined as two nucleons bonded twice to

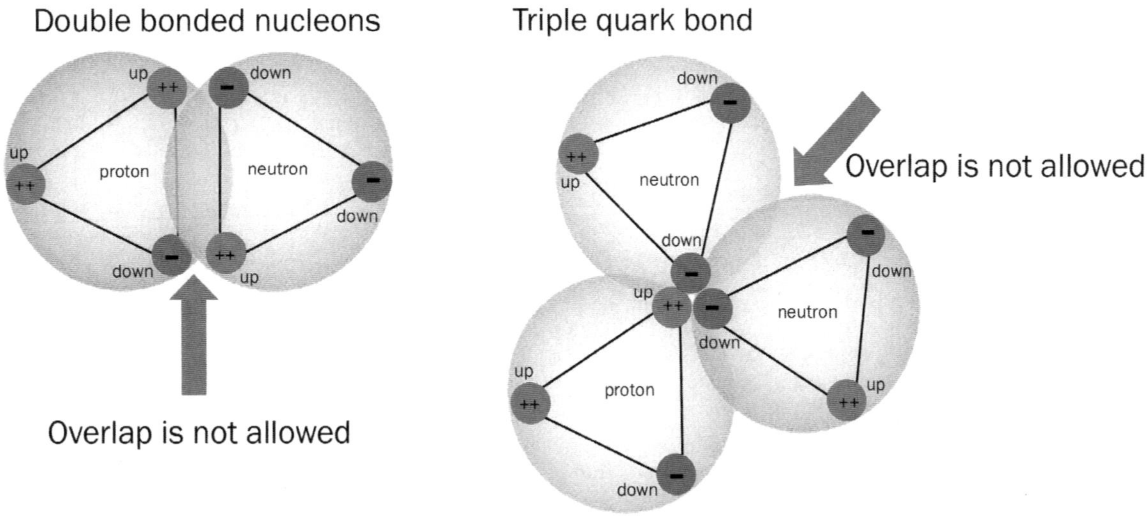


**Figure 4.** Left is an illustration of a proton double-bonded to a neutron. Right, is an illustration of a triple-quark bond.

each other, is not allowed. Similarly a triple-quark bond, defined as three quarks attempting to bond together, is not allowed. These conditions are illustrated in Fig. 4.

Double-bonded nucleons and triple-quark bonds are not allowed by quantum physics due to the overlapping that would result. As discussed above, it is assumed that the proton and neutron are disk-shaped, with a finite thickness that prevents the double-bonded quarks and the stacking one on top of another.

## 2.10. Simplified Representations of Protons and Neutrons

Given the considerations above, a representation of an atomic nucleus can be made for its lowest energy configuration. These representations of the atomic nuclei are not intended to be detailed physically accurate representations, but rather short-hand simplified graphics. In this first representation, the simplest atomic nuclei are shown, the individual proton and neutron. Fig. 5 illustrates four representations of the proton and neutron. The leftmost graphic shows the disk shape of the proton and neutron, with the symbols ++ and − for the up and down quarks. The up quarks are red and the down quarks are blue. (These colors are not related in any way to the colors of the Quantum Chromodynamic Force Model.) The next representation is more simplified, showing only the triangle for the nucleons, with the quarks as red and blue dots. (The dot sizes are not meant to be the actual relative quark size; rather the dots are sized to be easily viewed.) The next representation is even more simplified red-and-blue diagram of the proton and neutron, with an implied quark at each corner. The rightmost illustration shows a black and white short-hand simplified representation.

For every atomic nucleus in this paper, the position of each quark is defined in a matrix with an xyz spatial coordinate and electric charge value of either $-1/3$ or $+2/3$ of an elementary charge. The value and vector direction of the magnetic moment are also included in the matrix for every quark. This matrix file is used to calculate electromagnetic energy and the kinetic energy of the quantum angular momentum, using the equations above. The result is then used to calculate the nuclear binding energy.

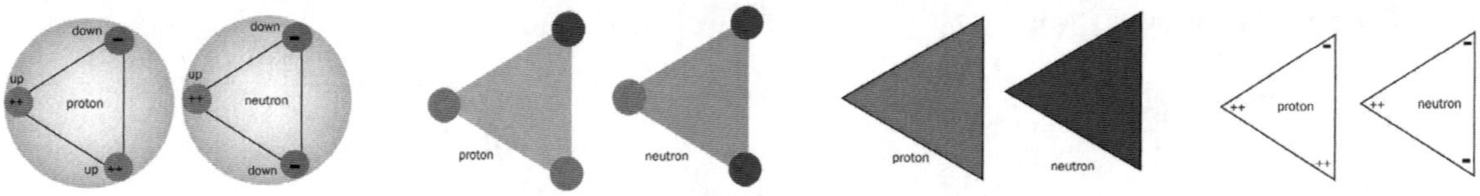


**Figure 5.** Four short-hand simplified representations of protons and neutrons.

### 2.11. Summary of the Electromagnetic Force within the Atomic Nucleus

The following is a review of the basic concepts for the Electromagnetic Model of the Nuclear Force.

- The Electromagnetic Force is valid inside the atomic nucleus.
- The electric charge and the magnetic moment of the nucleons are contained within the quarks.
- There are three possible bonds per nucleon, one for each of the three valence quarks. After forming three inter-nucleon quark-to-quark bonds, the nucleon cannot bond to a fourth nucleon.
- A pair of quarks, one from each of the two different nucleons, is needed for one bond.

All of the above considerations are used in this Electromagnetic Model, along with the following quantum considerations:

- The kinetic energy of the quantum angular momentum of the nucleus is properly taken into account.
- There is a minimum distance between two quarks of two different nucleons, consistent with the Pauli Exclusion principle and the hard core repulsion.
- The hard core repulsion of the nucleons prevents a nucleon from bonding more than once to another same nucleon. Any nucleon that attempts to bond twice to another same nucleon is involved in a double-nucleon bond. A double-nucleon bond is not allowed.
- The hard core repulsion of the nucleons prevents a quark from bonding to more than one other internucleon quark. Any quark that attempts to bond to two other internucleon quarks is involved in a triple-quark bond. A triple-quark bond is not allowed.
- Inside the protons and neutrons, the quarks have a spatial quantum probability distribution determined by the Expectation Value of their time averaged probability function. (An equilateral triangle is assumed for simplicity.)

## 3. The Quantum Angular Momentum Energy

The calculations for the change in the angular momentum of the configuration are also properly considered in this model. The change in the angular momentum energy is due to the overall change in the quantum angular momentum of the atomic nucleus as compared to the component parts.

The energy for the quantum angular momentum [32], [33] of spinning nucleus is shown in Eq. (9):

$$U_{angular\ momentum} = \frac{\hbar^2}{2I} J\,(J+1) \tag{9}$$

where $J$ is the angular momentum quantum number and $I$ is the total moment of inertia for the object. The quantity $\mathrm{U}_{\mathrm{angular\ momentum}}$ represents the total energy due to the total quantum angular momentum of the nucleus. The angular momentum quantum number $J$ is the number found in the nuclear tables for nuclear spin, and it represents the total angular momentum of the atomic nucleus, both intrinsic and orbital, for all the particles within the atomic nucleus.

For a non-spherical object, the calculation of the moment of inertia $I$ is slightly complicated. Specifically, the moment of inertia must not be oversimplified by approximating its shape as a sphere. However, this is what previous calculations and models in the past have done. Previous calculations and models have erroneously assumed a spherical shape when determining the nuclear moment of inertia, simply to make the calculations easier. An overly simplified calculation of the moment of inertia for an atomic nucleus causes errors in the angular momentum energy, and these errors have resulted in conundrums and erroneous conclusions, specifically with regard to the collective motion of the nucleus. By using a more accurate calculation for the moment of inertia, much of the previous historical confusion regarding collective motion can be explained and better understood.

A detailed explanation to show the calculations of the quantum angular momentum is done in Part I of this series [1], and it will not be repeated here. Calculating the correct quantum angular momentum of a non-spherical object is more complicated for a non-spherical object. In order to determine the proper moment of inertia, the calculation must include the position and mass of each proton and neutron, the center of mass for the configuration, and the axis of the spin. This calculation requires a computer, especially for the larger atomic nuclei. However, the effect of the change in the binding energy due to the quantum angular momentum is more pronounced for the smaller atomic nuclei, since it is a much larger percentage of the overall energy.

## 4. Brief Review of the Cluster Model and Nuclear Bonds

### 4.1. Introduction

Recent research regarding the clustering model of nuclear physics has shown that clustering-type structures do indeed exist experimentally within atomic nuclei [34], [35], [36], confirming the concept of structure inside an atomic nucleus. Clustering is observed as general phenomena at high excitation energies in light alpha-like atomic nuclei. Also, clustering is a general feature not only observed in light atomic nuclei, but also in less common systems, such as $^{11}$Li and $^{14}$Be. The Alpha Cluster Model has recently been extended beyond the alpha-only atomic nuclei, with much study and research in the field of cluster structure and of "nuclear molecules". These nuclear molecules can be thought of as being built from building blocks, called segments, linked together to form the nuclear molecules. These nuclear molecules are illustrated graphically in Cluster Model diagrams [37]. This recent research in the clustering structure of the atomic nucleus further corroborates the experimentally-observed collective motion of atomic nuclei, strongly indicating that protons and neutrons do not move independently inside an atomic nucleus. A sample of Cluster Model diagrams are shown in Fig. 6, to familiarize the reader with this type of diagram for nuclear structure.

As can be seen from the diagrams, the nuclear segments form chain-like configurations, to create the atomic nucleus. Again, this long chain-like shape is verified both experimentally and theoretically within the Cluster Model. (It is possible that this chain-like shape curls up to form a more compact overall shape, like a curve or a helix.) The Cluster Model claims that protons and neutrons cluster into various types of segments. Alpha segments predominate as the most common type of segment. These segments then link together to form a chain-like structure.

The Electromagnetic Model also claims that protons and neutrons cluster into segments; there are, however, a few differences. The Electromagnetic Model takes into consideration the electric dipole moments of the clusters, and how the segments bind to each other. Also, there are more types of segments in the Electromagnetic model. However, the most important difference is that the Electromagnetic Model explains why clustering occurs and why the alpha segment is predominant.

### 4.2. Explaining Why Clustering Occurs

Clustering is caused by the Electromagnetic Force putting the atomic nucleus in its lowest energy configuration. The lowest energy configuration is one that:

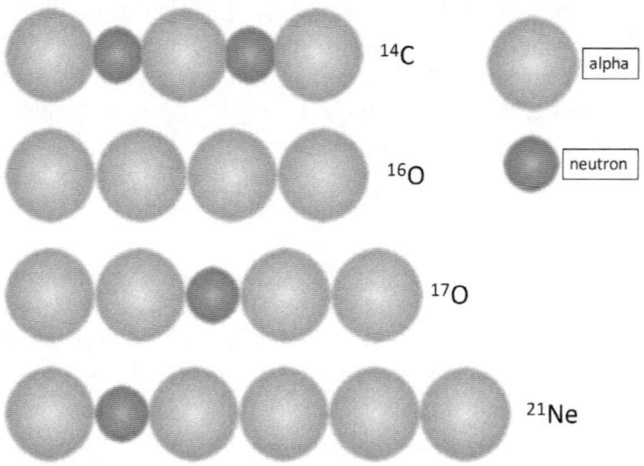


**Figure 6.** Cluster Model diagrams of $^{14}C$, $^{16}O$, $^{17}O$, and $^{21}Ne$.

- maximizes the number of quark-to-quark bonds. The Electromagnetic Forces within the nucleus will attempt to make as many bonds as possible and allowable.
- spreads out the net positive charge as far as possible, while still maintaining the bonds. This will reduce the Coulomb energy.
- situates any negative charge, such as an unbonded down quark, in a position what would otherwise be the highest concentration of positive charge. Again, this will reduce the Coulomb energy.

When these three factors are taken into account, the resulting nuclear shape is a chain-like structure of nuclear segments. This is because a chain-like structure utilizes the maximum number of allowed quark-to-quark bonds, and also allows the charge to be spread out as much as possible while maintaining the bonds. The negative charges can be inserted into the chain-like structure where they are most needed to mitigate and spread out the net positive charge. On the other hand, a spherical shape violates all three of these electromagnetic considerations. A spherical shape does not maximize the number of bonds that can be made, since there would be a significantly large number of unbonded quarks around the periphery of the sphere. A spherical shape does not spread out the charge, but rather it compresses it into the tightest possible package, which is the exact opposite of what is needed for the lowest energy. Also spherical configuration does not utilize the negative charge in such a way as to mitigate the strong electric field of the net positive charge.

### 4.3. Explaining Why Alpha Segments Predominate

The nuclear bond is a bond between nucleons. For the Electromagnetic Model, it is a bond between quarks from two different nucleons, an internucleon quark-to-quark bond. This bond can be conceptualized as an energy well formed between the up quark and the down quark of two different nucleons. Alternatively, it can be conceptualized as an attractive electromagnetic force between the up quark in one nucleon and the down quark in another nucleon. Regardless of how it is conceptualized, this bond lowers the overall energy of the atomic nucleus, similar to the way that chemical bonds between the atoms lower the overall energy of a molecule. The number of times that an atom can bond to another atom is limited by the number of valence electrons available for bonding. Similarly, the number of bonds for each nucleon is limited by the number of valence quarks available for bonding. Thus with only three valance

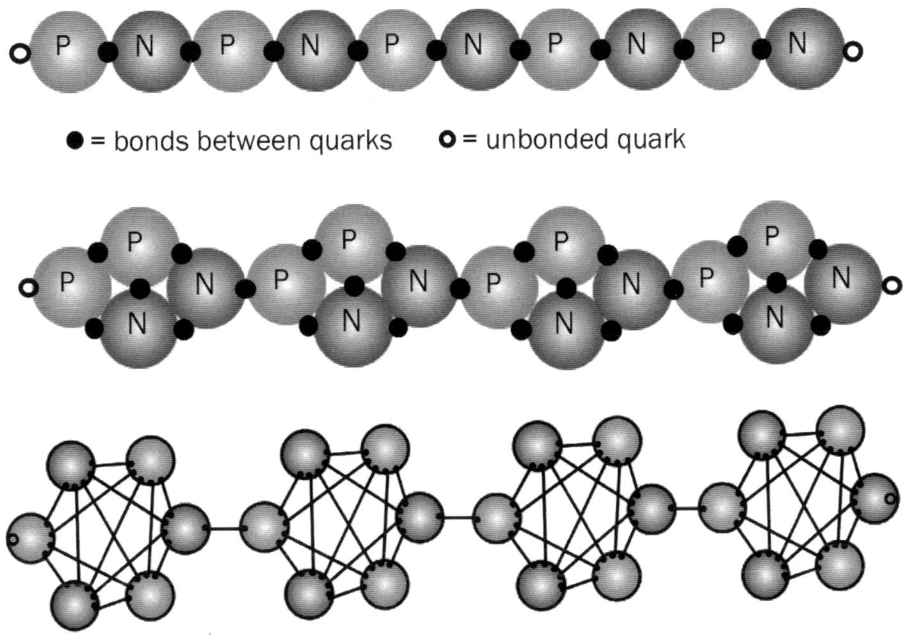


**Figure 7.** a) Hypothetical nuclear bonding if there were only two bonds for each proton and neutron. b) Nuclear bonding with three bonds for each proton and neutron. c) Hypothetical nuclear bonding if there were five bonds for each proton and neutron.

quarks, each nucleon can only bond three times. Two quarks, one from each of the two nucleons, are needed to form one bond. Thus, if every nucleon were fully bonded in an atomic nucleus, there would be three bonds for every two nucleons.

The alpha segment is predominant because there are only three bonds for each nucleon. The Electromagnetic Model claims that each nucleon can bond a maximum of three times, and this causes the nucleons to cluster into alpha segments. If there were only two bonds within each nucleon, they would simply string together as shown in the top illustration of Fig. 7a. As can be seen in that illustration, each hypothetical proton and neutron can bond only twice. A structure similar to a string of beads results. With three bonds allowed for each nucleon, the protons and neutrons form alphas segments, linked together, as shown in middle illustration of Fig. 7b. This results in clusters of alpha segments, linked together in a chain-like structure. If there were more bonds within each proton or neutron, the predominant cluster would be something larger than the alpha segment. For example, with five possible bonds per nucleon, the predominant structure inside such a hypothetical atomic nucleus would be $^6$Li, as seen in the bottom illustration of Fig. 7c.

Experimentally, however, it is known that other unusual segments such as a $^6$Li segment, are not indicated as being the predominant structure of atomic nuclei. This can be seen in the scallop-like pattern of the curve of binding energy per nucleon vs. A. This curve shows that there is a peak for every atomic nuclide made only alpha segments, this peak being more apparent for the smaller atomic nuclei. Also, radioactive alpha decay implies that the alpha segments are pre-formed within the atomic nucleus. Also, the binding energy and separation energies of $^4$He is the largest for any stable atomic nucleus. This implies that the alpha particle is energetically unique. Other segments, like $^6$Li, do not appear as common types of particle decay; however, alpha particles do. Thus, the predominant structure is the alpha segment, with two protons and two neutrons. This experimental evidence strongly implies that there are three bonds within each proton and neutron, exactly as might be expected with three quarks.

If there were more than three bonds for each nucleon, the segments would be larger. For example, clusters of $^{6}Li$ would form when there are five bonds per nucleon. Also note that the segments form a chain; they would not form a sphere.

## 5. The Nuclear Electric Quadrupole Moment and the Misconception of Spherical Atomic Nuclei

### 5.1. Understanding Quadrupole Moments

The quantum electric quadrupole moment of a distribution of charge is a frequently misunderstood topic that deserves review before proceeding.

First, the electric monopole moment and the electric dipole moment will be described. An electric monopole moment is the net charge of an object, assuming the object has a spherical distribution of charge with radial symmetry. Any distribution of charge with spherical symmetry will have only an electric monopole moment. If the distribution has a polarity, with positive charges on one side and negative charges on the other, then the object will also have an electric dipole moment. For example, consider a charge distribution along a line, with a total net charge, but with a positive charge on one side and a negative charge on the other; this type of distribution would have an electric dipole moment.

If the distribution of charge has an ellipsoidal shape, instead of being perfectly spherical, then the object will have an electric quadrupole moment [38], [39]. The quadrupole moment is related to the eccentricity of the charge distribution, and it can be either prolate (positive, cigar shaped) or oblate (negative, disk shaped). The quadrupole moment of a non-spinning object is referred to as its "intrinsic quadrupole moment", $Q_0$, and this is different from the "measured quadrupole moment" $Q_m$, if the object is spinning. If a prolate object with an intrinsic quadrupole moment $Q_0$ is spinning, then depending on the axis of spin, the measured quadrupole moment may be smaller than the intrinsic quadrupole moment, and it can even be negative. However, it can never be larger in absolute value. In other words, for a spinning prolate object the measured quadrupole moment can be smaller, in absolute value, than the intrinsic quadrupole moment, and it can be either positive or negative. This is shown in Fig. 8.

If an *oblate* object, with an intrinsic quadrupole moment $-Q_0$, is spinning, then depending on the axis of spin, the measured quadrupole moment can be smaller in absolute value, all the way to zero. It cannot be positive, and it cannot be larger in absolute value. This is shown in Fig. 9.

When comparing the measured quadrupole moment with the intrinsic quadrupole moments, another factor to consider is the uniformity of the alignment of the atomic nuclei. For a given sample of atomic nuclei, if the spin axis of the atomic nuclei are not precisely aligned to each other, then a smaller measured value will occur as a result. To summarize, the absolute value of the measured quadrupole moment of a spinning object can never be larger than the absolute value of the intrinsic quadrupole moment.

Taken into the quantum realm, similar principles apply. The measured quadrupole moment, $Q_{\text{measured}}$, is related to the intrinsic quadrupole moment, $Q_{\text{intrinsic}}$, as shown in Eq. (10).

$$Q_{measured} = Q_{instrinsic}\left(\frac{3K^2}{(J+1)(2J+3)}\right) \tag{10}$$

where: K is the projection of the spin axis onto the symmetry axis, and J is the total nuclear angular momentum of the atomic nucleus.

For quantum quadrupole moments, similar to regular quadrupole moments, the absolute value of the measured quadrupole moment will always be less than or equal to the intrinsic quadrupole moment. Therefore, if a nuclear model predicts a small inherent quadrupole moment, smaller than the measured quadrupole moment, then this model

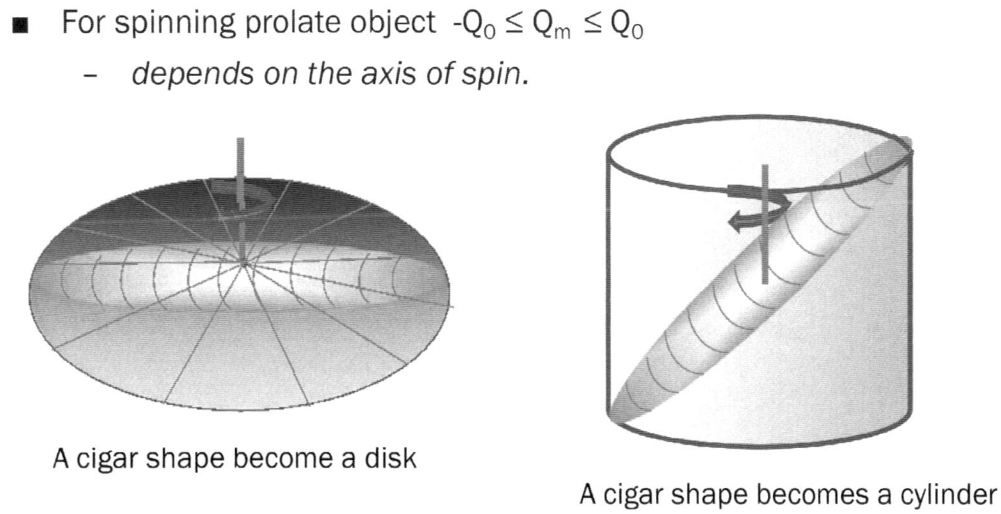


**Figure 8.** A spinning prolate object can have a measure quadrupole moment, $Q_m$, such that $-Q_0 \leq Q_m \leq Q_0$.

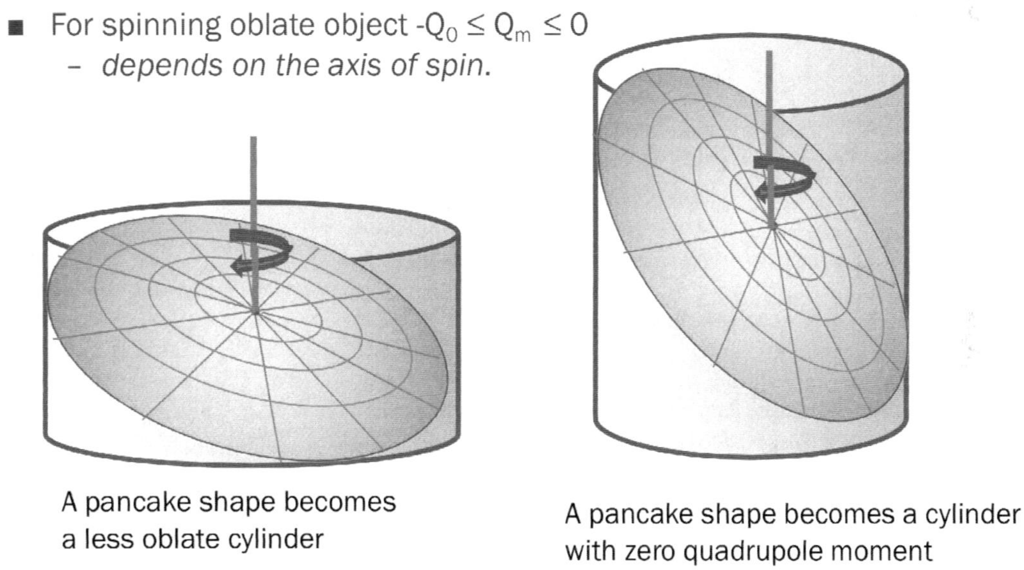


**Figure 9.** A spinning oblate object can have a measure quadrupole moment, $Q_m$, such that $-Q_0 \leq Q_m \leq 0$.

is inherently flawed. Conversely, if a nuclear model predicts a large inherent quadrupole moment, one that is larger than the measured quadrupole moment, then this inconsistency can easily be explained as being due to the angle of the spin axis or due to the imprecise alignment of the atomic nuclei.

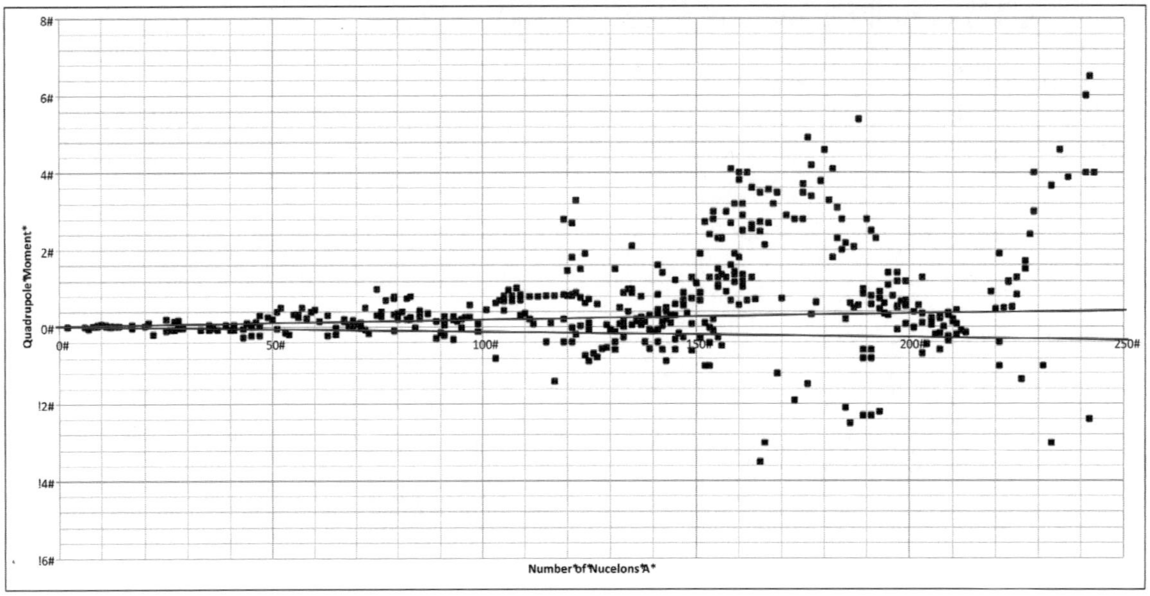


**Figure 10a.** Experimental quadrupole moment. The predicted deformation parameter of the Shell Model is shown by the blue lines.

## 5.2. The Misconception of Spherical Nuclei

With the exception of analyzing quadrupole moments, when researchers interpret the data from an experimentally probed atomic nucleus, the shape of the atomic nucleus is generally assumed to be spherical. Then, for this pre-assumed spherical shape, the best value of its radius R is forced-fit to the data, without any attempt to examine different possible shapes [40], [41]. For example, the interpretation of the electron scattering data only considers the electric monopole moment—in other words, the spherical shape. The interpretation of this data is then normalized over the charge density to force-fit the charge inside a sphere. A Gaussian curve is assumed as the skin of the atomic nucleus, and this process is applied without justification or even considering an alternative shape. Thus, regardless of the actual shape, the forced-fit normalized data makes the atomic nucleus appear to be spherical in shape with a Gaussian skin. Even the so-called "model-independent" interpretations of scattering data still assume a spherical shape [42]. This assumption of a spherical atomic nucleus does not compare well to the experimental data regarding the nuclear shape.

Fig. 10a shows the quadrupole moment for atomic nuclei with spin. (All data for Fig. 10a are extracted from reference [43].) Fig. 10b shows the quadrupole moment for zero-spin even-even atomic nuclei. (All data for Fig. 10b are extracted from reference [44].) The blue lines represent the maximum value predicted by the Shell Model. As can be readily seen, atomic nuclei are not spherical in shape. Almost all atomic nuclei have a measured quadrupole moment and deformation parameter that are larger than predicted by the Shell Model. The distortion is not just in a few isolated regions of the nuclear table, but covers the entire range.

The large experimentally-measured quadrupole moments for the vast majority of atomic nuclei is in opposition to the concept of a spherical atomic nucleus. Conversely, the inherent quadrupole moments may be much larger than the measured quadrupole moments, without posing any theoretical problems, as explained previously in section 5.1.

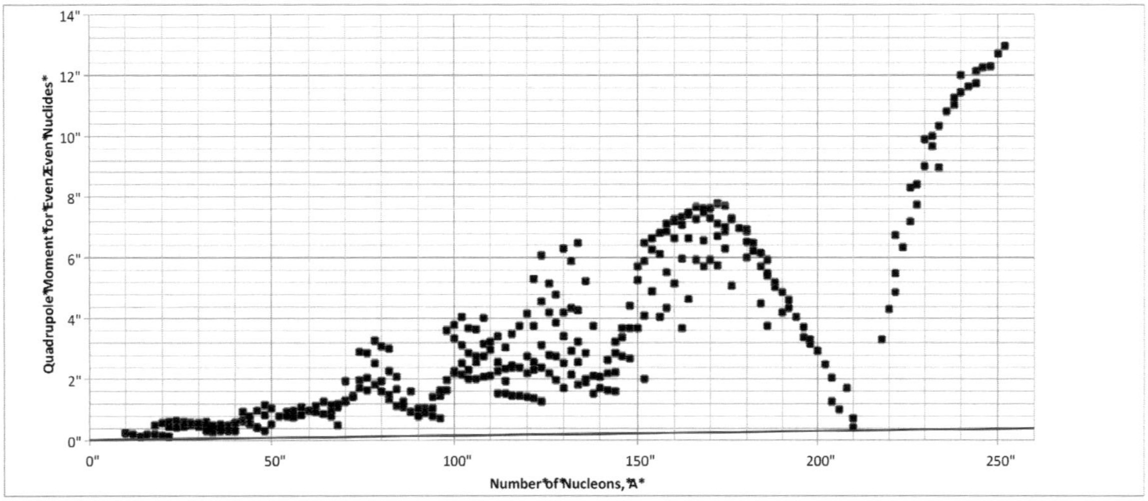


**Figure 10b.** Experimental quadrupole moment. The predicted deformation parameter of the Shell Model is shown by the blue line.

## 6. Summary of the Electromagnetic Considerations of the Atomic Nucleus

The following is a review of the basic concepts for the Electromagnetic Model.

(1) The Electromagnetic Force is valid inside the atomic nucleus.
(2) The electric charge and the magnetic moment of the protons and neutrons are contained within the quarks.
(3) The quarks inside of the protons and neutrons have a spatial distribution associated with their quantum probability and their resulting subsequent Expectation Value. (An equilateral triangle is assumed for simplicity.)
(4) There is a minimum distance between two quarks of two different nucleons, consistent with the Pauli Exclusion principle and the hard core repulsion.
(5) A pair of nucleons that are doubly-bonded to each other is called a double-bonded nucleon pair. The hard core repulsion of the nucleons prevents this situation. A doubly-bonded nucleon pair is not allowed.
(6) A triple-quark bond is defined as three internucleon quarks attempting to bond together. The hard core repulsion of the nucleons prevents a triple-quark bond, and it is not allowed.
(7) There are three possible bonds within each nucleon, one for each valence quark. After three bonds, a nucleon cannot bond to a fourth nucleon.
(8) A pair of quarks, one from two different nucleons, is needed for one bond.
(9) The protons and neutrons are assumed to be disk shaped, with a finite thickness.
(10) The kinetic energy of the quantum angular momentum of the atomic nucleus is properly taken into account.
(11) The lowest energy configuration, which only includes the Electromagnetic energies and the kinetic energy of the angular momentum, is assumed for the ground state. No other energies are included in this calculation.

## 7. The Configurations of the Lowest Energy States for Stable Atomic Nuclei

### 7.1. The Determination of the Lowest Energy Configurations

Given this understanding of the electromagnetic energy, and assuming the configuration will naturally fall into the lowest energy state, some analytic predictions about the shape of the segments can easily be made. The lowest energy

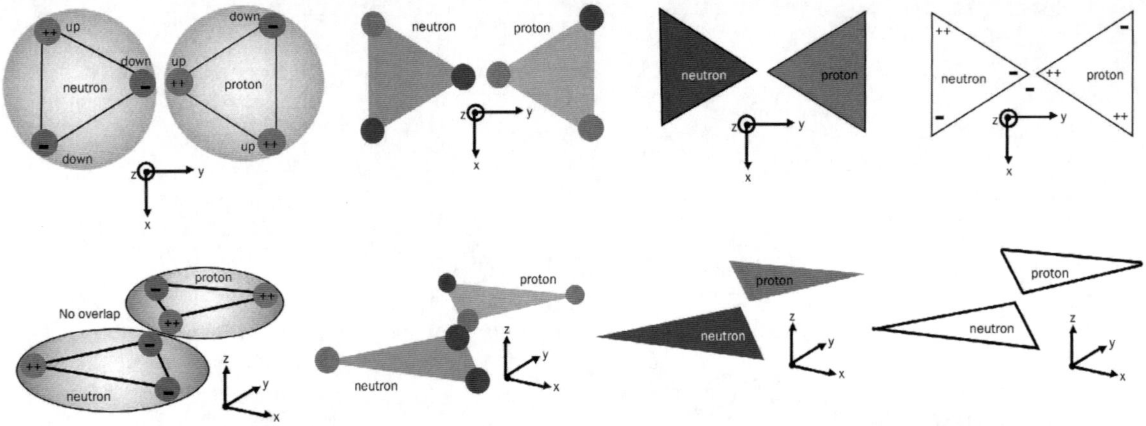


**Figure 11.** Eight short-hand simplified representations of $^2$H, shown from two different viewpoints.

configuration will be with the most number of bonds and with the positive charge spread out as far as possible, while still maintaining the bonds. Any negative charge will be situated at a place that would otherwise have the highest concentration of positive charge.

The lowest energy configurations of the atomic nuclei are determined by using the laws and equations of electromagnetics applied to the quarks. Thermodynamics insists that the stable configuration of the atomic nuclei are in their lowest energy states, and using these lowest energy states, a better understanding of nuclear behavior can be achieved. Any accurate model of the Nuclear Force must consider the lowest energy configuration of that nuclear structure. For a genuine and credible understanding of any structured atomic nucleus, the lowest energy configuration must first be determined. Since the lowest energy configuration is entirely dependent upon the forces and energies within the atomic nucleus, the equations for these energies and forces must be known. Any type of nuclear structure cannot be properly determined unless there is a fundamental mathematical formulation of the forces and energies that are internal to the atomic nucleus, between the nucleons. In this paper the lowest energy configuration of the structured atomic nucleus is determined by using the electromagnetic equations as applied to the quarks, and also to the angular momentum equations as applied to the nucleons.

As previously mentioned, for every atomic nucleus in this paper, the position of each quark is defined in a matrix with xyz spatial coordinates and an electric charge value of either $-1/3$ or $+2/3$ of an elementary charge. The value of the magnetic moment, for either the up or down quark, and the three dimensional vector direction of each magnetic moment for each quark are also included in the matrix. Based on the physical constraints of the configuration, a determination is made as to whether the magnetic bond is stacked, angled, or side-by-side.

The various possible lowest energy configurations are tested to determine which configuration is the lowest state. As a result, every atomic nucleus in this paper is in its lowest energy state, in accordance with the rules for electromagnetic energies and spin energies. Also, the configurations are in accordance with the quantum considerations that double-nucleon bonds or triple-quark bonds are not allowed.

### 7.2. Simplified Representations of $^2$H, $^3$H, $^3$He and $^4$He

Shown in Fig. 11 are six different representations of $^2$H. As before, this is not meant to be a physically accurate representation, but rather a short-hand simplified graphic. Leftmost is $^2$H, showing the three-dimensional disk shape

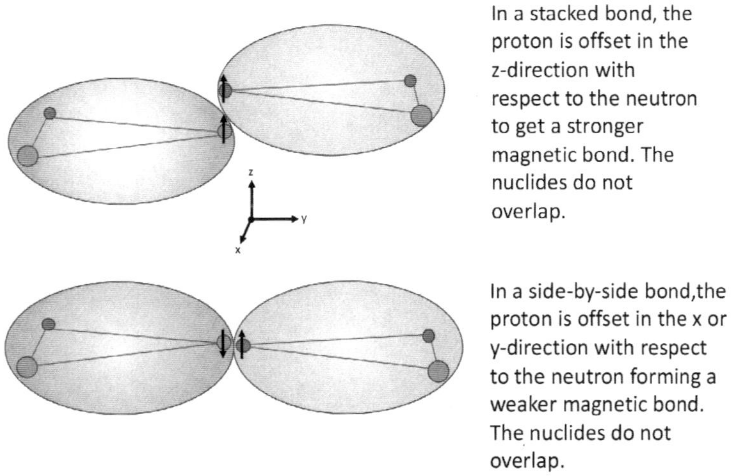


**Figure 12.** The stacked and side-by-side magnetic bonds between a proton and a neutron.

of the proton and neutron, with the ++ and – charges of the quarks inside, and with the up quarks as red and the down quarks as blue. Below it is the same configuration, but with a different viewpoint. To the right of that, the next representation is more simplified, showing the proton as a light red triangle and the neutron as a light blue triangle, and with the quarks as red and blue dots. In this second representation, the disk-like graphic has been removed in order to make this a more simplified graphic, but the disk-like shape is still implied. The third representation is an even more simplified red-and-blue diagram of the proton and neutron, with an implied quark at each corner. To the very right, is a simplified black and white representation. These simplified diagrams are very useful for representing the larger atomic nuclei.

Please note that these simplified graphic representations are not intended to be actual physical representations. Nor are they intended to show the relative size of the quarks. Rather, the dot size is chosen to be easily viewable. Also, the colors are not related to the colors in the Quantum Chromodynamic Force Model.

There is an interesting quantum aspect about the configuration of $^2$H, which explains the very low binding energy of $^2$H. It is known experimentally that the spin of $^2$H is 1, rather than 0. Thus, the spins of the neutron and proton add rather than cancel. In other words, the spins of the proton and neutron are parallel rather than anti-parallel. This parallel alignment of nucleon spins is also in accordance with the Pauli Exclusion Principal of quantum physics. This parallel spin alignment can only occur if the magnetic bond between the corresponding up and down quarks are a side-by-side bond, rather than a stacked bond. The side-by-side bond, however, is not as strong magnetically as a stacked bond. This is the reason why the bonding energy of $^2$H is so small; it is because the magnetic bond must be a weaker side-by-side bond in order to accommodate the quantum rules of spin alignment in the Pauli Exclusion Principal.

The stacked and side-by-side magnetic bonds are shown with more detail and from another viewpoint in Fig. 12. Note that there is no overlap of the nucleons. Although the stacked magnetic bond is not present in deuteron, is it present in most other configurations.

Shown in Fig. 13 is the configuration for $^3$H. Two bonds are stacked bonds, with the up and down quarks offset in the z dimension to prevent an overlap; the third bond is side-by-side. (In the more simplified red-and-blue diagrams, the reader is to understand that there is always a bond between an up quark and a down quark at the connection point where the triangles meet.) The last diagram to the right shows the disk-shaped proton and neutron, in a different

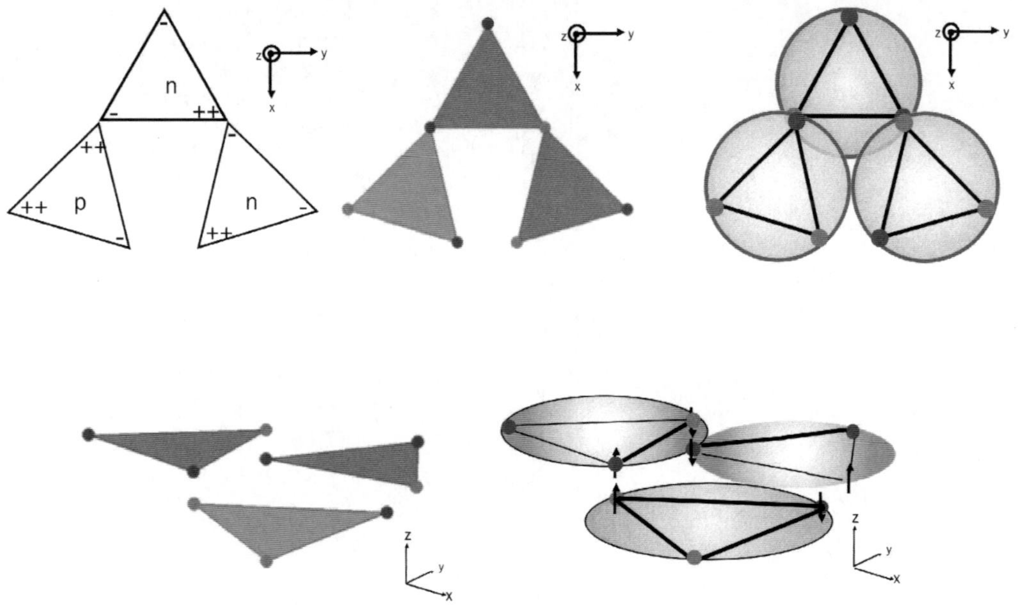

**Figure 13.** Short-hand simplified graphics of $^3$H, showing five different representations.

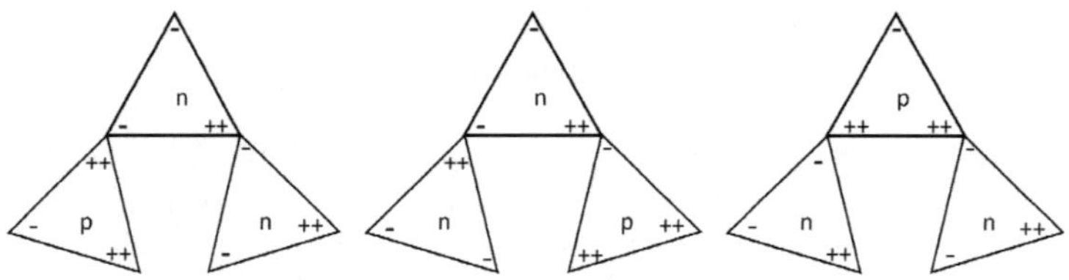

**Figure 14.** Alternative configurations of $^3$H, showing different quark positions within the bonds.

viewpoint, to illustrate the side-by-side magnetic bond and the two stacked bonds. As before, there is no physical overlap of the nucleons. (Even though the two lower-most up and down quarks in the diagrams have a further separation due to the disk shape of the nucleons preventing a closer connection, these two quarks are still considered to be bonded.)

Shown in Fig. 14 are three variations of the positions for the up and down quarks for $^3$H. By comparing them in detail, the differences in the exact placement of the up and down quarks can be seen. However, electromagnetically, these are practically the same. For the remaining atomic nuclei in this paper, these numerous alternative configurations will not be shown, but the reader should be aware that they exist.

The configuration for $^3$He is similar to $^3$H, but with two protons and one neutron. Fig. 15 shows four different representations of $^3$He. As mentioned previously, it is assumed that double-bounded nucleons and triple-bonded quarks are not allowed. It is assumed that the quarks are in an equilateral triangle within the nucleon, and that the

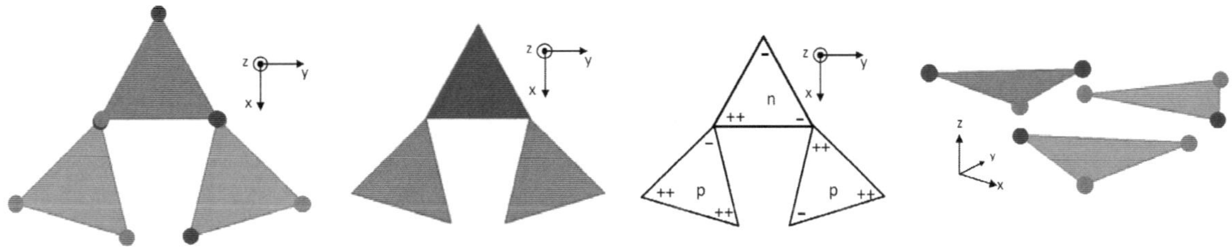


**Figure 15.** Short-hand simplified graphics of $^3$He, shown in four different graphic representations.

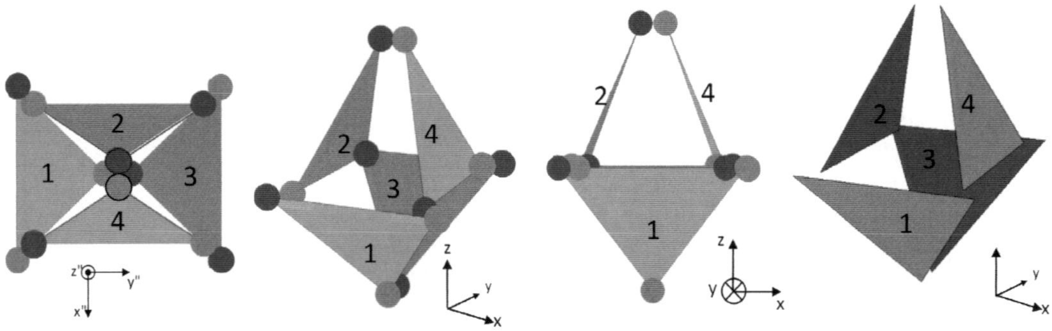


**Figure 16.** Short-hand simplified graphics of $^4$He, shown in four different representations.

quarks are within the oblate ellipsoidal shape of the nucleon. Plus, there is a finite thickness that prevents the stacking of the nucleons on top of one another. There is no physical overlap of the nucleons.

The configuration for $^4$He, shown in Fig. 16, is unique because it is the only atomic nucleus with all of its quarks bonded. This characteristic is what gives $^4$He such high binding energy and high stability; it is because it has the highest number of bonds per mass number A. This is also why the thermal neutron cross section for $^4$He is zero; a thermal neutron cannot form a triple-quark bond with a quark that is already bonded in the configuration.

Another interesting point about $^4$He is that it has 6 bonds, as compared to $^3$He with only three bonds. This explains why there is such a large jump in the binding energy, a difference of 20.58 MeV from $^3$He to $^4$He. The large jump is because three more bonds are added in the configuration of $^4$He compared to $^3$He. This also explains why the neutron and proton separation energies, $S_n$ and $S_p$, are so exceptionally high for $^4$He; it is because to remove either a proton or a neutron, three bonds must be broken. The $^4$He is not a segment in any nuclear configuration. Rather, it is the alpha segment that predominates in the nuclear molecules.

### 7.3. Simplified Representations of the Alpha Segment, H4 Segment, and Star Segment

A segment similar to $^4$He is called the alpha segment. The alpha segment is also made up of two neutrons and two protons, but unlike $^4$He, it can bond to other segments. The alpha segment is the predominant segment in other atomic nuclei. This segment is illustrated in Fig. 17 in four different representations.

Another segment similar to the alpha segment is called an H4 segment, made up of three neutrons and one proton. This is illustrated in Fig. 18 in four different representations.

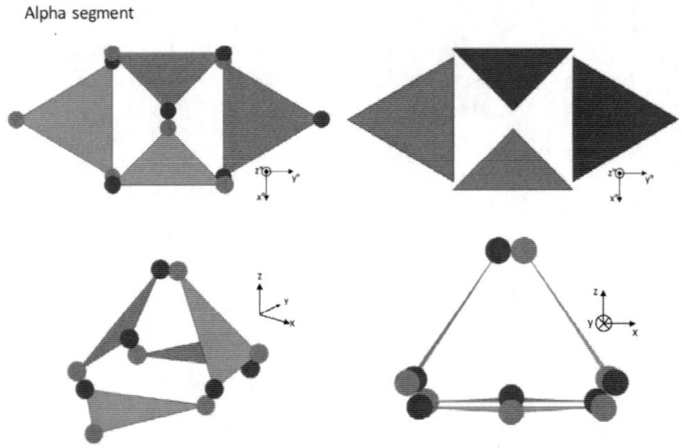


**Figure 17.** Shorthand symbolic graphics of an alpha segment, shown in four different representations.

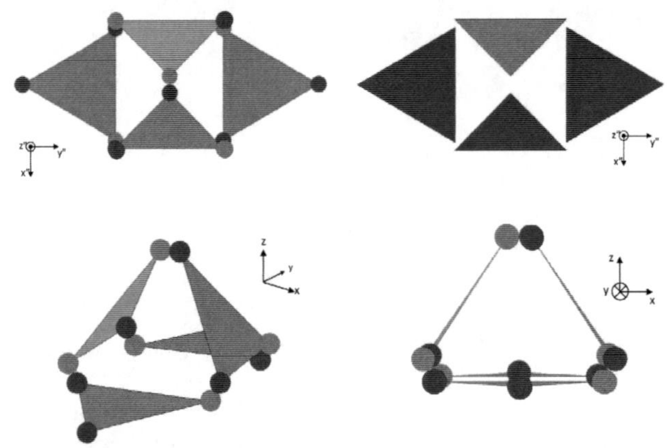

**Figure 18.** Short-hand simplified graphics of Hydrogen, $^4$H, shown in four different representations.

Another segment similar to an alpha segment is called the star segment. The star segment is similar to an alpha, but with one less bond. This is shown in four different representations in Fig. 19.

The star segment is similar to an alpha segment in that it has two protons and two neutrons. When this segment is part of a larger configuration, the two up quarks will bond to the other segments, and the two down quarks will remain unbonded. The star segment is required in order to remove the electric dipole moment from the configuration. If the star segment were not in the configuration, an electric dipole moment would exist with a line of charge having a positive unbonded quark on one end and a negative unbonded quark on the other end. By including the star segment, there are two down quarks in the middle that mitigate the high electric field there. Furthermore, the two unbonded up quarks both are on the ends, where their high positive electric charge density is more readily dissipated on the end. The star segment is thus able to reduce or remove the dipole moment associated with the y-dimension. Also, the star segment can remove or reduce the dipole moments associated with the x and z dimensions as well, by rotating the

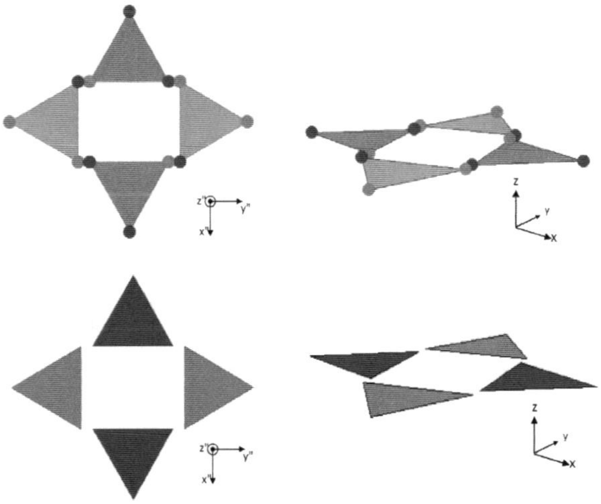


**Figure 19.** Short-hand simplified graphics of a star segment, shown in four different representations.

two unbonded down quarks as needed. (The two neutrons of the star segment, with their unbonded down quarks, may rotate in three dimensions, but they are drawn always as being flat, for simplicity.)

A reminder should be made here that the drawings are not meant to be physically realistic in that the nucleons are not really red or blue, or flat triangles. Rather, the drawings for these nucleons are shorthand symbolic graphical representations to show the quarks, the bonds, and the segments, and to also show how these segments form into chains-like nuclear molecules.

With regard to the cluster formation that the nucleons generally follow when they form into segments, here are some simple considerations to bear in mind.

- When a bond is formed the net charge of the bond is $+1/3$ of an elementary charge. This net positive charge is present at every bond. Similarly, there is a net magnetic moment which is the vector sum of the two magnetic moments of the two quarks involved in the bond.
- The protons and neutrons cluster into segments, as described in the Electromagnetic Model. This clustering is due to the Electromagnetic Force situating the atomic nucleus into its lowest energy configuration.
- Due to the consideration that each proton or neutron can only bond a maximum of three times, the protons and neutrons will tend to cluster into alpha segments, with each segment consisting of two protons and two neutrons. The alpha segment is the predominant type of segment within atomic nuclei.
- Any additional protons and neutrons that are not part of an alpha segment will form other types of segments.
- A tritium segment consists of one proton and two neutrons.
- An He3 segment consists of two protons and one neutron.
- A star segment is an alpha segment with one less bond. When part of a configuration, it has two unbonded down quarks and it bonds to the rest of the segments with two up quarks.
- An H4 segment is made of one proton and three neutrons, and it is similar in shape to the alpha segment.
- A single neutron that is not part of any other segment is called a single-neutron segment.
- A single proton that is not part of any other segment is called is called a single-proton segment.
- In the lowest energy state, each segment can bond to either one or two other segments to form a chain-like structure. There is no branching.

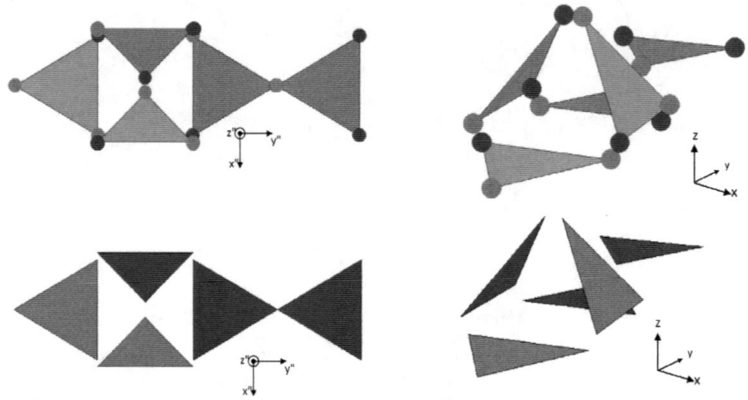


**Figure 20.** Short-hand simplified graphics of Helium $^5$He, shown in four different representations.

### 7.4. Simplified Representations of $^5$He, $^6$Li, $^7$Li, $^8$Be, $^9$Be, $^{10}$B, and $^{11}$B

The smaller stable atomic nuclei are $^6$Li, $^7$Li, $^9$Be, $^{10}$B, and $^{11}$B. (Although there are no stable isobars for $A = 5$ or $A = 8$, the figures for $^5$He and $^8$Be are also shown for consistency.) From $^{12}$C upwards, the stable atomic nuclei follow a relatively simple pattern. The atomic nuclei smaller than $^{12}$C cannot follow this pattern simply because they do not have enough protons and neutrons to do so. A discussion of the pattern for the stable atomic nuclei with $A \geq 12$ will be made after these the smaller atomic nuclei are first discussed.

The next atomic nucleus to discuss is $^5$He. The lowest energy configuration for Helium $^5$He is an alpha-neutron configuration. This is shown in Fig. 20 in four different representations.

The next stable atomic nucleus to discuss is $^6$Li. The configuration for $^6$Li is a combination of $^3$H and $^3$He, bonded together, as seen in Fig. 21. These are referred to as tritium segments and He3 segments, respectively. (Note that in the top view, both quarks of the stacked bonds are not readily seen, since they are stacked on top of each other, offset in the z-dimension. Even though the top viewpoint obscures the lower quark, the reader should know that both quarks are there, as can be easily seen in the angled viewpoint. There is no physical overlap of the nucleons.) The bond linking the two segments together is a stacked magnetic bond.

The next stable atomic nucleus is $^7$Li. The configuration for $^7$Li is a combination of a tritium segment, a neutron segment, and an He3 segment, bonded together in a tritium-neutron-He3 configuration, as seen in Fig. 22.

The next atomic nucleus to consider is Beryllium $^8$Be. There are no isobars with $A = 8$ that are stable, but $^8$Be is shown for consistency. The lowest energy configuration for Beryllium $^8$Be is a star-alpha configuration, shown in Fig. 23 in two representations.

The next stable atomic nucleus is Beryllium $^9$Be. The lowest energy configuration for $^9$Be is a triton segment, a triton segment, and an He3 segment, as shown in Fig. 24. This is one of the very few stable atomic nuclei that does not contain an alpha segment.

The next stable atomic nucleus is Boron, $^{10}$B. The lowest energy configuration for $^{10}$B is a triton-alpha-He3, shown in Fig. 25. Recall that the electrical charge is located in the quarks, rather than being homogenously distributed throughout the protons. Each bond has a net charge of $+1/3$ an elementary charge, each unbonded up quark has a

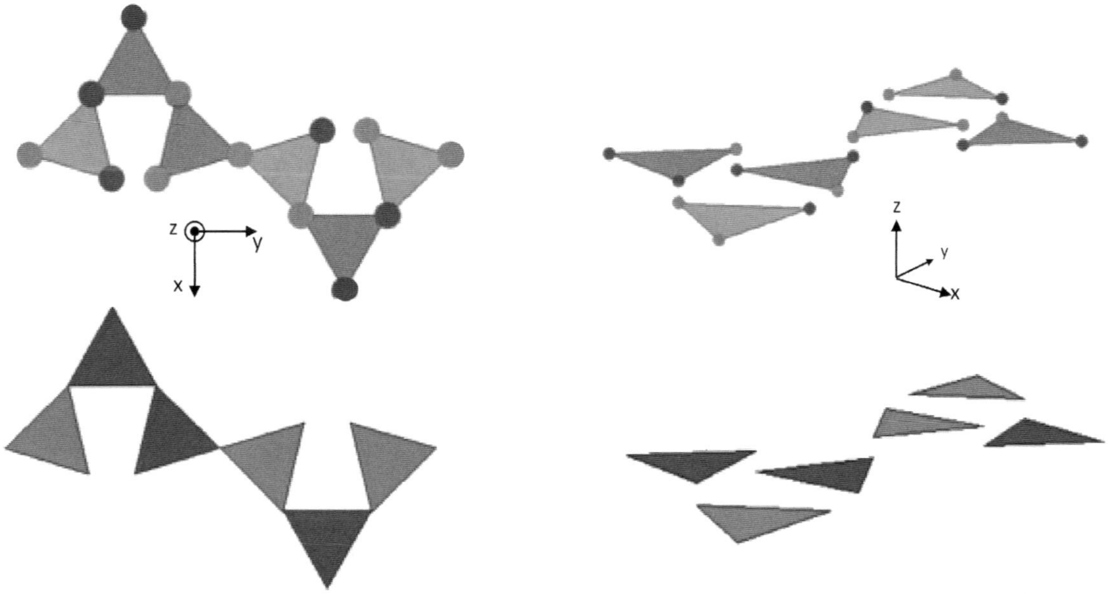


**Figure 21.** Short-hand simplified graphics of $^{6}$Li, shown in four different representations.

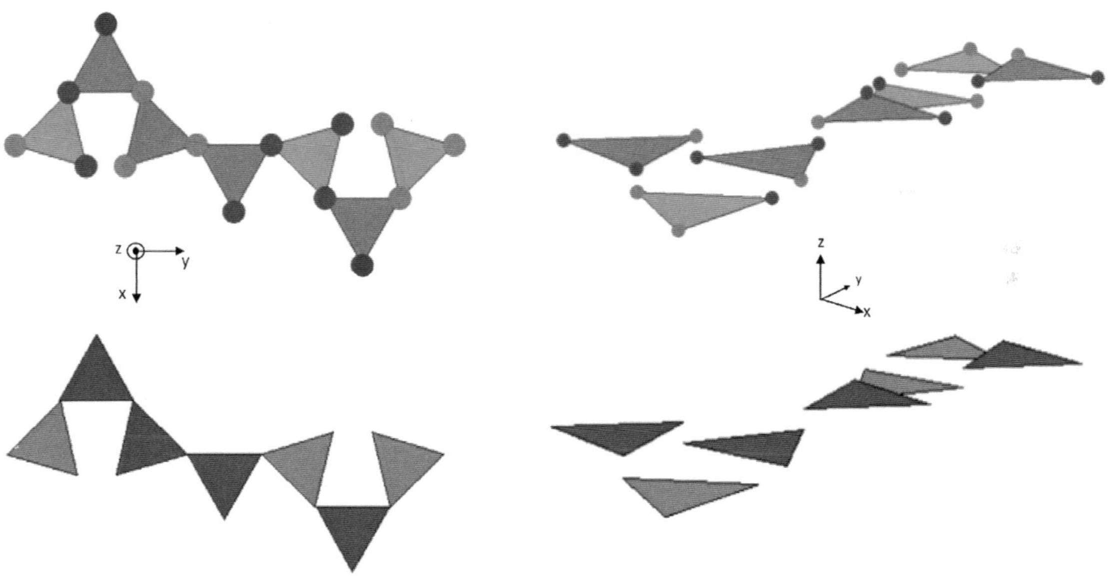


**Figure 22.** Short-hand simplified graphics of $^{7}$Li, shown in four different representations.

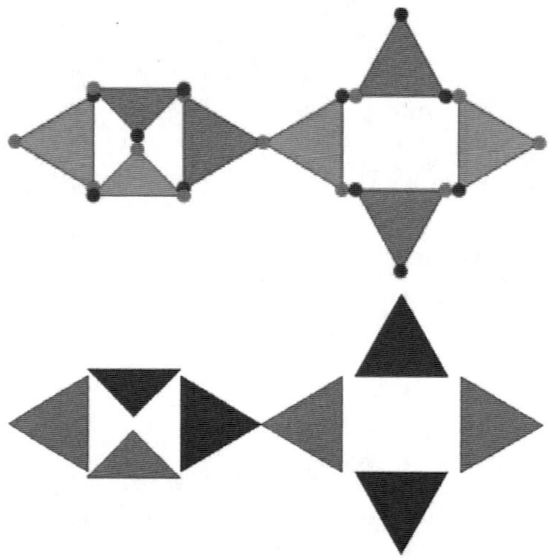

**Figure 23.** Short-hand simplified graphics of Beryllium $^{8}$Be, shown in two different representations.

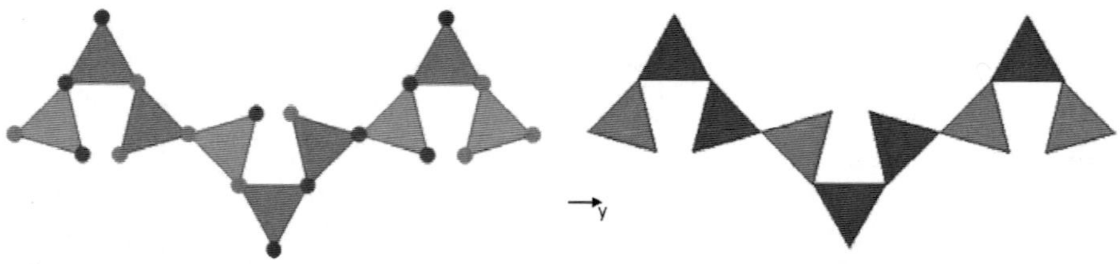


**Figure 24.** Short-hand simplified graphics of $^{9}$Be, shown in two different representations.

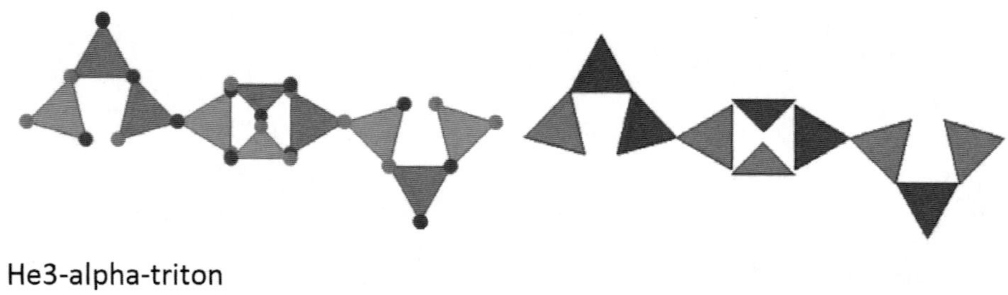


**Figure 25.** Short-hand simplified graphic representations of $^{10}$B.

charge of +2/3 an elementary charge, and each unbonded down quark has a charge of −1/3 an elementary charge. In Fig. 25, there is also a verbal representation, using the segment names, triton-alpha-He3.

The next stable atomic nucleus is Boron $^{11}$B. The lowest energy configuration for $^{11}$B is a tritium-star-alpha configuration, shown in Fig. 26, with two graphic representations and one verbal representation.

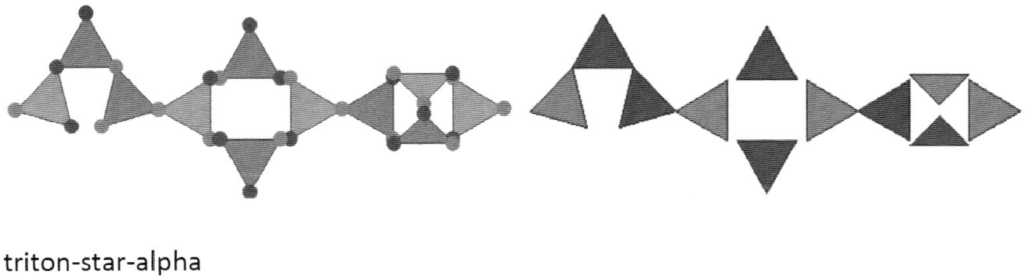


**Figure 26.** Short-hand simplified representations of $^{11}$B.

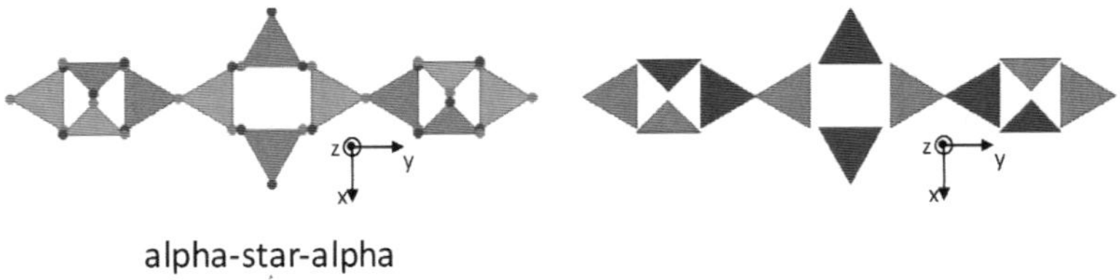


**Figure 27.** Short-hand simplified representations of $^{12}$C.

### 7.5. Basic Pattern for the Configurations of the Stable Atomic Nuclei with A$\geq$12

The Electromagnetic Model of the Nuclear Force has successfully modeled both stable and unstable atomic nuclei, all the way up to Californium $^{250}$Cf, using the electromagnetic equations. As previously mentioned, the stable atomic nuclei follow a relatively simple pattern, from $^{12}$C upwards. This basic pattern is now presented, describing the stable atomic nuclei from $^{12}$C and larger.

The configuration of $^{12}$C is shown in Fig. 27, in an alpha-star-alpha configuration.

This is the lowest energy state of $^{12}$C, with spin 0. On each end of $^{12}$C is an unbonded up quark, and in the middle, there are two unbonded down quarks. Other physical configurations are possible; however, if $^{12}$C is configured differently, it would either have fewer bonds, more Coulomb energy, or more kinetic spin energy. Therefore as a result, different configurations would be in a higher energy state.

For the larger atomic nuclei, the diagrams become too complex and awkward to show all the red and blue dots representing the quarks in the configuration. Thus, only the red-and-blue representation and the verbal representation will be shown for the atomic nuclei larger than $^{12}$C. (When looking at the illustrations, the reader is to understand that any unbonded quark on a neutron is a down quark and any unbonded quark on a proton is an up quark.)

Adding a neutron to $^{12}$C creates $^{13}$C, shown in Fig. 28 in both the red-and-blue and verbal representations. The additional neutron in $^{13}$C adds an additional unbonded down quark in the configuration. This additional unbonded quark lowers the average number of bonds per A, as compared to $^{12}$C. This, in turn, gives $^{13}$C a lower binding energy per A, as compared to $^{12}$C.

The configuration for nitrogen $^{14}$N has the addition of another proton compared to $^{13}$C. For $^{14}$N, shown in Fig. 29, the additional proton adds an additional unbonded up quark into the configuration, such that there are 6 unbonded quarks. This gives $^{14}$N an even lower binding energy per A, as compared with $^{12}$C or $^{13}$C. As with all atomic nuclei, there are numerous different ways to physically configure $^{13}$C and $^{14}$N, and these different configurations are the

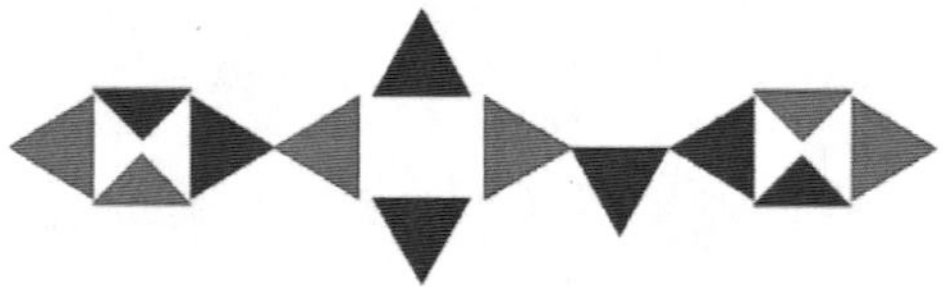


**Figure 28.** Two short-hand simplified representations of $^{13}$C.

**Figure 29.** Two short-hand simplified representations of $^{14}$N.

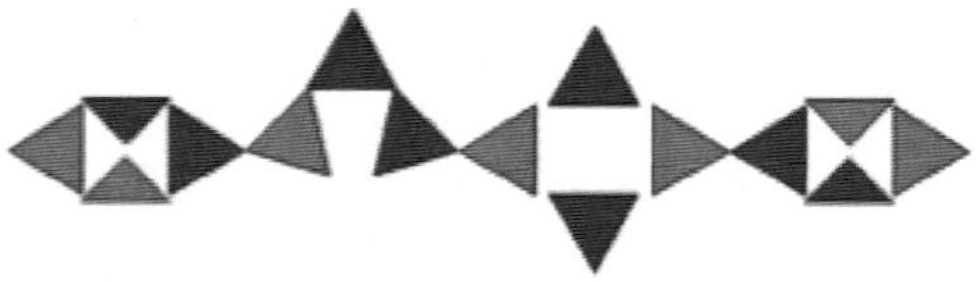


**Figure 30.** Short-hand simplified representations of $^{15}$N.

higher energy states. (The single proton segment and the single neutron segment cannot be situated next to each other in a stable atomic nucleus, because they would attempt to form a double-nucleon bond, which is not allowed.)

The configuration for $^{15}$N has an additional neutron compared to $^{14}$N, but it is able to form more bonds as a result of the added neutron. The ground state configuration for this atomic nucleus is shown in Fig. 30 in both the red-and-blue and verbal representations. For $^{15}$N, there are only 5 unbonded quarks. Thus, when compared to $^{14}$N with 6 unbonded quarks, the binding energy per A for $^{15}$N is slightly increased, as is experimentally observed.

The configuration for $^{16}$O has an additional proton, compared to $^{15}$N. The lowest energy configuration for this atomic nucleus is shown in Fig. 31, and is comprised of three alpha segments and one star segment, in an alpha-star-alpha-alpha configuration.

Once an understanding of the pattern from $^{12}$C to $^{16}$O has been gained, the configurations for the lowest energy states of the atomic nuclei are easily reproduced, with only a few simple energy-related rules. This pattern continues from $^{12}$C to $^{250}$Cf with the following energy-related rules. The first four rules apply to all atomic nuclei.

1. Due to the repulsion of the net electric charge, the particles separate themselves as far away as possible, while still maintaining as many bonds as possible.
2. There is one alpha segment for every two protons and two neutrons.

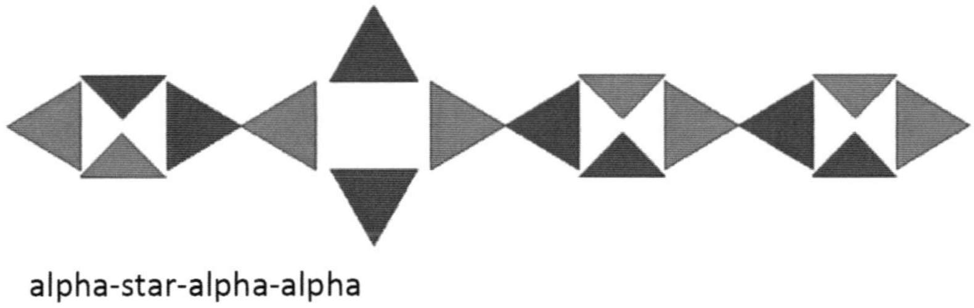


**Figure 31.** Short-hand simplified representations of $^{16}$O.

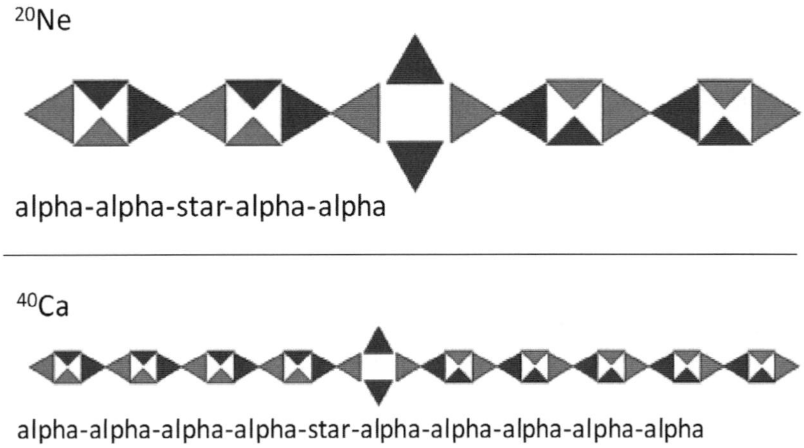


**Figure 32.** Short-hand simplified representations of $^{20}$Ne and $^{40}$Ca.

3. A star segment is in the middle, with two unbonded down quarks. The unbonded down quarks offset the high Coulomb energy in the middle of the atomic nucleus, and they help to separate the net positive charges. Good examples of these first three rules of the pattern are $^{20}$Ne and $^{40}$Ca, shown in Fig. 32.
4. If Z is odd and N = Z, there is a single neutron segment and a single proton segment. A good example of rule 4 is illustrated in $^{14}$N (shown previously in Fig. 29).

The next three rules relate to atomic nuclei with more neutrons than protons. These rules are:

5. For even Z and N = Z + 1, N = Z + 2, or N = Z + 3, the extra neutrons will be single neutron segments spaced between the alpha segments. The single neutron segments tend to be situated in what would otherwise be the highest concentration of positive charge, and they are situated between the alpha segments. Examples of the rule 5 are $^{13}$C (shown previously in Fig. 28) for one extra neutron and $^{42}$Ca for two extra neutrons, shown in Fig. 33.
6. For odd Z and N = Z + 1, there is one tritium segment, as in $^{31}$P. A good example of rule 6 is Phosphorous $^{31}$P, shown in Fig. 33.
7. If there are enough extra neutrons to do so, N = Z + 4 or greater, there will be two H4 segments, one on each end. These segments have an unbonded down quark on each end, as seen in $^{56}$Fe. Any additional extra

neutrons will be single neutron segments spaced between the alpha segments, as in $^{100}$Ru and $^{208}$Pb. Fig. 34 shows the configurations of $^{56}$Fe, $^{100}$Ru and $^{208}$Pb.

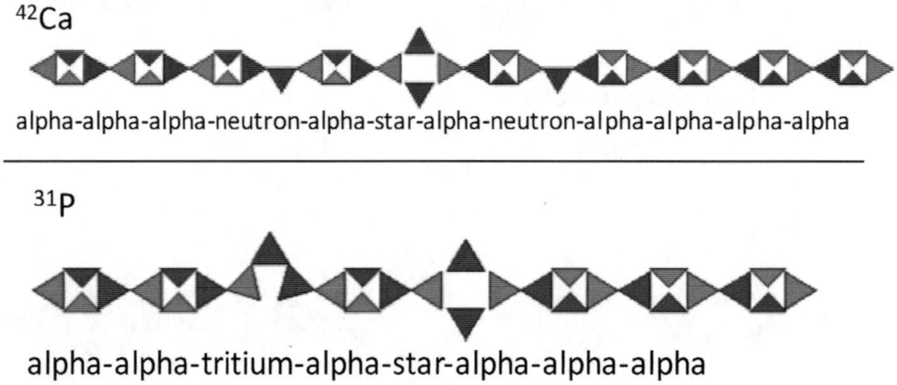


**Figure 33.** Short-hand simplified representations of $^{42}$Ca and $^{31}$P.

From Fig. 34, it is easy to understand why iron $^{26}$Fe has such a high binding energy per A. It is the largest stable atomic nucleus with only 4 unbonded quarks. Also, as can be noticed in Fig. 34, for the representations of the larger nuclei, such as $^{208}$Pb, it is difficult to see the graphical representation due to its large size. It is easier to view the verbal representation of it.

### 7.6. Summary of the Basic Pattern for the Configurations of Stable Atomic Nuclei

The Electromagnetic Model has successfully computer simulated both stable and unstable atomic nuclei, all the way up to californium $^{250}$Cf, using the electromagnetic equations and the equations for quantum angular momentum. From carbon $^{12}$C upwards, the stable atomic nuclei follow a relatively simple pattern. This basic pattern is not mere speculation, but rather the pattern is due to the lowest energy configuration of the protons and neutrons, when taking into consideration the laws of electromagnetics, as applied to the quarks. The pattern is as follows:

- There is one alpha segment for every two protons and two neutrons.
- There is one star segment with two unbonded down quarks, near the middle of the configuration. These two unbonded down quarks have a negative charge, which offsets the high Coulomb energy in the middle of the atomic nucleus, and also allows the net positive charge to spread out slightly.
- When there are more neutrons than protons, the extra neutrons are either combined in a tritium segment or in single-neutron segments. These segments are interspersed within the net positive charge of the alpha segments, in such a way as to lower the overall positive electric field of the atomic nucleus.
- If there are enough extra neutrons, there will be two H4 segments, one on each end of the configuration. The inclusion of these H4 segments changes the end-most unbonded quarks from positive to negative charges, thereby eliminating the high positive charge density associated with the unbonded up quarks.
- This pattern continues until there are more single neutron segments than alpha segments. That occurs at $^{209}$Bi.

$^{56}$Fe

H4-alpha-alpha-alpha-alpha-alpha-star-
alpha-alpha-alpha-alpha-alpha-alpha-H4

$^{100}$Ru

H4-alpha-alpha-neutron-alpha-alpha-neutron-alpha-alpha-neutron-alpha-alpha-neutron-alpha-alpha-star-alpha-alpha-neutron-alpha-alpha-neutron-alpha-alpha-neutron-alpha-alpha-neutron-alpha-alpha-H4

$^{208}$Pb

H4-neutron-alpha-neutron-alpha-neutron-alpha-neutron-alpha-neutron-alpha-neutron-alpha-neutron-alpha-neutron-alpha-neutron-alpha-neutron-alpha-neutron-alpha-neutron-alpha-neutron-alpha-neutron-alpha-neutron-alpha-neutron-alpha-neutron-alpha-neutron-alpha-neutron-alpha-neutron-alpha-neutron-alpha-neutron-alpha-neutron-star-alpha-neutron-alpha-neutron-alpha-neutron-alpha-neutron-alpha-neutron-alpha-neutron-alpha-neutron-alpha-neutron-alpha-neutron-alpha-neutron-alpha-neutron-alpha-neutron-alpha-neutron-alpha-neutron-alpha-neutron-alpha-neutron-alpha-neutron-alpha-neutron-alpha-neutron-alpha-neutron-alpha-neutron-alpha-neutron-H4

**Figure 34.** Short-hand simplified representations of $^{56}$Fe, $^{100}$Ru, and $^{204}$Pb.

These configurations of the lowest electromagnetic energy are based on detailed computer simulations, with the numerous possible configurations being tested and compared. These configurations are not just mere speculation.

To summarize, the pattern is as follows. The various segments are linked together in a chain-like structure. There is one alpha segment for every two protons and two neutrons. There is one star segment near middle of the configuration. If N>Z and Z is odd, there is one triton segment in the configuration. Any extra neutrons with their unbonded down quarks are interspersed between the alphas. These unbonded down quarks have a negative charge which offsets and separates the net positive charge of the atomic nucleus. The unbonded down quarks tend to be situated in a region that would otherwise be a high concentration of positive charge. If there are enough extra neutrons in the atomic nucleus, they form H4 segments on both ends of the atomic nucleus. This removes any unbonded up quarks, with their high positive charge density, from the configuration. For the largest stable atomic nuclei, more unbonded down quarks, with their negative charges, are needed to offset the increasing net positive charge. These single neutron segments, with their unbonded down quarks, are situated between the alpha segments. This pattern continues up to $^{209}$Bi, which is the largest stable atomic nucleus.

While these short-hand simplified graphic representations are drawn out as straight chains of segments, it should be mentioned that if the actual shape of the protons and neutrons interferes with the magnetic bond in such a way as to

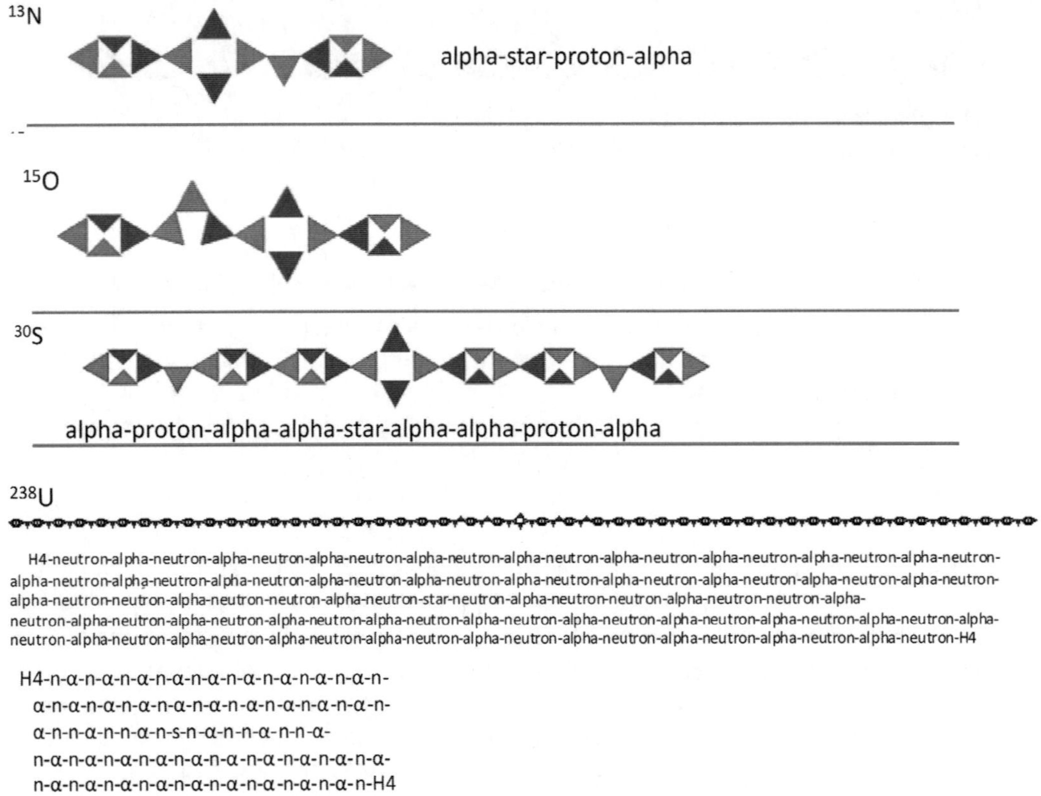


**Figure 35.** Short-hand simplified representations of $^{13}$N, $^{15}$O, $^{30}$S, and $^{238}$U.

prevent a perfectly straight stacked bond between the segments, then these chains may be bent, curled, or spiraled. If so, this effect would not only decrease the intrinsic quadrupole moment, but it would also decrease the bonding energy very slightly, due to the Coulomb energy. This effect would be more pronounced for the larger atomic nuclei, which have a large excess of neutrons.

## 8. Basic Pattern for Radioactive Atomic Nuclei

For the radioactive atomic nuclei, there are two more considerations for their lowest energy configurations. Again, these patterns are based on the patterns that emerge as a direct result of the electromagnetics energies.

- Any atomic nucleus with more single neutron segments than alpha segments will have the single neutron segments doubled-up, with no alpha segments in between them. These doubled-up single neutron segments tend to be near the middle of the chain, due to the electric force. For example, Uranium $^{238}$U, which is shown in Fig. 35.

- When there are more protons than neutrons, the protons will either be single proton segments as in $^{13}$N, or there will be in an He3 segment made of two protons and one neutron, as in $^{15}$O and $^{30}$S. Examples of these two additional rules for the radioactive atomic nuclei are shown in Fig. 35.

For $^{238}$U, the neutron segments are spaced between the alpha segments, with the star segment in the middle. Also note that the adjacent double-neutron segments are also near the middle. It is interesting to note that there are no stable atomic nuclei with doubled-neutron segments.

### 8.1. Summary of the Basic Pattern for Radioactive Atomic Nuclei

Above $^{209}$Bi, there are more extra neutron segments than alphas segments. For any atomic nuclei larger than $^{209}$Bi, the extra neutron segments but must be doubled up in pairs, with no alpha segments in between, for example as seen in $^{238}$U.

For atomic nuclei in which Z>N, there is only one stable atomic nucleus and that is $^{3}$He. For all other atomic nuclei in which Z>N, they are radioactive. For the radioactive atomic nuclei with Z>N, there are either single proton segments or He3 segments in the configuration. Both of these segment types have unbonded up quarks associated with them, and as a result, they also have a high positive electric field in the immediate vicinity of the unbonded up quarks. Thus, the unbonded up quarks, with their high positive charge, tend to be at or near the end of the configuration in order to disburse this excess positive field as much as possible.

## 9. Results

### 9.1. Binding Energy Results

There remains only one parameter to be selected for the best fit to the binding energy data, the minimum distance between the two quarks of the two different nucleons—the "minimum internucleon quark-to-quark distance". This model has only one parameter to determine. The value for the minimum internucleon quark-to-quark distance is selected to be $2.11082 \times 10^{-16}$ meters. The semi-empirical binding mass formula, which is also known as the Weizsäcker formula, uses five variables and a conditional logic statement to achieve its mathematical curve-fitting. The Electromagnetic Model of the Nuclear Force is able to get comparable results with only one variable. This is a significant achievement, and strongly implies the correctness of this model.

Hundreds of atomic nuclei have been modeled using this method to calculate the binding energy based on electromagnetics. This has been done for every atomic nucleus shown in Fig. 35. Each atomic nucleus is placed in the lowest energy configuration. Using this configuration, the electromagnetic energy and spin energy are then calculated, from which the binding energy is determined.

Using this value for the minimum distance between quarks, the resulting bonding energy curve is shown in Fig. 35. For comparison, two sets of values are shown: the calculated binding energy, shown below in red squares, and the experimental binding energy, show in light blue circles. (All experimental data has been extracted from reference [44].) Fig. 36a shows all the points from A = 2 to A = 208. Fig. 36b shows the detail of the points for the smaller atomic nuclei, from A = 2 to A = 60, for ease of viewing the details of this curve.

For the larger atomic nuclei, the kinetic energy due to nuclear spin is not a significant contributor to the binding energy. However, the smaller atomic nuclei, their binding energy is more strongly dependent on spin, as well as on the axis angle of the spin. For consistency, the axis angle has been set to 45° with respect to the y-axis, for all atomic nuclei. However, in actuality, the true axis angle of the spin could vary from this artificially fixed 45° setting. For the smaller atomic nuclei, this actual variation can appreciably affect the calculated binding energy, and this variation could easily explain the larger differences seen in the calculated binding energy for the smaller atomic nuclei.

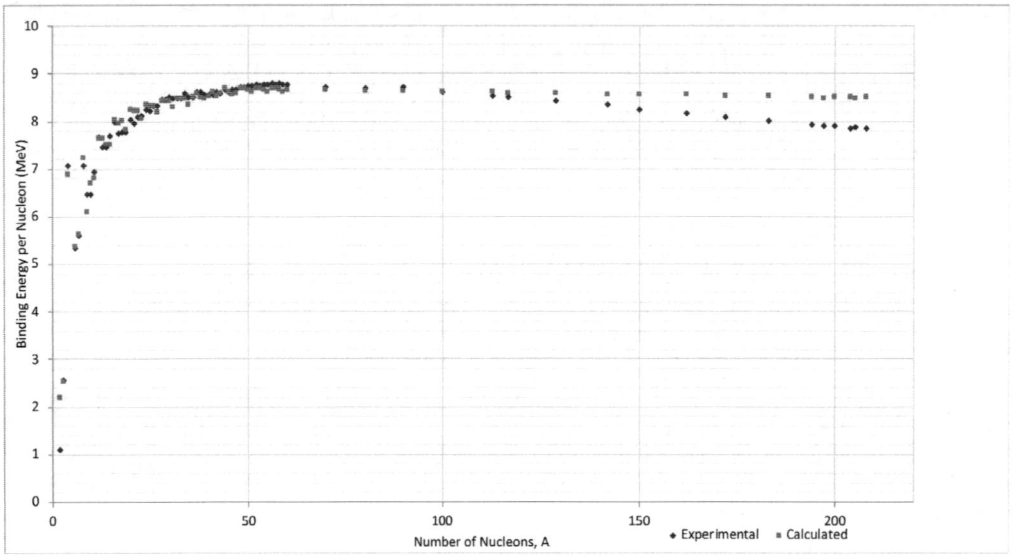


**Figure 36a.** Binding energy per A versus A, for the stable atomic nuclei, for both calculated and experimental data, from A = 2 to A = 208.

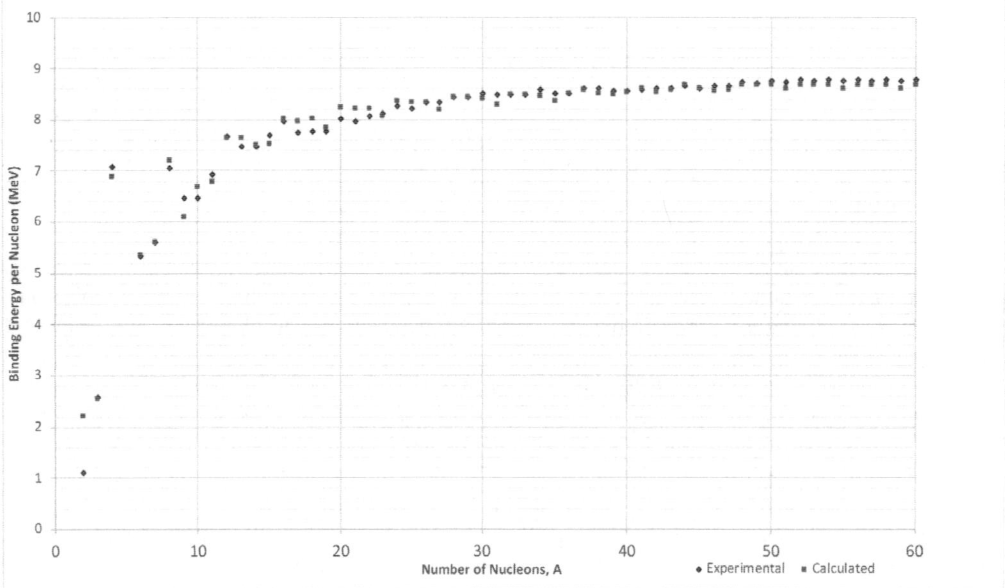


**Figure 36b.** Binding energy per A versus A, for both calculated and experimental data, showing detail for the smaller atomic nuclei.

This data show excellent reproduction of the experimental binding energy; most of the atomic nuclei fall within a one or two percent error. For the atomic nuclei with large A, the predicted downward slope is not as much as the experimentally observed downward slope. The largest error is 8.3%.

No other model has been able to predict such a tight correlation of binding energy. (Empirical numerical curve fitting is not a model.) Nor can any other model reproduce the general shape of this curve with only *one* variable. This is an unprecedented success in this result.

### 9.2. Results for Other Nuclear Behaviors

The Electromagnetic Model is also able to explain numerous results of other nuclear behaviors, behaviors that other models of the Nuclear Force have difficulty explaining. For example, excited energy states are very easily explained as being different configurations for the structure. Previous models do not incorporate structure into the models, rather they prefer to model the nucleus as independent particles in an energy well, or as a liquid drop, or as an energy shell. As a result, these models do not consider the concept of nuclear structure. Thus, the concept of different configurations of the atomic nucleus is not considered or explored. Because the Electromagnetic Model incorporates the concept of structure, excited states are easily understood as being a configuration that is not the lowest energy configuration for the Electromagnetic Force. The high energy states may or may not have a different spin, but they will always have a different configuration from the ground state configuration.

Particle decay is defined as when an atomic nucleus changes its mass number A, such as when a proton, neutron, or alpha particle is ejected from it. The previous models have difficulty explaining why particle decay occurs and why the Nuclear Force should suddenly turn itself off. Mathematically valid and coherent answers are not forthcoming from the previous older models, except to say that experimentally, particle decay is energetically allowed. However, such experimentally based answers do not prove the validity of any model. Hence, there remains many unanswered questions and incongruities with these previous older models.

The Electromagnetic Model can easily explain particle decay as being either an attempt of the nucleons to double bond or triple-quark bond. Both of these attempts are not allowed due to the Pauli exclusion principal and the hard core repulsion. When the Electromagnetic Force attempts to double bond or triple-quark bond, the bonds break, and particle decay results. (This will be discussed in further detail in subsequent papers.) The concept of nuclear structure is necessary for the understanding of particle decay.

Regarding the experimentally measured electric quadrupole moments, they are much larger than what older previous models, such as the Shell Model, can explain. These large quadrupole moments are simply due to the inherent chain-like structures of the atomic nuclei in their lowest energy state. There are several legitimate explanations as to why the measured quadrupole moment may be smaller than the inherent quadrupole moment. The converse, however, is not easily explained at all. Thus, the previous older models of the nuclear force cannot explain the large quadrupole moments. The Electromagnetic Model can readily explain the large measured quadrupole moments as being due to the inherent chain-like structure of the lowest energy state.

Thus, the techniques of incorporating a structure into the nucleus and determining the lowest energy configuration are extremely important for the understanding of nuclear behaviors. Structure and configuration are attributes that are able to explain much about nuclear behavior that the other previous older models can not explain.

### 9.3. The Intended Endeavor for this Paper

As discussed in the introduction, the intended endeavor of this paper was to explore how much of the nuclear behavior can be explained by the Electromagnetic Force. The Electromagnetic Model can predict numerous aspects of nuclear behavior, and it can duplicate the binding energy curve better than the previous older models of the Nuclear Force. This model can also explain many other aspects of nuclear behavior, better than any other singular model can do.

In the intended endeavor of this paper, the Electromagnetic Model is very successful. The answer to this investigation is that almost all of the nuclear behavior can be explained by electromagnetism. As intended, this investigation is done without having to include the mathematical equations or complicated simulations of other models—-such as the Shell Model, Independent Particle Models, or the Residual Color Force Model—-in order to understand nuclear behavior. This statement, that almost all nuclear behavior can be explained by electromagnetism, is a significant finding.

## 10. Conclusions

The laws of electromagnetics are valid inside an atomic nucleus, and these laws should not be disregarded. It is the electromagnetic properties of the quarks within the nucleus that create the Nuclear Force, the force holding the atomic nucleus together. Due to the electromagnetic properties of the quarks, the electromagnetic energies configure the protons and neutrons into the lowest energy state. This electromagnetic energy and the specific lowest energy configurations of the atomic nuclei are what control the nuclear behaviors. A better understanding of nuclear behaviors can be gained through this model by applying this knowledge and insight. Nuclear behaviors are a direct consequence of the Electromagnetic Forces acting within that nuclear structure. Thus, by recognizing the Electromagnetic Forces within the nuclear structure, a better understanding of nuclear behavior can be obtained. This model has directly unified the Nuclear Force to the Electromagnetic Force.

## Appendix: More Information about the Residual Color Force Model

### History

The Residual Color Force is also known as the "Residual Strong Force", the "Residual Strong Interaction", and the "Residual Chromodynamic Force". These are just different names for the same concept. The Residual Color Force Model is one of the many theoretical models of the Nuclear Force.

The Nuclear Force was thought, in 1935, to be mediated by virtual mesons. This theoretical development included a description of the Yukawa potential [1a] as an early representation of the nuclear potential. By the 1970s, particle physicists determined that nucleons were composed of quarks, and that the bond between nucleons was actually a bond between the internucleon quarks for neighboring nucleons. The Residual Color Force Model of the Nuclear Force hypothesizes that this internucleon quark-to-quark bond is related to the quark's color. Also, the Residual Color Force is presumed to be a residual effect of the Quantum Chromodynamic Force. However, this hypothesis is not confirmed experimentally, and it remains as a strictly theoretical concept. Because color forces cannot be directly measured, there is no direct evidence to corroborate the concept that the internucleon quark-to-quark bond is caused by quark color. Thus, it is strictly a hypothetical supposition that the Nuclear Force is based on quark's color rather than on the quark's electromagnetic properties.

### Introduction

Quantum Chromodynamics (QCD) is the theory of the color interactions inside the hadrons [2a]. It deals with quarks, gluons and their interactions. QCD is also part of the Standard Model of Particle Physics. QCD is a non-Abelian gauge field theory with color SU(3) as the underlying gauge group.

The Residual Color Force Model, which is based on QCD, is one of the many numerous models of the Nuclear Force, and it theorizes that the force between nucleons is a residual of the QCD color interaction, similar to the van der Waals force between electrically neutral molecules. The Residual Color Force Model is an extremely complex problem that can only be solved with brute computing power on a discretized, Euclidean space-time lattice, known as lattice QCD. In other words the quarks are positioned in a lattice of three dimensions of space and one dimension of time.

Similar to the Yukawa Model, the hypothesis of the Residual Color Force Model claims that the Nuclear Force is mediated by the exchange of mesons, in particular virtual pi-mesons, or pions. There is no closed mathematical form for the Residual Color Force. There are relatively few published scientific papers that attempt to perform simulations of the Residual Color Force, and these simulations use extremely complex lattice QCD calculations. The simulations of the Residual Color Force have been relatively unsuccessful in duplicating nuclear behavior, with extremely large errors in the resulting simulated quantities. Also, these computer models use quark masses which are much larger than what the quark mass is estimated to be, by as much as a factor of 50 or more. Thus, the computer simulations for the Residual Color Force are not only sparse, but also lacking credibility.

The hypothesis of the Residual Color Force Model is that most of the Quantum Chromodynamic Forces, which are due to the gluons and the quarks inside of the nucleon, cancel each other outside of the nucleon. In other words, when viewed from the outside, the nucleons are color neutral. The interaction between the various colors of the gluons and quarks inside the protons and neutrons is called the "Quantum Chromodynamic Force". The interaction between the various colors outside the protons and neutrons is called the "Residual Color Force". Not only does the name of the force change, but the range of the force also changes. Outside the proton or neutron, the force becomes a short range force, of only 1 to 2 femtometers. Also, the mediating particle of the force changes from a massless gluon to a virtual pi-meson.

## Virtual Particles

Note that these mediating pi-mesons are virtual particles; they are not actual physical particles. Virtual particles are believed to be a result of quantum energy fluctuations, existing for energies and time durations that are below the detection of the quantum uncertainty principles. Thus they can be conceptualized as going in and out of existence as the quantum fluctuations occur. These virtual particles are not an actual physical or measurable entity. As seen in lattice field theory, it is possible to perform quantum field theory calculations that are completely absent of the concept of virtual particles. As a result, it is often said that virtual particles are simply a conceptual or mathematical tool [3a].

## Computer Simulations and Equations from the Researchers, Aoki et al

In the computer simulations done by Aoki et al [4a], the Residual Color Force has a potential V that is the sum of four components, including a central potential $V_C(r)$ and a tensor potential $V_T(r)$. Also included in the sum are $V_{LS}$ and $O(v^2)$, where $V_{LS}$ is the potential due to spin orbit, and O is the energy due to its velocity v. This sum is written mathematically, as shown in Eq. (1a):

$$V = V_C\,(r) + V_T\,(r)\,S_{12} + V_{LS}\,(r)\,\boldsymbol{L}\cdot\boldsymbol{S} + O\left(v^2\right) \tag{1a}$$

where r is the center-to-center distance, and L and S are vector angular momentum quantum numbers.

The expression for the central potential $V_C$ is shown below, in Eq. (2a):

$$V_C\,(r) = E - \frac{1}{\Delta\,(r)}\left([QS_{12}\psi]_{21}\,(r)\,H_0[\mathfrak{p}\psi]_{21}\,(r) - [\mathfrak{p}S_{12}\psi]_{21}\,(r)\,H_0[Q\psi]_{21}\,(r)\right) \tag{2a}$$

And the tensor potential $V_T$ is shown in Eq. (3a):

$$V_T\,(r) = \frac{1}{\Delta(r)}\left([Q\psi]_{21}\,(r)\,H_0[\mathfrak{p}\psi]_{21}(r) - [\mathfrak{p}\psi]_{21}\,(r)\,H_0[Q\psi]_{21}(r)\right) \tag{3a}$$

In both of these equations, Δ(r) is also a function of r, and is defined by Eq. (4a):

$$\Delta(r) \equiv [\mathrm{p}\psi]_{21}(r)\,[QS_{12}\psi]_{21}(r) - [Q\psi]_{21}(r)\,[\mathrm{p}S_{12}\psi]_{21}(r) \tag{4a}$$

Once these potentials have been determined, they are used in the Schrödinger equation in order to consider the quantum wave-like properties of the system. This is done in an attempt to duplicate the Reid diagrams of the Nuclear Force—diagrams that plot energy and force as a function of distance, for a single nucleon interacting with another nucleon. However, very large quark masses must be used in order to facilitate convergent of these computer simulations.

Computer Simulations and Equations from the Researchers, Beane et al.

The scattering behavior between two nucleons has been calculated using the theoretical model of the Residual Color Force Model by Beane et al [5a]. By theoretically examining two nucleons contained within a box of length L, with their center of mass at rest, the energies of the low lying energy levels can be computed in terms of p cot $\delta$. The values of $p^2$ that solve Eq. (5a) give the locations of all of the energy eigenstates in the box, including bound states, for which $p^2 < 0$.

$$p\cot\delta(p) = \frac{1}{\pi L}\,S\left(\left(\frac{Lp}{2\pi}\right)^2\right) \tag{5a}$$

where the value of S is a function of $\eta$, as shown in Eq. (6a).

$$S(\eta) \equiv \sum_{j}^{\Lambda_j} \frac{1}{|j|^2 - \eta} - 4\pi\Lambda_j \tag{6a}$$

The authors state that this derivation is valid within the radius of convergence for the Effective Field Theory without pions. In the limit $L \gg a$ (where a is the scattering length), an approximation for the attractive binding energy of the two nucleon system can be obtained. The approximate energy of the lowest lying state is shown in Eq. (7a):

$$E_{-1} = \frac{\gamma^2}{M}\left[1 + \frac{12}{\gamma L}\frac{1}{1 - 2\gamma(pcot\delta)'}e^{-\gamma L} + \ldots\right] \tag{7a}$$

where: $(p\cot\delta)' = \frac{d}{dp^2}p\cot\delta$ is evaluated at $p^2 = -\gamma^2$ and the quantity $\gamma$ is the solution of the Eq. (8a):

$$\gamma + p\cot\delta|_{p^2=-\gamma^2} = 0 \tag{8a}$$

This quantity $\gamma$ is the bound state binding energy, in the infinite-volume limit. The scattering behavior caused by the Residual Color Force between two nucleons is then simulated, to extract the desired quantitative information.

In a subsequent follow-up paper [6a], the researchers determined the binding energy for both exotic hyper-nuclei (which contain strange quarks) and non-exotic nuclei. The results of this computer calculation for the non-exotic nuclei are shown in Table 1.

As can be seen in this table, the calculated binding energies have extremely large errors when compared to the actual binding energies. It should also be noted that in these simulations, the quarks masses are set to values much larger than the currently accepted value. The results are then extrapolated to the smaller quark masses, a process that could introduce very large errors.

**Table 1a.** Calculated binding energies from the Residual Color Force, compared to the actual binding energies for 4 nuclei. Note this is binding energy, not binding energy per A.

| Nuclide | Calculated Binding Energy as calculated in this referenced report, in MeV | Actual Binding Energy | %error |
|---|---|---|---|
| H-2 | 19.5 | 2.22 | 878% |
| H-3 | 53.9 | 8.48 | 636% |
| He-3 | 53.9 | 7.72 | 698% |
| He-4 | 107 | 28.3 | 378% |

Computer simulations from the researchers, Beane et al [6a].

In another lattice QCD calculation, the nucleon-nucleon (NN) potential was studied by Hatsuda et al [7a]. For this study, it must again be noted that unrealistically large quark masses were used. This paper only studied exotic nucleons, which are thought to exist theoretically only in neutron stars, and are not normally found in nature. Thus, the results of this paper are more of interest to cosmology than to the topic of this paper.

There are other papers related to binding energy of nuclei and lattice QCD, however they are not considering quarks nor the Residual Color Force in the simulations. Rather they examine only the Effective Field Models.

An Inherent Problem with the Residual Color Force Model

In Table 1a, the calculated binding energy of the $^3$H and the $^3$He are the same. (Also in the referenced paper, the exotic nuclei with A = 4 have the same binding energy as $^4$He.) This result is because the Residual Color Force Model considers only the color of the quarks and not the flavor of the quarks. The nucleons all have three colors—red, green, and blue—inside of them. Nucleons are distinctive from each other due the flavor of the quarks—up, down, top, bottom, charm or strange.

For example, consider isobar with the same A but with different Z and N values. A computer simulation of the Residual Color Force might look at the nuclear binding energy of an isobar, with a given mass number A, and it would erroneously conclude that the binding energy for all the isobars are the same. This is because the Residual Color Force Model does not consider the flavor of the quarks, ignoring the up and down attributes of the quarks. With only the quark color as a consideration, all isobars are essentially identical. (This is especially true if the Coulomb energy is also ignored.) This is an inherent problem for the Residual Color Force Model, and this effect is seen in the binding energy simulations of this force. For example, the binding energies of $^3$H and $^3$He are identical [5a]. More tellingly, for the A = 4 nuclei in the referenced paper, the binding energies for all the exotic nuclei with A = 4 and the non-exotic nuclide of $^4$He are also exactly the same.

If the bonds between the nucleons are bonds based only on the color of the quark, as the Residual Color Force Model speculates, then all isobars for a given A would have the similar binding energy and similar behavior, differing only slightly due to Coulomb energy. Empirically it is known that such behavior is not representative of reality.

Summary of Research about the Residual Color Force

The simulations of the Residual Color Force Model are complex, difficult, and have yielded only a relatively small amount of data. Furthermore, there has been little recent progress in this field. Thus the Residual Color Force Model is not a viable or practical solution for the understanding nuclear structure, nuclear behavior, or nuclear physics.

**References for Appendix:**

1a.) Yukawa, H. (1935). "On the interaction of elementary particles". *Proc. Phys. Math. Soc. Jpn.* 17: 48.

2a.) http://www.scholarpedia.org/article/Nuclear_Forces#QCD_and_the_nuclear_force
3a.) https://en.wikipedia.org/wiki/Virtual_particle
4a.) Sinya Aoki et al., (2010). "Theoretical Foundation of the Nuclear Force in QCD and its Applications to Central and Tensor Forces in Quenched Lattice QCD Simulations" *Progress of Theoretical Physics,* Vol. 123. No. 1.
5a.) S.R. Beane, et al. "Two Nucleons on a Lattice" arXiv:hep-lat/0312004v1. *Phys. Lett.* B585:106-114 (2006.)
6a.) S.R. Beane, et al. "Light nuclei and hyper nuclei from quantum chromodynamics in the limit of SU(3) flavor symmetry" *Physics Review* 87, 034506 (2013.)
7a.) T. Hatsuda, (2012). "Hadron interactions from lattice QCD." *Prog. Part. Nucl. Phys.* 67: 122.

## References

[1] N. L. Bowen "The Electromagnetic Considerations of the Nuclear Force", *J. Condensed Matter Nucl. Sci.* 33, (2020), p 194 www.iscmns.org/CMNS/JCMNS-Vol33.pdf
[2] Andrew Watson, The Quantum Quark, Unviersity Press, Cambridge, 2004, pp. 172–183
[3] F. Halzen and A. Martin, *Quarks and Leptons: an Introductory Course Modern Particle Physics*. New York, Wiley, 1984.
[4] R. Eisberg, R. Resnick, Quantum Physics of Atoms, Molecules, Solids, Nuclei, and Particles, Wiley, New York, pp 562-567, 1985.
[5] Gell-Mann (1964). "A Schematic Model of Baryons and Mesons". Physics Letters. 8 (3):214-215. doi: 10.1016/S0031-9163(64)92001-3
[6] G. Zweig (1964). "An SU(3) Model for Strong Interaction Symmetry and its Breaking" (PDF). CERN-TH-401.
[7] G. Zweig (1964). "An SU(3) Model for Strong Interaction Symmetry and its Breaking: II". CERN-TH-412.
[8] K. N. Mukhin, Experimental Nuclear Physics, Volume 1, Physics of Atomic Nucleus. Mir Publishers, Moscow, p. 120, 1987.
[9] R. Eisberg, R. Resnick, Quantum Physics of Atoms, Molecules, Solids, Nuclei, and Particles, Wiley, New York, pp 509-549, 1985.
[10] Cook, N. D. (2010), Models of the Atomic Nucleus, 2nd edition, Springer Verlag, Heidelberg.
[11] R. Machleidt, Nuclear Forces. Scholarpedia, 9(1):30710, 2014. doi: 10:4249/scholarpedia.30710. (Large pion (quark) mass in Residual calculations, then extrapolated down, and the mass is still too large to be meaningful. Also, it is very difficult calculation and no practical.)
[12] U.G. Meisner, Modern theory of nuclear forces A.D. 2006, Eur. Phys. J. A 31: 397-402, 2007. doi: 10.1140/epja/i2006-10170-1.
[13] R. Machleidt, Nuclear Forces. Scholarpedia, 9(1):30710, 2014. doi: 10:4249/scholarpedia.30710.
[14] N. L. Bowen, A Simple Calculation of the Inter-Nucleon Up-to-Down Quark Bond and its Implications for Nuclear Binding, J. Condensed Matter Nucl. Sci. 29 (2019) 249–259.
[15] D. Halliday and R Resnick, *Fundamentals of Physics*. New York, Wiley, 1974, p 477.
[16] P. Lorrain, and D. Corson, *Electromagnetic Fields and Waves*. San Francisco, Freeman and Company, 1970, p72-76.
[17] D. Giancoli, *Physics Principles with Applications*. Upper Saddle River, New Jersey, Prentice Hall, 1998, pp. 509–510.
[18] P. Lorrain, and D. Corson, *Electromagnetic Fields and Waves*. San Francisco, Freeman and Company, 1970, p292-299.
[19] G. E. Owen, *Electromagnetic Theory*. Boston, Allen and Bacon, Inc., 1963, p 205.
[20] K. Yosida, Theory of Magnetism. Springer, New York, 1996, p 13.
[21] F. Halzen and A. Martin, *Quarks and Leptons: an Introductory Course Modern Particle Physics*. New York, Wiley, 1984, p. 188.
[22] R E Marshak, *Conceptual Foundations of Modern Particle Physics*, World Scientific Publishing Co, Singapore, 1993, p. 37.
[23] A. J. Buchmann, et al., Intrinsic quadrupole moment of the nucleon, arXiv:hep-ph/0101027, Feb. 2008.
[24] Martin M. Block and Francis Halzen. New experimental evidence that the proton develops asymptotically into a black disk, Phys. Rev. D 86, 051504(R) – Published 25 September 2012. DOI: https://doi.org/10.1103/PhysRevD.86.051504
[25] Martin M. Block and Francis Halzen, Experimental confirmation that the proton is asymptotically a black disk Phys. Rev. Lett. 107, 212002 – Published 15 November 2011. DOI: https://doi.org/10.1103/PhysRevLett.107.212002
[26] H. P. Robertson, Phys. Rev. A 35, (1930), 667

[27] E. Schrödinger, Ber. Kgl. Akad. Wiss. 24, (1930), 296;
[28] Silas R. Beane et al., Hadronic interactions from lattice QCD. arXiv:0805.4629v6 Jun 2008
[29] N. Ishii, S. Aoki and T. Hatsuda, Nuclear force from lattice QCD, *Phys. Rev. Lett.* 99-2:022001, (2007). doi: 10.1103/PhysRevLett.99.022001
[30] S. Aoki, T. Hatsuda, N. Ishii, Theoretical Foundation of the Nuclear Force in QCD and Its Applications to Central and Tensor Forces in Quenched Lattice QCD Simulations, *Progress of Theoretical Physics*, 123: 1, 2010.
[31] Hatsuda, T (2012). Hadron interactions from lattice QCD. Prog. Part. Nucl. Phys. 67: 122.
[32] K. S. Krane, *Introductory Nuclear Physics*, Wiley, New York, p.144, 1988.
[33] J. D. McGervey, *Introduction to Modern Physics*, Academic Press, New York, p.498, 1971.
[34] D. M. Brink. "History of cluster structure in nuclei," *J. Phys.: Conf. Ser.,* vol. 111, p. 012001, 2008
[35] M. Freer, "The clustered nucleus—cluster structures in stable and unstable nuclei," *Reports on Progress in Physics*, vol. 70, p. 2149, 2007.
[36] B. R. Fulton, "Clustering in nuclei: nuclear chains, nuclear molecules and other exotic states of nuclear matter," *Contemporary Physics,* vol. 40, pp. 299–311, 1999.
[37] K. Ikeda, N. Tagikawa and H. Horiuchi, "The Ikeda Diagram". *Prog. Theo. Phys. Suppl.* extra number: p 464, 1968.
[38] P. Lorrain, and D. Corson, Electromagnetic Fields and Waves. San Francisco, Freeman and Company, 1970, p 64-70.
[39] K Mukhin, Experimental Nuclear Physics, Moscow, Mir Publishers, 1974, pp. 101–108.
[40] K. N. Mukhin, Experimental Nuclear Physics, Volume 1, Physics of Atomic Nucleus. Moscow: Mir Publishers, 1987, pp. 62–68.
[41] K. S. Krane, Introductory Nuclear Physics. New York, Wiley, 1988, p 49-57.
[42] H. De Vries, C. W. De Jager, and C. De Vries, "Nuclear charge-density-distribution parameters from elastic electron scattering". Atomic Data and Nuclear Data Tables, vol. 36, pp. 495–536, 1987.
[43] N. E. Holden, "Table of the Isotopes", in D.R. Lide, Ed., *CRC Handbook of Chemistry and Physics*, 84th Ed., CRC Press, Boca Raton, FL, 2004.
[44] S. Raman, C.W.Nestor, P. Tikkanen, "Transition Probability from the Ground to the First 2+ state of Even-Even Nuclides", in B. Pritychenko. Ed., Atomic Data and Nuclear Data Tables, vol. 78, Elsevier Inc., Cambridge, MA. pp1-128 (2001).
[45] National Nuclear Data Center, information extracted from the NuDat 2 database, http://www.nndc.bnl.gov/nudat2/

J. Condensed Matter Nucl. Sci. 36 (2022) 184–202

Research Article

# Discrepancies with the Recent Models of Nucleons

N. L. Bowen*

*Colorado Mountain College, Glenwood Springs, Colorado, USA*

**Abstract**

Recent theoretical calculations about the mass of the partons within protons have raised questions, specifically regarding what comprises the bulk of the mass of the nucleons. As a result of these recent theoretical calculations and the questions they precipitate, there are many new models of the nucleons that claim to have solved the question of the missing nucleon mass. This paper is an examination these recent nucleon models, exploring the accuracy of the science, the mathematics, and the concepts described therein.



*Keywords:* Nucleons, theoretical calculations, missing nucleon mass, quarks, relativistic mass

## 1. Introduction

In 1964, the existence of quarks was proposed independently by Gell-Mann and Zweig [1], [2], [3], [4], changing the concept of the proton and neutron from homogeneously-charged particles to particles having electrical inhomogeneity. Since that time, much experimental evidence of quarks has confirmed their existence, mainly through the experimental evidence of deep inelastic scattering [5], [6], [7], a process used to probe protons and neutrons. Deep inelastic scattering provided the first convincing evidence of the reality of quarks. This is an extension of Rutherford scattering, but with higher energies and finer resolution to detect the inner components of the nuclei.

There are six different types of quarks: up, down, charm, strange, top, and bottom [8]. This is shown in Table 1.

Only two of these quarks are present in ordinary stable matter, the up and down quarks, and the remaining four quarks are short lived, with lifetimes much less than a second. These stable up and down quarks are the composite parts of stable matter, the proton and neutron. An up quark has a charge of 2/3 e and the down quark has a charge of $-1/3$ e, where e is an elementary charge of a proton.

Recent theoretical calculations of the mass of quarks imply that the mass of the quarks is much smaller than the mass of the nucleon [9]. These theoretical calculations posit that the mass of the quarks is only a small percentage of the mass of the nucleon, roughly 1 percent.

However, the mass of the up and down quarks cannot be measured directly. This inability to measure the quark mass directly is because quarks cannot be isolated from their nucleon, regardless of how much energy is used to try

*Corresponding author: nbowen@coloradomtn.edu

**Table 1.** Quark characteristics, including the quark name, charge, spin, and mass, which is either theoretically estimated or measured.

| Quark Flavor | Charge | Spin | Mass in MeV/$c^2$ | Mass in kg | Measured or Theoretical Esimates of Mass |
|---|---|---|---|---|---|
| up | 2/3 e | 1/2 | 2.2 | 3.92211E-30 | Theoretical Estimate |
| down | -1/3 e | 1/2 | 4.7 | 8.37906E-30 | Theoretical Estimate |
| charm | 2/3 e | 1/2 | 1,280.0 | 2.28196E-27 | Theoretical Estimate |
| strange | -1/3 e | 1/2 | 96.0 | 1.71147E-28 | Theoretical Estimate |
| top | 2/3 e | 1/2 | 172,100.0 | 3.06816E-25 | Measured |
| bottom | -1/3 e | 1/2 | 4,180.0 | 7.45201E-27 | Theoretical Estimate |

to unbind these quarks from each other [10], [11], [12]. In various experiments, large amounts of externally-applied energy has been focused at protons, using particle accelerators and colliders in an attempt to break the bonds holding the quarks together. The goal of these experiments is to add external energy to the proton to separate the quarks, and thereby force the proton to break apart as a result of this added energy, in order to obtain an isolated quark as a result.

However, the external energy/mass, which must be added to a proton to break the chromodynamic bonds between the quarks, is larger than the energy/mass of the quarks. As a result, this added energy creates a quark/anti-quark pair, known as a pi-meson. Thus, the desired effect of the chromodynamic bonds being broken does not occur. Note, the mass of the pi-meson is 139.6 MeV/$c^2$ [13].

Although isolated quarks have not been detected experimentally to measure its actual mass, theoretical calculations have been made using chiral perturbation theory. These theoretical calculations are done by first experimentally measuring the mass of a pi-meson. Then by applying the theoretical predictions of the quantum chromodynamics chiral perturbation theory, a calculation can be done to determine the estimated mass of the quarks.

These calculations are extremely complicated, requiring the lattice quantum chromodynamics theory and contour improved perturbation theory to evaluate the ratio of two different types of quarks. In particular, using the theoretic assumptions of quantum chromodynamic chiral symmetry breaking, it is hypothesized that the mass of the up and down quarks are related to the square root of the mass of the pion [14], [15]. Using these assumptions, the ratio of the charmed quarks as compared to the strange quarks was first determined, and then this result was used to subsequently determine the ratio of the charmed quark to the up and down quarks. Comparing this value to the mass of the proton, which is 938.3 MeV/$c^2$, implies that the mass of the three quarks inside is roughly about 1% of the nucleon mass. The remaining 99% of the mass remains unexplained.

These numbers have sparked many new and recent theories regarding what makes up the missing 99% of the mass of the nucleons. Unfortunately, these new theories contain mistakes, errors, and unsubstantiated assumptions.

Here is an example, quoted directly, of what these recent nucleon models claim [16].

> "In addition to mass that comes from quarks, another 32 percent comes from the energy of the quarks zipping around inside the proton, Liu and colleagues found. (That's because energy and mass are two sides of the same coin. Albert Einstein described that in his famous equation, $E=mc^2$. E is energy, m is mass and c is the speed of light.) Massless particles called gluons, which help hold quarks together, contribute another 36 percent of a proton's mass via their energy."

Here is another example, quoted directly [17]:

> "In quantum chromodynamics, the modern theory of the nuclear force, most of the mass of protons and neutrons is explained by special relativity. The mass of a proton is about 80–100 times greater than the sum of the rest masses of the quarks that make it up, while the gluons have zero rest mass. The extra energy of the quarks and gluons in a region within a proton, as compared to the rest energy of the quarks alone in the QCD vacuum, accounts for almost 99% of the mass."

Another example, quoted directly [18]:

"Quantum chromodynamic binding energy (QCD binding energy), gluon binding energy or chromodynamic binding energy is the energy binding quarks together into hadrons. It is the energy of the field of the strong force, which is mediated by gluons. Motion-energy and interaction-energy contribute most of the hadron's mass."

"Most of the mass of hadrons is actually QCD binding energy, through mass-energy equivalence. This phenomenon is related to chiral symmetry breaking. In the case of nucleons – protons and neutrons – QCD binding energy forms about 99% of the nucleon's mass. That is if assuming that the kinetic energy of the hadron's constituents, moving at near the speed of light, which contributes greatly to the hadron mass, is part of QCD binding energy."

And a final example, quoted directly [19]:

"Protons are not made of quarks the way that a wall is made of bricks but rather like the way that a fire is made of flames. They are seething balls of spontaneously forming and annihilating quarks. (...)"

"The number of quarks plus the number of antiquarks depends on how closely you look. Just as a coastline seems to get longer as you zoom in (because the true coastline winds around every grain of sand on the beach), the number of quarks and antiquarks increases at finer scales."

Thus, depending on the model you look at, there are three concepts that make up the missing 99% of the proton mass:

1. The binding energy of the gluon particle field has a mass associated with it, due to the energy/mass equivalence.
2. The quark masses are inflated relativistically as they move at relativistic speeds inside the proton.
3. The additional of non-valance quarks.

These three concepts will be explored in Sections 2, 3, and 4 of this paper.

## 2. Binding Energy and Conservation of Energy

### 2.1. Overview of the Concept that Missing Mass Comes from Binding Energy

One of the biggest mistakes is a sign error made when accounting for binding energy, and in particular, this is a violation of conservation of energy.

This erroneous concept, that mass comes from binding energy, claims the binding energy holding the quarks together within the proton has mass. Furthermore, this erroneous concept claims that this mass of this binding energy contributes substantially to the proton's missing mass [20]. The models go into complex mathematical detail, performing lattice quantum chromodynamic calculations by using super computers, to determine the binding energy. However, the model fails to understand that binding mass must be subtracted from the mass of the component parts; it is not added to it. And herein lies the sign error.

This is a blatant violation of conservation of energy. This statement is not due to a naivete or incomprehension of the complexities of Quantum Chromodynamics. Rather, this statement is a fact of thermodynamics that the QCD models of the nucleon must, at some point, admit is correct. The rest of this section 2 will demonstrate this fact to be true.

### 2.2. Review of Correct Binding Energy Concepts

Before proceeding, a brief review of the actual correct binding energy concept is in order. Consider several isolated parts, not interacting with each other, and with no force or energy shared among them. These parts come together and form a bound object, an object that is made up of these component parts. These parts are now bonded together by some type of force, holding the component parts together in a stable bound object. This is represented in the Fig. 1.

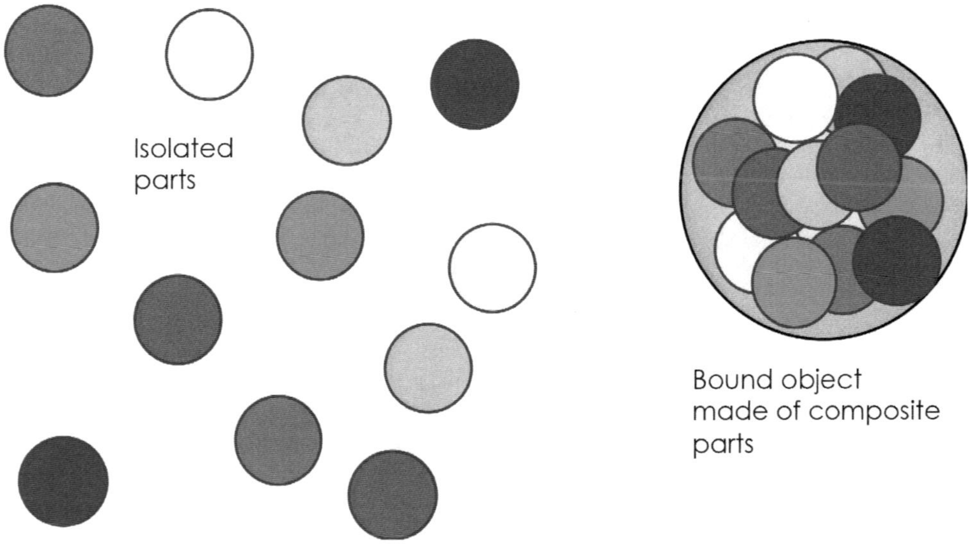


**Figure 1.** A bound object and its composite parts, before and after they are bound.

The total overall energy and mass of the bound object is less than the total overall energy and mass of the isolated component parts. The binding mass is defined as the difference between the mass of the bonded object and the mass of the isolated component parts. Similarly, the binding energy is defined as the difference between the energy of the bonded object and the energy of the isolated component parts. The binding mass and energy are related by $E=mc^2$, where c is the speed of light. By definition, that binding mass is positive, as is the binding energy. However, that binding mass is subtracted from the mass of the isolated component parts, in order to get the reduced and lower mass of the bonded object. Similarly, the binding energy is subtracted from the energy of the isolated component parts, in order to get the reduced and lower energy of the bonded object. This is shown in Eq. (1a) and (1b).

$$Mass_{bonded\ object} = Mass_{isolated\ component\ parts} - Mass_{binding} \tag{1a}$$

$$Energy_{bonded\ object} = Energy_{isolated\ component\ parts} - Energy_{binding} \tag{1b}$$

From thermodynamics, it is known that all systems in nature have a tendency to seek their lowest energy state. If there exists an attractive force among a group of separate particles, then this manifests itself as a tendency of the particles to bond together into a lower energy object, an object that is also less massive. If allowed, a group of particles sharing an attractive force among them, will transform into a bound object, whose total mass is less than the mass of the isolated composite particles. The mass difference is determined by $E = mc^2$, where E is the energy difference between the system of isolated particles and the bound object.

Stated again for clarity, the mass of the bonded object is less than the sum of the masses of the isolated component parts. This is true for any stable bonded object, regardless of the type of force that holds the parts together. The bonded object is in a lower energy than the isolated parts. Explicitly, this is a known fact because energy was released into the external environment when the component parts bonded together. The energy that was released to the external environment lowers the overall energy of the system, and puts the bonded object into a lower energy than the isolated component parts. These same arguments apply similarly for the mass, because of the energy/mass equivalence.

### 2.3. Energy/Mass Equivalence

Energy and mass are equivalent, and can be regarded as simply different units of the same thing. They are related by the constant $c^2$. If the energy of the object goes up, the mass of the object goes up. If the energy goes down the mass goes down. They are not traded off like a teeter totter, wherein one goes up and the other one goes down. Rather they are simply different units of the same thing, a concept that is defined as the energy/mass equivalence.

If there is confusion about this concept, that mass and energy are not traded off for one another, then it must be remembered that the system of bound particles is not a closed system. Rather, when the binding occurs, energy is removed from the system and released to the external environment. Similarly, when the unbinding occurs, the addition of energy is required from the external environment.

### 2.4. Example of Nuclear Bonding Energy for Fusion

One example of binding energy is that of nucleons that fuse together to form a nucleus. Energy is released from the system into the environment when fusion occurs, thus the energy of the system is less. An external source of energy, such as the energy of a collider or particle accelerator, is required to break apart a nucleus.

Nuclear science readily confirms that binding energy lowers the mass and energy of a stable nuclide. An atom, made up of protons, electrons, and neutrons, is at a lower mass than its component parts. This is shown in Eq. (2), for the nuclear force [21].

$$Mass_{atom} = (Z \times Mass_{1_H}) + (N \times Mass_n) - (Mass_{binding}) \tag{2}$$

The subscript $^1$H refers to the mass and energy of a hydrogen atom; this is used to properly and accurately include both the energy and mass of the electron as well as that of the proton. The Z is the number of proton and electrons, and N is the number of neutrons. $\mathrm{Mass_{1H}}$ is the mass of the proton and electron, and N is the mass of the neutrons. The first two addends in this equation are the total masses of the isolated component parts. The subtrahend is the binding mass. Note that this binding mass, which is a positive number, is subtracted. It is not added. Thus, the mass of the bound nuclide is always lower than the masses of the isolated composite parts.

Similarly for binding energy, the energy of the bound nuclide is less than the energy of the isolated composite parts, by the amount of the binding energy. This is shown in Eq. (3).

$$Energy_{bound\ object} = c^2 \left(Z \times Mass_{H1} + N \times Mass_n - Mass_{binding}\right) \tag{3}$$

Consider a proton and a neutron, isolated from each other, with no binding energy between them. If they get near enough to each other, they will bond via the nuclear force, to form $^2$H. This nuclear force is *not* like the gravitational field, which gradually decreases to zero as the distance increases. Nor is the nuclear force like a stretchy spring, wherein the further the distance, the harder it pulls back. Rather the nuclear force can be compared to being more like a snap. If the nucleons touch each other, they snap together, forming a strong bond. And when the proton and neutron form this bond, energy is released into the environment in the form of gamma radiation. For a neutron and proton forming $^2$H, this amount of released energy $\mathrm{E_{released}} = 2.224$ MeV. The energy of the bound object (the $^2$H) is less than the energy of the composite particles (the neutron and proton) by the amount of energy that was released into the environment. Furthermore, this amount of released energy is defined as the binding energy. The mass associated with this energy, the binding mass, is subtracted from the original masses of the neutron and proton. It would require 2.224 MeV or more of externally applied energy to pull the proton and neutron apart. This energy, supplied externally, is added back into the bonded object, in order to separate the two parts. Energy is always conserved throughout this process. Quantum physics, which is involved in this process, does not violate conservation of energy.

In nuclear physics, separation energies are the energies that must be added to the nuclide, by an external source of energy, in order to remove a nucleon. The separation energy for removing a neutron from a nuclide, $\mathrm{S_n}$, is seen in Eq. (4). The final product of this neutron removal is the isotope with one less neutron. Energy must be added to the original nuclide from an external source, in order to remove a neutron.

$$S_n = \left[m\left({}^{A-1}_{Z}X\right) - m\left({}^{A}_{Z}X\right) + m_n\right]c^2 \tag{4}$$

where: X represents a generic nuclide.

Similarly for the separation energy of removing a proton, $S_p$, is the energy that must be added to the nuclide in order to remove a proton, as in Eq. (5).

$$S_p = \left[m\left({}^{A-1}_{Z-1}X\right) - m\left({}^{A}_{Z}X\right) + m_{H1}\right]c^2 \tag{5}$$

The mass of a stable bound object is always smaller than the mass of the isolated component parts. The difference is defined as the binding mass. That binding mass multiplied by $c^2$ is defined as the binding energy.

The laws of quantum physics obey the laws of conservation of energy; when nuclear fusion occurs energy is released. The bound system is at a lower energy and at a lower mass than the unbound system of parts.

### 2.5. Atomic Example of Binding Energy

Consider the binding energy of a chemical bond, formed by the interaction of electrons and atoms. For example, an external source of electric energy is required to separate water, a bound molecule, into the component atoms of oxygen and hydrogen. This external energy is absorbed by the system when the chemical bond is broken; therefore the energy of the system of unbound particles is more. Similarly, energy is released into the external environment when a bond is made. For example, when the gases of oxygen and hydrogen combine to form water, there is a release of light and heat into the external environment. Energy is lost from the system, therefore the energy of the system is less. The chemical bonds of electrons and atoms obey the laws of conservation of energy; when a chemical bond occurs, energy is released. The bound system is at a lower energy and at a lower mass than the unbound system of parts, by the amount of the binding energy/mass.

### 2.6. Electromagnetic Example of Binding Energy

Another example of binding energy is electromagnetic. Quantum physics says that the electromagnetic interaction is mediated by photons, much the way that the chromodynamic interaction is mediated by gluons. Consider the binding energy of two glass beads, one with a positive charge and one with a negative charge. If these beads are isolated from each other, there is no force or interaction between them. If they are brought closer together, there is an attractive force that will draw them even closer together; this is a force that is mediated by the photons going back and forth between these two charged glass beads. Assuming the beads are free to move, they will eventually contact each other, emitting energy in the form of a noise, a vibration, friction, and heat into the external environment. Although this may not be a lot of energy, it is nonetheless energy that is released by the system. The beads are bonded together by the electromagnetic force, and by the photons that mediate this force. Energy is lost from the system, and thus the two beads, in their bonded state, are at a lower energy. To pull the beads apart would require an external energy from an external source. This external added energy would bring their unbonded isolated state back up to where it was at the beginning.

The important thing to note about this particular example is that massless particles, the photons, are mediating the interaction. With regards to conservation of energy/mass, this mediation due to the massless photon particles is a process that conserves energy/mass.

### 2.7. Gravitational Example of Binding Energy

As another example, consider the force of gravity, a force that has both a field and binding energy. Consider several massive rocks that come together in outer space to form an asteroid. That asteroid has a gravitational field, and that gravitational field has energy. The gravitational field has binding energy, which has bound the rocks together to form the asteroid. When the rocks come together, they crash into each other and emit energy in the form of light and heat, into the external environment. (Any small debris that scattered away will eventually return, thus this is not the mass that is lost.) The released energy, which was released in the form of light and heat, is the energy that must be subtracted from the overall energy of the asteroid, as compared to its isolated composite parts. Thus, the overall energy/mass of the asteroid is slightly less than the sum of its isolated composite energy/masses. To unbind the asteroid back to the separate isolated rocks, a large external explosive energy would be needed to do so. This large explosive energy would provide the external energy to unbind the asteroid; however, that external energy source would have to be bigger than the gravitational binding energy of the asteroid in order to successfully dissociate it.

In this example, there is a force (gravity), and that force has a field (the gravitational field). That field has binding energy and binding mass. The overall energy/mass of the asteroid was reduced because a quantity of energy was lost into space when the asteroid was formed. Thus, the binding energy/mass is subtracted from the energy/mass of the composite parts. In order to return the asteroid to its composite isolated rocks, external energy must be added to it, to dissociate it.

The laws of gravity obey the laws of conservation of energy; when a gravitational bond occurs, energy is released. The bound system is at a lower energy and at a lower mass than the unbound system of parts.

### 2.8. Summary of Correct Binding Energy Concepts

Regardless of the force—electromagnetic, chromodynamic, nuclear, weak nuclear, gravity, or some new unusual force that has not yet been discovered—if two or more particles are bound together by a force, then the binding energy of that force is subtracted from the sum of the energies of the component particles. This subtraction of energy is due to the law of conservation of energy. This required subtraction of the binding energy is independent of the particular force involved. Table 2 [22] lists the three interactions of the microscopic world, i.e., of quantum physics, and all of the quantities that are conserved in certain interactions. It should be emphasized that no force is allowed to violate

**Table 2.** Applicability of the Conservation Laws to the observed interactions. "yes" means conserved, "no" means not conserved.

| Quantity Conserved | Strong | Electromagnetic | Weak |
|---|---|---|---|
| Energy | yes | yes | yes |
| Linear momentum | yes | yes | yes |
| Angular momentum | yes | yes | yes |
| Charge | yes | yes | yes |
| Electronic lepton number | yes | yes | yes |
| Muonic lepton number | yes | yes | yes |
| Tauonic lepton number | yes | yes | yes |
| Baryon number | yes | yes | yes |
| Isospin magnitude | yes | yes | yes |
| Isospin z component | yes | yes | yes |
| Strangeness | yes | yes | yes |
| Parity | yes | yes | yes |
| Charge conjugation | yes | yes | yes |
| Time reversal (or CP) | yes | yes | yes |

conservation of energy. All of these relations among conservation laws, invariance principles, and symmetries can be proved classically or quantum mechanically [23]. Stated specifically, the Strong Force of Quantum Chromodynamics does not violate conservation of energy.

To summarize, particles held together by binding energy are in a lower energy and a lower mass than the isolated component particles. That is because, when a bond is formed, energy is released to the external environment. This energy is removed from the system of particles. The energy is less as a result of this released energy. Similarly, with mass. Conversely, to break a bond, an external energy must be added to the bound object to dissociate it. As a result, the isolated dissociated particles are in a higher energy state, due to the external energy that was added to the system. Thus, the bound particles in an object are at a lower energy state and a lower mass than the isolated component particles. The isolated component particles are at a higher energy state and a higher mass than the bound particles.

Using the complex quantum chromodynamic calculations, the recent models for nucleons have calculated the amount of the binding energy. However for proper conservation of energy, this binding energy must be subtracted from the rest mass of the isolated quarks, not added to it.

### 2.9. Gluons

The binding force holding the quarks together is postulated to be mediated by a massless particle called a gluon. These models do not claim that the massless gluons contribute to the mass of the proton. The models do claim, however, that the gluons have a "gluon particle field", and that the particle field has energy. This gluon particle field acts similar to the force of a stretchy spring–the further it is stretched, the harder and stronger it pulls back. However, it does not matter that the chromodynamic force is caused by massless gluons inside the proton. The electromagnetic force is thought to be caused by massless photons, and we know that the electromagnetic force abides by conservation of energy. The nuclear force is thought to be mediated by pions, and we know that the nuclear force abides by conservation of energy. Similarly, gravity is thought to be caused by massless gravitons, and we know that gravity still abides by the laws of conservation of energy. It is also known that the gravitational field has energy associated with it. However the energy of that gravitational field does not add to the mass of a bound object, as we saw in our examples. Similarly, the gluon particle field has energy associated with it, and, similarly, the energy of that gluon field does not add to the mass of the bound object.

If the proton is stable, the quantum chromodynamic binding energy is subtracted from the mass of the three component quarks, and overall net mass is less than the rest mass of the three isolated quarks.

### 2.10. Heisenberg Uncertainty Principle

The Heisenberg Uncertainty Principle is shown in Eq. (6).

$$\Delta E \Delta t > \frac{\hbar}{2} \tag{6}$$

This uncertainty principle states that the uncertainty in the energy, $\Delta$E, of a particle times the uncertainty in the lifetime, $\Delta$t, of the particle can only be measured and discernible when that product is greater than Planck's reduced constant divided by two. Thus, this uncertainly principle allows for extremely small variations in energy that may violate conservation of energy for extremely short durations of time. These violations must be so small and so short in their duration, that they are indiscernible and unmeasurable. Specifically, according to the Heisenberg Uncertainty Principle, energy conservation may be violated as long as the product of the energy of the violation multiplied by the time duration of the violation is *less than* the reduced Planck constant divided by two, which is an extremely small value, $5.237 \times 10^{-35}$ in MKS units. In other words, violations of energy are allowed only if the violation is so small that it is not detectable and not measurable. For this situation, the inequality sign usually associated with the Heisenberg Uncertainty Principle, is reversed to show the condition of this energy abnormality being undetectable and indiscernible. This is shown in Eq. (7), with the inequality sign properly and correctly reversed to demonstrate this condition of indiscernibility.

$$\Delta E \Delta t < \frac{\hbar}{2} \tag{7}$$

It should be noted that this uncertainty principle, depicted in Eqs. (6) and (7), is valid only for uncorrelated states of the considered particle. In particular, if either or both of the conditions exist:

- "phase memory"
- cross-correlation for different states of the same particle

then, for a given quantum system, the states are considered to be correlated. When such a correlation exists, then the Heisenberg uncertainty relation of Eq. (6), is replaced by the Robertson-Schrödinger uncertainty relation, shown in Eq. (8).

$$\Delta E \Delta t > \frac{\hbar}{2} \frac{1}{\sqrt{1-r^2}} \tag{8}$$

where: r is the correlation coefficient, and $|r| < 1$.

As seen in Eq. (8), the Robertson-Schrodinger uncertainty relation [24], [25] allows for an uncertainty that is different from that of Eq. (6), specifically that the product of the uncertainties may be larger that that allowed by Heisenberg.

The inequality sign is reversed for the condition of indiscernibility, as shown in Eq. (9).

$$\Delta E \Delta t < \frac{\hbar}{2} \frac{1}{\sqrt{1-r^2}} \tag{9}$$

In such cases, it is shown in [26], [27] that the existence of larger energy fluctuations over longer time intervals is possible.

However, even with this advanced understanding of the distinctions of the various Uncertainty Principles, this uncertainty allows for only a very small violation of energy to exist for only a very short duration of time. When the effects become macroscopic in size and time, quantum physics theories, including the uncertainty principles, adhere to the laws of conservation of energy. Uncertainty principles cannot violate conservation of energy for 99% of the visible matter in the universe, for billions of years.

As we have seen, electrons around an atom and molecule, such as water, are within the quantum realm of physics, and their binding energy obeys energy conservation laws. Similarly, nucleons in a nucleus are within the quantum realm of physics, and the binding energy of a nucleus obeys energy conservation laws. Consequently, a valid model of the partons inside a nucleon cannot naively claim that a violation of the laws of conservation of energy are allowed to occur simply because the theory is quantum.

### 2.11. Mass Creation and Chiral Symmetry Breaking

The second rationale sometimes used as to why energy/mass conservation can be violated is due to the QCD chiral symmetry breaking. Below is a quote making this claim [28].

> "Chiral symmetry breaking is most apparent in the mass generation of nucleons from more elementary light quarks, accounting for approximately 99% of their combined mass as a baryon. It thus accounts for most of the mass of all visible matter."

This quote claims that the quantum chromodynamic chiral symmetry breaking is responsible for the generation of 99% of the energy/mass of the visible universe. However, this is an erroneous claim and a misinterpretation of QCD chiral symmetry breaking.

### 2.12. A Brief Description of Chiral Symmetry Breaking Theory

Experimentally, it is observed that the masses of the pseudo-scalar mesons, such as the pi-meson, are less massive than the masses of the vector mesons, such as the rho-meson. This mass difference between these two types of mesons, according to the theory, is due to spontaneous symmetry breaking of chiral symmetry within the fermion sector of QCD, specifically the broken chiral symmetry of the up, down, and strange flavors of quarks.

Here is an explanation, from a science textbook, of chiral symmetry breaking [29]. In this description, the chiral symmetry breaking is compared to an analogy of a large magnet, inside of which the numerous smaller atomic magnets are aligned. The term symmetry means "sameness", and in this particular ferromagnetic analogy, it refers to the "sameness" of the rotational direction of the atomic magnets.

> "This is an example of what physicists term spontaneous symmetry breaking: the symmetry of the lowest energy state is less than that of the interaction in question. Some people say that the expression 'spontaneous symmetry breaking' is a little misleading. The lowest energy state, or ground state, of the magnet, that in which all atomic magnets are aligned, does not share the rotational symmetry of the basic interaction between those atomic magnets. So the symmetry of the interaction is just hidden, but hasn't gone away. Hidden symmetry, they argue, is maybe a better expression than spontaneous broken symmetry. But it's the latter that has taken hold."

In this ferromagnetic example, the term "spontaneous symmetry breaking" means that although the atomic magnets are aligned, the large magnet can rotate in all directions. As a result, the large magnet is less "symmetrical" than the atomic magnets. However, even though the alignment of these atomic magnets is hidden, that alignment is still there. The term, "spontaneous broken symmetry" is really a misnomer, the symmetry is not broken; it is hidden. In this ferromagnetic example, the alignment and the symmetry of the atomic magnetics are hidden within the large magnet.

The book goes on to describe chiral symmetry breaking for QCD [30].

> "The source of that symmetry breaking is the vacuum. The vacuum fails to respect the chiral symmetry of the theory, and thus the vacuum's symmetry is less. In the ferromagnetism example above, the aligned multitudinous atomic magnets gave a lowest energy configuration having a reduced symmetry relative to the interaction between those magnets, yielding Goldstone bosons. The chiral QCD case runs exactly parallel, the vacuum provides the analog of the lowest energy state, and its disregard for chiral symmetry mimicking the aligned atomic magnets' disdain for the rotation-independence of the interaction.
>
> "In the ferromagnetic case, symmetry breaking results in magnons. In QCD, the result of breaking symmetry results is a fistful of mesons: three pions, $\pi^0$, $\pi^+$, $\pi^-$, four Kaons, $K^0$, anti-particle $K^0$, $K^+$, and $K^-$, and last but not least $\eta$. When the symmetry that is broken is exact, then the Goldstone bosons are massless. If the symmetry is not exact, but only approximate, then the result is, instead, low-mass spinless particles called pseudo-Goldstone bosons, instead of massless Goldstone bosons. Real up, down, and strange quark masses, though small on the energy scale of strong interactions, are not exactly zero, so real QCD does not have an *exact* chiral symmetry only an *approximate* chiral symmetry. So chiral symmetry breaking in QCD gives particles having a small mass, the eight mesons listed above. Through this mechanism, meson masses are linked to the (current) quark masses.
>
> "Indeed, chiral symmetry breaking is the route to estimating the up quark mass as 4 MeV and the down quark mass of 7 MeV. The effective masses of quarks bound into observable particles such as pions, protons, and neutrons, in other words the constituent masses of around 300 MeV, are due to the interaction of

quarks with the surrounding chiral-symmetry-breaking vacuum. In this way, spontaneous symmetry breaking is responsible for effective quark masses. The chiral symmetry aspects of QCD are also the launch pad for another calculational tool that works over longer distance scales where conventional QCD perturbation theory breaks down. This chiral perturbation theory, which takes as its starting point the perfect chiral symmetry of a hypothetical QCD with massless quarks, then allows actual real masses to enter as perturbations."

In other words, the theory of spontaneous chiral symmetry breaking, along with QCD chiral perturbation theory, can calculate the masses of the up and down quarks, as perturbations around an ideal situation of zero-mass quarks.

Here is another quote from a scientific peer-reviewed journal, showing the mathematical relationship of chiral symmetry breaking and the mass of the pions [31]:

"In the standard understanding of the strong force interaction as defined by quantum chromodynamics, pions are loosely portrayed as Goldstone bosons of spontaneously broken chiral symmetry. That explains why the masses of the three kinds of pions are considerably less than that of the other mesons, such as the scalar or vector mesons. If their current quarks were massless particles, it could make the chiral symmetry exact and thus the Goldstone theorem would dictate that all pions have a zero mass.

"In fact, it was shown by Gell-Mann, Oakes and Renner (GMOR) that the square of the pion mass is proportional the quarks condensate $M_\pi^2{=}(m_u + m_d)B + \vartheta(m^2)$, with $B = |\langle 0|\overline{u}u|0\rangle / f_\pi^2|_{m_q \to 0}$ the quark condensate. This is often known as the GMOR relation and it explicitly shows that $\mathrm{M}_\pi = 0$ in the massless quark limit."

As is noted in the quote above, the theory of chiral symmetry is idealized for massless quarks; if these quarks were massless, then this chiral symmetry breaking would not occur. However, since the quarks are not idealized massless quarks, they break this symmetry.

The masses of the pseudo-scalar mesons are specified by the of chiral perturbation theory. This chiral perturbation theory provides an internally-consistent argument, in which the pseudo-scalar meson masses are related to the square-root of the quark masses, but for only the smallest three quarks, the up, down, and strange. For the charm, bottom, and top quarks, their masses are much larger than the scale for the QCD spontaneous chiral symmetry breaking. Hence, the three larger quarks cannot be treated as small perturbations around the explicit symmetry limit of zero, and the chiral perturbation theory cannot be applied to these larger quarks.

What is important to highlight is that the theories of QCD chiral symmetry breaking do not claim that chiral symmetry breaking is able to generate energy/matter out of nothing. Rather, the models simply claim that chiral symmetry breaking is the basis for the predictions of the small masses of the up and down quarks. The misconstrued claim that chiral symmetry breaking is responsible for the creation of 99% of all visible energy/mass in the universe is an erroneous concept. It is a concept not made by the peer-reviewed scientific journals, but rather by a naïve misinterpretation of the model.

In conclusion, the creation of energy/mass due to chiral symmetry breaking should not be considered as more valid or more credible than the laws of thermodynamics, such as conservation of energy.

### 2.13. Experiments to Break Apart a Proton

The proponents of these recent nucleon models claim that QCD gluon binding energy creates a positive mass, thereby making the proton heavier. However, if these proponents actually did believe that the binding energy of the quarks somehow makes the proton heavier, then the experiments attempting to separate the quarks would have to pull that binding energy out of the proton. However they do not do this in their experiments. Rather they add energy into the proton, energy from an external source, such as a particle acceleration or collider. They add this energy in an attempt to break the QCD bonds and dissociate the proton into its component parts. This is an extremely important statement and bears repeating.

Experiments attempt to separate the quarks by adding external energy into the proton. This is done in order to put the proton into an excited energy state, and to use that added external energy to break the chromodynamic bonds within the proton.

In other words, the particle physicists cannot deny that they do, indeed, acknowledge that the object of a bound proton is at a lower energy state than its composite parts. This is why they add external energy into the proton, to try to increase the energy of the proton, to break the chromodynamic bonds. If the particle physicists truly thought that the binding energy was some hypothetical thermodynamically-impossible "substance" that increased the mass of the proton, they would not try to break the chromodynamic bonds by putting energy into it.

The mathematics of this chromodynamic bond is very complicated and difficult to understand. From this very complicated mathematics and computations, the models have derived a value for the binding energy of the chromodynamic force between the quarks. Those calculations, and that complexity behind them, is not being questioned here. The deserved respect of the particle physicists who have made these calculations is not being questioned here; they are highly respected people and they remain respected. The calculations of the absolute value of the binding energy are not being questioned here.

What is being questioned, and called out as incorrect, is that the binding energy is being erroneously added, rather than being properly subtracted.

### 2.14. Conclusions about Binding Energy

Since these quarks are bound together inside a nucleon, held together by the extremely strong binding energy of the chromodynamic force, the resulting collection of three quarks bound together in a nucleon is less massive than the mass of the three isolated quarks, as seen in Eq. (10).

$$Mass_{nucleon} < \sum_{i=1}^{3} Mass_{quark\_i} \tag{10}$$

Regardless of the force involved, regardless of quantum uncertainty, regardless of quantum chiral symmetry breaking–the QCD binding energy of the three quarks does not increase the mass of the nucleon. As mentioned previously, this statement is not due to a naivete or incomprehension of the complexities of Quantum Chromodynamics. Rather, this statement is a fact of thermodynamics that is incontrovertibly correct.

## 3. Relativistic Mass Inflation

### 3.1. Overview of Concept that Missing Mass is Due to Relativistic Mass Inflation

Another concept to make up the missing 99% mass is to make the three quarks move extremely fast, such that they inflate their mass relativistically and become much more massive, enough so to account for the 99% of the missing proton mass. This relativistic speed increases the mass of the quarks due the relativistic mass inflation from Einstein's special theory of relativity, as shown in Eq. (11).

$$Mass_{relativistic} = \frac{Mass_{rest}}{\sqrt{1 - \left(\frac{v}{c}\right)^2}} \tag{11}$$

As can be ascertained from this equation, when the velocity v is close to the speed of light c, the denominator in this equation becomes less than 1, and the relativistic mass become larger than the rest mass, as a result. To increase the velocity of an object, external energy is added to the object, in the form of kinetic energy. This added energy thus increases the overall energy and mass of the object. Thus, Eq. (11) is simply another way of expressing the equation of E=mc$^2$. In this case, the energy added to the object is in the form of kinetic energy, enough to bring its velocity up to relativistic speeds. This process of increasing the mass of an object, by adding kinetic energy to the object, is known as relativistic mass inflation.

### 3.2. Relativistic Mass Inflation Applied to Quarks in a Proton

If this relativistic mass inflation is to account for 99% of the mass of the proton, then the quarks need to go 0.9999349 times the speed of light, as shown in (Calc. 1).

$$0.99 \times Mass_{proton} = \frac{(2.5 + 2.5 + 5.6)}{\sqrt{1 - \left(\frac{0.9999349\ c}{c}\right)^2}}$$

$$0.99 \times 938.3\ \mathrm{MeV}/c^2 = \frac{(10.6)}{\sqrt{1 - (0.9999349)^2}}$$

$$928.9 \frac{\mathrm{MeV}}{c^2} = 928.9\ \frac{MeV}{c^2}\quad check \qquad (Calc.\ 1)$$

However, if the quarks are moving at 0.9999349 c, the quarks would instantly be hurled from the proton. This added kinetic energy, which is needed to inflate the relativistic mass of the quarks, would give the quarks so much momentum that they would instantly explode out of the proton.

That is, unless the binding energy is stronger than this relativistic kinetic energy.

In order to prevent the quarks from instantly exploding out of the proton, we need a binding energy that is stronger than the relativistic kinetic energy of the quarks, as shown in Eq. (12).

$$E_{binding} > E_{kinetic}\ where:\ E_{kinetic} = (m_{relativistic} - m_{rest})\, c^2 = (\Delta m)\, c^2 \qquad (12)$$

To clarify, the binding mass must be greater than the relativistic mass increase of the quarks, if the proton is to be a stable particle. Thus the two energies, the binding energy and the relativistic kinetic energy, are in an energetic tug-of-war with each other.

As is evident in the quotes above for these new models, they are claiming that the kinetic energy adds to the binding energy. However, this concept is incorrect. Binding energy and kinetic energy oppose each other; they subtract from each other, they do not add to each other. In order for the proton to be stable, with super-speed relativistic quarks zooming around inside it, the binding energy of the chromodynamic force would have to be stronger than the kinetic energy of the relativistic mass inflation. Otherwise, the proton would explode from its own internal kinetic energy. Thus, if we use relativistic speeds to increase the mass of the quarks, the binding energy of the proton must reduce that mass by more than that amount. Herein lies another mistake, and another sign error, of these current nucleon models. Relativistic mass inflation cannot be used to account for the missing mass of the proton.

If the quarks are to be bound inside the proton, then the binding energy has to be more than their kinetic energy, in order to contain them inside the nucleon. If the proton is to be stable, then the relativistic mass increase of the quarks cannot be more than the mass decrease of the binding energy. In other words, when comparing the mass of the proton to the rest mass of the isolated quarks, there must be an overall net decrease in the mass–a decrease which is caused by the binding energy being greater than the kinetic energy. The net mass of the three quarks inside a stable proton cannot be increased by kinetic energy more than the mass of the proton is decreased by the binding energy.

### 3.3. Example of Binding Energy vs. Kinetic Energy

For example, imagine several glass beads sitting on a flat smooth turntable, connected together by the thin threads of a spider web. The threads of the web are the binding force holding the beads together. If that turntable starts to spin and gives the beads some kinetic energy, then the beads will fly off the turntable, because the thin spider webs are too weak to hold the beads together against the centrifugal forces. However, if the spider webs were replaced by sturdier wires glued to the beads, then the beads would not be able to fly off the turntable. This simple analogy is similar to the kinetic energy and binding energy of the quarks inside a proton. The binding energy must be stronger than the kinetic energy for a stable object.

### 3.4. Summary of Mass Inflation Concept

The proton is stable only if the binding energy of the chromodynamic force is stronger than the internal kinetic energy of the component parts; otherwise, the proton would explode due the internal kinetic energy of its own quarks. The binding energy of the composite particles is in direct conflict with their kinetic energy. The binding energy of the object reduces the mass of the object. Conversely, the kinetic energy of its internal particles increases its energy. In order for the object to be stable and not explode, the binding energy of an object must be greater than its internal kinetic energy. Thus, the decrease in mass due to binding energy must be more than the increase in mass due to kinetic energy.

To summarize, within a stable bound object, the two energies–the kinetic energy of the composite particles and binding energy holding them together–oppose each other and are in direct conflict with each other. These two energies, the internal kinetic energy and the binding energy, are not added together, rather they are in conflict with each other. In order for an object to be stable, the binding energy must be more than the kinetic energy. This means that the overall mass of a stable bound object is less than the rest mass of its composite particles, regardless of the relativistic velocity of the composite particles.

Hence the idea that the missing mass of the proton is due to the relativistic mass inflation of the quarks is not a valid concept for a stable nucleon. If the quarks are to be bound together inside a stable nucleon, the overall mass of the bound object must be at a lower mass than the rest mass of the original isolated composite parts.

## 4. The Addition of More Non-Valance Quarks

### 4.1. Overview of Concept that Missing Mass Comes Additional Quarks

Another concept for adding mass to the proton is to add more non-valance quarks into the proton, hundreds of them, to make up for the missing 99% of the mass. The idea behind this concept is simply to add more quarks, called non-valence quarks or "sea" quarks, into the nucleons, to give the nucleons more mass. The quarks must be added in groups of six to keep the colors, spins, and electric charges neutral. At the very least, this gives nine quarks inside a nucleon: three valance quarks and six non-valence quarks.

### 4.2. Additional Non-Valance Quarks and the Pauli Exclusion Principle

According to the Pauli Exclusion Principle, there cannot be particles in a quantum object with the same set of quantum numbers. Quarks are quantum objects. Historically, this is why colors were assigned to the quarks, in order to distinguish the three quarks from one another. Hence the quarks were assigned colors–red, green and blue–to distinguish their properties and provide them with different quantum states. With regards to color, there cannot be two red quarks or two green quarks or two blue quarks inside the same nucleon, otherwise they would conflict with the Pauli Exclusion Principle [32]. All of these additional non-valance quarks would violate the Pauli Exclusion Principle.

### 4.3. Additional Non-Valance Quarks and the Heisenberg Uncertainty Principle

All of these additional non-valance quarks would also violate the Copenhagen Interpretation of the Heisenberg Uncertainty Principle. The Heisenberg Uncertainty Principle claims that the uncertainty in the momentum of an object times the uncertainty of its position cannot be smaller than the reduced Planck constant divided by two. This is similar to the equation as discussed in section 2, but with the two conjugate-pair uncertainties converted to momentum and position. This is shown in Eq. (13).

$$\Delta p \Delta x \geq \frac{\hbar}{2} = 5.237 \times 10^{-35}\ m^2 kg/s \tag{13}$$

This uncertainty relationship for quarks is now examined, using a non-relativistic calculation first. The uncertainty in the position of the quarks is the diameter of the proton, $D_{\text{proton}} = 1.682 \times 10^{-15}$. The mass of the up quark as $3.922 \times 10^{-30}$ kg. For non-relativistic non-inflated mass, the largest number we can reasonably use for the uncertainty in the velocity of the quarks is about one tenth the speed of light, or $2.99 \times 10^7$ m/sec. This gives us a small enough mass inflation that we can reasonably consider it to be non-relativistic. Using no mass inflation and 0.1c as the uncertainty in the velocity, and putting these numbers in Eq. (9) in MKS units, we get (Calc. 2):

$$\Delta p \Delta x = \left(3.922 \times 10^{-30}\right)\left(2.99 \times 10^{7}\right)\left(1.682 \times 10^{-15}\right) = 1.97 \times 10^{-37}\ m^2 kg/s \qquad (Calc.\ 2)$$

The Heisenberg Uncertainty Principle insists that this product of uncertainties must be bigger than $5.237 \times 10^{-35}\ m^2kg/s$; clearly it is not. Thus, the addition of hundreds of non-valence quarks into the proton now violates not just one, but two, of the foundational pillars of quantum physics.

Without mass inflation, the mere existence of a 2.6 MeV/$c^2$ quark confined inside the proton violates the Heisenberg Uncertainty Principle. Relativistic mass inflation must be used to increase the mass of the quarks, such that they do not violate the Heisenberg Uncertainty Principle. Indeed, this relativistic mass inflation is how the theoretical models attempt to prevent this violation [33]. This possibility is explored below.

### 4.4. Relativistic Mass Inflation to Avoid Violating Uncertainty Principles

Again, it is assumed that the uncertainty in the quark position is the diameter of the proton, $D_{\text{proton}}$. The mass of the quark is now the relativistically inflated mass. Combining Eqs. (8) and (9), results in Eq. (14).

$$\left(\frac{m_{quark\_rest}}{\sqrt{1-\left(\frac{v_{quark}}{c}\right)^2}}\right)(v_{quark})(D_{proton}) \geq \frac{\hbar}{2} \tag{14}$$

Since this is a transcendental equation for the quark velocity $v_{\text{quark}}$, this is solved by iteration. Using the values of $D_{\text{proton}} = 1.682 \times 10^{-15}$ meters, $m_{\text{quark\_rest}} = 3.922 \times 10^{-30}$ kg, and $h/4\pi = 5.275 \times 10^{-35}$ m$^2$kg/s, and $c = 2.9979 \times 10^8$ m/s, we get the required mass of the quark as $1.04689 \times 10^{-28}$ kg, and the uncertainty in the velocity is 0.999298c. This is checked below, in (Calc. 3).

$$\left(\frac{3.922 \times 10^{-30}}{\sqrt{1-\left(\frac{0.999298\ c}{c}\right)^2}}\right)(0.999298)(2.9979 \times 10^8)(1.682 \times 10^{-15}) \geq 5.275 \times 10^{-35}$$

$$\left(1.04689 \times 10^{-28}\right)(0.999298)(2.9979 \times 10^8)(1.682 \times 10^{-15}) \geq 5.275 \times 10^{-35}$$

$$5.275 \times 10^{-35}\frac{m^2kg}{s} \geq 5.275 \times 10^{-35}\frac{m^2kg}{s}\ \ check \qquad (Calc.\ 3)$$

These values narrowly allow the up quark to fit within the confines of the Heisenberg Uncertainty Principle. This required minimum mass is $1.04689 \times 10^{-28}$ kg $= 58.72$ MeV/c$^2$. This is 26.7 times the original rest mass of the quark, and for the one quark, it is 6.258% of the mass of a proton.

If the quarks are in correlated states inside the proton, the Robertson-Schrodinger uncertainty principle of equation (8) would apply. Similar modified uncertainty relations take place for the uncertainty in spatial coordinates and momentum for particles in a coherent correlated state, shown in Eq. (15).

$$\Delta p \Delta q \geq \frac{\hbar}{2} \frac{1}{\sqrt{1-r^2}} \tag{15}$$

This would give a different value for the required inflated mass needed to satisfy the uncertainty principle, depending on the correlation coefficient. Regardless, these quarks would instantly explode from the proton, unless the binding mass is more than the relativistic mass inflation, as was discussed in Section 3. Even if hundreds of quarks are added, and their mass is inflated with relativistic speeds, they must be held together by a binding energy that is stronger than their kinetic energy. And recall from section 2, that binding energy is subtracted, not added, and as a result, all of that inflated mass is negated by the binding. What remains is a proton that is less massive than the sum of the rest masses of the quarks inside it.

### 4.5. Possible Ways to Revise These Recent Nucleon Models with Hundreds of Quarks

One possible way to revise these models, to remove this violation of the Pauli Exclusion Principle could be done conceptually by adding more colors, in groups of six, to the theory of chromodynamics. However, we need hundreds of colors as determined in Section 4.4. This would require that quantum chromodynamics be rewritten to accommodate hundreds of colors. Not only is this option theoretically unattractive, another issue is that the experimental data strongly implies that there are only three colors [34].

In order to fit within the confines of the Copenhagen Interpretation of the Heisenberg Uncertainty Principle, the new models of the nucleons would also have to be revised. One possible way to revise the models in order to remove this violation of the Uncertainty Principle, conceptually, would be to claim that the Copenhagen Interpretation of the Heisenberg Uncertainty Principle does not apply to quarks since quarks are not isolated particles. Or perhaps it could be claimed that Einstein was right all along about the Copenhagen Interpretation of the Heisenberg Uncertainty Principle. Again, these options are unattractive to most physicists.

### 4.6. Fractals, Antimatter, and the Residual Quantum Chromodynamic Model

The nuclear force is the force which holds the nucleons together in a nuclide. The most widely accepted model of the nuclear force is the Residual Quantum Chromodynamic model. This Residual Quantum Chromodynamic Model would have serious difficulty reconciling itself with these recent nuclide models and the "seething balls of spontaneously forming and annihilating quarks" purported to exist inside each nucleon. Specifically, the residual quantum chromodynamic bond that exists between the inter-nucleon quarks would have extreme difficulty surviving the continuous creation and annihilation of its quarks. The fractal concept would add another absurdly complicated dimension into these already complex models. Any reconciliation between these recent models of the nucleon and the Residual Quantum Chromodynamic Model would be extremely difficult.

### 4.7. Conclusion about Adding More Quarks

Thus, adding more quarks does not work well as a concept–requiring revisions not only of the recent nucleon models, but also of Quantum Chromodynamics and the Heisenberg Uncertainty Principle. However, such revisions are not necessary if the binding energy and kinetic energy are simply accounted for correctly, without sign errors. The addition of more non-valance quarks is an unattractive and troublesome idea for numerous reasons. It seems to be a better idea to maintain the concept of three quarks and to simply use proper accounting for the kinetic and binding energies.

Thus, we are back to the conundrum of having 99% of the mass of the proton unexplained. However, recall that this 99% missing mass is based on the theoretical calculations. It is not based on actual experimental measurements of the quark mass.

## 5. Theory versus Experiment

### 5.1. The Standard Model

The complex calculations of the Standard Model have theoretically predicted that quarks have an unexpectedly small mass. As detailed in the 2005 article "Beyond the Standard Model" [35], [36], it is stated that the Standard Model is inherently an incomplete theory, with revisions and changes being applied as needed. The Standard Model should not be considered, at this point in the progress of physics, as the absolute truth. For example, it does not explain gravity, dark matter, dark energy, neutrino mass, or the anti-matter/matter asymmetry of the universe. Thus, a revision of one small part of it, specifically the chiral symmetry breaking, might be appropriate to adjust for the masses of the quarks.

### 5.2. Lower Limit of Quark Mass

Although it is not the main focus of this paper, an interesting result becomes obvious, and should be mentioned. With the understanding that binding energy is subtracted from the mass of the isolated component particles, this information readily gives us a lower limit to the mass of the up and down quarks. If the quarks inside a proton cannot be unbound by the addition of 139.6 MeV/c$^2$ of energy, then we know the binding energy is <u>at least</u> this much. And, this same binding energy has decreased the mass of the three isolated quarks by at least that amount. Therefore, the rest mass of the partons is at least the mass of the proton plus the amount of 139.6 MeV/c$^2$. This can be determined by rearranging Eq. (1). Including the inequality, this is shown in Eq. (16). The calculation is shown in (Calc. 4).

$$Mass_{component\ parts} \geq Mass_{bonded\ object} + Mass_{binding} \tag{16}$$

$$Mass_{component\ parts} \geq 938.3\frac{MeV}{c^2} + 360\frac{MeV}{c^2} \geq 1077.9\ \frac{MeV}{c^2} \qquad (Calc.\ 4)$$

Assuming there are no other massive particles, as yet undiscovered, inside the proton, this gives the sum of the masses of the three quarks to be at least 1077.9 MeV/c$^2$. Thus, each of the masses of the quarks are at least 350 MeV/c$^2$ $= 6.4 \times 10^{-28}$ kg. However, because of the sign error made in accounting for the energies, this simple determination of the lower limit for the mass of the quarks has not previously been recognized.

Since the up and down quarks have a much larger mass than was previously thought, their masses are too large for perturbation theory to effectively be applied to them. This means that the up and down quarks should not be treated as small perturbations around the explicit symmetry limit of zero mass, and the chiral perturbation theory should not be applied to them.

Also, the experimental attempts to dissociate the proton are done by <u>adding</u> energy/mass into the proton. There are no experimental attempts to <u>pull</u> energy out of the proton in order to dissociate it. Because of these experiments, scientists already know that the quarks confined inside the proton are in the lowest energy state, specifically the proton is in a state of lower energy than the dissociated quarks would be. As is demonstrated by the experiments that attempt to separate the quarks by adding external energy into the proton—it is known that the binding energy/mass of the bound system of a proton is at a lower energy/mass than the isolated quark components.

By correctly accounting for the additional energy needed to dissociate a proton, the experimental data directly contradicts the theoretical estimates of the chiral perturbation theory of small quark masses for the up and down quarks.

### 5.3. Conclusions about Theory vs. Experiment

"It doesn't matter how beautiful your theory is, it doesn't matter how smart you are. If it doesn't agree with experiment, it's wrong ..." –Richard Feynman.

## 6. Overall Conclusions

There are sign errors in the new models of the proton, errors that are made when trying to account for the various energies and masses. Also there are violations of Thermodynamics, Energy Conservation, Pauli Exclusion Principle, and the Heisenberg Uncertainty Principle. Furthermore, these new models cannot be reconciled with the currently accepted models of the nuclear force.

By applying proper accounting of the energy/mass within the nucleon, the following conclusions can be made. There are only three quarks inside each nucleon. Assuming there are no other as-yet-undiscovered massive particles inside the proton, then each of these three quarks inside a nucleon has a mass of at least 350 MeV/$c^2$. None of the quarks violate the Pauli Exclusion Principle. None of the quarks violate the Heisenberg Uncertainty Principle. There is no violation of Thermodynamics. There is no violation the Conservation of Energy. And most importantly, there are no sign errors.

Nucleon models are the building blocks for Nuclear Force models. Similarly, the nucleon models are used in Cosmology for a better understanding of the beginnings of the universe. For a better understanding of nuclear physics and cosmology, nucleon models must be free of blatant violations of sign errors, violations of thermodynamics, violations of conservation of energy, and violations of the foundational pillars of modern physics. When trying to account for the nucleon masses, the recent models of the nucleons are severely flawed. These recent models of nucleons should not be endorsed by the scientific community as correct.

## References

[1] Gell-Mann (1964). "A Schematic Model of Baryons and Mesons". Physics Letters. 8 (3):214-215. doi: 10.1016/S0031-9163(64)92001-3

[2] G. Zweig (1964). "An SU(3) Model for Strong Interaction Symmetry and its Breaking" (PDF). CERN-TH-401.

[3] G. Zweig (1964). "An SU(3) Model for Strong Interaction Symmetry and its Breaking: II". CERN-TH-412.

[4] M. Gell-Mann (2000) [1964]. "The Eightfold Way: A Theory of Strong Interaction Symmetry". In M. Gell-Mann, Y. Ne'eman (ed.). The Eightfold Way. Westview Press. p. 11. ISBN 978-0-7382-0299-0.Original: M. Gell-Mann (1961). "The Eightfold Way: A Theory of Strong Interaction Symmetry". Synchrotron Laboratory Report CTSL-20. California Institute of Technology. doi:10.2172/4008239

[5] E. D. Bloom; et al. (1969). "High-Energy Inelastic e–p Scattering at 6° and 10°". Physical Review Letters. 23 (16): 930–934. Bibcode:1969 PhRvL..23..930B. doi:10.1103/PhysRevLett.23.930.

[6] M. Breidenbach; et al. (1969). "Observed Behavior of Highly Inelastic Electron–Proton Scattering". Physical Review Letters. 23 (16): 935-939. Bibcode: 1969PhRvL..23..935B. doi: 10.1103/PhysRevLett.23.935. OSTI 1444731.

[7] J. I. Friedman. "The Road to the Nobel Prize". Hue University. Archived from the original on 2008-12-25. Retrieved 2012-02-25.

[8] R. Eisberg, R. Resnick, *Quantum Physics of Atoms, Molecules, Solids, Nuclei, and Particles*, Wiley, New York, pp 653, 645 1985.

[9] C. T. H. Davies, C. McNeile, K. Y. Wong, E. Follana, R. Horgan, K. Hornbostel, G. P. Lepage, J. Shigemitsu, and H. Trottier (HPQCD Collaboration) "Precise Charm to Strange Mass Ratio and Light Quark Masses from Full Lattice QCD". Phys. Rev. Lett. 104, 132003 – Published 31 March 2010.
[10] Barger, V.; Phillips, R. (1997). *Collider Physics.* Addison–Wesley. ISBN 978-0-201-14945-6.
[11] Greensite, J. (2011). "An introduction to the confinement problem." *Lecture Notes in Physics.* 821. Springer. Bibcode:2011LNP...821.....G. doi:10.1007/978-3-642-14382-3. ISBN 978-3-642-14381-6.
[12] Wu, T.-Y.; Pauchy Hwang, W.-Y. (1991). "Relativistic quantum mechanics and quantum fields." World Scientific. p. 321. ISBN 978-981-02-0608-6.
[13] P.A. Zylaet al. (Particle Data Group), Prog. Theor. Exp. Phys.2020, 083C01 (2020) and 2021 update
[14] https://en.wikipedia.org/wiki/Pion
[15] Gell-Mann, M.; Renner, B. (1968). "Behavior of current divergences under SU3×SU3" (PDF). Physical Review. 175 (5): 2195–2199. Bibcode:1968PhRv..175.2195G. doi:10.1103/PhysRev.175.2195.
[16] https://www.sciencenewsforstudents.org/article/much-protons-mass-comes-from-energy-of-particles-inside
[17] https://en.wikipedia.org/wiki/Proton
[18] https://en.wikipedia.org/wiki/Quantum_chromodynamics_binding_energy
[19] https://news.fnal.gov/2014/01/how-many-quarks-in-a-proton/
[20] Y.B. Yang et al. "Proton mass decomposition from the QCD energy momentum tensor." Physical Review Letters. Vol. 121, November 23, 2018, p. 212001. doi: 10.1103/PhysRevLett.121.212001.
[21] K. S. Krane, *Introductory Nuclear Physics*, Wiley, New York, p.65, 1988.
[22] R. Eisberg, R. Resnick, *Quantum Physics of Atoms, Molecules, Solids, Nuclei, and Particles*, Wiley, New York, p 654, 1985.
[23] R. Eisberg, R. Resnick, *Quantum Physics of Atoms, Molecules, Solids, Nuclei, and Particles*, Wiley, New York, p 655, 1985.
[24] H. P. Robertson, Phys. Rev. A 35, (1930), 667
[25] E. Schrödinger, Ber. Kgl. Akad. Wiss. 24, (1930), 296;
[26] V.I.Vysotskii, S.V.Adamenko, M.V.Vysotskyy. Annals of Nuclear Energy, 2013, 62, pp. 618–625;
[27] Vysotskii, V.I., Vysotskyy, M.V. Journal of Experimental and Theoretical Physics, 2019, 128(6), pp. 856–864
[28] https://en.wikipedia.org/wiki/Chiral_symmetry_breaking
[29] A. Watson, *The Quantum Quark*, Cambridge University Presss, Cambridge, p 374, 2004.
[30] A. Watson, *The Quantum Quark*, Cambridge University Presss, Cambridge, p 376-377, 2004.
[31] https://en.wikipedia.org/wiki/Pion
[32] K. Rith and A. Schafer, "The Mystery of Nucleon Spin", Scientific American, July 1999, vol 281, #1,p 60.
[33] R. Eisberg, R. Resnick, *Quantum Physics of Atoms, Molecules, Solids, Nuclei, and Particles*, Wiley, New York, p 683-684, 1985.
[34] Womersley, J. (February 2005). "Beyond the Standard Model" (PDF). Symmetry Magazine. Archived from the original (PDF) on 2007-10-17. Retrieved 2010-11-23.
[35] Lykken, J. D. (2010). "Beyond the Standard Model". CERN Yellow Report. CERN. pp. 101–109. arXiv:1005.1676. Bibcode:2010arXiv1005.1676L. CERN-2010-002.

J. Condensed Matter Nucl. Sci. 36 (2022) 203–209

Research Article

# Formation of Hydrogen Miniatoms in the Medium of Free Electrons–the Key to the Mechanism of Low-energy Nuclear Reactions

A. I. Goncharov*

*Institute of Digital Technologies, Electronics and Physics, Altai State University, Prospect Lenina, 61, Barnaul, 656099, Russia*

V. A. Kirkinskii

*Institute of Geology and Mineralogy, Siberian Branch of the Russian Academy of Sciences, Prospect Akad. Koptyuga, 3, Novosibirsk, 630090, Russia E-mail: kirkinsky2011@yandex.ru*

---

**Abstract**

We have modeled the behavior of hydrogen atoms in the flow of free electrons in metals by the molecular dynamic method. The trajectories of the particles were calculated by numerically solving a system of differential equations of mechanics. Relativistic equations were used, and the interaction of particles was considered Coulombic without taking into account magnetic effects. About $10^4$ stories were modeled, each of them containing up to 100 collisions of free electrons with a hydrogen atom. The total number of simulated atoms that experienced collisions was $\sim 10^6$. Dynamic modeling revealed the formation of neutral particles consisting of protons (deuterons) with an electron rotating around them in nonstationary, close to elliptical orbits with an apogee to a distance of less than $10^{-11}$ cm and to a perigee of $\sim 10^{-12}$ cm. These particles, which are continuously changing in size and shape, are up to 3 to 4 orders of magnitude smaller than ordinary hydrogen atoms, and 1-2 orders of magnitude larger than neutrons. Such nonstationary hydrogen miniatoms can exist for a short time (on our estimate, up to $\sim 10^{-12}$ sec.) in the environment of free electrons of metals, easily move in them and, like neutrons, approach the nuclei of isotopes of hydrogen or other elements at a distance at which nuclear fusion reactions or transmutation of elements are possible due to the tunneling effect. Taking into account the formation of such hydrogen miniatoms the previously calculated rate of low-energy nuclear reactions in metals[1−6] increases more than by 5 – 6 orders of magnitude, that is, to values corresponding to experimental data. Formation of hydrogen miniatoms in the medium of free electrons is of primary importance in the mechanism of low-energy nuclear reactions.


*Keywords:* LENR, Molecular dynamics, Hydrogen mini-atoms

---

*Corresponding author: goncharov.ai@mail.ru

© 2022 ICCF. All rights reserved. ISSN 2227-3123

## 1. Introduction

In computer simulations using molecular dynamics methods, it was previously shown that the electrons of the outer shells of metals have a shielding effect on deuterons and significantly accelerate nuclear fusion reactions. The results of quantitative calculations of the reaction rates of hydrogen isotopes in a number of metals are published in Refs. [1]–[6]. In this paper, we model the behavior of hydrogen atoms in the flow of free electrons in metals.

### 1.1. Description of the Model

At the first stage, within the frame of classical (non-quantum) mechanics, the behavior of a single deuterium atom is simulated when acted upon by a flow of free electrons[7]. It is assumed that the particles do not radiate electromagnetic waves. The system consists of a deuteron and two electrons. At the initial moment, one of the electrons (orbital) is in the circular orbit of Bohr radius 0.529 Å. The flow of incoming electrons is isotropic; the impact parameter is limited by the value 0.53 Å.

The incoming electrons appear at the 1 Å distance from the deuteron; the kinetic energy of the incoming electrons follows the Fermi distribution with the parameters characteristic of free electrons in a Pd crystal at room temperature[8]:

$$f(\varepsilon) = const\sqrt{\varepsilon}/\{1 + \exp[(\varepsilon - \varepsilon^*)/kT]\},$$

where $\varepsilon^* = h^2(3n_e/\pi)^{3/2}/8m_e$ is the so-called characteristic energy, $n_e$designates the free electrons concentration, $h$ is Planck's constant, $k$ is the Boltzmann constant; and the absolute temperature $T$ is assumed to be equal to 350 K. According to Ref. [3], we assume that for every atom there is a single free electron. Then $n_e \approx N_A \cdot \rho/A$. For palladium, $n_e \approx 6.8 \times 10^{22}\text{cm}^{-3}$, $\varepsilon^*/k \approx 7 \times 10^4$ K. The average kinetic energy of free electrons in palladium is $3\varepsilon^*/5 \approx 3.6$eV, the average electron velocity $v \approx 1.13 \times 10^8$ cm/s.

After random initial conditions have been specified, particle trajectories are calculated by numerically solving a system of differential equations of mechanics. Relativistic equations are used but without allowance for magnetic effects, i.e. the particle interaction is considered purely Coulomb interaction. The system of equations is solved by Runge-Kutta method of the fourth order with the relative error less than $10^{-5}$.

As has already been mentioned, at any instant three particles can be tracked, i.e., two electrons and one deuteron. When one of the electrons travels at a distance exceeding 10 Å, the remaining system (unsteady state of deuterium D* atom) is bombarded by a new free electron, etc. $N_{\max} = 10^4$ model experiments have been conducted. Number $l$ of free electrons collisions with the D* "atom" could reach up to $l_{\max} = 100$.

When one of the electrons approached the deuteron at a distance less than $10^{-13}$ cm, trajectories were no longer calculated. This implementation of incoming electron parameters was excluded from consideration and a transition to a new implementation of a system took place with an initial Bohr orbit of the deuterium atom. Therefore, the number of simulated D* "atoms", which underwent $l$ collisions with electrons, $N(l) < N_{\max}$. For instance, $N(10) = 9101$, $N(50) = 3331$, $N(80) = 1489$, $N(100) = 847$. The total number N* of simulated D* "atoms"

$$N* = \sum_{l=1}^{l_{\max}} N(l) = 420748 \quad < N_{\max} \times l_{\max} = 10^6.$$

A high proportion of electrons collisions with D* results in the decreased energy of D* and its size. For the D* size the distance r from the electron to the deuteron is taken at the apogee of the orbit. The size distribution density of D*, which underwent $l$ collisions with electrons, will be designated as $f(r|l)$.

We had to exclude from our consideration trajectories in the region of small distances in order to prevent the degradation of the trajectories calculation accuracy and to enhance the program response. Special calculations showed that for a significant proportion of neglected trajectories there was a pronounced trend for the electron impingement on the deuteron, which could have resulted in the formation of a bineutron. The program, however, was not yet sufficient to calculate this process accurately enough. Part of the neglected trajectories could have also lead to the formation of D* of even smaller sizes. Those two possibilities potentially can significantly increase the theoretically estimated value of the fusion reaction rate and require supplementary research. For the time being, we assume that under more accurate trajectories calculation all the excluded implementations result in the formation of D* whose size obey the same distributions $f(r|l)$. In this connection, distributions are normalized similarly for any $l$:

$$\int_0^\infty f(r|l)dr = 1.$$

The number $l$ of the D* "atom" collisions with free electrons obeys Poisson distribution $P_a(l)$ with the parameter $a$; the average number of collisions $a$ is proportional to the electron flow density and time.

The main contribution to the reaction rate make small D* ($r \sim 10^{-11}$ cm), that have a high mobility. So, the crystal lattice does not strongly interfere with D*-D* collisions. Let $n$ and $N$ be the concentration and full quantity of D* atoms in the working volume of crystal. The proportion of deuterons $\lambda$, which reacted in a unite time (fractional reaction rate $\lambda$) is equal to:

$$\lambda(a) = -\frac{1}{N}\frac{dN}{dt} = \sum_{l=0}^{l_{\max}}\sum_{m=0}^{l_{\max}} P_a(l)P_a(m)\lambda(l,m),$$

$$\lambda(l,m) = \frac{n}{2}\int_0^\infty dr_1 \int_0^\infty dr_2 f(r_1|l)f(r_2|m)\int_0^\infty dEv\sigma(v)F_M(E),$$

where $v$ is the relative speed of deuterons, $E = \mu v^2/2$; $\mu$ – reduced mass of deuterons pair; $F_M$(E) – Maxwell distribution at room temperature normalized as:

$$\int_0^\infty F_M(E)dE = 1.$$

The cross-section of D*–D*-collisions resulting in a fusion reaction in the quasiclassical approximation is $\sigma(v) = S_0 P(E)/E$, where $S_0 = 0.55 \times 10^{-19}$ cm$^2$ eV;

$$P(E) = \exp\left(-\frac{4\pi}{h}\int_0^R \sqrt{2\mu[V(r)-E]}dr\right),$$

where $R = V^{-1}(E)$ is the turning point of a classical trajectory.

It will be assumed that D* "atoms" with the shell sizes $r_1$, $r_2$ move freely if the "shells" do not overlap, i.e., if the distance between the D* "atoms" is $r > r_1 + r_2$. If the "shells" overlap, a purely Coulomb repulsion of the D* "atoms" is observed. The potential deuterons interaction energy in this approximation is equal to

$$V(r) = \begin{cases} e^2/r - E_s & \text{if } r < r_1 + r_2, \\ 0 & \text{if } r > r_1 + r_2, \end{cases}$$

where $E_s = e^2/(r_1 + r_2) = m_e c^2 r_e/(r_1 + r_2)$ is the screening energy, $r_e = 2.82 \times 10^{-13}$ cm is the classical radius of an electron. By integration, we obtain

$$P(E) = \exp\left(-993.09 \text{ eV}^{1/2}/\sqrt{E+E_s}\right).$$

Since at any energy $E$ contributing to $\lambda$ there holds a condition $E << E_s$, the approximation $P(E) \approx P(0)$, $F_M(E) = \delta(E - \overline{E})$ can be used where $\overline{E} = 3kT/2$. Then

$$\lambda(l, m) = C \int_0^\infty dr_1 \int_0^\infty dr_2 f(r_1|l) f(r_2|m) W(r_1 + r_2),$$

where $C = (nS_0/2)(2/\mu\overline{E})^{1/2}$, $W(r_1 + r_2) = \exp\left(-1.39\sqrt{(r_1 + r_2)/r_e}\right)$.

Taking into account the probabilistic sense of $f(r|l)$, the reaction rates $\lambda(l, m)$, $\lambda(a)$ could be approximately calculated by the formulas

$$\lambda(l, m) = \frac{C}{N(l)N(m)} \sum_{i=1}^{N(l)} \sum_{j=1}^{N(m)} W(r_i^{(l)} + r_j^{(m)}), \quad \lambda(a) = C \sum_{i=1}^{N*} \sum_{j=1}^{N*} \frac{P_a(l(i)) P_a(m(j))}{N(l(i))N(m(j))} W(r_i + r_j),$$

where $r_i^{(l)}$, $i$ = 1, 2, 3,…, $N(l)$ are random implementations of the D* "atom" size with $l$ collisions of electrons; $r_i$, $i$ = 1, 2, 3…, $N^*$ denotes the complete set of implementations. Let every octahedral site contain a deuterium atom. Then the deuterium concentration $n = 6.8 \times 10^{22}$ cm$^{-3}$; at T = 350 K, $C = 1.2 \times 10^{10}$ s$^{-1}$.

## 2. Calculation Results

Recall that $f(r|l)$ is the size distribution of the atom that has experienced $l$ collisions with electrons. The simulation showed that the distributions of $f(r|l)$, starting from about $l$ = 50, are almost the same with a further increase in $l$ (see Figure 1). The equilibrium distribution can be close to that found if the characteristic time of the quantum relaxation of the "miniatom" exceeds $10^{-13}$ s.

Table 1 presents the number of events with the size $r$ and the number of collisions $l$ from the different ranges; $\Delta N*$ is the total number of events with $l$ from these ranges.

Table 2 shows the dimensions of the 10 smallest "miniatoms" of deuterium D*, which have $r \leq 2 \times 10^{-11}$cm. The "orbits" are usually very elongated: $r_{\min} << r$, so using the value $r$ as the size of the "atom" should lead to a lower estimate of $\lambda$.

Figure 2 shows the functions $\lambda(l, l)$ and $\lambda(a)$. Averaging $\lambda(a)$ over $a \leq l_{\max}$, which is equivalent to averaging over time, leads to the result $\lambda \approx 2 \times 10^{-4}$ s$^{-1}$, which is $10^6$ times greater than the value obtained in Refs. [2] and [4]

**Table 1.** Number of events with $r$ and $l$ from the indicated ranges

| Ranges of $l$ | $\Delta N*$ | Ranges of size $r$ of the D* "atom" in units of $10^{-11}$ cm | | | | | | | | |
|---|---|---|---|---|---|---|---|---|---|---|
| | | 0.5 – 1. | 1. – 1.5 | 1.5– 2. | 2. – 2.5 | 2.5 – 3. | 3. – 3.5 | 3.5 – 4. | 4. – 4.5 | 4.5 – 5. |
| 1 – 20 | 177731 | 0 | 0 | 0 | 0 | 2 | 4 | 7 | 2 | 0 |
| 21 – 40 | 114983 | 0 | 0 | 1 | 5 | 4 | 14 | 19 | 25 | 43 |
| 41 – 60 | 66531 | 0 | 0 | 2 | 10 | 9 | 16 | 23 | 25 | 36 |
| 61 – 80 | 39042 | 0 | 0 | 1 | 3 | 7 | 13 | 17 | 10 | 22 |
| 81 – 100 | 22461 | 1 | 3 | 2 | 2 | 2 | 3 | 8 | 11 | 18 |

**Table 2.** Sizes of D* "atoms" at the apogee of $r$ and at the perigee $r_{min}$ in units of $10^{-11}$ cm. $l$ is the number of electron collisions with D*

| | | | | | | | | | | |
|---|---|---|---|---|---|---|---|---|---|---|
| $l$ | 88 | 83 | 84 | 82 | 25 | 89 | 79 | 56 | 98 | 51 |
| $r$ | 0.87 | 1.17 | 1.27 | 1.38 | 1.64 | 1.69 | 1.78 | 1.81 | 1.82 | 1.99 |
| $r_{min}$ | 0.22 | 0.41 | 0.19 | 0.88 | 0.12 | 0.11 | 0.08 | 0.19 | 0.88 | 0.38 |

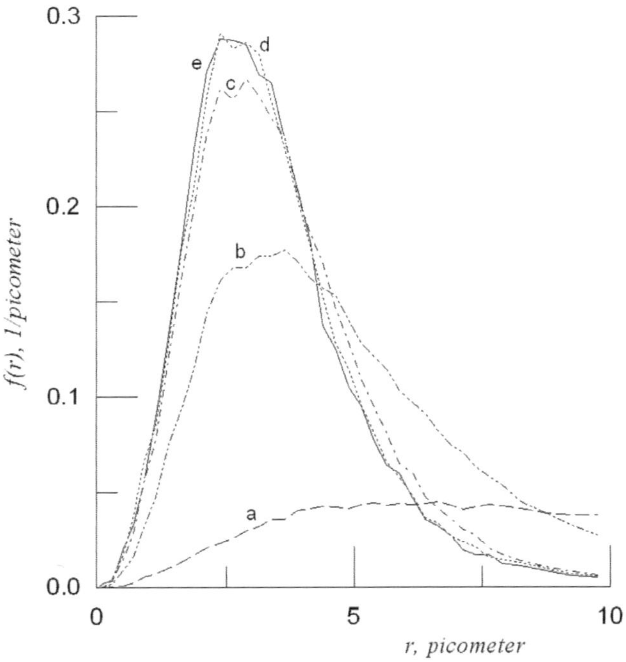


**Figure 1.** The size distributions of "atoms" D*, averaged over the number $l$ of collisions with electrons: a – $l = 1 \div 20$; b – $l = 21 \div 40$; c – $l = 41 \div 60$; d – $l = 61 \div 80$; e – $l = 81 \div 100$.

on the basis of the dynamic model of electron orbital deformation (EODD). Taking into account that $\lambda(a)$ somewhat increases with the increase in a, the obtained average $\lambda$ should be treated as a lower-bound estimate. The contribution of 370 events presented in Table 2 to the obtained value of $\lambda$ is 99%, including the contribution of 10 events from Table 1 amounting to about 80%.

## 3. Discussion

The application of classical mechanics usually provokes objections because classical mechanics enables one to put a shielding charge at any point and impart it any initial velocity (e.g., zero velocity). We deem it therefore important to emphasize that the equiprobable initial conditions we employed correspond to the *real-life* situation. In addition, the applied approach, which could be called the Bohr approach, is not a purely classical one, since the radiation has been "turned off". Quantum-mechanical models allowing for elastic processes only give a sufficiently accurate description of real physical processes for a very wide range of phenomena.

It should be noted that from the quantum-mechanical viewpoint, $D^*$ is not a kind of a hypothetical atomic state with energies lower than that of the ground state. $D^*$ should be treated as an unsteady structure within many-particle nonradiating system.

We indicate a possible mechanism for the occurrence of such configurations under the action of a flow of free electrons, having previously made an estimate of the collision frequency of the "atom" D* with electrons. It was said above that the maximum value of the impact parameter of the incoming electrons in our calculations was equal to $r_B$, and any outcome was counted as a collision. The average number of such collisions of an "atom" with electrons

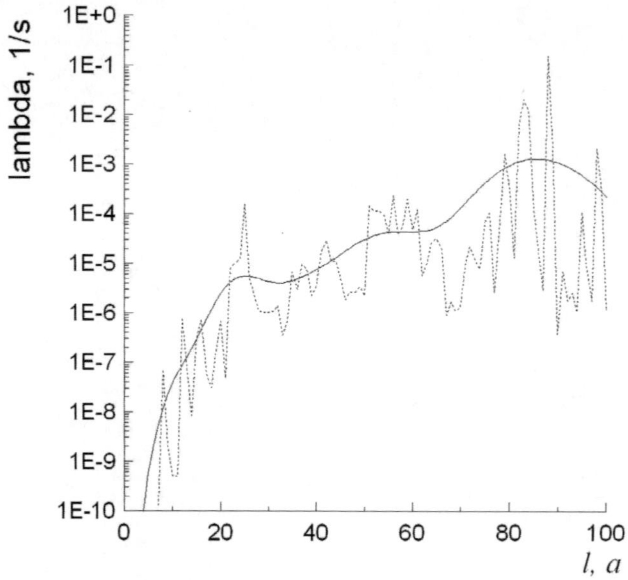


**Figure 2.** Reaction rate. Dashed line denotes $\lambda(l, l)$; solid line designates $\lambda(a)$.

during time t is equal to $a = \Phi\pi\ r_B^2 t$, where $\Phi = n_e v \approx 7.7 \cdot 10^{30}\ \mathrm{cm}^{-2} s^{-1}$ is the density of the flow of free electrons in a palladium crystal. The average time between two successive collisions of an "atom" with free electrons is equal to $\Delta\tau = 1/\Phi\pi\ r_B^2 \approx 1.5 \cdot 10^{-15}$s. In some cases, as a result of the interaction of electrons, one of them (flying or orbiting, let's call it "electron 1") can fly a sufficiently large distance (several Å) from the deuteron; for a while ($\Delta t$) it seems to be out of the game, since it is far away and does not determine the size of the "atom". But it takes away some of the energy, which leads to the fact that the remaining electron ("electron 2") will make revolutions around the center along an open trajectory, which we will call a "non-stationary orbit"; its average dimensions are smaller than the original dimensions of the atom and during $\Delta t$ these determine the actual dimensions of the "atom". The returned electron 1 returns part of the energy to electron 2, which as a result flies away to infinity. If the final energy of the atom is not less than the initial energy, then such a classical movement of electrons does not contradict the fact of the existence of a lower energy stationary state of the atom. Since electron 1 can fly far enough away, a non-stationary orbit may well exist for a time of order $\Delta t = 10^{-15} s$; then it is very likely that it will be exposed to the next free electron, and this may lead to the formation of a non-stationary orbit of an even smaller size, etc.

## 4. Conclusions

Dynamic modeling without taking into account quantum effects predicts the formation of non-stationary particles consisting of protons (deuterons) with an electron rotating around them in orbits close to elliptical, with an apogee of $\sim 10^{-11}$ cm and a perigee of $\sim 10^{-12}$ cm. These "mini-hydrogen atoms", which vary in size and shape, are up to 3-4 orders of magnitude smaller than ordinary hydrogen atoms, but 1-2 orders of magnitude larger than neutrons. They can exist and move in the environment of free electrons of metals, and, like neutrons, approach the nuclei of isotopes of hydrogen or other elements at a distance at which, due to the tunneling effect, nuclear reactions of synthesis or transmutation of elements are possible. The formation of such atoms increases the previously calculated rate of nuclear reactions in metals[1-6] by 5-6 orders of magnitude, that is, to values corresponding to experimental data.

## References


[1] V. A. Kirkinskii, Yu. A. Novikov. A new approach to theoretical modelling of nuclear fusion in palladium deuteride, Europhysics Letters, vol. 46, No.4, pp. 448–453, 1999.

[2] V. A. Kirkinskii, Yu. A. Novikov. Theoretical modeling of cold fusion. Novosibirsk State University, Novosibirsk, 105 p, 2002.

[3] V.A. Kirkinskii, Yu. A. Novikov. Numerical calculations of cold fusion in metal deuterides. In the book: "Condensed Matter Nuclear Science", Proceedings of the ICCF-9, ed. by Xing Z. Li, Beijing, pp. 162–165, 2003.

[4] V. A. Kirkinskii, Yu. A. Novikov. Modelling of dynamic screening effects in solid state, Europhysics Letters, vol. 67, N 3, pp. 362–367, 2004.

[5] V.A. Kirkinskii, Yu. A. Novikov. Calculation of nuclear reaction probability in iron hydride and the problem of nucleosynthesis in the Earth's interior, in the book "Condensed matter nuclear science", Proc. of ICCF-9, ed. by Xing Z. Li, Beijing, pp. 166–169, 2003.

[6] V. A. Kirkinskii, Yu. A. Novikov. Calculation of nuclear reaction probabilities in a crystal lattice of titanium deuteride. In the book "Condensed matter nuclear science", Proc. of the ICCF-10, 2003, Cambridge Massachusets, ed. by P. L. Hagelstein and Scott R. Chubb, World Scientific, New Jersey, London, Singapore, Beijing, 2006, pp. 681–685.

[7] A. I. Goncharov and V. A. Kirkinskii. Theoretical modeling of electron flow action on probability of nuclear fusion of deuterons. In the book: Progress in condensed nuclear science, World Scientific, Proc. of ICCF-12, Yokohama 2005, pp. 582–586, 2007.

[8] R. H. Fowler and E. A. Guggenheim. Statistical Thermodynamics (Cambridge University Press, 1939).

J. Condensed Matter Nucl. Sci. 36 (2022) 210–246

# Recent Progress on Phonon-Nuclear Theoretical Models

Peter L. Hagelstein
*Massachusetts Institute of Technology, Cambridge, MA, USA*

**Abstract**

Excess heat in the Fleischmann-Pons experiment presents numerous challenges for theorists, the biggest of which is to account for the absence of energetic nuclear particles in amounts commensurate with the energy produced. Our focus has been on the development of such a model, with recent progress reviewed. To account for excess heat production we contemplate a complicated model based on nuclear excitation transfer, subdivision and the down-conversion of keV nuclear quanta to a large number of phonons. The phonon-nuclear interaction that provides the foundation has been derived from a many-nucleon Dirac formalism, and we discuss recent efforts to confirm and improve upon the derivation. The quantification of the models under consideration require the evaluation of phonon-nuclear matrix elements, with recent progress discussed. The first step in the excess heat model involves the excitation transfer of a 24 MeV quantum associated with the $D_2/^4He$ transition, which motivates a consideration of modeling and connections with experiments showing low-level energetic nuclear emissions. The analogous first step in the case of the 5.5 MeV quantum associated with the $HD/^3He$ transition is considered, where there is the possibility of low-level energetic nuclear emission, and where the issue of the availability of near-resonant energy levels can be studied. A nearby level in $^{137}Cs$ is identified, which suggests a possible mechanism for the acceleration of the decay rate. We have recently revisited the issue of the null reaction, in which a $D_2/^4He$ transition leads to the production of a compact $D_2$ state from $^4He$, with the conclusion that this may be important in models for excess heat production and for low-level nuclear emission. Progress on subdivision is reviewed, including connections with experiments showing low-level nuclear emission, and candidates for laboratory tests. The possibility of a new $^{57}Co/^{57}Fe$ resonant excitation transfer experiment in a near single crystal sample is presented. A numerical approach for the evaluation of the rate for excess heat production in the models under discussion is discussed.


*Keywords:* LENR theory, Phonon-nuclear coupling, Excess heat model, Review, Excitation transfer

## 1. Introduction

For more than three decades following the announcement of excess heat in the Fleischmann-Pons experiment there has been the promise of a small scale clean nuclear energy source [1], [2]. Progress has been hindered due to a lack of resources and manpower, largely due to the belief within the mainstream physics community that the basic effect is inconsistent with nuclear physics and solid-state physics – in essence that there is no possibility that the excess heat effect could be real [3].

One approach to changing things is to develop a (useful) commercial product based on Fleischmann-Pons excess heat, which would once and for all provide an unambiguous demonstration that the effect is real. Quite a few efforts along these lines have over the years met with financial or technical headaches, but the prospects at present seem good that some of these efforts (such as Technova [4]–[6] and Brillouin [7], [8]) will succeed.

We have long focused on the question of how it works [9]–[11] which if solved could help move the experiments and technology development forward. To engage the mainstream scientific community in taking an interest in theoretical and experiment problems in the field has always been a major goal, but in our view this has always been more of a social rather than a scientific issue.

The theoretical challenges associated with how it works have been intimidating from the start. Critics and skeptics initially drew attention to the lack of neutrons and 24 MeV gammas as evidence that dd-fusion did not occur [3], arguing that the source of the energy could not be nuclear. At this point it seems clear that the critics and skeptics did not go far enough – in many experiments considerable energy is detected but there is no evidence for the presence of *any* energetic nuclear particles in amounts consistent with the energy produced (and also no evidence for the much larger amount of chemical reaction products that would be needed to account for the energy). According to the textbooks, energetic nuclear particles must be produced in connection with energy generation as a direct consequence of energy and momentum conservation; hence, their absence would seem to require a violation of energy and momentum conservation at the outset. From our perspective this was the biggest theoretical problem, even though it was not the main focus of the critics early on.

As has been often repeated for more than 30 years the Coulomb barrier between deuterons would be expected to prevent deuteron-deuteron interactions and fusion at the fermi scale in the case of collisions near room temperature, or due to tunneling in molecular $D_2$ [12]. This argument has been used effectively to rule out excess heat in the Fleischmann-Pons experiment. A focus in connection with this argument was on the rate for incoherent $D_2$ fusion, which would result in energetic nuclear products such as neutrons – hence, the theoretical argument that this process is too slow to account for the excess heat effect should be understood as being in agreement with experiment. We know that excess heat is not due to tunneling in $D_2$ leading to incoherent dd-fusion, since the corresponding dd-fusion neutrons are not seen. A better question to have been asked is whether the fusion rate for a mechanism more closely connected with experiment, in particular one which did not result in neutrons or other energetic nuclear particles, could be on the order of what is observed in experiment.

Aside from being the source of theoretical headaches, the absence of commensurate energetic nuclear emissions prevents the reaction mechanism from being studied using conventional nuclear diagnostics. For example, in conventional fusion or fission reactions the energetic nuclear products are detected [13], and from measurements we know a lot about what goes on internally during the reactions. We are not able to make use of the same diagnostics to study the nuclear processes which occur to produce heat in the Fleischmann-Pons experiment, which has prevented an experimental elucidation of the reaction mechanism. For experimental nuclear physicists this gives rise to doubt that excess heat generation has anything to due with nuclear processes.

The theoretical picture which has emerged from our efforts is at once both simple in concept [9]–[11], but complicated in implementation. The theoretical starting place has to be the absence of commensurate energetic nuclear particles, which implies that nature has found a way to take a very large 24 MeV nuclear quantum and split it up into an enormous number of lower energy quanta. Such a thing is unprecedented – while down-conversion as a physical process is certainly known, we have no examples where a large quantum is split up into on the order of $10^9$ smaller quanta; yet that seems to be what is needed to account for the experimental results.

This immediately leads to more theoretical headaches. We might expect that a large down-conversion effect might be possible in a strongly coupled quantum system. However, models for nuclei embedded in a solid view the internal nuclear degrees of freedom as largely independent of condensed matter degrees of freedom (such as vibrations or plasmons), with only a weak second-order interaction resulting from perturbations of local electric or magnetic fields. On the face of it the prospects for strong coupling are not encouraging.

All of this emphasizes just how daunting the theoretical issues are, with little help from mechanisms studied in the mainstream literature.

A positive step toward a potential solution came first in the case of new mechanisms for down-conversion. If we consider down-conversion in the case of generalized spin-boson models (or sometimes referred to as Dicke models), an analysis of up-conversion and down-conversion indicates that the indirect coupling between an initial state and final state (where the conversion has been completed) under normal conditions is hindered by destructive interference. Hence, any mechanism capable of reducing or eliminating this destructive interference will result in a dramatic many-orders-of-magnitude increase in the up-conversion and down-conversion rate. In 2002 we noticed that loss asymmetric around the resonant energy point could greatly reduce the destructive interference and dramatically enhance up-conversion and down-conversion [14]–[18]. Models developed to quantify the effect suggested that a down-conversion of a large quantum into on the order of a million smaller quanta was within the realm of possibility. For many years our focus was on this kind of model, until we recognized relatively recently that there needed to be another mechanism (from an initial comparison of theory and experiment from results of excitation transfer experiments) [10], [11]. In this case there is an obscure effect in which the nuclear binding energy varies off of resonance, which can greatly reduce the destructive interference.

Another positive step was the recognition that there could be an alternative coupling mechanism between internal nuclear states and condensed matter modes beyond the weak second-order interaction based on electric and magnetic fields. The idea is that a nucleus embedded in the lattice must be consistent with relativity, undergoing a Lorentz contraction as well as spin and isospin rearrangement when in motion, so that changes in the internal nuclear state when the nucleus moves in connection with lattice vibrations implies a (low-order) relativistic phonon-nuclear interaction [19]. This provides a stronger phonon-nuclear interaction than results from local electric and magnetic fields. It also addresses the conceptual issue of how the small fermi-scale internal nuclear degrees of freedom can interact with lattice vibrations organized on the meso- or macro- scale involving a large number of atoms.

At some point theory needs to connect in a meaningful way with experiment. Ideally, we would like a clean demonstration of the phonon-nuclear interaction in the simplest way possible, which would be in a process where the exchange of a single phonon results in the change of an internal nuclear state. Unfortunately, this doesn't work since relevant phonon energies are much less than 1 eV and relevant nuclear transition energies are greater than 1 keV. Consequently, the lowest-order physical effect that we might study is excitation transfer, in which the excitation in one nucleus is transferred to a different nucleus, with the exchange of (at least) one phonon for each nuclear transition [20]. This second-order process then by necessity becomes the focus of basic experimental studies to clarify the phonon-nuclear interaction. In our lab we have evidence for resonant phonon-mediated nuclear excitation in observations of dynamic angular anisotropy, and observations consistent with nonresonant excitation transfer in delocalization experiments [21]–[23].

The mechanism for accelerated up-conversion or down-conversion mentioned above operates through a large number of sequential non-resonant excitation transfer steps which must occur before the system decoheres. If excitation transfer can be rapid, then there is the possibility of a subdivision effect, in which a few excitation transfer steps starting from an initial excited nuclear state leads to excitation in two or more nuclei at lower energy where the total energy is conserved [10], [11]. An unambiguous demonstration of subdivision in the lab has not yet been accomplished, but it may be that some of the low-level neutron emission seen in experiments with deuterated metals might be a result of subdivision.

As another positive step in the development of a picture for excess heat generation, subdivision provides a bridging mechanism, allowing for some of the down-conversion to take place largely in the nuclei (through subdivision), and the rest of the down-conversion to take place as a multi-phonon process through a coupling of a few nuclear transitions with a highly-excited vibrational mode.

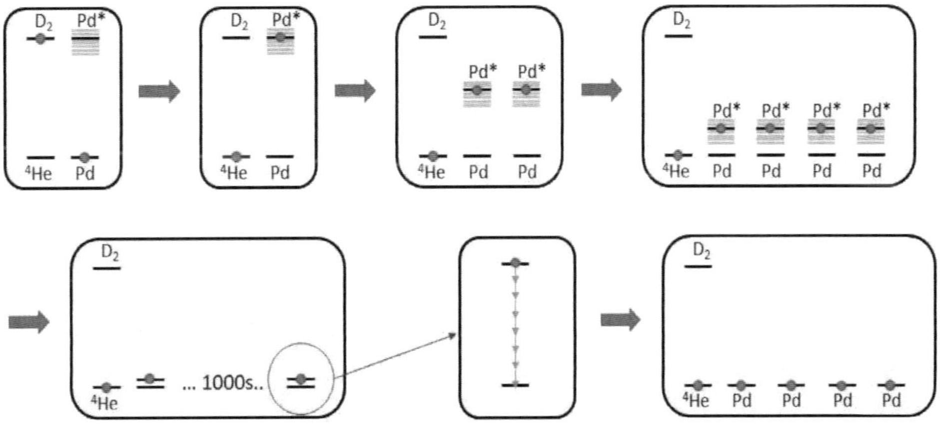


**Figure 1.** Schematic of the model for excess heat production under consideration. The first step is excitation transfer of 24 MeV from the $D_2/^4He$ transition to produce excitation in a Pd nucleus (or other nucleus in the lattice). This is quickly followed by a subdivision step resulting in two excited nuclei with roughly 12 MeV each. More subdivision steps leads ultimately to a thousands of nuclei with much lower levels of excitation – likely involving excited impurity nuclei since the Pd isotopes do not have long-lived low-lying excited states. Finally, this lower nuclear excitation is down-converted into phonons.

The overall picture for excess heat production consistent with these mechanisms is illustrated in Figure 1. It is clear that the overall process is pretty complicated, being made up of many individual theoretical pieces. If the associated dynamics can proceed coherently (in the sense of a coherent quantum process analogous to Rabi oscillations), then there is the possibility of great acceleration in the associated reaction rate. The decay associated with a quantum mechanical incoherent process is governed by the Golden Rule, and the associated rate is quadratic in the coupling matrix element. The dynamics associated with a quantum mechanical coherent process can be linear in the coupling matrix element. For incoherent fusion, the coupling matrix element is hindered by the Gamow factor $e^{-G}$ resulting from tunneling through the Coulomb barrier, so that the reaction rate is proportional to $e^{-2G}$. If the mechanism outline above proceeds as a coherent quantum process, then the rate can be proportional to $e^{-G}$. Since $e^{-G}$ is on the order of $10^{-40}$, there is the potential for a dramatic acceleration of the reaction rate (and hence fusion rate) due to quantum coherence.

In what follows we review some recent advances in the theory, models, picture, and applications.

## 2. Lowest-Order Phonon-Nuclear Interaction From the Dirac Model

We consider first the lowest-order phonon-nuclear interaction derived from the Dirac model. As discussed briefly above, the idea is that the internal degrees of freedom of a nucleus are modified as the nucleus moves in connection with lattice vibrations due to relativity, which means that there must be an associated relativistic phonon-nuclear interaction. Our preference would be to make use of a fully covariant model for the derivation of the effect, which the many-body Dirac model with realistic interactions is not. However, the Dirac model is relativistic, it is cast in a Hamiltonian formulation which make a derivation simple, and over many decades of use we know that the Dirac model is capable of capturing a significant part of the relativistic contribution to the quantum problem.

We can start with a many-particle Dirac model in the form

$$\hat{H} = \sum_j \beta_j mc^2 + \sum_j \boldsymbol{\alpha}_j \cdot c\hat{\mathbf{p}}_j + \sum_{j<k} V_{jk} \tag{1}$$

and then make use of a partial Foldy-Wouthuysen transformation to obtain [19]

$$\begin{aligned} e^{i\hat{S}}\hat{H}e^{-i\hat{S}} &= \frac{|\hat{\mathbf{P}}|^2}{2M}\sum_j \frac{m_j}{M}\beta_j + \sum_j \beta_j mc^2 + \sum_j \boldsymbol{\alpha}_j \cdot c\hat{\boldsymbol{\pi}}_j + \sum_{j<k} V_{jk} \\ &+\frac{1}{M}\sum_j \beta_j \hat{\mathbf{P}} \cdot \hat{\boldsymbol{\pi}}_j + \frac{1}{2Mc}\sum_j \left[(\beta_j\boldsymbol{\alpha}_j + \beta_k\boldsymbol{\alpha}_k)\cdot\hat{\mathbf{P}}, V_{jk}\right] \\ &+ \cdots \end{aligned} \tag{2}$$

What this transformation does is to provide for an approximate separation of the center of mass degrees of freedom (as a nonrelativistic problem) and relative degrees of freedom (as a relativistic problem). On the first line are terms that can be interpreted as the center of mass kinetic energy, a relative Hamiltonian for the internal nuclear states, and low-order terms that provide a coupling between the center of mass and relative degrees of freedom.

The lowest-order phonon-nuclear interaction describing single phonon exchange from this model is

$$\hat{H}_{int} = \hat{\mathbf{a}} \cdot c\hat{\mathbf{P}} \tag{3}$$

where the center of mass momentum $\hat{\mathbf{P}}$ for a nucleus in a condensed matter environment is an operator made of up phonon creation and annihilation operators, and where $\hat{\mathbf{a}}$ is an operator which acts on the internal nuclear degrees of freedom for consistency with the Dirac model version of relativity. The $\hat{\mathbf{a}}$ operator is given by

$$\hat{a} = \frac{1}{Mc}\sum_j \beta_j \hat{\boldsymbol{\pi}}_j + \frac{1}{2Mc^2}\sum_j \left[(\beta_j\boldsymbol{\alpha}_j + \beta_k\boldsymbol{\alpha}_k), V_{jk}\right]. \tag{4}$$

## 3. Unsuccessful Effort to Correct the Many-Particle Dirac Model

The relativistic phonon-nuclear interaction mentioned above provides one foundation for the models under discussion. Consequently, a concern is that this interaction does not appear to be known in the mainstream literature, and we have been concerned that it may not be a real physical effect. A headache is that our derivation of the effect is from a model based on a many-particle Dirac model, and it is well known that the many-particle Dirac model with realistic interaction is not covariant.

One way to proceed might be through the development of a modified many-particle Dirac model which is corrected in order to make it more covariant. The thought was that this might clarify whether the proposed phonon-nuclear interaction is real, or an artifact of the many-particle Dirac model. A substantial effort was put into the problem, ultimately without success. It seems useful to give an account of the ideas and issues here.

To make a version of the many-particle Dirac model that is more nearly covariant, what is needed is a consistent set of operators for the many-particle system $\hat{H}$, $\hat{\mathbf{P}}$, $\hat{\mathbf{J}}$ and $\hat{\mathbf{K}}$ associated with translations in time, space, rotational angle and momentum. Poincaré invariance requires that these operators satisfy [24]–[28]

$$[\hat{P}_i, \hat{P}_j] = 0 \quad [\hat{P}_i, \hat{H}] = 0 \quad [\hat{J}_i, \hat{H}] = 0$$

$$[\hat{J}_i, \hat{J}_j] = i\hbar\epsilon_{ijk}\hat{J}_k \quad [\hat{J}_i, \hat{P}_j] = i\hbar\epsilon_{ijk}\hat{P}_k \quad [\hat{J}_i, \hat{K}_j] = i\hbar\epsilon_{ijk}\hat{K}_k$$

$$[\hat{K}_i, \hat{P}_j] = i\hbar\delta_{ij}\frac{1}{c^2}\hat{H} \quad [\hat{K}_i, \hat{K}_j] = -i\hbar\epsilon_{ijk}\frac{1}{c^2}\hat{J}_k \quad [\hat{K}_i, \hat{H}] = i\hbar\hat{P}_i \tag{5}$$

Strictly speaking, such coupled commutation relations are quite difficult to solve in general. For the special case of a many-particle Pauli model, a solution by construction was found for a Pauli type of formulation (in which spins are included, but not involving 4-vectors) by Foldy [29], and by Krajcik and Foldy [30].

Krajcik and Foldy noted that the most important of the commutation relations was the one involving boost and Hamiltonian, which we can write as

$$[\hat{\mathbf{K}}, \hat{H}] = i\hbar\mathbf{P} \tag{6}$$

In one attack on the problem we sought to develop a solution based on matching orders according to

$$\hat{H} = \hat{H}_{+2} + \hat{H}_0 + \hat{H}_{-2} + \cdots$$

$$\mathbf{K} = \mathbf{K}_0 + \mathbf{K}_{-2} + \mathbf{K}_{-4} + \cdots \tag{7}$$

Guided in part by the work of Krajcik and Foldy, after some experimentation we worked with Hamiltonian and boost operators for the two-body version of the problem, trying a variety of different operators. A specific example makes use of a Hamiltonian operator according to

$$\hat{H}_2 = m_1c^2\beta_1 + m_2c^2\beta_2$$

$$\hat{H}_0 = \boldsymbol{\alpha}_1 \cdot c\mathbf{p}_1 + \boldsymbol{\alpha}_2 \cdot c\mathbf{p}_2 + V_{12}$$

$$\hat{H}_{-2} = -\left[\frac{\beta_1\boldsymbol{\alpha}_1 \cdot c\mathbf{p}_1}{2m_1c^2} + \frac{\beta_2\boldsymbol{\alpha}_2 \cdot c\mathbf{p}_2}{2m_2c^2}, V_{12}\right]$$

$$\hat{H}_{-4} = \frac{1}{2}\left[\frac{\beta_1\boldsymbol{\alpha}_1 \cdot c\mathbf{p}_1}{2m_1c^2} + \frac{\beta_2\boldsymbol{\alpha}_2 \cdot c\mathbf{p}_2}{2m_2c^2}, \left[\frac{\beta_1\boldsymbol{\alpha}_1 \cdot c\mathbf{p}_1}{2m_1c^2} + \frac{\beta_2\boldsymbol{\alpha}_2 \cdot c\mathbf{p}_2}{2m_2c^2}, V_{12}\right]\right]$$

$$-\frac{1}{8c^2}\left(\frac{|\mathbf{p}_1|^2V_{12} + V_{12}|\mathbf{p}_1|^2}{m_1^2} + \frac{|\mathbf{p}_2|^2V_{12} + V_{12}|\mathbf{p}_2|^2}{m_2^2}\right) \tag{8}$$

and a boost operator given by:

$$\mathbf{K}_0 = \mathbf{r}_1\beta_1m_1 + \mathbf{r}_2\beta_2m_2 \tag{9}$$

$$\mathbf{K}_{-2} = \frac{1}{2c^2}\left(\mathbf{r}_1(\boldsymbol{\alpha}_1\cdot c\mathbf{p}_1) + (\boldsymbol{\alpha}_1\cdot c\mathbf{p}_1)\mathbf{r}_1 + \mathbf{r}_2(\boldsymbol{\alpha}_2\cdot c\mathbf{p}_2) + (\boldsymbol{\alpha}_2\cdot c\mathbf{p}_2)\mathbf{r}_2\right) + \frac{1}{4c^2}\left(\mathbf{r}_1 V_{12} + V_{12}\mathbf{r}_1 + \mathbf{r}_2 V_{12} + V_{12}\mathbf{r}_2\right)$$

$$\mathbf{K}_{-4} = -\frac{1}{2c^2}\left[\frac{\beta_1\boldsymbol{\alpha}_1\cdot c\mathbf{p}_1}{2m_1c^2} + \frac{\beta_2\boldsymbol{\alpha}_2\cdot c\mathbf{p}_2}{2m_2c^2}, V_{12}(\mathbf{r}_1+\mathbf{r}_2)\right]$$

$$\mathbf{K}_{-6} = \frac{1}{4}\left[\frac{\beta_1\boldsymbol{\alpha}_1\cdot \mathbf{p}_1}{2m_1c^2} + \frac{\beta_2\boldsymbol{\alpha}_2\cdot \mathbf{p}_2}{2m_2c^2}, \left[\frac{\beta_1\boldsymbol{\alpha}_1\cdot \mathbf{p}_1}{2m_1c^2} + \frac{\beta_2\boldsymbol{\alpha}_2\cdot \mathbf{p}_2}{2m_2c^2}, V_{12}(\mathbf{r}_1+\mathbf{r}_2)\right]\right] \tag{10}$$

For the commutation relation at different orders we found

$$[\mathbf{K}, \hat{H}]_{+2} = [\mathbf{K}_0, \hat{H}_{+2}] = \left[\mathbf{r}_1\beta_1 m_1 + \mathbf{r}_2\beta_2 m_2, m_1c^2\beta_1 + m_2c^2\beta_2\right] = 0 \tag{11}$$

$$[\mathbf{K}, \hat{H}]_0 = [\mathbf{K}_0, \hat{H}_0] + [\mathbf{K}_{-2}, \hat{H}_{+2}] = 0 \tag{12}$$

$$[\mathbf{K}, \hat{H}]_{-2} = [\mathbf{K}_0, \hat{H}_{-2}] + [\mathbf{K}_{-2}, \hat{H}_0] + [\mathbf{K}_{-4}, \hat{H}_{+2}] = i\hbar\mathbf{P} \tag{13}$$

$$\begin{aligned}
[\mathbf{K}, \hat{H}]_{-4} &= [\mathbf{K}_0, \hat{H}_{-4}] + [\mathbf{K}_{-2}, \hat{H}_{-2}] + [\mathbf{K}_{-4}, \hat{H}_0] + [\mathbf{K}_{-6}, \hat{H}_2] = \\
&+\frac{1}{4m_1c^2}\beta_1\left[\boldsymbol{\alpha}_1\cdot\hat{\mathbf{p}}_1\mathbf{r}_1\boldsymbol{\alpha}_1\cdot\hat{\mathbf{p}}_1, \hat{V}\right] + \frac{1}{4m_2c^2}\beta_2\left[\boldsymbol{\alpha}_2\cdot\hat{\mathbf{p}}_2\mathbf{r}_2\boldsymbol{\alpha}_2\cdot\hat{\mathbf{p}}_2, \hat{V}\right] \\
&-\frac{1}{4m_1c^2}\beta_1\mathbf{r}_2\left[|\mathbf{p}_1|^2, \hat{V}\right] - \frac{1}{4m_2c^2}\beta_2\mathbf{r}_1\left[|\mathbf{p}_2|^2, \hat{V}\right]
\end{aligned} \tag{14}$$

For this model we assumed that the potential satisfies

$$[\beta_1, \hat{V}] = [\beta_2, \hat{V}] = 0 \tag{15}$$

as well as

$$[\mathbf{r}_1, \hat{V}] = [\mathbf{r}_2, \hat{V}] = 0 \tag{16}$$

We were not able to develop a consistent matching of terms at all orders using this approach. Interestingly enough, it is possible to obtain models which are more covariant than the two-body Dirac model as measured by deviations from the commutators from zero at higher order (as this one is). The corrections to the potential are similar in this kind of construction to those found by Krajcik and Foldy.

A two-body Dirac model with improved covariance would seem to be an important result, and worth further consideration. We were very interested as a first application of the corrected model to see what happened to the phonon-nuclear coupling terms. Consistent with our worst fears, the lowest-order phonon-nuclear interaction involving the potential is cancelled out by the new correction terms. This appeared to suggest that the phonon-nuclear coupling interaction obtained from the many-particle Dirac model is an artifact, and that the interaction goes away when corrected to be more covariant.

However, there is a problem in this argument that needs to be thought about. If we view the problem as a mathematical problem, then we have succeeded in developing versions of the two-body Dirac model which are objectively more covariant, and in a physical system governed by the resulting Hamiltonian the lowest order potential contribution to the phonon-nuclear interaction vanishes. However, if we view the problem as a physics problem, then through this construction we have added corrections to the potential, and the question might be asked whether the new terms correspond to things that are measured in the lab. In this case the corrections needed to improve covariance correspond to low-order interactions that are not observed experimentally.

The conclusion at this point is that two-body Dirac models with improved covariance that we found developed using this approach involve interactions that are not physical.

There are additional subtleties with this approach. One is that even if operators are constructed that satisfy the commutation relations, it is possible for the solutions to be non-covariant. An example of this is for non-interacting particles with different masses, where in the $+-$ and $-+$ sectors the energy-momentum relation is not covariant. This might provide additional support for not including not including mixed sector solutions in general in connection with physical systems for the many-particle Dirac model. It is possible that the Poincare commutation relations overconstrain the model, since at the end of the day covariance is only needed in the positive energy sector.

## 4. Thinking About the Foldy and Krajcik-Foldy Transformations

Even though the approach outlined above did not pan out, the development of two-body and many-body Dirac models corrected to be more nearly covariant remains of interest. This led to a consideration of making use of a transformation similar to what was pioneered by Foldy [29] and by Krajcik and Foldy [30] (but applying it to the two-body Dirac model).

The basic idea is to start with a Hamiltonian $\hat{H}$ expressed in terms of individual particle positions and momenta, and then rotate to produce a rotated Hamiltonian $\hat{H}'$ expressed in terms of center of mass and relative positions and momenta:

$$e^{i\hat{S}}\hat{H}(\mathbf{r}_1,\mathbf{r}_2,\hat{\mathbf{p}}_1,\hat{\mathbf{p}}_2)e^{-i\hat{S}} \;=\; \hat{H}'(\mathbf{r},\mathbf{R},\hat{\mathbf{p}},\hat{\mathbf{P}}) \tag{17}$$

where $\mathbf{r}_j$ and $\hat{\mathbf{p}}_j$ are individual particle position and momentum operators, where $\mathbf{r}$ and $\hat{\mathbf{p}}$ are relative operators and where $\mathbf{R}$ and $\hat{\mathbf{P}}$ are center of mass operators. Ideally the Hamiltonian $\hat{H}$ would be the two-body Dirac model with a correction

$$\hat{H} \;=\; \beta_1 mc^2 + \beta_2 mc^2 + \alpha_1\cdot c\hat{\mathbf{p}}_1 + \alpha_2\cdot c\hat{\mathbf{p}}_2 + V_{12} + \delta\hat{H} \tag{18}$$

where $\delta\hat{H}$ would make the resulting model more nearly covariant. The rotated Hamiltonian $\hat{H}'$ would ideally involve a covariant combination of the at rest Hamiltonian (perhaps with a rotation) and center of mass energy.

In the event that the potential is set to zero, it is possible to develop this kind of transformation exactly, since the two-body Dirac equation is covariant (if the masses are equal). We have developed a construction of this kind at different orders based on

$$e^{i\hat{S}}\left(\beta_1 mc^2 + \beta_2 mc^2 + \alpha_1\cdot c\hat{\mathbf{p}}_1 + \alpha_2\cdot c\hat{\mathbf{p}}_2\right)e^{-i\hat{S}} \;=\; \hat{H}' \tag{19}$$

where the part of the rotated Hamiltonian in the positive energy sector is consistent with [31]

$$\left(\hat{H}'\right)_{++} = \sqrt{(Mc^2)^2 + 4c^2|\hat{\mathbf{q}}|^2 + c^2|\hat{\mathbf{P}}|^2} \tag{20}$$

with

$$\hat{\mathbf{q}} = \hat{\mathbf{p}} - \frac{1}{2}\frac{\hat{\mathbf{p}}\cdot\hat{\mathbf{P}}}{(Mc)^2 + |\hat{\mathbf{p}}|^2}\hat{\mathbf{P}} + \cdots \tag{21}$$

An interpretation is possible that $\hat{\mathbf{q}}$ is the Dirac model's version of the relative momentum in the moving frame. After experimenting with a large number of approaches that uniformly failed, it was encouraging to see that this construction could be carried out reasonably for this simple model.

The generalization of the approach to include a potential on the face of it appears to be straightforward – we have carried out such a construction for several orders. The initial conclusion based on this exploratory calculation is that a rotation is possible resulting in a transformed Hamiltonian consistent with covariance up to higher order than we were expecting (and higher order than is obtained from numerical calculations). The headache is that at each point in the construction terms appear which could be treated as covariant corrections to the Hamiltonian, or which can be rotated out (so that their impact is deferred to higher order) – and we are not yet able to identify what the lowest-order correction should be.

The conclusion for now is that it should be possible to develop a version of the two-body Dirac model where the lowest-order covariant correction can be identified. An issue is that it is possible that corrections which improve covariance may also be nonphysical. Another issue is that requiring overall covariance in this case may over-constrain the system as once again covariance is only required for the positive energy sector.

## 5. Exploratory Numerical Solutions

In connection with the models investigated above, we developed a finite basis code to solve the two-body Dirac model with a scalar SHO potential in order to gain some insight as to how much the lack of covariance of the model impacted the eigenvalues for a composite in motion. The results were interesting.

We found that the energy-momentum relation for the ground state in the positive energy sector was very nearly consistent with covariance (the numerical and analytic covariant solutions cannot be distinguished for the parameters studied as shown in Figure 2), which is encouraging [31]. We also found that (minor) nonphysical changes in the level splitting occurred in the moving frame. Of concern is that the deviations from covariance occur at the same order ($|\mathbf{P}|^2$) as would occur with the term giving rise to the phonon-nuclear interaction.

What remains to be done is to examine potential candidates for correction terms, and then to verify numerically whether they are effective or not. At this point we are beginning to suspect that the lack of covariance is a consequence of using a Hamiltonian formulation, and that corrections that improve covariance are nonphysical.

## 6. Phonon-Nuclear Interaction From a Covariant Model

From the discussion above it might be concluded that the many-particle Dirac model is good enough to allow a derivation of the phonon-nuclear interaction at low order, but the lack of covariance probably precludes a treatment at high order and doubt remains as to whether the low-order interaction derived is physical. This motivates interest in whether the interaction can be derived from a covariant model.

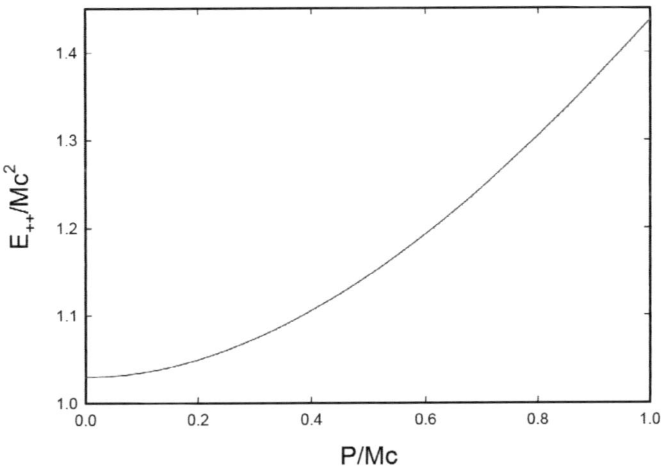


**Figure 2.** Numerical (blue) and analytic covariant (red) ground state energy as a function of the center of mass momentum for a two-body Dirac model with an SHO potential with $\hbar\omega_0/Mc^2 = 0.02$.

One such model is the covariant two-body Bethe-Salpeter model, which for us is an obvious candidate with which to examine the issue. We worked with a version of this model in a form close to the two-body Dirac model described above, and make use of a partial Foldy-Wouthuysen transformation which leads to an approximate separation of the (nonrelativistic) center of mass degrees of freedom and (relativistic) relative degrees of freedom. The model that results includes terms which are close to those derived from the two-body Dirac model. In this case the kernel describing the interaction is transformed according to [31]

$$V'(\mathbf{k},\epsilon) = V(\mathbf{k},\epsilon) + \frac{1}{2Mc}\left[\beta_1\boldsymbol{\alpha}_1 + \beta_2\boldsymbol{\alpha}_2, V(\mathbf{k},\epsilon)\right]\cdot\hat{\mathbf{P}} + \cdots \tag{22}$$

This compares well with the lowest-order interaction due to the potential obtained from the two-body Dirac model

$$\hat{H}_{int} = \frac{1}{2Mc}\left[\beta_1\boldsymbol{\alpha}_1 + \beta_2\boldsymbol{\alpha}_2, V(|\mathbf{r}_2 - \mathbf{r}_1|)\right]\cdot\hat{\mathbf{P}} \tag{23}$$

It is very encouraging that the phonon-nuclear interaction can be derived from a covariant model, and that the lowest-order interaction is essentially the same as obtained from the two-body Dirac model.

## 7. Progress on Matrix Element Calculations

A quantification of the models under discussion in this paper requires the evaluation of phonon-nuclear matrix elements, as well as the estimation of off-resonant energy shifts. Years ago we made use of Pauli reduction to develop an estimate for the phonon-nuclear matrix element for the $D_2/^4He$ transition [32]. A better approach to this problem would be to develop a reduction for the phonon-nuclear interaction, and to develop a more standard calculation. An early attempt as this kind of model was carried out for the spin-orbit contribution to the phonon-nuclear matrix element in the case of the 6.2 keV transition of $^{181}Ta$ [33].

### 7.1. Nonrelativistic Reduction

The phonon-nuclear interaction discussed above is given in terms of relativistic operators; however, most nuclear structure calculations these days are done in the context of a nonrelativistic description with nonrelativistic interactions that are derived from a covariant field theory (chiral effective field theory) model [34]–[37]. It would be possible to make use of the relativistic phonon-nuclear interaction in connection with the relativistic interactions from this model, and then reduce the expressions that result to a form appropriate for a nonrelativistic calculation.

This has recently been done for the relativistic one-pion exchange potential based on the spatial part of the pseudovector interaction, which is consistent with the lowest-order interaction from chiral effective field theory [38]. For the one-phonon exchange contribution this can be written as

$$\left(\hat{H}'_{int}(j,k)\right)_{NR} = -\frac{1}{4Mmc^2}\left[\boldsymbol{\sigma}_j\cdot\mathbf{p}_j,\left[\boldsymbol{\sigma}_j\cdot\mathbf{P},\hat{V}_{jk}^{(NR)}\right]\right] - \frac{1}{4Mmc^2}\left[\boldsymbol{\sigma}_k\cdot\mathbf{p}_k,\left[\boldsymbol{\sigma}_k\cdot\mathbf{P},\hat{V}_{jk}^{(NR)}\right]\right] \tag{24}$$

where $\hat{V}_{jk}^{(NR)}$ is the nonrelativistic one-pion exchange interaction.

### 7.2. Progress on the HD/$^3$He Phonon-Nuclear Matrix Element

One of the more straightforward calculations to be done is the evaluation of the phonon-nuclear matrix element in the case of the HD/$^3$He transition, which is important for models under discussion for excess heat in the NiH system. A construction for the molecular HD states and doublet S component for the $^3$He state is given in [39], along with a reduction of the spin, isospin, and angular degrees of freedom. The resulting matrix elements are in terms of relative radial coordinates, which are convenient for numerical evaluation. The plan is to develop estimates based on simple approximate radial wave functions, and afterward to attempt a more sophisticated calculation based on an optimized three-body solution for the $^3$He nucleus.

There are many issues associated with such models. In an ideal situation one would like to work with a phonon-nuclear operator that is consistent with the potential used. This would suggest that we should start with molecular HD and $^3$He states evaluated with the chiral effective field theory model, and then develop reductions for the LO, NLO and NNLO potentials [37] needed for the structure calculations. This would constitute a major theoretical project. It would take much less work to make use of the same structure calculation, but then use only the lowest-order (LO) interaction for the phonon-nuclear matrix element, with an associated loss of accuracy. Less work would be required to develop wave functions based on older empirical potentials for the matrix elements, but then there is the issue of developing a consistent phonon-nuclear interaction (since the older empirical potentials were not relativistic). In this case one could make do with the LO contribution to the phonon-nuclear interaction in the large $r$ regime, which would pick up some of the contributions to the matrix element.

### 7.3. Progress on Models for Transitions in Heavier Nuclei

We would like to use or develop codes that can provide wave functions for heavier nuclei that could be used for the systematic evaluation of phonon-nuclear matrix elements. Developing single configuration wave functions is straightforward based on effective interactions (such as the Skyrme interaction [40]). However, correlation is important in structure calculations, and single configuration models in general do not give good results for E1, M1 or E2 matrix elements for low-energy transitions. Much progress has been made including correlation through the development of multiconfigurational structure models based on effective interactions (see [41] as an example of a substantial literature on this problem). Codes (such as ANTOINE [41]) are available for this type of calculation, and there is the possibility of developing a post-processor for phonon-nuclear matrix element calculations.

We have considered the development of this kind of code for some exploratory calculations. In the case of the LO chiral effective field theory interaction we can write the potential in coordinate space as

$$v_{LO}(\mathbf{r}) = \frac{\mu_\pi c^2}{\kappa^3}\left(C'_S + C'_T(\boldsymbol{\sigma}_1\cdot\boldsymbol{\sigma}_2) + \frac{1}{3}f^2_{\pi NN}(\boldsymbol{\tau}_1\cdot\boldsymbol{\tau}_2)(\boldsymbol{\sigma}_1\cdot\boldsymbol{\sigma}_2)\right)v_{contact}(r)$$

$$-f^2_{\pi NN}\frac{\mu_\pi c^2}{\kappa^3}(\boldsymbol{\tau}_1\cdot\boldsymbol{\tau}_2)\frac{\boldsymbol{\sigma}_1\cdot\boldsymbol{\sigma}_2}{3}v_{central}(r)$$

$$-f^2_{\pi NN}\frac{\mu_\pi c^2}{\kappa^3}(\boldsymbol{\tau}_1\cdot\boldsymbol{\tau}_2)\left(\frac{(\boldsymbol{\sigma}_1\cdot\mathbf{r})(\boldsymbol{\sigma}_2\cdot\mathbf{r})}{r^2} - \frac{\boldsymbol{\sigma}_1\cdot\boldsymbol{\sigma}_2}{3}\right)v_{tensor}(r) \quad (25)$$

For multiconfigurational calculations we require a reduction of the interaction in the case of simple ($m$-scheme [41]) basis states. For example, in the case of the central potential we can write for the central contribution to the general matrix element

$$\left\langle a,b\left| - f^2_{\pi NN}\frac{\mu_\pi c^2}{\kappa^3}(\boldsymbol{\tau}_1\cdot\boldsymbol{\tau}_2)\frac{\boldsymbol{\sigma}_1\cdot\boldsymbol{\sigma}_2}{3}v_{central}(r)\right|a,b\right\rangle =$$

$$-\frac{1}{3}f^2_{\pi NN}\frac{\mu_\pi c^2}{\kappa^3}\left\langle |\tau,m_{\tau a}\rangle_1|\tau,m_{\tau b}\rangle_2\Big|(\boldsymbol{\tau}_1\cdot\boldsymbol{\tau}_2)\Big||\tau,m_{\tau c}\rangle_1|\tau,m_{\tau d}\rangle_2\right\rangle$$

$$\sum_l\int_0^\infty\int_0^\infty P_a(r_1)P_b(r_2)v_l^{central}(r_1,r_2)P_c(r_1)P_d(r_2)dr_1dr_2$$

$$\sum_{m_a}\sum_{m_b}\sum_{m_c}\sum_{m_d}\sum_m 4\pi(-1)^m\langle l_a,m_a|Y_{l,-m}|l_c,m_c\rangle\langle l_b,m_b|Y_{l,m}|l_d,m_d\rangle$$

$$\sum_{m_{sa}}\sum_{m_{sb}}\sum_{m_{sc}}\sum_{m_{sd}}\left\langle |s,m_{sa}\rangle_1|s,m_{sb}\rangle_2\Big|(\boldsymbol{\sigma}_1\cdot\boldsymbol{\sigma}_2)\Big||s,m_{sc}\rangle_1|s,m_{sd}\rangle_2\right\rangle$$

$$\langle j_a,m_{ja}|l_a,m;s,m_{sa}\rangle\langle j_b,m_{jb}|l_b,m;s,m_{sb}\rangle\langle l_c,m;s,m_{sc}|j_c,m_{jc}\rangle\langle l_d,m;s,m_{sd}|j_d,m_{jd}\rangle \quad (26)$$

At this point reductions have been developed for the Skyrme interaction, for the one-pion exchange potentials, and for the LO and NLO potentials from chiral effective field theory.

With luck we will have some numerical results for phonon-nuclear matrix elements for problems of interest in the coming year.

## 8. Excitation Transfer of a 24 MeV Quantum

As discussed in the introduction, the overall process for excess heat production in the picture under consideration is quite complicated. Since energetic particles are not seen in amounts commensurate with the energy produced, we are not able to study the excess heat process when it is producing excess heat using standard nuclear particle detection.

Consider the first step of the proposed reaction process outlined in the introduction, which involves the transfer of 24 MeV to a (reasonably stable) highly excited state of a nucleus in the host lattice (either a Pd nucleus, or an impurity nucleus). When excess heat is produced, the thought is that excitation in this highly excited state is rapidly transferred elsewhere through excitation transfer processes associated with subdivision. If these excitation transfer processes are fast compared to the lifetime of the highly excited nuclear state, then very little in the way of disintegration products would be seen. It may be that the excitation is rapidly transferred to a longer-lived highly excited state prior to subdivision, perhaps one with more angular momentum, which would also hinder the detection of decay products.

### 8.1. Excitation Transfer to Pd

Suppose now that there is a problem so that the reaction proceeds much more slowly. For example, suppose that the highly-excited phonon mode is insufficiently excited, or that the number of $D_2$ dideuterium species is low, or perhaps that there are insufficient impurity nuclei present in the lattice with reasonably stable low-lying excited states. In this case it may be that the highly excited nucleus decays a larger fraction of the time [20], as illustrated for $^{106}$Pd in Figure 3.

We conjectured previously [20] that the energetic alphas produced this way may be the source of the signals seen in Ref. [42], and that the 14 MeV neutrons from this process may be the source of energetic neutrons commented on in Ref. [43].

### 8.2. Systematics

Whether this mechanism is responsible for some of the energetic nuclear particles observed in experiments is at present not confirmed; however, a confirmation (or disproof) is possible since the alpha energies have a strong dependence on nuclear mass [20] (as shown in Figure 4).

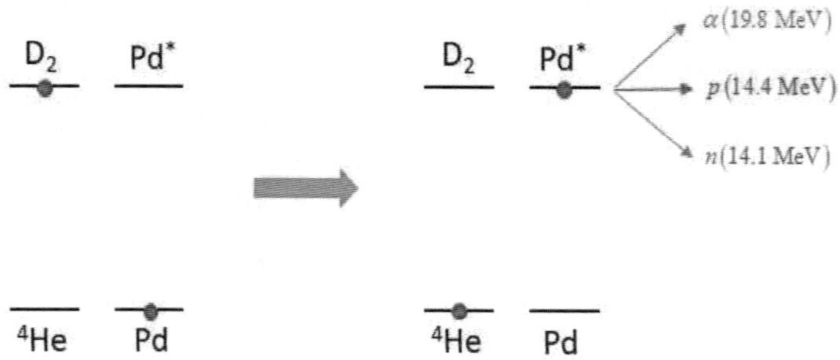


**Figure 3.** Schematic of excitation transfer of 24 MeV from the $D_2/^4$He transition resulting in a highly-excited state in $^{106}$Pd that subsequently decays.

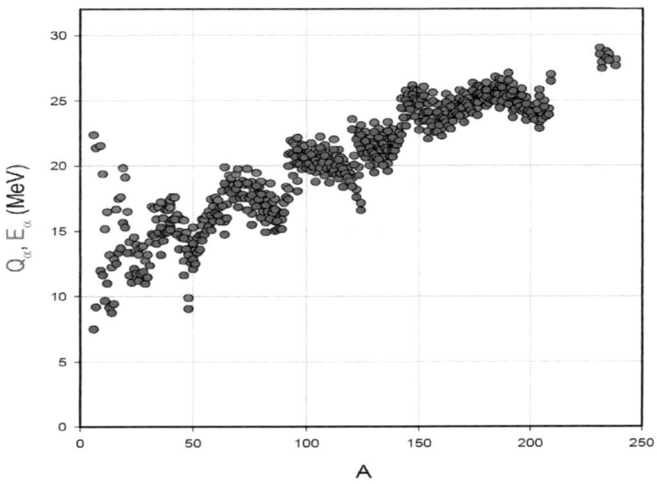


**Figure 4.** Reaction energy $Q$ (red) and alpha energy $E_\alpha$ (green) for $D_2/^4He$ excitation transfer to isotopes with different nuclear mass $A$.

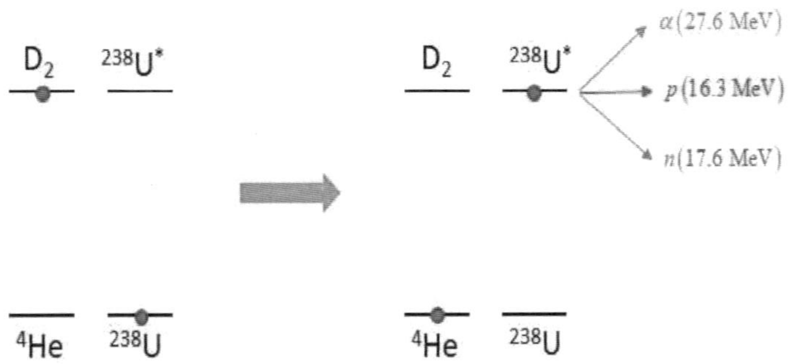


**Figure 5.** Schematic of excitation transfer of 24 MeV from the $D_2/^4He$ transition resulting in a highly-excited state in $^{238}U$ that subsequently decays.

This suggests that looking for energetic alpha emission from elements with high Z might be an important goal, as the confirmation of an energetic 27.6 MeV alpha from $^{238}U$ (see Figure 5) would strongly support the proposed mechanism.

## 8.3. Excitation Transfer to Ti

Chambers and co-workers reported the observation of charged particles near 5 MeV in the case of deuteron ion bombardment of Ti at low energy (350 eV) [44], [45]. In Table 1 is listed the various candidate reaction particles possible associated with decay of highly excited states in the Ti isotopes following the excitation transfer of a 24 MeV quantum from the $D_2/^4He$ transition.

**Table 1.** Energies (MeV) for neutrons, protons, deuterons, tritons, $^3$He, and $^4$He (including recoil) expected from the decay of the different stable Ti isotopes following excitation transfer of a 24 MeV quantum from the $D_2/^4$He transition.

| isotope | $E_n$ | $E_p$ | $E_d$ | $E_t$ | $E_{3He}$ | $E_{4He}$ |
|---|---|---|---|---|---|---|
| $^{46}$Ti | 10.42 | 13.21 | 4.21 | 0.89 | 2.99 | 14.46 |
| $^{47}$Ti | 14.64 | 13.09 | 6.55 | 1.66 | 5.10 | 13.62 |
| $^{48}$Ti | 11.96 | 12.14 | 3.81 | 1.38 | 1.16 | 13.20 |
| $^{49}$Ti | 15.38 | 12.24 | 6.22 | 1.97 | 3.28 | 12.55 |
| $^{50}$Ti | 12.65 | 11.45 | 3.63 | 1.69 | – | 12.08 |

Note that $^{47}$Ti has the possibility of emitting a $^3$He at 5.10 MeV resulting from this kind of excitation transfer. Not much is known about the specific highly-excited nuclear states in the titanium isotopes in the vicinity of 24 MeV; however, we would expect that reasonably long-lived states that are nearly resonant with the transition energy would be favored. This suggests a scenario in which $^{47}$Ti might be special in this sense, which might the emission in this channel to dominate.

This possibility could be pursued by making use of an isotopically pure sample made of $^{47}$Ti, and perhaps a sample with $^{47}$Ti excluded, which may be able to help sort whether the reaction mechanism proposed here is consistent with experiment. Also, it would be helpful to attempt charged particle detection at higher energy to look for alpha and proton emission. In the data there seems to be no sign of a deuteron signal at 6.55 MeV (the upper energy for which data is given is about 6.2 MeV), which should be clarified by extending range over which ions are detected.

## 9. Excitation Transfer of a 5.5 MeV Quantum

The picture outlined above for excess heat in PdD can be extended to excess heat in NiH systems under the assumption that the process starts with an HD/$^3$He transition instead of a $D_2/^4$He transition. If so, then $^3$He should be observed in amounts commensurate with the energy produced. At this point we are not certain that $^3$He is the product of light water excess heat, as there are at present no reports where appropriate high-resolution quadrupole mass spectrometry has been used to assay for $^3$He. From a theoretical perspective we consider this proposed modification of the picture to be a strong candidate for excess heat in light water systems.

### 9.1. Systematics

The possibility of the excitation transfer of a 5.5 MeV quantum, which is implied as a first step in an excess heat model, can be tested experimentally by looking for decay products from highly-excited nuclear states that are produced following excitation transfer from the HD/$^3$He transitions in a system where excess heat production is frustrated [20]. Once again the expected alpha energies depend strongly on the nuclear mass $A$ as illustrated in Figure 6 – although in this case the more energetic alphas potentially produced at high Z are hindered by the Coulomb barrier. Consequently, alpha emission is expected only in the case of excitation transfer to low $A$ target nuclei; for high $A$ nuclei gamma emission should be sought, unless neutron decay is energetically allowed.

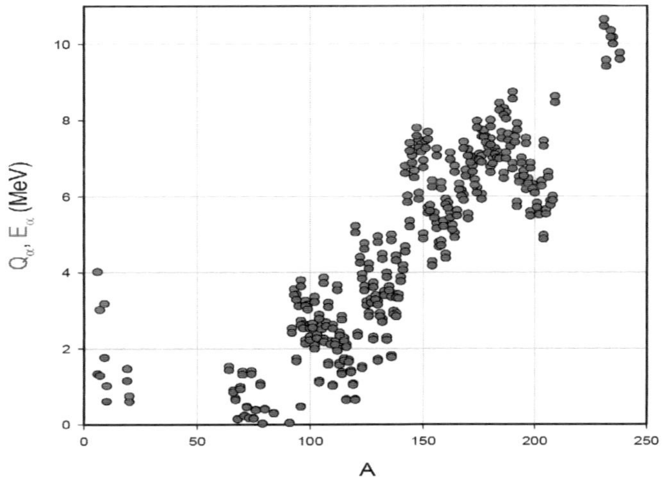


**Figure 6.** Reaction energy $Q$ (red) and alpha energy $E_\alpha$ (green) for HD/$^3$He excitation transfer to isotopes with different nuclear mass $A$.

### 9.2. Proton Emission From $^6$Li

Of special interest in connection with this discussion is excitation transfer to $^6$Li according to [20]

$$\left[ \text{H+D} \rightarrow {}^3\text{He} + 5.49 \text{ MeV} \right] + \left[ {}^6\text{Li} \rightarrow \text{p} + {}^5\text{He} - 4.43 \text{ MeV} \right] + 1.06 \text{ MeV}$$

which is the only example of a proton decay channel starting with a stable ground state nucleus available for the transfer of a 5.5 MeV quantum. We have proposed that this reaction is responsible for the proton signal reported in [46], [47].

## 10. Resonance Issues for the HD/$^3$He Transition

For excitation transfer from the HD/$^3$He transition near 5.5 MeV the issue of resonance becomes more important, as there are far fewer nuclear states available than for the $D_2$/$^4$He transition near 24 MeV. At this lower energy there are available detailed lists of nuclear states which have been observed, and which might be candidates for excitation transfer.

### 10.1. Excited Nuclear States in the Stable Ni Isotopes

Consider excess heat production in and Ni experiment where the first step in the associated reaction picture might be the transfer of a 5.5 MeV quantum to a highly-excited state in a nickel isotope. In this case we can make use of the excited states of the nickel isotopes tabulated in the NUDAT2 database [48], [49] (that extends above 5.5 MeV) to create Figure 7. There are many states in the general vicinity of the HD/$^3$He transition energy, as can be seen in this figure, with the highest density of states associated with $^{61}$Ni.

A detailed comparison leads to the conclusion that the nearest state is off of resonance by more than 5 keV (the closest is $^{62}$Ni(5488) which compares with the HD/$^3$He transition at 5493.4780 keV), a mismatch which is much too large to be made up through the exchange of a few phonon or plasmon quanta.

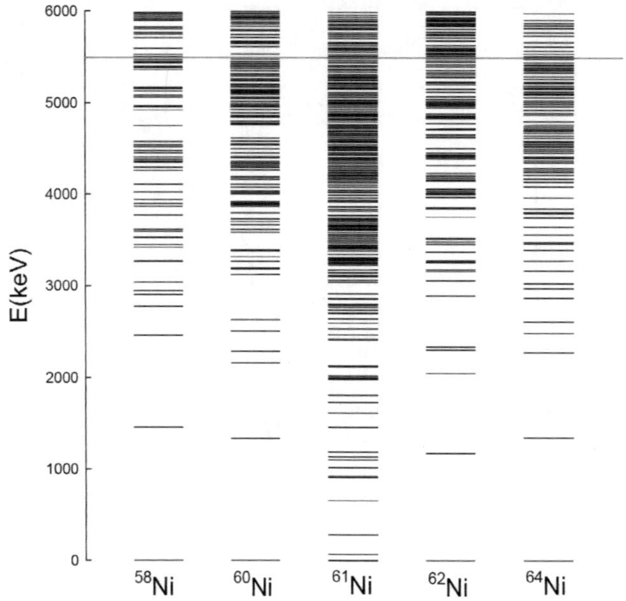


**Figure 7.** Known energy levels of the stable Ni isotopes (dark blue) in the vicinity of the HD/$^3$He transition near 5.5 MeV (red).

### 10.2. Impurity Nuclei

This draws attention to the start up problem within the model. When excess heat is being produced, then there is the possibility of making up energy mismatches at the keV or 10s of keV scale via up-conversion and down-conversion, in which case if some of the nearby levels of the nickel isotopes are reasonably stable then they can be candidates to participate in the overall reaction. However, before excess heat is being made, the question can be asked as how the process starts, since substantial up-conversion and down-conversion under these conditions seems unlikely. What is needed are long-lived states more nearly resonant, presumably from impurities in the nickel.

We can make use of the NUDAT2 database to ask what energy levels of the stable nuclei are known that are more closely resonant with the HD/$^3$He transition energy. Results from a search of this kind are shown graphically in Figure 8. Closest is the $^{116}$Sn(5493.2) state, which looks like it may be within 300 eV (there are uncertainties involved since the known energies are reported to fewer digits than the HD/$^3$He transition is known). Note that there are many more nuclear energy levels in this region than what are tabulated in the database, as not all of the stable isotopes have been studied in this way in detail.

The conclusion is that within the overall proposed reaction scheme, excess heat in the NiH system would require some deuterium to be present, and would not be able to start up without impurity nuclei present in the Ni with nuclear states close to the HD/$^3$He transition energy. If correct, at some point in the future studies would be of interest in which specific impurity isotopes were present, from which it could be inferred have reasonably stable states near resonance at 5493.4780 keV.

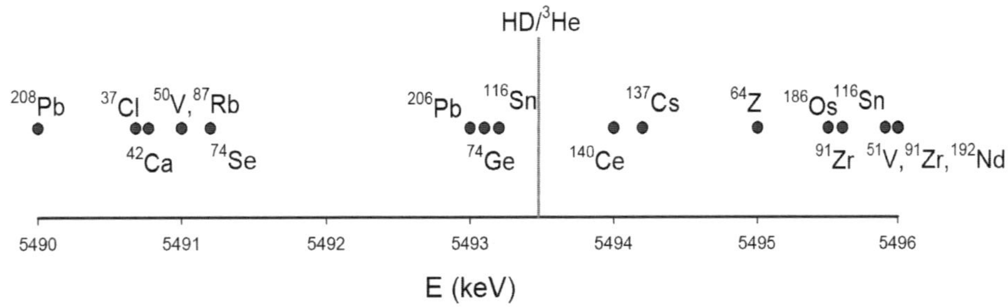


**Figure 8.** Known energy levels of stable isotopes (and $^{137}$Cs) from the NUDAT2 database (blue circles) compared with the HD/$^3$He transition energy (red line).

## 11. $^{137}$Cs and the HD/$^3$He Transition

The radioactive isotope $^{137}$Cs beta decays to $^{137}$Ba with the emission of an electron and an antineutrino with a half-life of 30.08 years. Most of the time this decay populates the excited state of $^{137}$Ba at 661.659 keV (94.5%), and the rest of the time ground state $^{137}$Ba is produced. The $^{137}$Ba excited state decays produce a 661.657 keV gamma, which makes $^{137}$Cs important as a calibration source for gamma spectroscopy.

### 11.1. Multiple Excitation Transfers

In the discussion above we found that $^{137}$Cs has an excited state at 5494.2 keV, which is nearly resonant with the HD/$^3$He transition at 5493.4780 keV. The energy mismatch is less than 1 keV. Due to the phonon-nuclear interaction, indirect coupling would be expected between HD + $^{137}$Cs(0), and $^3$He+$^{137}$Cs(5494). We know almost nothing about this excited state in $^{137}$Cs other than its energy. Most likely the coupling with this state would lead to gamma decay to some of the $^{137}$Cs levels.

We might speculate that further excitation transfer can induce a coupling with longer-lived states of higher spin, for which beta decay is dominant (illustrated in Figure 9). In this case there is the possibility of seeing an overall acceleration of the beta decay of $^{137}$Cs. Depending on the details of which states are produced from the beta decay of this highly-excited state, it is possible that an increase in the intensity of the 661.657 keV gamma might occur, as well as emission at other energies.

### 11.2. Possible Connection With Kornilova and Vysotskii Experiment

Kornilova, Vysotskii and coworkers have claimed a factor of 40 increase in the decay rate for $^{137}$Cs in biological cultures measured from the time dependence of the 661 keV gamma [50]. It seems natural to consider whether the mechanism discussed in this section might be relevant.

Years ago we noted that under conditions where anomalous energy exchange is possible, that beta decay had the potential to be accelerated assuming that phonon exchange occurs. However, it seems unlikely that anomalous energy exchange would be possible in a biological culture, which would seem to rule out such a mechanism. In this case the near-resonance noted here may be a better candidate.

From the discussion above, it seems likely that if the near resonance is important then there may be additional gamma lines produced from highly-excited $^{137}$Cs states that gamma decay rapidly. To account for a substantial acceleration in the beta decay rate, then excitation transfer to highly-excited $^{137}$Cs states which beta decay faster than gamma decay likely occurs. There is the question as to what states are produced from the beta decay from these

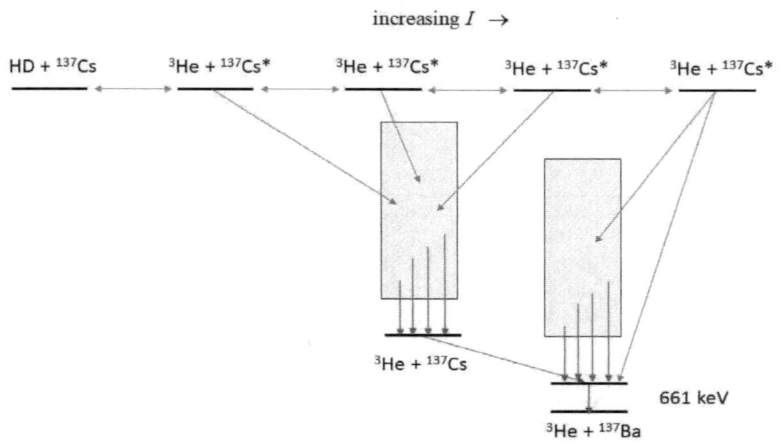


**Figure 9.** Schematic of how the HD/$^3$He transition might increase the decay rate for $^{137}$Cs. The first excitation transfer step results in HD transitioning to $^3$He and ground state $^{137}$Cs transitioning to a nearby excited state, which would likely gamma decay to lower energy levels in $^{137}$Cs (the grey box on the left). Repeated excitation transfer steps can lead to highly-excited states with higher total orbital and spin angular momentum, eventually leading to states where beta decay dominates over gamma decay, resulting in gamma decays to highly-excited states in $^{137}$Ba (the grey box on the right) and possibly to the 661 keV state.

highly-excited $^{137}$Cs. One might expect this beta decay to result in highly-excited states in $^{137}$Ba, which would give rise to other gammas. What the yield is for the 661 keV gamma is an interesting question – one could imagine working with an initial known amount of $^{137}$Cs and measuring the 661 keV emission before it is added to the biological culture, and then measuring again after it has been taken up into the cells. If a change in the 661 keV emission occurs, then this might be due to a finite yield of 661 keV emission resulting from the beta decay from highly-excited $^{137}$Cs. If the mechanism under discussion is relevant, one might expect the decay rate to increase with a small addition of $D_2O$ into the solution.

### 11.3. Possible Connection With the Piantelli Experiment

Piantelli and coworkers have described seeing gamma emission at $661.5 \pm 0.8$ keV in connection with their NiH excess heat experiments [51], [52]. There are issues in connection with these observations: whether the emission correlates with excess heat production or not; and whether the line is due to the well known transition in $^{137}$Ba produced from $^{137}$Cs or not.

In Table 2 we consider all known gamma transitions in the stable isotopes up to $^{137}$Ba in the range 661-662 keV from the NUDAT2 database. There do not seem to be compelling candidates in general among this list, and no compelling candidates were proposed by the Piantelli group.

In connection with whether the line correlates with excess heat or not, during a visit to the Piantelli lab in 2011 the group demonstrated excess heat for Mike McKubre and I, and showed that the 661 keV line turned on when the excess heat turned on.

We would expect that when excess heat is produced, energy exchange beyond the few phonon or plasmon level becomes possible, so that the energy mismatch between the HD/$^3$He transition and the $^{137}$Cs level at 5494.2 keV (or other levels) can be made up (this is illustrated in Figure 10). This has the potential to produce a stronger version of

**Table 2.** Excited state energies and gamma energies for all known gamma transitions between 661 keV and 661 keV among the stable isotopes up to $^{137}$Ba.

| isotope | $E_j$ | $E_\gamma$ |
|---|---|---|
| $^{48}$Ti | 5974 | 662 |
| $^{50}$Cr | 3825.7 | 661.6 |
| $^{76}$Ge | 3409.19 | 661.4 |
| $^{75}$As | 860.0 | 661.4 |
| $^{77}$Se | 911.5317 | 661.719 |
| $^{88}$Sr | 8094.8 | 661.3 |
| $^{98}$Mo | 2419.63 | 661.12 |
| $^{98}$Ru | 3852.3 | 661.6 |
| $^{99}$Ru | 3536.29 | 661.76 |
| $^{102}$Pd | 5094.11 | 661.3 |
| $^{114}$Cd | 2525.420 | 661.21 |
| $^{120}$Te | 1863.29 | 662.0 |
| $^{126}$Xe | 3252.1 | 661 |
| $^{128}$Xe | 2941.9 | 661.0 |
| $^{137}$Ba | 661.659 | 661.657 |

the effect. The conjecture is that a small amount of $^{137}$Cs was initially present in the nickel bar used in the experiment, and that its decay was dramatically increased leading to a large increase in the intensity of the 661 keV line.

### 11.4. $^{137}$Cs as an Indicator of Anomalous Energy Exchange

If it is true that the intensity of the 661 keV line is sensitive to excess heat being produced, then this might be important for several reasons. It has the potential to provide for a very different kind of test for phonon-mediated and plasmon-mediated nuclear excitation transfer. It has the potential to provide information as to how much anomalous energy exchange occurs in connection with excitation transfer, under conditions where up-conversion and down-conversion occurs. The effect also has the potential to provide an independent diagnostic for heat production in light water excess

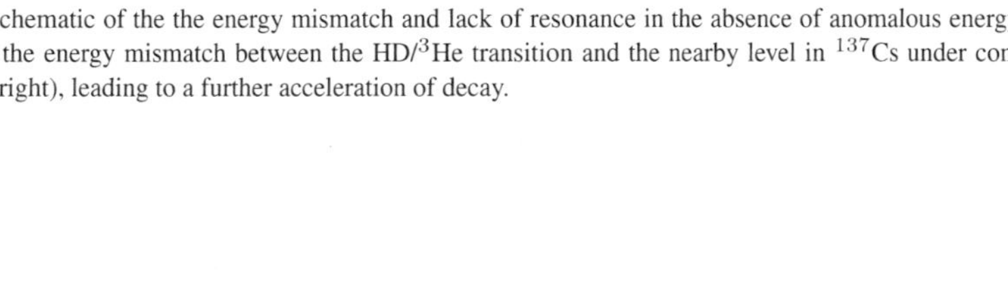


**Figure 10.** Simplified schematic of the the energy mismatch and lack of resonance in the absence of anomalous energy exchange (left); and the possibility of making up the energy mismatch between the HD/$^3$He transition and the nearby level in $^{137}$Cs under conditions where anomalous energy exchange occurs (right), leading to a further acceleration of decay.

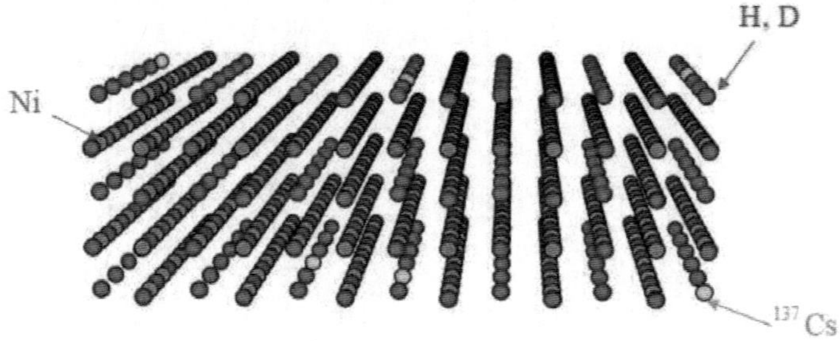


**Figure 11.** Illustration of $^{137}$Cs going into vacancies in Ni co-deposited using the Letts protocol, assuming that locally we end up with a super-abundant vacancy phase.

heat experiments, which would be interesting and perhaps useful. There is the prospect for a mechanism that can be associated with the acceleration of radioactive decay, supporting the effort of Kornilova, Vysotskii and coworkers to remediate harmful radioactive isotopes. Although this mechanism has been discussed specifically in the context of the HD/$^3$He transition and $^{137}$Cs, it is likely to work for other unstable isotopes as well.

How might this be tested? There are many possibilities. One approach might be to work with the Letts co-deposition with either Ni or Pd [53], which according to our earlier conjectures may make host metal atom vacancies [54]. If so, then the task would be to find a way to get $^{137}$Cs into some of these vacancies (as illustrated in Figure 11). One possibility might be to arrange for $^{137}$Cs to be in the electrolyte during the Ni or Pd co-deposition, hoping for a some of it to deposit. Another possibility might be to use different steps, where the Ni or Pd co-deposition is first done (in acid) following the Letts protocol. Then the surface could be rinsed off and placed in base, with electrolysis involving $^{137}$Cs in the electrolyte. According to [55] Cs is found to go into the cathode at low levels during electrolysis in base. Then the surface could be rinsed, and then placed into a new cell for live tests. For this perhaps a small amount of Ni or Pd could be co-deposited, hoping to produce excess heat along with an increase in the 661 keV gamma line.

## 12. The Null Reaction

One of the first proposals for phonon-mediated excitation transfer years ago was for the null reaction [56], [57], in which a $D_2$ making a transition to $^4$He at one site produced a new $D_2$ from a $^4$He at another site (as illustrated in Figure 12). This process seemed of interest since it could be resonant, so that it could proceed as a coherent process. At the time this seemed to be of academic interest only, since it would seem to be hard to distinguish experimentally whether a $D_2$ at one site and $^4$He at another site had undergone excitation transfer.

Our attention has recently returned to this problem motivated by the interpretation of the enhanced fusion rate observed in low-energy deuteron beam experiments, where the yield as a function of energy is fitted successfully to a model that includes a resonance at low energy [58]. The probability that a long-lived excited state of the alpha particle is nearly resonant at the eV level with $D_2$ seems low. On the other hand, excitation transfer from the $D_2$/$^4$He transition to produce p+t or n+$^3$He final state from $^4$He seems like it might be a candidate explanation to account for the excess fusion rate observed at low beam energy.

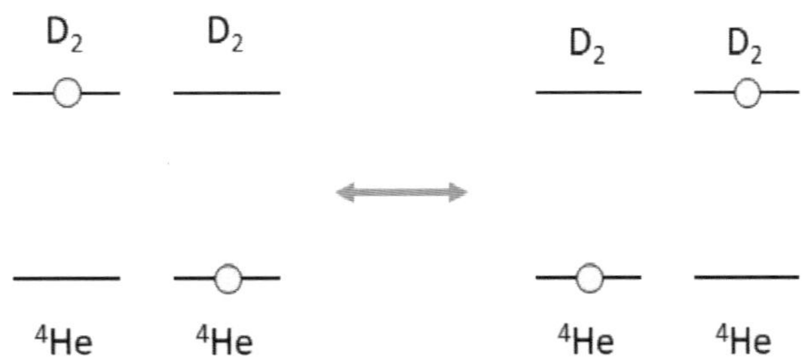


**Figure 12.** Concept of phonon-mediated nuclear excitation transfer in the case of the null reaction.

### 12.1. Simple Model

Let us consider a naive model for excitation transfer in the null reaction, where we might use a Hamiltonian of the form

$$\hat{H} = \hbar\omega_0 \hat{a}^\dagger \hat{a} + \sum_j \sum_k \frac{|\hat{\pi}_{k|j}|^2}{m} + \sum_j \sum_{k<k'} V_{k,k'|j}$$

$$+\frac{1}{\Delta E_-} \sum_{j<j'} \Big| \mathrm{D}_2(j),\ {}^4\mathrm{He}(j') \Big\rangle \langle \mathrm{D}_2(j)|\hat{\mathbf{a}}_j \cdot c\hat{\mathbf{P}}_j|{}^4\mathrm{He}(j)\rangle \langle {}^4\mathrm{He}(j')|\hat{\mathbf{a}}_{j'} \cdot c\hat{\mathbf{P}}_{j'}|\mathrm{D}_2(j')\rangle \Big\langle\ {}^4\mathrm{He}(j),\ \mathrm{D}_2(j') \Big|$$

$$+\frac{1}{\Delta E_+} \sum_{j<j'} \Big| {}^4\mathrm{He}(j), \mathrm{D}_2(j') \Big\rangle \langle {}^4\mathrm{He}(j)|\hat{\mathbf{a}}_j \cdot c\hat{\mathbf{P}}_j|\mathrm{D}_2(j)\rangle \langle \mathrm{D}_2(j')|\hat{\mathbf{a}}_{j'} \cdot c\hat{\mathbf{P}}_{j'}|{}^4\mathrm{He}(j')\rangle \Big\langle \mathrm{D}_2(j), {}^4\,\mathrm{He}(j') \Big| \tag{27}$$

where we assume one highly-excited phonon mode, the relative nucleon kinetic energies (where $k$ refers to a nucleon and $j$ refers to a nucleus), the nucleon-nucleon interactions, and excitation transfer terms involving single-phonon exchange for all nuclear transitions.

There are technical issues associated with solving for the states of the system in this kind of model; however, after some thought it becomes clear that if a $^4$He nucleus makes a transition to the $D_2$ channel it will initially be localized and not able to tunnel through the Coulomb barrier in a reasonable time. An approximation (not satisfying $P \to 0$ for $r \to 0$) for the resulting compact state spatial distribution is illustrated compared with the molecular $D_2$ state. The compact $D_2$ state is localized at the few fm scale. More work remains to develop a self-consistent compact state solution.

### 12.2. Compact $D_2$ State and the Kasagi Experiment

It might be asked whether such a compact $D_2$ state might be connected with experiment. An immediate application is to the three-body fusion experiment of Kasagi and coworkers [59], in which an incoming deuteron leads to energetic alpha and proton particles with an energy distribution consistent with a three deuteron fusion process. A confirmation of this result was reported in [60]. The compact $D_2'$ state on the fermi scale discussed above could be involved in a reaction of the form

$$\mathrm{d} + \mathrm{D}_2' \longrightarrow \alpha + \mathrm{p} + \mathrm{n} + 21.62\ \mathrm{MeV}$$

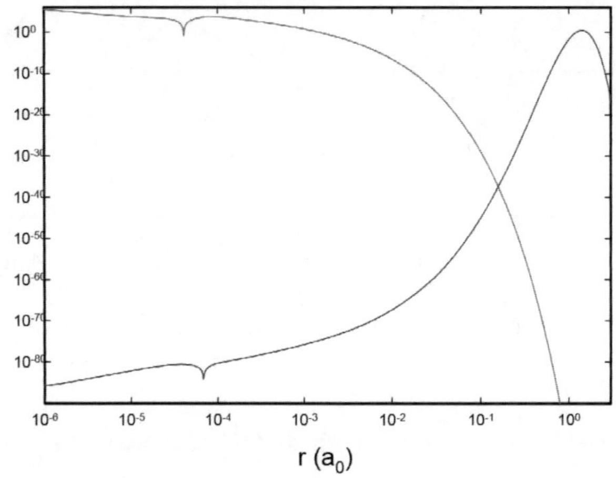


**Figure 13.** Approximate radial squared probability densities ($P^2(r)$) for an $l = 1$ $D_2$ state (blue, outer) and for an $l = 1$ compact state (red, inner).

A better notation for the same reaction in this case might be

$$\mathrm{d} + (\mathrm{dd})_{compact} \longrightarrow \alpha + \mathrm{p} + \mathrm{n} + 21.62 \text{ MeV}.$$

### 12.3. Deuteron-Deuteron Fusion Products

Based on the discussion above, there appear to be multiple ways in which dd-fusion products could be produced. The route simplest in concept could be written as a two-site excitation transfer reaction

$$\mathrm{D}_2 +^4\mathrm{He} \begin{cases} {}^4\mathrm{He} + n +^3 \mathrm{He} + 3.27 \ \ \mathrm{MeV} \\ {}^4\mathrm{He} + \mathrm{D}_2' \\ {}^4\mathrm{He} + p + t + 4.03 \ \ \mathrm{MeV} \end{cases}$$

where the $\mathrm{D}_2'$ state is compact as discussed above. We might expect this process to be inefficient at producing low-level p+t and n+$^3$He fusion products since the final states associated with this kind of excitation transfer process are so unstable. The compact $\mathrm{D}_2'$ state is unstable against 24 MeV gamma decay (how unstable depends on the details of precisely which $LSJ$ states are involved). The longer-lived the state, the more likely that excitation transfer will result in occupation of the state.

There is a related but more complicated route to the formation of dd-fusion products where the first step in the process is the formation of a compact $\mathrm{D}_2'$ species. We might write this as a three-site excitation transfer reaction

$$\mathrm{D}_2 + {}^4\mathrm{He} + {}^4\mathrm{He} \rightleftarrows {}^4\mathrm{He} + \mathrm{D}_2' + {}^4\mathrm{He} \rightleftarrows \begin{cases} {}^4\mathrm{He} + {}^4\mathrm{He} + n + {}^3\mathrm{He} + 3.27\ \mathrm{MeV} \\ {}^4\mathrm{He} + {}^4\mathrm{He} + \mathrm{D}_2' \\ {}^4\mathrm{He} + {}^4\mathrm{He} + p + t + 4.03\ \mathrm{MeV} \end{cases}$$

The idea here is that the formation of dd-fusion products starting from a compact state $D_2'$ may be more efficient, since the relevant phonon-nuclear matrix element for the $D_2'/^4He$ transition is much larger.

Testing whether these processes are responsible for the production of low-level dd-fusion products would seem to involve controlling how much $^4He$ is in the sample, and then checking on the dependence of the reaction rate on $^4He$ concentration.

### 12.4. First Step in the Excess Heat Process

In the Introduction we described a complicated coherent reaction scheme for excess heat production in the PdD system (Figure 1). A weakness of this proposed scheme is that the first excitation transfer step requires a relatively long-lived highly-excited state in the vicinity of 24 MeV, which on the face of it seems to be unlikely. In light of the discussion in this section it seems clear that this first step could be replaced with an excitation transfer that results in a compact $D_2'$ state instead, which is expected to be much more stable. A modification of the coherent reaction scheme to include this is shown in Figure 14.

### 12.5. Low-Level Energetic Particle Emission

The phonon-nuclear coupling matrix element is orders of magnitude stronger for the compact state $D_2'/^4He$ transition than for the molecular $D_2/^4He$ transition, which would be expected to impact the transfer to highly-excited states of

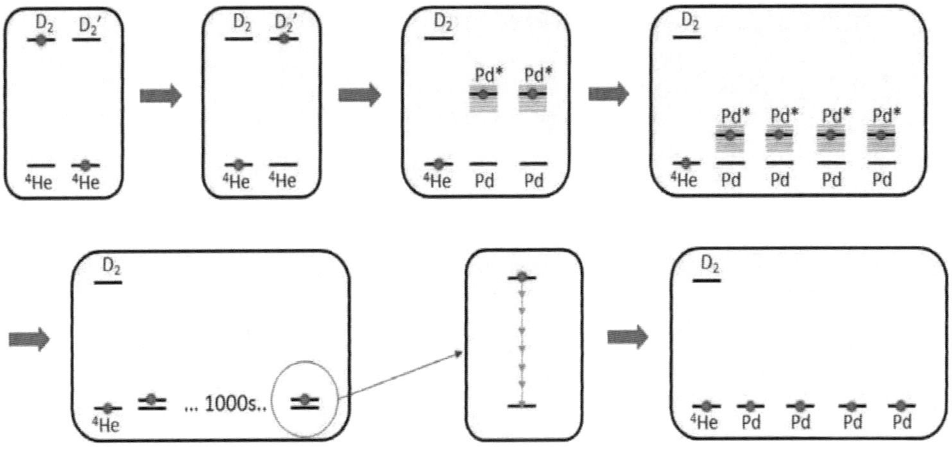


**Figure 14.** Schematic of the revised version of the model for excess heat production where the first excitation transfer step results in a compact $D_2$ state (indicated as $D_2'$ here).

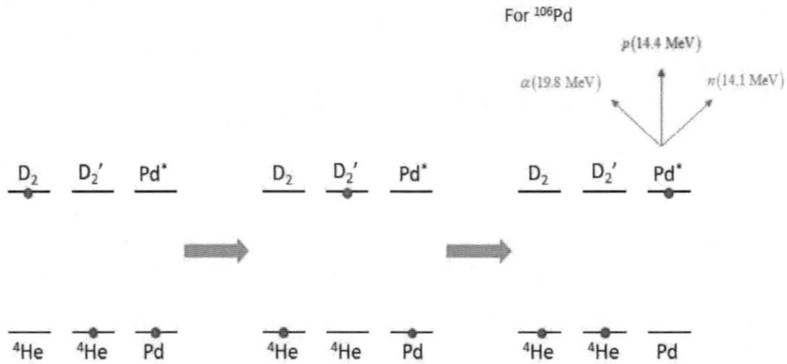


**Figure 15.** Schematic of excitation transfer of 24 MeV from the $D_2/^4He$ transition to a compact $D_2'/^4He$ transition resulting in a highly-excited state in $^{106}Pd$ that subsequently decays.

nuclei in the host lattice. The issue is that excitation transfer involving a small coupling matrix element to an unstable highly-excited state that decay rapidly is hindered; while excitation transfer from the compact $D_2'/^4He$ transition would involve a much larger coupling matrix element, and hence lead to a much larger probability of transfer and decay. As an example, we consider the associated modification of Figure 3 to include excitation transfer to produce a compact $D_2'$ intermediate state prior to the excitation of a Pd nucleus in Figure 15.

## 13. Subdivision

Subdivision as we consider it involves multiple excitation transfer events that start with a single excited nucleus and result in two or more excited nuclei where the total excitation energy is (approximately) conserved [10], [11].

From the discussions above we see that subdivision is a key element in the overall reaction schemes proposed for excess heat in the Fleischmann-Pons experiment. In this section we are interested three issues: is there evidence in experiments done so far that subdivision occurs; can states be avoided in the first subdivision step in the proposed excess heat process; and is there a possibility that subdivision could be demonstrated in the lab, so that we could be certain that it occurs and so that we might be able to study it in simpler controlled laboratory experiments.

### 13.1. Subdivision and the $D_2/^4He$ Transition

The simplest reaction scheme for subdivision starting from the $D_2/^4He$ transition is

$$D_2 + {}^{A_1}Z_1 + {}^{A_2}Z_2 \longleftrightarrow {}^4He + {}^{A_1}Z_1^* + {}^{A_2}Z_2^*$$

where the 24 MeV transition energy is split to produce excitation in two lattice nuclei. Ideally the initial and final state energies are resonant, or can be made up by the exchange of a few phonons or plasmons. We would expect the highly-excited states produced to be unstable, which may help in arranging for the overall multi-step excitation transfer process to be on resonance.

In light of the discussion in the previous section, it may be that a compact $D_2'$ state is formed as an intermediate state according to

$$D_2 + {}^4\text{He} + {}^{A_1}Z_1 + {}^{A_2}Z_2 \longleftrightarrow {}^4\text{He} + D_2' + {}^{A_1}Z_1 + {}^{A_2}Z_2 \longleftrightarrow$$
$${}^4\text{He} + {}^4\text{He} + {}^{A_1}Z_1^* + {}^{A_2}Z_2^*$$

In both schemes the two excited nuclei that result subsequently decay incoherently. Extensions are also possible where subdivision occurs following more excitation transfers of the compact $D_2'$ state among $^4$He nuclei.

### 13.2. Simple Empirical Model for Subdivision

The evaluation of subdivision rates and products would require computational models for the highly-excited states/resonances including energies and lifetimes, as well as phonon-nuclear coupling matrix elements. We would expect that the overall subdivision rate will be larger if the two highly-excited final states are nearly energy resonant with the initial state, and if the decay rates are slow.

An accurate first principles calculation of the associated rates and distribution of products is hindered by limitations in our ability to carry out accurate structure, energy, and decay rates for the large number of highly-excited nuclear states in the continuum. Under such conditions is seem dubious that we should be able to make relevant predictions at all.

We defer these difficult accurate subdivision calculations, and consider here a much simpler approach. If it happens that a reasonably long-lived highly-excited state occurs with exactly half of the transferred energy, then a Dicke-enhanced subdivision would be possible. To expect that such a coincidence might be a common occurrence would not be reasonable; however, with such an ansatz it is possible to develop a very simple model with which we are able to begin a theoretical discussion. For example, if there are energetic particles close to what is predicted from such a model, then perhaps further work might be worthwhile to following up to see if the isotopes implicated contribute the products.

A specific empirical model can be developed according to

$$C(E) = \sum_j f_j e^{-(E-E_j)^2/\Delta E^2} \tag{28}$$

where $C(E)$ is the associated simple model prediction, and where the (optimized) $f_j$ coefficients keep track of the weight of the contribution to the spectrum from isotope $j$. The energies $E_j$ can be determined simply from the ansatz, and $\Delta E$ is a free parameter (potentially connected with the detector resolution) to be optimized.

### 13.3. Possible Subdivision in the Pd Isotopes

We first consider the Pd isotopes, where "ideal" subdivision to produce matched excited final states would lead to the alpha, proton and neutron energies listed in Table 3.

Neutron emission above 2.45 MeV was reported by Takahashi and coworkers [61], with evidence for neutrons between 4 MeV and 5 MeV. We digitized the data of Figure 2 in [61] using the WebPlotDigitizer of Ankit Rohatgi, and fit to the simple model. Results from the optimization, augmented with a contribution at 2.45 for the dd-fusion neutrons, is shown in Figure 16. The model in this case suggests that the neutrons in the 4-5 MeV band may be a result of subdivision in $^{105}$Pd, and that some of the lower energy neutrons may be due to subdivision in other Pd isotopes.

An obvious experiment suggested by this model is to try looking for neutron emission in the 4-5 MeV region of the spectrum using a Pd sample made entirely of mass 105, and alternatively with a Pd sample in which no mass

**Table 3.** Alpha energies $E_\alpha$, proton energies $E_p$ and neutron energies $E_n$ (including recoil) expected from decay of the Pd excited state formed under idealized subdivision where the final excited state energy is exactly half of the transferred energy.

| isotope | $E_\alpha$ (MeV) | $E_p$ (MeV) | $E_n$ (MeV) |
|---|---|---|---|
| $^{102}$Pd | 9.41 | 4.07 | 1.34 |
| $^{104}$Pd | 8.97 | 3.24 | 1.92 |
| $^{105}$Pd | 8.69 | 3.14 | 4.78 |
| $^{106}$Pd | 8.37 | 2.55 | 2.34 |
| $^{108}$Pd | 7.77 | 1.96 | 2.67 |
| $^{110}$Pd | 7.22 | 1.29 | 3.10 |

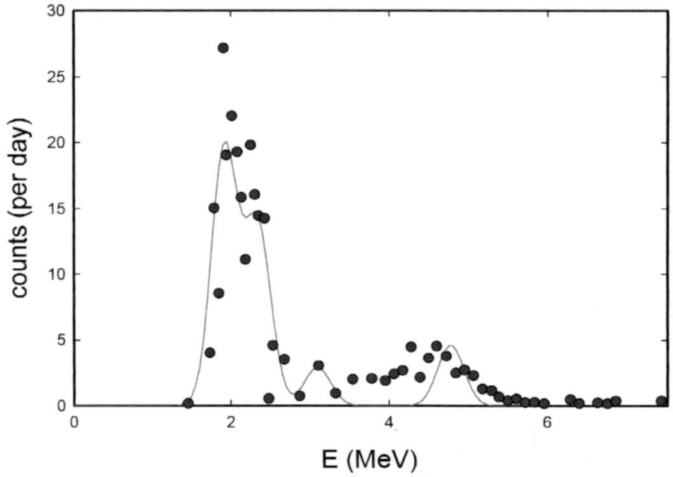


**Figure 16.** Plot of neutron data reported by Takahashi et al compared with the simple model described in the text for neutron emission from subdivision.

105 isotopes are present. If energetic neutrons in this part of the spectrum were correlated with $^{105}$Pd this would be supportive of the proposed interpretation that subdivision is responsible for them.

## 13.4. Possible Subdivision in the Er Isotopes

A similar approach might be taken in the case of subdivision in the Er isotopes. In this case Er has six stable isotopes with alpha, proton and neutron decay energies given in Table 4 for ideal subdivision into identical excited states.

In this case we consider neutron emission for a deuterated Eu sample irradiated by gammas in the range 2.5-2.9 MeV [62], where the protons resulting from deuteron disintegration would be expected to produce THz vibrations capable of stimulating nuclear excitation transfer. We have digitized the data in Figure 5(a), and fitted it using

**Table 4.** Alpha energies $E_\alpha$, proton energies $E_p$ and neutron energies $E_n$ (including recoil) expected from decay of the Er excited state formed under idealized subdivision where the final excited state energy is exactly half of the transferred energy.

| isotope | $E_\alpha$ (MeV) | $E_p$ (MeV) | $E_n$ (MeV) |
|---|---|---|---|
| $^{162}$Er | 13.24 | 5.46 | 2.72 |
| $^{164}$Er | 12.91 | 5.04 | 3.06 |
| $^{166}$Er | 12.45 | 4.58 | 3.43 |
| $^{167}$Er | 12.29 | 4.39 | 5.45 |
| $^{168}$Er | 12.18 | 3.90 | 4.13 |
| $^{170}$Er | 11.69 | 3.30 | 4.64 |

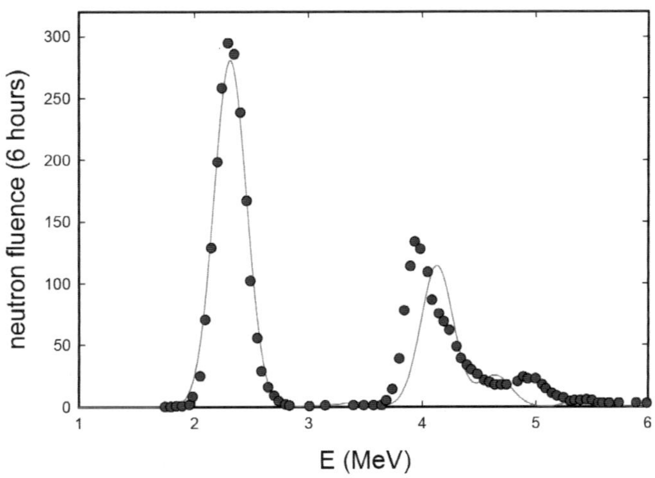


**Figure 17.** Plot of neutron data reported by Steinetz et al compared with the simple model described in the text for neutron emission from subdivision. For this model we allowed the dd-fusion energy to be determined from the data rather than fixing it at 2.45 MeV.

the empirical subdivision model, augmented by a contribution for 2.45 MeV dd-fusion neutrons (allowing the center energy to be determined from the data). Results are shown in Figure 17.

A conclusion from this is that there are subdivision candidates in the vicinity of the observed neutron peaks at higher energy, and that symmetric subdivision in $^{168}$Er is a candidate to account for the peak near 4 MeV. As above, to provide clarification of this point we would like to have similar experimental results involving a sample composed of $^{168}$Er, and a sample with no $^{168}$Er.

### 13.5. Avoiding Very Unstable Excited States

Note that there may also be a potential problem finding reasonably stable states after one subdivision resulting in two excited states each with roughly half the transferred 24 MeV energy quantum. The speculation here is that if the first subdivsion from the compact $D_2'$ state results in four excited states with lower excitation, then the resulting excited states may be much more stable. A schematic of how this might work in the case of the initial subdivision resulting in four excited states is shown in Figure 18.

### 13.6. Subdivision at Lower Energy

The discussion above has been concerned with subdivision in the case of the large 24 MeV quantum associated with the $D_2/^4He$ transition; however, we note that in the various proposed schemes for coherent reactions for excess heat production that successive subdivisions are required for part of the down-conversion of the large quantum. This suggests that a study of subdivision at lower energy would be useful. Very convenient would be subdivision implemented in experiments similar to those which have been studied for excitation transfer [21]–[23].

We have considered possible subdivision reactions in which an initial nuclear excited state is produced following the decay of a (commercially available) radioactive isotope that decays to produce excited states in two other nuclei. If the process were resonant, then the initial excitation energy would match the sum of the two excitation energies of the final state nuclei. A computer code was developed to search low-lying (below 250 keV) excited states of the stable nuclei for resonances, and a modest number of candidates were identified [63]. We do not have confidence that the mismatch could be made sufficiently small in any of the candidate sets of transitions that the mismatch could be made up through the exchange of a few phonons. However, it does seem possible to arrange for subdivision where the energy mismatch is made up through the exchange of a few photons (the idea here is that optical photons couple to plasmons which mix with phonons).

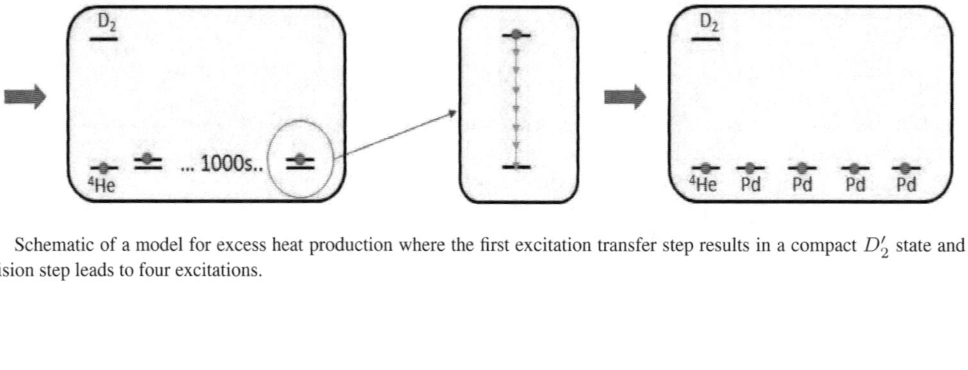


**Figure 18.** Schematic of a model for excess heat production where the first excitation transfer step results in a compact $D_2'$ state and where the first subdivision step leads to four excitations.

**Table 5.** Candidate photon energies around which to scan under the assumption that different numbers of photons are absorbed. In the second row we do not list the $\Delta n = \pm 6$ case since that is higher order since two of the transitions are E1 (one-phonon) and one is an E2 (two-phonon) transition.

| ${}^AZ_i$ | $E_i$ (keV) | ${}^AZ_{f1}$ | $E_{f1}$ (keV) | ${}^AZ_{f2}$ | $E_{f2}$ (keV) | $\hbar\omega_0$ $\Delta n = \pm 2$ | $\hbar\omega_0$ $\Delta n = \pm 4$ | $\hbar\omega_0$ $\Delta n = \pm 6$ |
|---|---|---|---|---|---|---|---|---|
| ${}^{123}$Te | 159.024 | ${}^{158}$Gd | 79.5143 | ${}^{158}$Gd | 79.5143 | 2.30 eV | 1.15 eV | 0.77 eV |
| ${}^{155}$Gd | 105.3106 | ${}^{181}$Ta | 6.237 | ${}^{183}$W | 99.0791 | 2.75 eV | 1.37 eV | – |
| ${}^{57}$Fe | 136.4743 | ${}^{155}$Gd | 60.0106 | ${}^{174}$Yb | 76.471 | 3.65 eV | 1.82 eV | 1.22 eV |
| ${}^{85}$Rb | 151.192 | ${}^{159}$Tb | 57.9964 | ${}^{178}$Hf | 93.1803 | 7.65 eV | 3.82 eV | 2.55 eV |

In Table 5 we list four candidate sets of transitions which are close to resonance, where the energy mismatch is balance by photon exchange according to

$$E_i = E_{f1} + E_{f2} + \Delta n\hbar\omega_0. \tag{29}$$

## 14. Phase Coherence With Resonant Excitation Transfer

We have been working on excitation transfer experiments in which ${}^{57}$Co decays to produce excited states in ${}^{57}$Fe, and then looking for changes in the gamma and x-ray emission following stimulation with (stress-induced) THz vibrations [21]–[23]. We were interested in developing a simulation model that could be used to model the experiment, and perhaps to begin a quantitative comparison of theory with experiment. Such a project is hindered from a basic lack of understanding of exactly where the THz vibrations are produced (a possibility is in the relative motion of the epoxy on top of the evaporated ${}^{57}$Co/${}^{57}$Fe deposit), what frequencies are involved, and how strong the vibrations are. What is really needed is a simpler experiment that is easier to model and to understand.

### 14.1. Phonon-Nuclear Interaction

We consider the first few steps in the development of a model for a different nucleus – one involving an electric dipole (E1) nuclear transition, so that only one phonon needs to be exchanged when a nuclear transition occurs (nuclear transitions in ${}^{57}$Fe involve two-phonon exchange, so the model will be more complicated). The key to modeling this new system is the phonon-nuclear interaction. The relevant part of the Hamiltonian can be written as

$$\hat{H}_{int} = \sum_j \hat{\mathbf{a}}_j \cdot c\hat{\mathbf{P}}_j \tag{30}$$

The $\hat{\mathbf{a}}_j$ operator acts on the internal nuclear degrees of freedom of nucleus $j$, which for this discussion allows for a transition from one nuclear state to another. In the case of resonant excitation transfer, this operator mediates a transition between the excited state and the ground state. The $\hat{\mathbf{P}}_j$ operator creates or destroys one phonon when a nuclear state change occurs.

### 14.2. Resonant Excitation Transfer

Consider now a resonant excitation transfer process for an excited nucleus that has just been formed as a result of the beta decay of an appropriate radioactive source nucleus. The interaction above can cause the excited nuclear state to make a transition to the ground state, coupling to an off-resonant state where there are no excited nuclei. A subsequent interaction can bring the system back into resonance by raising another ground state nucleus to the excited state, where the phonon exchange in both cases has to be from the same phonon mode, one creation and one annihilation so that there is no net change in the phonon energy. If so, then the new excited nucleus will be formed within a few hundred Angstroms of where the original one started, since the coherence domain of a THz phonon is not very large.

If phonon-mediated nuclear excitation transfer worked conventionally, then the contributions from all of the different phonon modes would destructively interfere so that excitation transfer would be restricted to nearest neighbor nuclei. However, since nuclear states shift off of resonance, this destructive cancellation can in part be removed, leading to a much larger delocalization if there happens to be a highly-excited phonon mode.

### 14.3. Beam Formation in an Ordered Lattice

If the highly-excited phonon mode has a long wavelength (on the order of the size of the coherence domain), then the excited nuclei at the different locations will have the same phase. Assuming local order, then this phase coherence will lead to beam formation very similar to what happens in crystal diffraction experiments. In the $^{57}$Fe experiments done in our experiments there was likely local order; however, the local crystal orientation likely varies over the different parts of the sample, so that it would not be expected for a macroscopic sample to produce a directed beam.

This suggests that a better way to do a resonant excitation transfer experiment is to work with a sample with macroscopic order, such as would occur with a single crystal sample (see Figure 19). For such a sample resonant excitation transfer in the case of a long wavelength phonon mode would produce beamlets as illustrated in Figure 20. If the vibrational mode wavelength is not long, then we might expect a modulation of the amplitude of the occupation probability for the excited nuclei, resulting in a modification of the angular emission pattern.

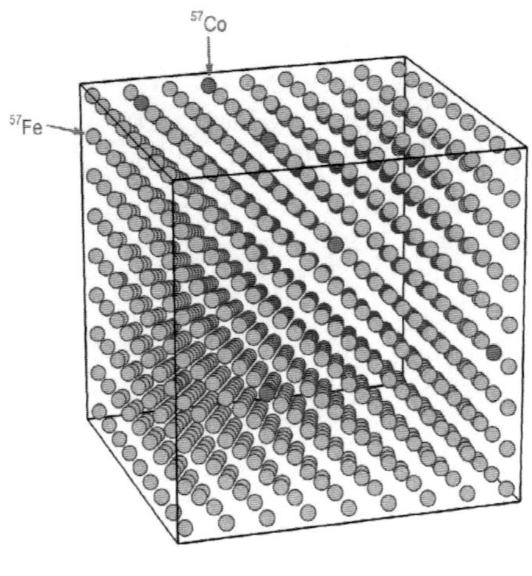


**Figure 19.** Proposal for a $^{57}$Co/$^{57}$Fe resonant excitation transfer experiment making use of a single crystal sample.

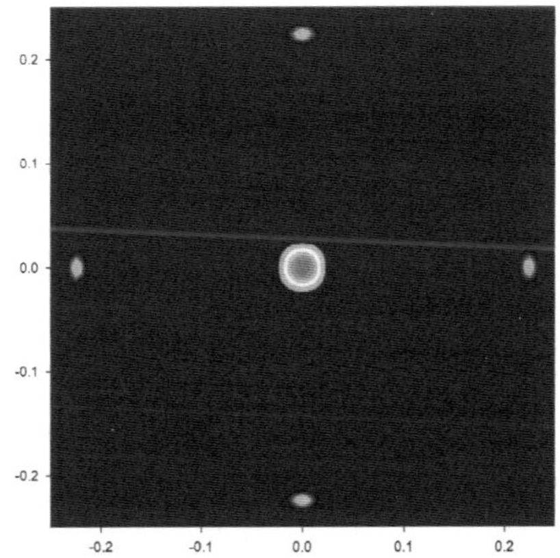


**Figure 20.** Beamlets produced around the [100] direction from the decay of 136 keV nuclei with phase coherence, with angles in radians.

## 15. Thoughts on a Model for Excess Heat Production

The discussion in the sections above provide a picture for how excess heat might be produced in the Fleischmann-Pons experiment. The question is whether this picture can be translated into a quantitative model that can be evaluated to give predictions for specific experiments.

### 15.1. Practical Issues

The number of issues involved is very large. As a practical matter, modeling sequential subdivision among the excited states of Pd nuclei (or impurity nuclei) will be hindered by a lack of precise state energies and wave functions with which to evaluate the phonon-nuclear matrix elements. Modeling down-conversion will be hindered in general by a lack of knowledge of precisely which low-energy nuclear transitions are involved, or in most cases which vibrational modes are involved.

Given this situation, what might be possible is to make use of lists of known nuclear levels with which to assemble a candidate set of subdivisions, and then develop estimates for the phonon-nuclear coupling matrix elements. Another approach might be to carry out moderately large nuclear structure calculations for the different isotopes, from which a set of (representative) phonon-nuclear matrix elements can be assembled.

For the down-conversion part of the model, it might be possible to work with an excess heat experiment in which an impurity is added which has a known low-energy transition, in which a down-conversion based on that transition may be relevant. This might be combined with a diagnostic (such as in Ref. [64]) to provide input about which phonon modes are involved.

### 15.2. One Approach to Putting Things Together

Assuming that an appropriate set of input data can be assembled for such a model, there remains more issues to be dealt with. A key motivation for pursuing this kind of model is the possibility that the dynamics will be coherent in the sense of Rabi oscillations. We need a formulation of the problem that works this way.

Conceptually simple might be an approach which starts with finite basis matrix equations of the form

$$E\mathbf{c} = \mathbf{H} \cdot \mathbf{c} \tag{31}$$

and decompose into

$$\begin{aligned} E\mathbf{c}_i &= \mathbf{H}_{ii} \cdot \mathbf{c}_i + \mathbf{H}_{ij} \cdot \mathbf{c}_j \\ E\mathbf{c}_j &= \mathbf{H}_{jj} \cdot \mathbf{c}_j + \mathbf{H}_{ji} \cdot \mathbf{c}_i + \mathbf{H}_{jf} \cdot \mathbf{c}_f \\ E\mathbf{c}_f &= \mathbf{H}_{ff} \cdot \mathbf{c}_f + \mathbf{H}_{fj} \cdot \mathbf{c}_j \end{aligned} \tag{32}$$

where $\mathbf{c}_i$ contains the initial state, where $\mathbf{c}_j$ includes intermediate states, and where $\mathbf{c}_f$ contains the final state. Then the intermediate states could be eliminated

$$\mathbf{c}_j = \left(E - \mathbf{H}_{jj}\right)^{-1} \cdot \left(\mathbf{H}_{ji} \cdot \mathbf{c}_i + \mathbf{H}_{jf} \cdot \mathbf{c}_f\right) \tag{33}$$

leading to

$$\begin{aligned} E\mathbf{c}_i &= \mathbf{H}_{ii} \cdot \mathbf{c}_i + \mathbf{H}_{ij} \cdot \left(E - \mathbf{H}_{jj}\right)^{-1} \cdot \left(\mathbf{H}_{ji} \cdot \mathbf{c}_i + \mathbf{H}_{jf} \cdot \mathbf{c}_f\right) \\ E\mathbf{c}_f &= \mathbf{H}_{ff} \cdot \mathbf{c}_f + \mathbf{H}_{fj} \cdot \left(E - \mathbf{H}_{jj}\right)^{-1} \cdot \left(\mathbf{H}_{ji} \cdot \mathbf{c}_i + \mathbf{H}_{jf} \cdot \mathbf{c}_f\right) \end{aligned} \tag{34}$$

In this kind of formulation an indirect coupling matrix between initial and final states could be defined according to

$$\mathbf{H}_{fi} = \mathbf{H}_{fj} \cdot \left(E - \mathbf{H}_{jj}\right)^{-1} \cdot \mathbf{H}_{ji} \tag{35}$$

The matrices involved in a realistic model would be expected to be large, but this might provide a way to think about the problem and to perhaps develop quantitative estimates.

### 15.3. Another Approach

The approach outlined above might be considered a standard and general approach in which intermediate states can be eliminated algebraically. We might expect there to be difficulties in the event that some of the intermediate states have the same energy as the initial and final state, as the matrix inverse would become nearly singular. In this case an alternate approach might be more appropriate. In this case we can decompose the vector $\mathbf{c}$ into a component $\mathbf{a}$ involving degenerate states and a component $\mathbf{b}$ involving nondegenerate states. For this separation we can write

$$\begin{aligned} E\mathbf{a} &= \mathbf{H}_{aa} \cdot \mathbf{a} + \mathbf{H}_{ab} \cdot \mathbf{b} \\ E\mathbf{b} &= \mathbf{H}_{bb} \cdot \mathbf{b} + \mathbf{H}_{ba} \cdot \mathbf{a} \end{aligned} \tag{36}$$

States off of resonance can be eliminated according to

$$\mathbf{b} = \left(E - \mathbf{H}_{bb}\right)^{-1} \cdot \mathbf{H}_{ba} \cdot \mathbf{a} \tag{37}$$

leading to

$$E\mathbf{a} = \mathbf{H}_{aa} \cdot \mathbf{a} + \mathbf{H}_{ab} \cdot \left(E - \mathbf{H}_{bb}\right)^{-1} \cdot \mathbf{H}_{ba} \cdot \mathbf{a} \tag{38}$$

If the shift of basis states off of resonance is important, it can be included in this formulation according to

$$E\mathbf{a} = \mathbf{H}_{aa} \cdot \mathbf{a} + \mathbf{H}_{ab} \cdot \left(E - \mathbf{H}_{bb}(E)\right)^{-1} \cdot \mathbf{H}_{ba} \cdot \mathbf{a} \tag{39}$$

For this kind of model there is no direct coupling matrix between initial and final states. Instead it would be possible to develop a dynamic model for the evolution of the nearly degenerate states starting with occupation of the initial state.

### 15.4. Loss

In the models outlined briefly above the basic description is Hamilonian based and loss-free. We would expect loss mechanisms to be important, including decay channels for the nuclear states and phonon loss from highly-excited THz phonon modes which drive the nuclear dynamics.

There are two approach that may be useful. In one approach we might work with a non-Hermitian model where loss is included by augmenting the real Hamiltonian matrices above with complex parts that model the loss. In this case the basic formulation remains the same, except that the linear algebra will be complex.

The other approach retains an underlying loss-free Hermitian model, but then adds loss to density matrix equations derived from the Hamiltonian for the degenerate states. The challenge in this case is that we expect the Hamiltonian models to be large, which means that the associated density matrix version of the model will be much larger.

## 16. Discussion and Conclusions

Excess heat in Fleischmann-Pons experiments is a real physical effect worthy of scientific investigation. In this paper we have summarized our approach to the development of theoretical models that are relevant. For excess heat production the approach which is most promising is illustrated in Figure 18. This scheme is pretty complicated, involving nuclear excitation transfer, nuclear subdivision, and the down-conversion of nuclear excitation into a great many phonons.

Understanding whether this approach is correct or not must involve understanding of the different parts of the model and comparing theory with experiment. In our view a foundation of the model is a novel relativistic phonon-nuclear interaction, one which has not at this point been of interest in the mainstream literature, and which needs experimental confirmation. Of interest is the development of new excitation transfer experiments that are more easily compared with theory, and corresponding quantitative theoretical models which can be tested in detail against experiment. From such tests it should be possible to tell whether this phonon-nuclear interaction is dominant, and whether we can rely on it for the development of predictive models.

We would like to develop quantitative predictions for models for excitation transfer, subdivision, down-conversion, and for excess heat, and then apply these models to experiment. As is clear from the discussions above this will involve considerable work. And as is also clear, some progress has been made in recent years.

### 16.1. On a comment from a reviewer

A reviewer made the comment: "It seems to me that the article would become more convincing if the author showed on the basis of an analysis of specific successful experiments, where exactly his theory is better for justifying these experiments than alternative theories." There is no question that quantitative analyses of many experiments based on the models under development is desirable, and would be compelling if agreement were obtained. The bottom line is that in spite of decades of development, more work remains to be done (both theoretically and experimentally) before we get there. One issue is that we still need nuclear matrix element and off-resonant energy shift calculations (or empirical estimates). We are also going to need models for the phonon distributions in the THz regime produced in electrochemical experiments, discharge experiments and gas loading experiments. And we are going to need simulation codes that implement the models and include the necessary physics to model important features of experiments.

## References


[1] M. Fleischmann, S. Pons and M. Hawkins, "Electrochemically induced nuclear fusion of deuterium," *Journal of Electroanalytical Chemistry and Interfacial Electrochemistry* **201** (1989) 301–307; errata **263** (1990) 187.

[2] M. Fleischmann, S. Pons, M. W. Anderson, L. J. Li and M. Hawkins, "Calorimetry of the palladium-deuterium-heavy water system," *Journal of Electroanalytical Chemistry and Interfacial Electrochemistry* **287** (1990) 293-348.

[3] J. R. Huizenga, *Cold Fusion. Fiasco of the century*, Oxford University Press (1994)

[4] A. Kitamura, A., A. Takahashi, K. Takahashi, R. Seto, T. Hatano, Y. Iwamura, T. Itoh, J. Kasagi, M. Nakamura, M. Uchimura, H.Takahashi, S. Sumitomo, T. Hioki, T. Motohiro, Y. Furuyama, M. Kishida, H. Matsune, "Excess heat evolution from nanocomposite samples under exposure to hydrogen isotope gases. Int," *J. Hydrogen Energy* **43** (2018) 16187–16200.

[5] A. Takahashi1, T. Yokose, Y. Mori, A. Taniike, Y. Furuyama, H. Ido, A. Hattori, R. Seto and J. Hachisuka, "Enhancement of excess thermal power in interaction of nano-metal and H(D)-gas,"*Proceedings of JCF20* (2019) 9–27.

[6] A. Takahashi, H. Ido, A. Hattori, R. Seto, A. Kamei, J. Hachisuka, T. Yokose, Y. Mori, A. Taniike and Y. Furuyama, "Latest progress in research on AHE and circumstantial nuclear evidence by interaction of nano-metal and H(D)-gas," *J. Condensed Matter Nucl. Sci.* **33** (2020) 14–32.

[7] F. Tanzella, R. George and R. Godes, "Nanosecond pulse stimulation in the Ni-$H_2$ system," *J. Condensed Matter Nucl. Sci.* **29** (2019) 202–210.

[8] F. Tanzella, R. Godes, J. Liu and R. George, "Mass and heat flow calorimetry in Brillouin's reactor," *J. Condensed Matter Nucl. Sci.* **33** (2020) 33–45.

[9] P. L. Hagelstein, "Bird's eye view of phonon models for excess heat in the Fleischmann-Pons experiment," *J. Condensed Matter Nucl. Sci.* **5** (2012) 169–180.

[10] P.L. Hagelstein, Models based on phonon–nuclear coupling, *Cold Fusion – Advances in Condensed Matter Nuclear Science*, J.P. Biberian (Ed.), Elsevier, Amsterdam (2020), pp. 283–297.

[11] P. L. Hagelstein, "Theory and experiments in Condensed Matter Nuclear Science," *J. Condensed Matter Nucl. Sci.* in press.

[12] S. E. Koonin and M. Nauenberg, "Calculated fusion rates in isotopic molecules," *Nature* **339** (1989) 690–691.

[13] M. L. E. Oliphant, P. Harteck and Lord Rutherford, "Transmutation effects observed with heavy hydrogen," *Proceedings of the Royal Society of London. Series A* **144** (1934) 692–703.

[14] P. L. Hagelstein and I. U. Chaudhary, "Energy exchange in the lossy spin-boson model," *J. Condensed Matter Nucl. Sci.* **5** (2011) 52–71.

[15] P. L. Hagelstein and I. U. Chaudhary, "Dynamics in the case of coupled degenerate states," *J. Condensed Matter Nucl. Sci.* **5** (2011) 72–86.

[16] P. L. Hagelstein and I. U. Chaudhary, "Second-order formulation and scaling in the lossy spin-boson model," *J. Condensed Matter Nucl. Sci.* **5** (2011) 87–101.

[17] P. L. Hagelstein and I. U. Chaudhary, "Local approximation for the lossy spin-boson model," *J. Condensed Matter Nucl. Sci.* **5** (2011) 102–115.

[18] P. L. Hagelstein and I. U. Chaudhary, "Coherent energy exchange in the strong coupling limit of the lossy spin-boson model," *J. Condensed Matter Nucl. Sci.* **5** (2011) 116–139.

[19] P. L. Hagelstein, "Quantum composites: A review, and new results for models for Condensed Matter Nuclear Science," *J. Condensed Matter Nucl. Sci.*, **20** (2016) 139–225.

[20] P.L. Hagelstein, Phonon-mediated nuclear excitation transfer, *J. Condensed Matter Nucl. Sci.* **27** (2018) 97–142.

[21] F. Metzler, P. L. Hagelstein and S. Lu, "Observation of non-exponential decay in x-ray and $\gamma$ emission lines from Co-57," *J. Condensed Matter Nucl. Sci.* **27** (2018) 46–96.

[22] S. Lu, "Exploring Possible Coupling between Phonons and Internal Nuclear States," PhD Thesis (MIT) (2018).

[23] F. Metzler, *Experiments to Test for Phonon-Nuclear Interactions*, Master's Thesis (MIT) 2019.

[24] E. Wigner, "On unitary representations of the inhomogeneous Lorentz group," *Annals of Mathematics* **40** (1939) 149-204.

[25] V. Bargmann, "Irreducible unitary representations of the Lorentz group," *Annals of Mathematics* **48** (1947) 568-640.

[26] E. Wigner, "Invariance in physical theory," *Proceedings of the American Philosophical Society* **93** (1949) 521-526.

[27] P. A. M. Dirac, "Forms of relativistic dynamics," *Reviews of Modern Physics* **21** (1949) 392.

[28] E. P. Wigner, "Relativistic invariance in quantum mechanics," *Il Nuovo Cimento* **10** (1956) 517-532.

[29] L. L. Foldy, "Relativistic particle systems with interaction," *Physical Review* **122** (1961) 275.

[30] R. A. Krajcik and L. L. Foldy, "Relativistic center-of-mass variables for composite systems with arbitrary internal interactions," *Physical Review D* **10** (1974) 1777.

[31] P. L. Hagelstein, "Lowest order relativistic interaction between lattice vibrations and internal nuclear degrees of freedom," submitted to *Phys. Rev C.*

[32] P. L. Hagelstein, "Central and tensor contributions to the phonon-exchange matrix element for the $D_2/^4$He transition," *J. Condensed Matter Nucl. Sci.* **11** (2013) 15–58.

[33] P. L. Hagelstein, "Calculation of the boosted spin-orbit contribution to the phonon-nuclear coupling matrix element of $^{181}$Ta," *J. Condensed Matter Nucl. Sci.* **29** (2019) 392–400.

[34] S. Weinberg, "Effective chiral Lagrangians for nucleon-pion interactions and nuclear forces," *Nuclear Physics B* **363** (1991) 3–18.

[35] D. R. Entem and R. Machleidt, "Accurate nucleon-nucleon potential based upon chiral perturbation theory," *Physics Letters B* **524** (2002) 93–98.

[36] D. R. Entem and R. Machleidt, "Accurate charge-dependent nucleon-nucleon potential at fourth order of chiral perturbation theory," *Physical Review C* **68** (2003) 041001.

[37] R. Machleidt and D. R. Entem, "Chiral effective field theory and nuclear forces," *Physics Reports* **503** (2011) 1–75.

[38] P. L. Hagelstein, "Nonrelativistic reduction of the phonon-nuclear interaction for the one-pion exchange potential with pseudovector interaction," *J. Condensed Matter Nucl. Sci.* in press.

[39] P. L. Hagelstein, "Reduction of the HD/$^3$He phonon-nuclear matrix element," *J. Condensed Matter Nucl. Sci.* in press.

[40] J. R. Stone and P. G. Reinhard, "The Skyrme interaction in finite nuclei and nuclear matter," *Progress in Particle and Nuclear Physics* **58** (2007) 587–657.

[41] E. Caurier, F. Nowacki, A. P. Zuker, G. Martinez-Pinedo, A. Poves and J. Retamosa, "Large scale diagonalization in the pf shell: Achievements and perspective," *Nuclear Physics* **A654** (1999) 747c–758c.

[42] G. P. Chambers, J. E. Eridon, K. S. Grabowski, B. D. Sartwell and D. B. Chrisey, "Charged particle spectra of palladium thin films during low energy deuterium ion implantation," *J Fusion Energy 9* (1990) 281–285.

[43] A. S. Roussetski, "Application of CR-39 plastic track detector for detection of DD and DT-reaction products in cold fusion experiments," *Proc. ICCF8* (2000) 253–257.

[44] G. P. Chambers, G. K. Hubler and K. S. Grabowski, "Search for energetic charged particle reaction products during deuterium charging of metal lattices," *AIP Conference Proceedings* **228** (1991) 383–396.

[45] G. P. Chambers, G. K. Hubler and K. S. Grabowski, "Evidence for MeV particle emission from Ti charged with low energy deuterium ions," *Naval Research Laboratory Report 6927* (1991).

[46] S A Lipinski and H U Lipinski, "Hydrogen-lithium fusion device, methods and applications," *United States Patent Application Publication* US 2009/0274256 A1 (2009).

[47] S A Lipinski and H U Lipinski, "Hydrogen-lithium fusion device," *WIPO/PCT Patent Application Publication* WO 2014/189799 A9 (2014).

[48] A. A. Sonzogni, "NuDat 2.0: Nuclear Structure and Decay Data on the Internet," *AIP Conference Proceedings* **769** American Institute of Physics (2005) 574–577.

[49] https://www.nndc.bnl.gov/nudat2/
[50] V. I. Vysotskii, A. B. Tashyrev, and A. A. Kornilova, "Experimental observation and modeling of Cs-137 isotope deactivation and stable isotopes transmutation in biological cells," *Low-Energy Nuclear Reactions Sourcebook*, American Chemical Society, Washington, DC. (2008) 295–309.
[51] S. Focardi, V. Gabbani, V. Montalbano, F. Piantelli and S. Veronesi, "Evidence of electromagnetic radiation from Ni-H systems," *Proc. ICCF11* (2004).
[52] S. Focardi and F. Piantelli, "Produzione di energia e reazioni nucleari in sistemi Ni-H a 400 C," *XIX Congresso Nazionale UIT* (2004).
[53] D. Letts and P. L. Hagelstein, "Modified Szpak protocol for excess heat," *J. Condensed Matter Nucl. Sci.* **6** (2012) 44–54.
[54] P. L. Hagelstein, "Molecular $D_2$ near vacancies in PdD and related problems," *J. Condensed Matter Nucl. Sci.* **13** (2014) 138–148.
[55] A. Czerwinski, M. Czauderna, G. Maruszczak, I. Kiersztyn, R. Marassi, S. Zamponi, "Influence of cesium cations on hydrogen and deuterium electrosorption in palladium," *Electrochimica acta* **42** (1997) 81–86.
[56] P. L. Hagelstein, "A unified model for anomalies in metal deuterides," *Proc. ICCF8* (2000) 363–374.
[57] P. L. Hagelstein, " A unified model for anomalies in metal deuterides," *Proc. ICCF9* (2002).
[58] K. Czerski, D. Weissbach, A. I. Kilic, G. Ruprecht, A. Huke, M. Kaczmarski, N. Targosz-Śleczka and K. Maass, "Screening and resonance enhancements of the $^2H(d, p)^3H$ reaction yield in metallic environments," *European Physics Letters* **113** (2016) 22001.
[59] J. Kasagi, T. Ohtsuki, K. Ishii and M. Hiraga, "Energetic protons and $\alpha$ particles emitted in 150 keV deuteron bombardment on deuterated Ti," *J. Phys. Soc. Japan* **64** (1995) 777–783.
[60] G. K. Hubler, K. S. Grabowski, D. L. Knies, R. A. Walker and P. L. Hagelstein, "Anomalous energetic proton emission during 170 keV deuteron bombardment of $TiD_2$," *J. Phys. Soc. Japan* **81** (2012) 124203.
[61] A. Takahashi, T. Takeuchi, T. Iida and M. Watanabe, "Emission of 2.45 MeV and higher energy neutrons from $D_2O$-Pd cell under biased-pulse electrolysis," *J. Nuclear Science and Technology* **27** (1990) 663–666.
[62] B. M. Steinetz, T. L. Benyo, A. Chait, R. C. Hendricks, L. P. Forsley, B. Baramsai, P. B. Ugorowski, M. D. Becks, V. Pines, M. Pines, R. E. Martin, N. Penney, G. C. Fralick and C. E. Sandifer II, "Novel nuclear reactions observed in bremsstrahlung-irradiated deuterated metals," *Phys. Rev. C* **101** (2020) 044610.
[63] P. L. Hagelstein, "Candidates for near-resonant subdivision experiments," *J. Condensed Matter Nucl. Sci.*, in press.
[64] M. R. Swartz, "Optical detection of phonon gain distinguishes an active cold fusion/LANR component," *J. Condensed Matter Nucl. Sci.* **20** (2016) 29–53.

J. Condensed Matter Nucl. Sci. 36 (2022) 247–265

Research Article

# Excess Energy from Heat-Exchange Systems

Bin-Juine Huang*, Ming-Li Tso, Ying-Hung Liu, Jong-Fu Yeh, Litu Wu, I-Fee Chen, Yu-Hsiang Pan, Ching-Kang Huang, Mou-Yung Liao, Yi-Chun Chen, and Po-Hsien Wu

*Department of Mechanical Engineering, National Taiwan University, Taipei, Taiwan*

**Abstract**

At ICCF-22, we presented a vapor compression machine (VCS) which can produce excess energy. The hot refrigerant vapor from the freon compressor is used to heat the water flowing through a tiny passage of a triple-pipe heat exchanger (THX). This can cause a violent cavitation of water and two-phase flow instabilities, and produce excess energy. The VCS machine was further modified and tested for two years since then. We improved the calorimetry in measuring COP. From the energy balance of THX, we defined COP as the net heat output from water divided by the net heat input to THX, which is denoted as $COP_X$. If the measured $COP_X$ was greater than 1, we can confirm the occurrence of excess energy generation or the low energy nuclear reaction (LENR). The test results show that the maximum $COP_X$ reaches 2.39 and $COP_X$ increases with decreasing inlet water temperature. We observed some peculiar phenomena. The output refrigerant vapor from the compressor is, instead, heated by water at some part of the heat exchanger. In addition, abnormally-high water line pressure damaged the pressure gauge, and also caused crack and leakage of the copper pipe. We also observed severe buckling and black appearance of the middle pipe. The buckling may result from a huge exerted pressure around 740 bar which is far beyond the system operating pressure. This may be caused by LENR. We also made another machine (DHX-1) using a double-pipe heat exchanger heated by a steam boiler. The maximum $COP_X$ reaches 2.55 and some peculiar phenomena similar to the VCS machine was observed. The inner copper pipe was deformed by buckling and the outer pipe wall was cracked and leaked. This may be caused by huge pressure (>225 bar) or a high temperature (>500°C) which results in buckling and rupture of the pipes. The surface of the ruptured inner pipe becomes shining black and looks like CuO nanowires produced by thermal oxidation of copper. The chemical element of the ruptured copper pipe increases by 3 to 4 folds in C, 10 folds in O, and 1.4 to 4.4 folds in Fe. Replacing the identical brand-new double-pipe heat exchanger (DHX-1B), we obtained the same results as DHX-1 with $COP_X > 2.0$, inner pipe rupture and blackening, and chemical element change. $COP_X > 2.0$ also occurred in a new heat exchanger (DHX-2) made from stronger copper tubes with a larger size. All of these excess-energy phenomena in VCS and DHX are probably induced by LENR. In the present study, we experimentally confirm that excess energy or LENR can be induced by a heat transfer process involving cavitation in heat-exchange systems. The excess energy process has been shown reproduceable and controllable.



*Keywords:* Excess energy, low-energy nuclear reaction, LENR, cavitation, cavitation-induced LENR

*Corresponding author: bjhuang38@gmail.com

© 2022 ICCF. All rights reserved. ISSN 2227-3123

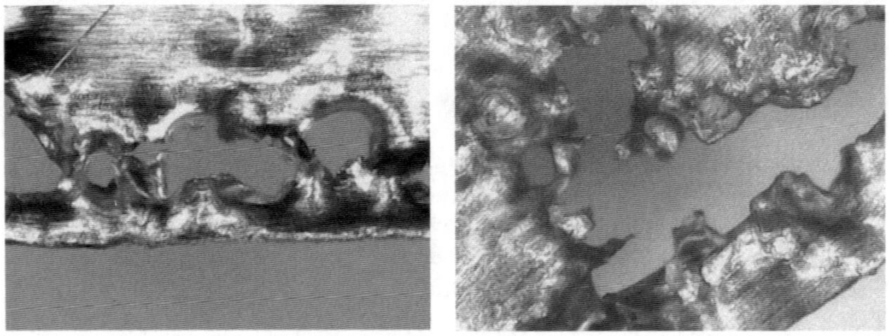

**Figure 1.** Damage of copper foil immersed in an ultrasonic vibrator for 7 minutes (left) and 21 minutes (right).

**Table 1.** Variation of water vapor bubble pressure with diameter.

| **Bubble diameter (*d*)** | **ΔP, N/m$^2$** | **ΔP, bar** |
|---|---|---|
| 1 mm | 287 | 0.00287 |
| 500 $\mu$m | 574 | 0.00574 |
| 100 $\mu$m | 2,869 | 0.0287 |
| 50 $\mu$m | 5,739 | 0.0574 |
| 1 $\mu$m | 286,926 | 2.87 |
| 100 nm | 2.87 x $10^6$ | 28.7 |
| 10 nm | 28.7 x $10^6$ | 287 |
| 2 nm | 143.5 x $10^6$ | 1,435 |
| 1 nm | 287 x $10^6$ | 2,870 |

## 1. Introduction

Cavitation is a process of liquid vaporization at a pressure lower than thermodynamic vapor pressure. This process generates bubbles and creates implosions with intense shock wave or micro-jet which can damage a propeller or container wall etc. Cavitation can also generate flash at several thousand Kelvin temperatures, during the collapse of bubble, called 'sonoluminescence' [1], [2]. In a simple experiment, we immersed a copper foil (0.08 mm thick, 10 cm x 1 cm strip) in the water of a 50 W ultrasonic vibrator for 7 and 21 minutes. The damage of the copper foil (Fig. 1) reveals that the cavitation induces force and heat that is larger than the ultimate stress and the melting point of copper. Similar phenomena were also found in other research using aluminium foil etc. [3]–[5].

From bubble dynamics, the pressure difference between the inside and outside of a bubble is proportional to surface tension ($\sigma$) and inversely proportional to bubble diameter (*d*), equation (1).

$$\Delta P = P_i - P_o = \frac{4\sigma}{d} \tag{1}$$

For a vapor bubble having a 1 nm diameter at 25°C, the pressure difference ($\Delta P = P_i - P_\mathrm{o}$) can reach 2870 bar (Table 1). If a vapor bubble of the size of 10 nm collapses, the bubble will explode at a huge pressure (287 bar) whose destructive power is incredible. It was found by many scientists [6]–[7] that water is capable of sustaining pressures up to 30 or 140 MPa (300$\sim$1400 bar) before it breaks by cavitation. This implies the force of cavitation possesses an incredibly high level of pressure which might induce a low-energy nuclear reaction.

Since cavitation is so powerful, it is thought that low-energy nuclear reaction (LENR) might be induced from the collapsing of vapor bubbles created during cavitation [8]. Some engineers also found that cavitation can be utilized

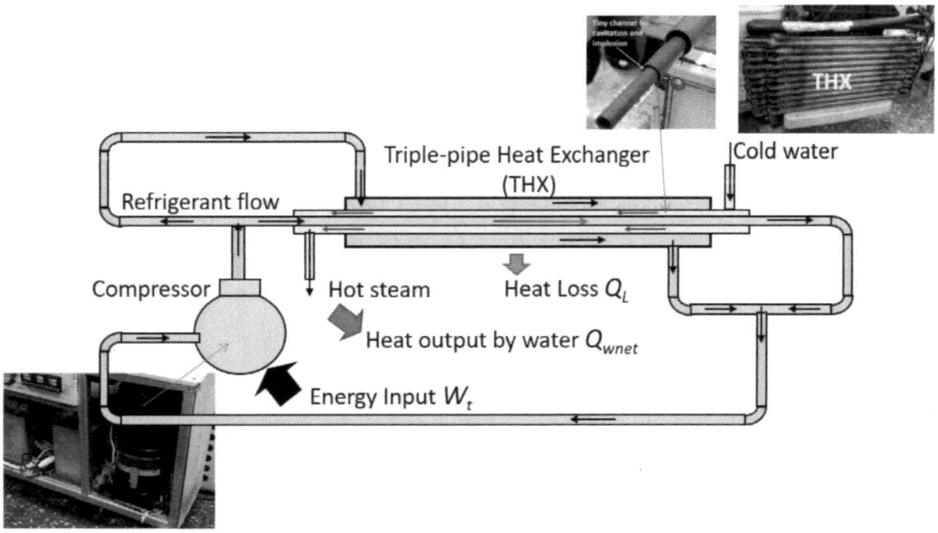


**Figure 2.** Schematic of the triple-pipe heat exchanger heated by a vapor compression system (VCS).

to produce heat for industrial application (called "cavitation heat generator") [9], [10]. The phenomenon of excess heat generation was found occasionally in some machines [11]–[13], but not much has been published in detail. Here, we would like to present two machines designed from multiple-pipe heat exchangers heated by a vapor compression system and by a steam boiler, respectively. We observed the phenomena of excess energy during heat transfer involving cavitation.

## 2. Excess Energy From a Triple-Pipe Heat Exchanger (THX) Heated by a Vapor Compression System (VCS)

At ICCF-22, we presented a vapor compression machine (VCS-1) using 2.75 RT freon (R22) compressor (Fig. 2) which can produce excess energy [14]. The hot superheated refrigerant vapor from a R22 compressor (around 150°C and 35 bar) is used to heat the water flowing through a tiny passage of a triple-pipe heat exchanger. This can cause a violent cavitation of water. The machine (VCS-1) has been tested for two years in various operating conditions.

*Design of triple-pipe heat exchanger*

The triple-pipe heat exchanger (THX) is made from copper pipes connected in series with a total length 30 meters (right upper corner of Fig. 2). The dimensions of the copper pipe are shown in Table 2. The water flow through THX is driven by the feed pump used in RO (reverse osmosis) machines which consumes about 30 W of power. The maximum flowrate is 1.5 L/min. The water passage clearance is 2.29 mm as shown in Fig. 3. City water is used in the tests of the VCS.

*Development of adiabatic calorimetry to measure net energy output by water ($Q_{wnet}$) and excess energy index $COP_x$*

To measure the excess energy, we developed the following method of calorimetry. Applying the first law of thermodynamics to the whole system (VCS) and assuming no internal energy generation such as heat source or excess energy, we derive the steady-state energy balance equation as

$$Q_{wnet} = W_t - Q_L \tag{2}$$

**Table 2.** Major dimension of the triple-pipe heat exchanger (THX).

| Inner copper pipe | | Middle copper pipe | | Outer copper pipe | |
|---|---|---|---|---|---|
| OD (mm) | thickness (mm) | OD (mm) | thickness (mm) | OD (mm) | thickness (mm) |
| 6.35 | 0.76 | 12.7 | 0.89 | 19.1 | 1.07 |

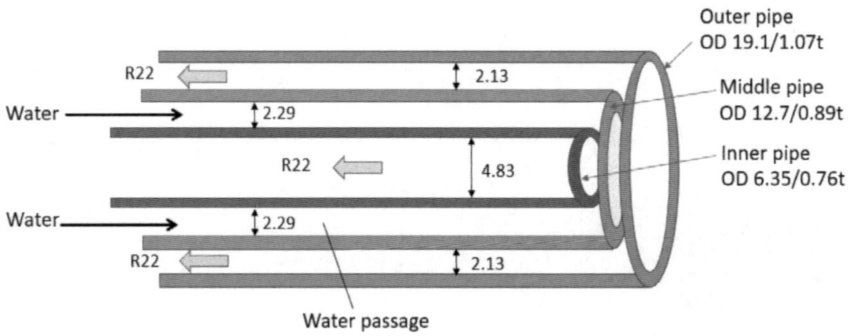


**Figure 3.** Flow paths and dimensions of triple-pipe heat exchanger (THX).

where $W_t$ is the total system power input including power consumptions of the R22 compressor, water pump (30 W), and controllers (around 30 W), $Q_{wnet}$ is the net heat output by water, $Q_L$ is the heat loss. We define the COP as the net heat output ($Q_{wnet}$) of water, i.e., the left-hand side of equation (1), divided by the net heat input ($W_t - Q_L$), i.e., the right-hand side of equation (2), denoted as COP$_x$ and called the "excess energy index":

$$\text{Excess Energy Index}: \; COPx = \frac{\mathrm{Q_{wnet}}}{\mathrm{W_t - Q_L}} \tag{3}$$

$COP_x$ is used as the criterion to detect the occurrence of excess energy generation. If the measured $COP_x$ was greater than 1, we conclude there is an excess energy.

We designed adiabatic calorimetry [23] to measure $Q_{wnet}$ and the excess energy index $COP_x$. A well-insulated water tank holds some cold water with temperature $T_i$ and mass $M_i$, at the beginning. The water tank is sitting on a scale to measure the weight change. See Fig. 4. Turn on the VCS machine, including the water flow through THX, for about an hour to reach a steady-state operation. The output hot water stream (containing steam) is suddenly fed into the water tank to mix with the prefeed cold water. The hot water stream stops when the mixed water temperature reaches near boiling point (100°C). By measuring the mass change ($\Delta M = M_f - M_i$) during the time interval ($\Delta t$) and the water temperature in the tank at the beginning ($T_i$) and at the end ($T_f$), we can determine the energy output ($Q_w$) from DHX using adiabatic calorimetry with the following energy equation:

$$Q_w = \frac{\mathrm{M_f u_f - M_i u_i}}{\Delta t} \tag{4}$$

where $u_i$ and $u_f$ are evaluated from the internal energy of water at the initial temperature $T_i$ and at the final temperature $T_f$, respectively. To reduce the energy loss by vapor that escapes from the tank, the tank is well sealed and insulated, and the test was terminated before the water temperature reached 90°C. The inevitable small energy loss is not taken into account in the calculation of COP$_x$. Therefore, COP$_x$ is under-estimated in order to obtain a more conservative conclusion.

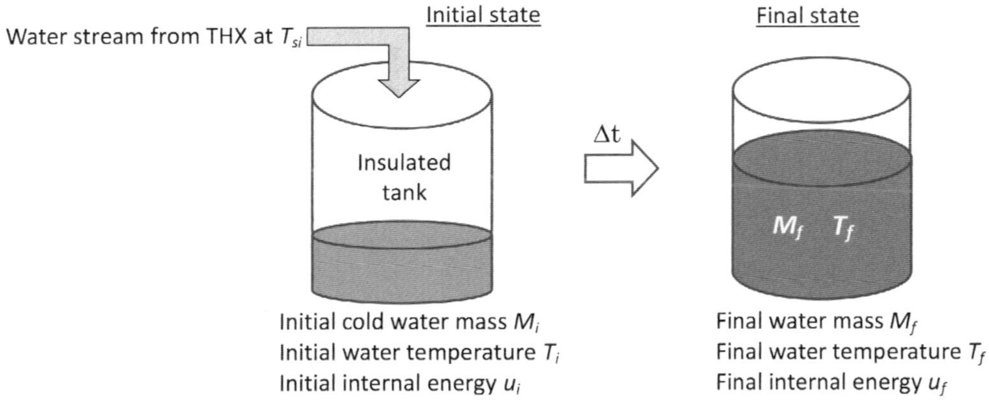


**Figure 4.** Schematic diagram of the calorimetry to measure the water output energy ($Q_w$) from THX.

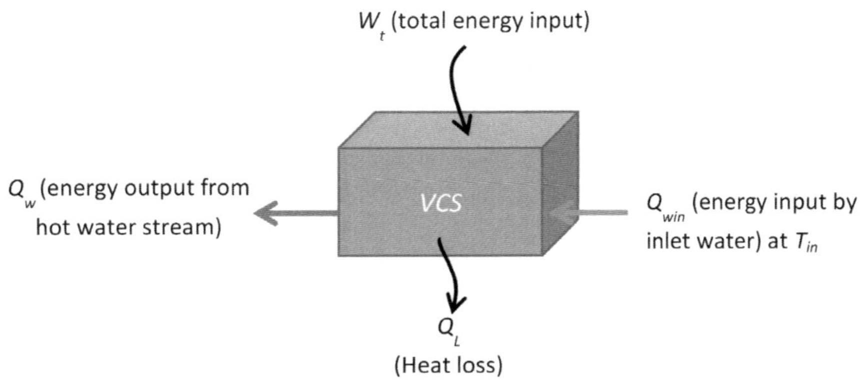


**Figure 5.** Overall energy balance of VCS at steady state.

From the energy balance diagram of VCS, Fig. 5, we obtain the net energy output ($Q_{wnet}$) of water from the heat exchanger:

$$Q_{wnet} = Q_w - Q_{win} = \frac{M_i C_v (T_f - T_i) + \Delta M C_v (T_f - T_{in})}{\Delta t} \tag{5}$$

Using equations (2) and (4) and the heat loss $Q_L$ which was determined from the measured surface temperatures at various parts of the VCS, we obtain $COP_x$. We can calibrate the error of $Q_L$ from checking whether $COP_x = 1$ when no excess energy appears. That is, all the data taken with $COP_x = 1$ can be treated as the baseline calibration of the test bed.

*Measurement errors*

To be more conservative, the measurement error ($e_{COPx}$) of $COP_x$ is evaluated for the worst case. According to the definition of $COP_x$ in equation (3), it is:

$$e_{COPx} = e_{Wt} + e_{Qwnet} + e_{QL} \quad (6)$$

where $e_{Wt}$ is the error of $W_t$; $e_{Qwnet}$ is the error of $Q_{wnet}$; $e_{QL}$ is the error of $Q_L$.

The digital power meter used to measure the energy input to VCS ($W_t$) has an error ($e_{Wt}$) of around 10%. The worst-case error ($e_{Qwnet}$) in measuring $Q_{wnet}$ can be estimated, according to equation (5):

$$e_{Qwnet} = e_M + e_{Ti} + e_{Tf} + e_t \quad (7)$$

$e_M$ is the error in measuring the mass of water using an electronic scale, which is around 1%; $e_{Ti}$ and $e_{Tf}$ are the errors in measuring the initial and final water temperatures using T-type thermocouple which is around 1% from a calibration at the ice and the boiling points; $e_t$ is the error in measuring the time interval using a stopwatch, which is around 1%. Therefore, the error of $e_{Qwnet}$ is around 4%.

The error ($e_{QL}$) in estimating the heat loss of VCS ($Q_L$) is around 5%. Hence, the worst-case error in measuring $COP_x$ is $e_{COPx} = e_{Wt} + e_{Qwnet} + e_{QL} = 10\% + 4\% + 5\% = 19\%$.

Therefore, the excess energy or LENR can be confirmed only when the measured $COP_x$ is greater than 1.20.

*Test results*

The VCS has been tested since May 29, 2019. The major operating parameters include water flowrate, inlet water temperature, and charging of refrigerant R22 and lubricating oil filled in the vapor compression system. Slight modifications in the triple-pipe heat exchanger (THX) and the compressor were also carried out during the tests.

Figure 6 shows that $COP_x$ of the VCS-1 machine was between 1.29 and 1.97 during the first test period before the end of 2020, which reveals the occurrence of excess energy. The temperature of the refrigerant vapor at the outlet of the compressor is around 150°C which is the upper limit of the compressor (Copeland CRN5-0500-PFV). The compressor overheated and broke at the end of 2020. It was replaced with a new compressor of a newer model CRNO-0500-PFV in the new prototype (VCS-2a), which was tested again. It took about 5 months to tune-up and reach a $COP_x$ higher than the VCS-1. This prototype was renamed "VCS-2b". The key factors in tuning VCS-2 for better performance include the water flow control, and the charged quantities of the refrigerant R22 and the compressor lubricating oil.

We found that the $COP_x$ increases with decreasing inlet water temperature for VCS-1, as shown in Fig. 7. We assume that colder inlet water provides a larger temperature difference between water and freon in the THX, which causes stronger boiling heat transfer and cavitation, which probably leads to two-phase flow instabilities and LENR [21]. For the second machine (VCS-2a) using a new compressor, $COP_x$ increases only slightly with decreasing inlet water temperature. And $COP_x$ was below 1.8 since VCS-2a was not tuned very well. But after good tuning (renamed as "VCS-2b"), $COP_x$ increases with decreasing inlet water temperature. The slope of $COP_x$ curve in this better-tuned prototype (VCS-2b) is nearly the same as VCS-1 but with higher $COP_x$. It is noted that the heat exchanger seems to experience a faster scale formation due to unknown phenomena and needs more frequent descaling. After descaling, $COP_x$ increases to 2.39 (the prototype is named as "VCS-2c") as shown in Fig. 7. The measured excess heat $Q_{ex}$, total heat output $Q_{tot}$ (= $Q_{wnet}$ +$Q_L$), total power input $W_t$ in kilowatt levels is shown in Fig. 8. It is seen that a machine producing several kilowatts of excess heat is feasible.

*Peculiar phenomena observed in the VCS machine*

We have noted that several peculiar phenomena take place in the VCS machine when $COP_x > 1$:

(1) The output water stream temperature ($T_{si}$) from THX is occasionally found around 150°C which is higher than the compressor output refrigerant temperature (around 145°C). VCS uses refrigerant vapor from the

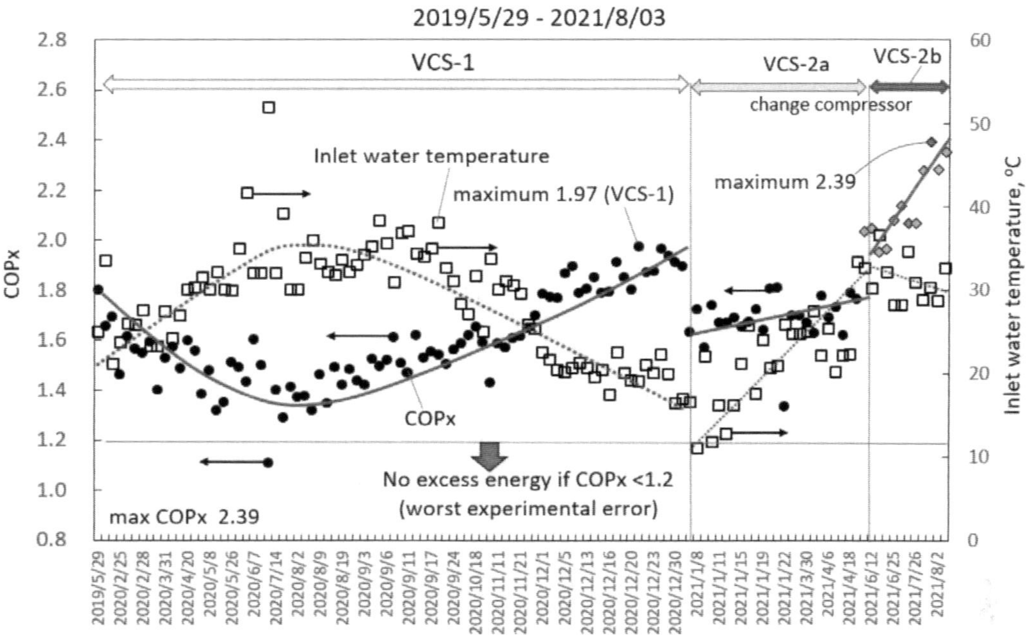


**Figure 6.** Measured $COP_x$ under various operating conditions.

compressor as the heat source to heat the water in the THX. The refrigerant vapor should be hotter than the water flow. Instead, the refrigerant vapor is heated by water at some part of the heat exchanger. This implies that excess energy or LENR is generated in water to raise the water temperature to be higher than the refrigerant vapor. The liquid water was pressurized by the RO pump up to around 17.6 bar before entering THX. Since the thermodynamic saturated pressure of water at 140°C (10°C below R22 temperature) is 3.6 bar, the maximum total water pressure inside the THX is around 21.2 bar (17.6 + 3.6). Since the water is discharged to the tank or open air, the THX outlet pressure is at atmospheric pressure (1 bar). Such a large pressure drop in THX with a long pipeline would induce two-phase flow instabilities [21] and probably LENR.

(2) Abnormally high water line pressure (> 50 bar) was observed, which broke the pressure gauge. Meanwhile the copper pipe was found cracked and leaking in the VCS machine at some portion of the THX (in the colder part) in which the flow passages of water and R22 as shown in Fig. 3 are interchanged. In this portion, water flows in the inner and outer rings and R22 flows in the middle. Figure 9 shows that the middle pipes deform seriously, and the surface looks black. The pipe buckling wraps around the inner pipe tightly and blocks the R22 flow path, and some of the inner pipes were deformed simultaneously. The critical pressure $P_{cr}$ required to produce the buckling of the middle coper pipe can be calculated by Equation (8) [22].

$$P_{cr} = \frac{2E}{1 - \upsilon^2} \left( \frac{t}{D} \right)^3 \tag{8}$$

where $E$ is the Young's modulus, $\upsilon$ is the Poisson's ratio, $t$ is the pipe thickness, $D$ is the pipe diameter. For the middle pipe with diameter $D = 12.7$ mm and thickness $t = 0.8$ mm, $P_{cr}$ is 740 bar, taking $\upsilon = 0.35$ and $E = 130$ GPa for copper pipe. Considering the R22 pressure inside the middle pipe (35 bar), the external

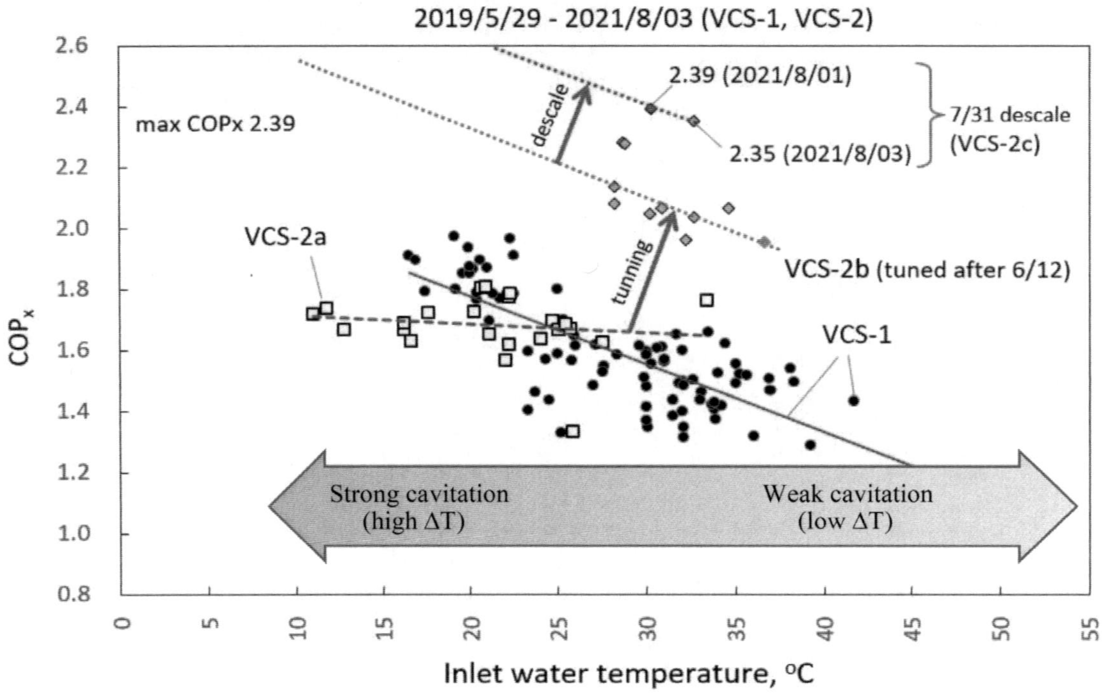


**Figure 7.** Variation of $COP_X$ with inlet water temperature.

pressure required to produce the buckling of the middle copper pipe is higher than 775 bar (740 + 35) which is far beyond the maximum total pressure of water (21.2 bar) in the outer ring. This also implies that the cavitation force is huge, probably caused by cavitation bubbles with a size of around 5 nm, according to bubble dynamics. Or, LENR might take place in the vicinity of pipe wall and create a local high temperature to reduce the strength and induce buckling of the middle copper pipe. Two-phase flow instabilities may play important role [21].

(3) Serious fouling happens often, even within a day, causing blockage of water passage. Scaling is common in boilers or heating equipment but is not so serious as it is in the VCS. This is probably induced by LENR.

(4) Lubricating oil in VCS plays some role in heat transfer inside the triple-pipe heat exchanger and causes interesting heat transfer phenomena. The R22 flow in THX behaves like a pulsating heat pipe [15].

## 3. Excess Energy From a Double-Pipe Heat Exchanger (DHX) Heated by Steam Boiler

The manufacture of triple-pipe heat exchanger (THX) in the VCS machine is quite complicated and expensive. Besides, the vapor compression system requires a huge compressor for large-scale applications. This makes the application of the triple-pipe heat exchanger heated by vapor compression system (VCS) limited to small scale. Hence, we developed a simpler machine using a double-pipe heat exchanger (DHX) which is heated by a steam boiler.

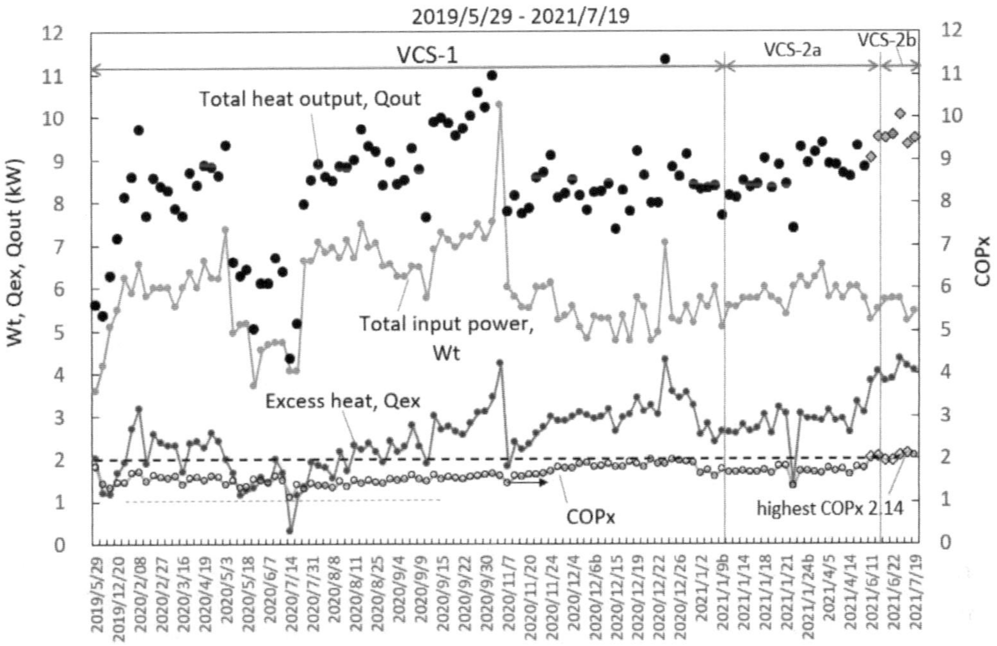


**Figure 8.** Measured excess energy $Q_{ex}$ and total heat output $Q_{out}(=Q_{wnet} + Q_L)$ in the kilowatt scale.

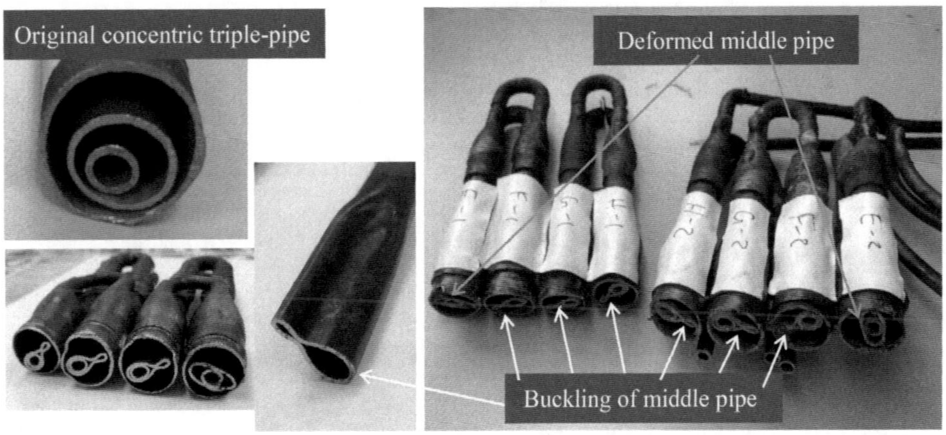


**Figure 9.** Buckling and deformation of middle pipes in VCS-1 when $COP_x > 1$.

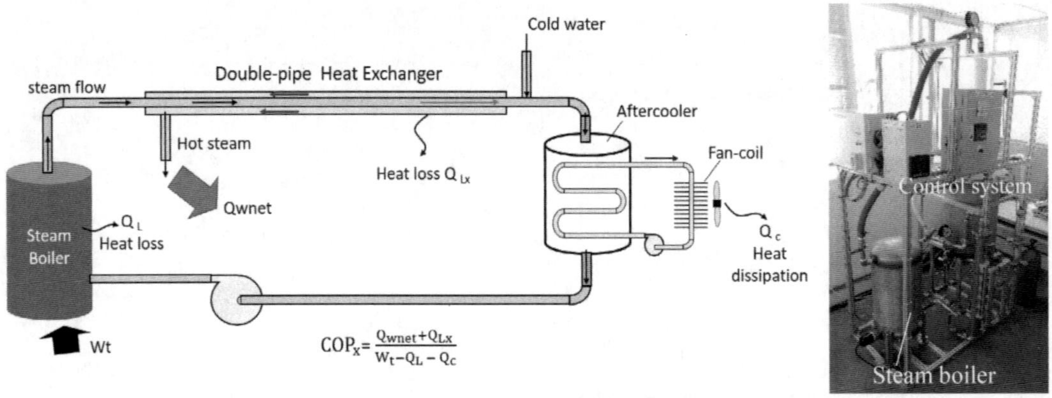


**Figure 10.** Schematic of double-pipe heat exchanger heated by steam boiler.

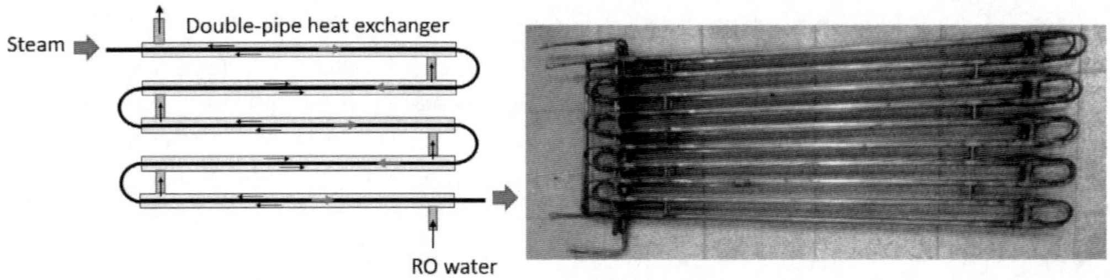


**Figure 11.** Double-pipe heat exchanger connected in series.

**Table 3.** Major dimension of double-pipe heat exchanger (DHX).

| **Inner copper pipe** | | **Outer copper pipe** | |
|---|---|---|---|
| OD (mm) | thickness (mm) | OD (mm) | thickness (mm) |
| 9.52 | 0.4 | 12.71 | 0.8 |

*Design of double-pipe heat exchanger*

The double-pipe heat exchanger (DHX) is heated by a steam boiler as shown in Fig. 10. The dimension of DHX is shown in Table 3. The gap of the water passage is 0.8 mm. Ten double-pipes are connected in series as a unit. Three units are connected in series to make the double-pipe heat exchanger 30 meters in total length (Fig. 11).

The maximum power input to the steam boiler is 7 kW. The temperature of the boiler output steam is controlled by a PID controller. The discharged steam/water flow from the DHX is fed into a water after-cooler to dissipate the remaining heat to the ambient air. The cooled water returns to the steam boiler by a feed pump to complete the cycle. The feed pump is the pump used in RO (reverse osmosis) equipment which consumes about 30 W. The water used in the DHX tests is purified by a RO machine.

Applying the energy balance to DHX and assuming no internal energy generation such as heat source or excess heat, we can derive the steady-state energy balance equation as

$$Q_{wnet} + Q_{Lx} = W_t - Q_L - Q_c \tag{9}$$

where $W_t$ is the total input power consumed by the electric heater, water pump (30 W), and controllers (around 30 W), $Q_{wnet}$ is the net power output from water flow, $Q_L$ is the heat loss of the boiler, $Q_{Lx}$ is the heat loss from the body of the double-pipe heat exchanger. We define the excess energy index ($COP_x$) of DHX as the total heat output ($Q_{wnet}$ $+Q_{Lx}$), the left-hand side of equation (9), divided by the net heat input ($W_t - Q_L - Q_c$), that is the right-hand side of equation (9):

$$COP_x = \frac{Q_{wnet} + Q_{Lx}}{W_t - Q_L - Q_c} \tag{10}$$

Therefore, the excess energy or LENR can be confirmed when the measured $COP_x$ is greater than 1 or 1.2 if taking into account the worst-case experimental error.

Adiabatic calorimetry and measurement errors

Similar to the previous treatment in a conservative manner, the measurement error ($e_{COPx}$) of $COP_x$ is evaluated at the worst-case condition. That is, according to the definition of $COP_x$ in equation (10),

$$e_{COPx} = e_{Wt} + e_{QL} + e_{Qc} + e_{Qwnet} + e_{QLx} \tag{11}$$

where $e_{Wt}$ is the error of $W_t$ (heating power of electric heater); $e_{Qwnet}$ is the error of $Q_{wnet}$; $e_{QL}$ is the error of boiler heat loss ($Q_L$). $e_{Qc}$ is the error of heat dissipation ($Q_c$); $e_{QLx}$ is the error of heat loss ($Q_{Lx}$) from DHX.

In the adiabatic calorimetry, the energy input to the electric heater ($W_t$) of the steam boiler was measured by a digital power meter (CP-500 by AIGO Tech Co.). The electric heater was connected to a temperature controller to fix the steady output temperature of steam from the boiler. The power consumptions by the CP-500 and the temperature controller are all included in the readings of the CP-500. Thus, the readings of the CP-500 need to be to converted into the actual heating power of the electric heater ($W_t$). We design a thermal balance system to calibrate the CP-500 as shown in Fig. 12. Water in an insulated tank was heated by a 5 kW electric heater which is connected to the temperature controller and the power meter (CP-500). The rate of heat absorbed by water ($Q_w$) was measured from the rates of increase in temperature and mass of the water.

The actual heating power ($Q_g$) of the heater is determined from $Q_g = Q_w + Q_{Lt}$ where $Q_{Lt}$ is the tank heat loss evaluated from the measured temperature difference between the outside surface of the tank and the ambient. It was found that the CP-500 reading is higher than the actual heating power of the heater by 5% ($\delta_{pm}$) on average. This deviation is a systematic error which comes from the power consumptions of the CP-500 and the temperature controller. This systematic error can be calibrated in $W_t$. The standard deviation (stdev) of the random errors in the heating power of heater ($W_t$) after calibration of the systematic error is 0.93%. The random error at 95% confidence ($e_R$) is 1.82%. We take the worst-case error in estimating the tank heat loss ($Q_{Lt}$) as 144 W (the largest) which corresponds to the maximum relative error of $Q_w$, i.e., $e_{hL} = 3.39\%$.

The worst-case error in measuring $W_t$ is thus $e_{Wt} = e_{hL} + e_R = 5.2\%$. The conversion of the CP-500 power meter reading into actual heating power of the electric heater is shown in Table 4 and Fig. 13.

The adiabatic calorimetry [23] of the net energy output by water ($Q_{wnet}$) follows the identical procedures in the VCS machine. Therefore, its measurement error $e_{Qwnet}$ is around 4%. The heat loss from the steam boiler ($Q_L$) is estimated from the measured surface temperature and ambient temperature. The error $e_{QL}$ is around 4%.

The heat dissipation $Q_c$ from the after-cooler connected downstream of the DHX is determined from measuring the temperature change of water flow through it and the flowrate. The error $e_{Qc}$ comes from the calibration errors of

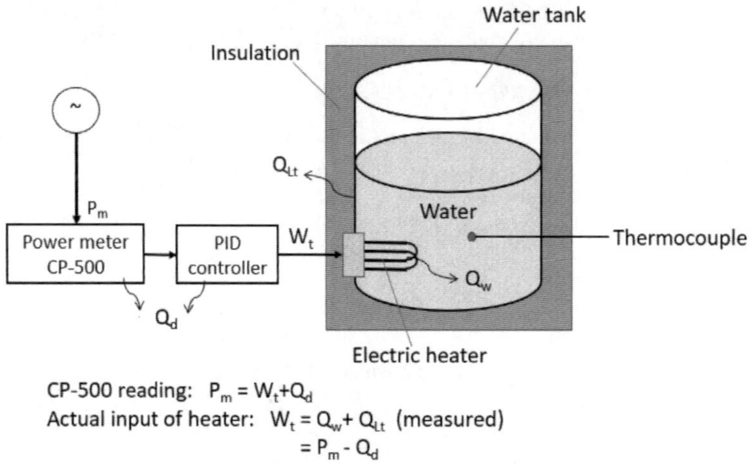


**Figure 12.** Schematic of power meter (CP-500) calibration device.

**Table 4.** Calibration of power meter CP-500 for the measurement of actual heating power of electric heater.

| **Ambient temp** (°C) | **Heat absorbed by water Qw (W)** | **Tank heat loss $Q_{Lt}$ (W)** | $Q_{Lt}/Q_w$ | **Actual input of heater $W_t=Q_w+Q_{Lt}$ (W)** | **CP-500 reading $P_m$ (W)** | **Deviation $d_{Pm}=(P_m-Q_g)/P_m$** | **Calibrated heating power of heater $W_t=(1-\delta_{pm})P_m$ (W)** | **Error of CP-500 reading for Wt, $(W_t-Q_g)/W_t$** |
|---|---|---|---|---|---|---|---|---|
| 25.2 | 4207 | 73 | 1.74% | 4280 | 4549 | 5.91% | 4323 | 0.99% |
| 26.3 | 4222 | 106 | 2.51% | 4328 | 4587 | 5.65% | 4359 | 0.71% |
| 25.8 | 4222 | 117 | 2.77% | 4339 | 4555 | 4.74% | 4328 | −0.24% |
| 28.2 | 4257 | 60 | 1.41% | 4317 | 4547 | 5.06% | 4321 | 0.09% |
| 28.3 | 4207 | 68 | 1.62% | 4275 | 4531 | 5.65% | 4306 | 0.71% |
| 28.5 | 4222 | 59 | 1.40% | 4281 | 4552 | 5.95% | 4326 | 1.03% |
| 25.6 | 4237 | 125 | 2.95% | 4362 | 4553 | 4.20% | 4327 | −0.82% |
| 25.6 | 4253 | 144 | 3.39% | 4397 | 4564 | 3.66% | 4337 | −1.38% |
| 25.8 | 4222 | 138 | 3.27% | 4360 | 4539 | 3.94% | 4313 | −1.08% |
| Worst-case heat loss estimation error $e_{hL}$ 3.39% | | | | Average deviation $\delta_{pm}$ | | 5.0% | stdev | 0.93% |
| Error in 95% confidence $e_R$ (=1.95 x stdev) | | | | | | | | 1.82% |
| Worst-case error of $W_t$ ($e_{Wt}=e_{hL}+e_R$) | | | | | | | | 5.2% |

temperature sensors and flowrate measurement and is 6.6%. Since the double-pipe heat exchanger is well insulated, the heat loss is very little and its error can be neglected. That is, $e_{Lx} \approx 0$.

Hence, the worst-case error in measuring COP$_x$ is $e_{COPx} = e_{Wt} + e_{QL} + e_{Qc} + e_{Qwnet} + e_{QLx} = 5.2\% + 4\% + 6.6\% + 4\% + 0 = 19.8\%$. Therefore, the excess energy or LENR can be confirmed only when the measured COP$_x$ is greater than 1.20.

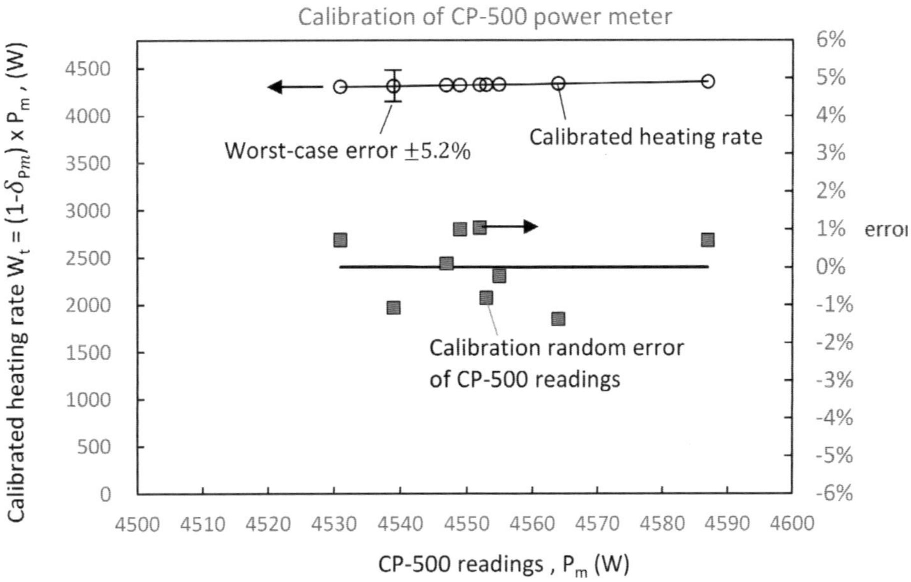


**Figure 13.** Conversion of CP-500 power meter reading into actual heating power of heater.

*Test results*

To perform a steady-state test, the DHX machine is turned on first, including the electric heater, the cooling water in the aftercooler, the cold-water flow through the double-pipe heat exchanger for at least an hour in order to reach a steady state. And then similar to the operation of the thermometry in the VCS machine (Fig. 4), the output hot water stream (containing steam) from the double-pipe heat exchanger is suddenly switched into the water tank to mix with the preloaded cold water to measure $Q_{\rm wnet}$. The energy input (Wt) from the electric heater is measured using the CP-500 calibrated power meter. The pressures and temperatures at various locations were recorded using a data acquisition system for further analysis.

The DHX-1 (the first prototype) was tested starting from July 9, 2020, under various operating conditions including boiler steam temperature, water flowrate through the DHX, and the flowrate from the aftercooler to the boiler, etc. During the first 4 months, the measured $COP_x$ all lied around 1.0 as seen from Fig. 14, which means that no excess energy can be confirmed. These results can only be treated as the base line calibration of the test bed. Then in mid-November 2020, some interesting results started to appear. $COP_x$ is over 1.0 and lies between 1.55 and 2.55 during the period from 11/14/2020 to 12/28/2020, as shown in Fig. 15. After that, $COP_x$ dropped to around 1.0 again until 2/02/2021 when water leakage was found, and the test was stopped. This means that DHX-1 was run with copper rupture and leakage for a while from 12/28/2020 to 2/02/2021. The highest $COP_x$ obtained with the DHX-1 is 2.55 on 12/19/2020. Several kilowatts of excess heat $Q_{ex}$ and total heat output $Q_{tot}$ $(=Q_{wnet} + Q_{Lx})$ are shown in Fig. 15. It is seen that the minimum $COP_x$ was observed at a boiler temperature around 123°C.

We built a new DHX (named DHX-2) using a thicker inner copper pipe to improve the strength. The pipe length was also increased from 30 to 40 meters for better heat transfer. The test continued for about 3 months from 3/11/2021, and $COP_x$ all lied around 1.0 again. This implies that the optimal operating conditions of DHX-1 has not been found

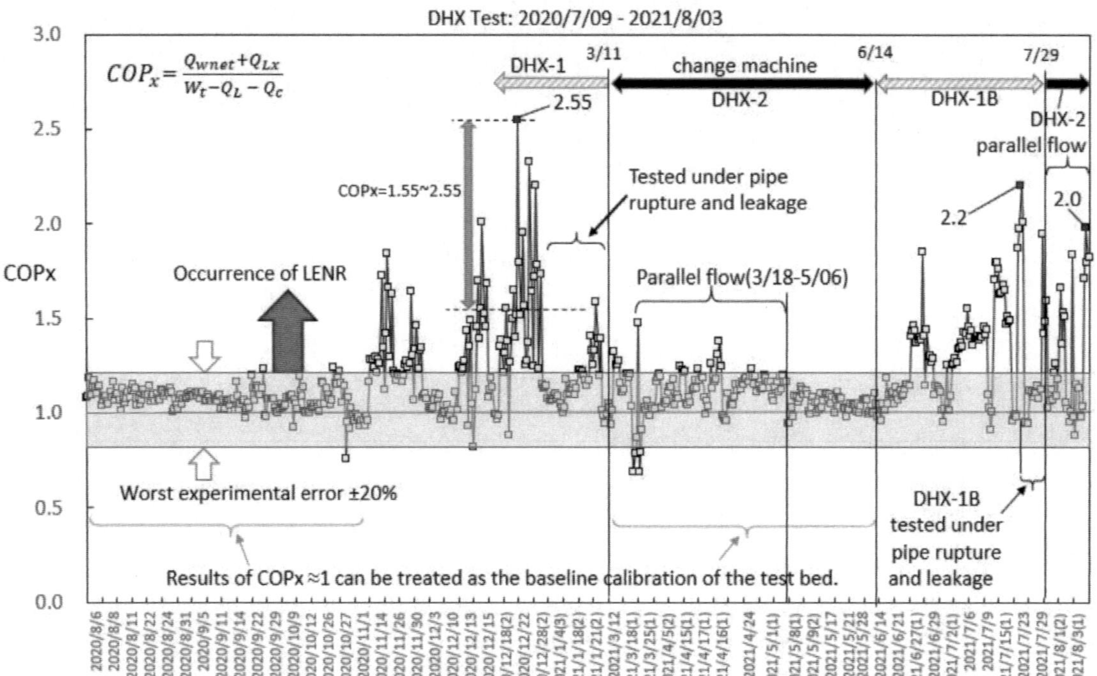


**Figure 14.** Performance of DHX-1 and DHX-2.

yet and no excess energy or LENR can be confirmed. This was probably due to the increase in the pipe length from 30 m to 40 m that reduces the heat flux in DHX-2 under the same input power of the electric heater.

We built another identical brand-new DHX-1 (named "DHX-1B") using the same design, and tested it starting June 14, 2021, under various operating conditions. Figure 14 shows that the excess-energy phenomenon can be repeated and $COP_x > 2$. However, after $COP_x$ reaching 2.20, the inner pipe was deformed and ruptured again, as was found in DHX-1. We went back to testing the DHX-2 again with a pipe length 30 m (the same as DHX-1) and parallel flow, starting July 29, 2021, under various operating conditions. It was found that $COP_x$ of DHX-2 increases up to 2.0 and the excess energy is confirmed again. This implies that we can reproduce and control the phenomenon of excess energy production.

*Peculiar phenomena observed in DHX*

The ruptured double-pipe heat exchanger of the DHX-1 was cut off piece by piece for inspection. Several peculiar phenomena were observed.

The inner copper pipe (steam side) of the DHX-1 was found deformed (Fig. 16) and the outer pipe wall (water side) was cracked and leaking. From the buckling theory of pipe, Equation (8) with $D = 9.52$ mm, $t = 0.4$ mm, $E =$ 130 GPa, and $\nu = 0.35$, the critical pressure $P_{cr}$ that can cause pipe buckling is 220 bar, assuming a vacuum inside the pipe. Since the steam flow inside the pipe has a pressure 4.76 bar at 150°C (maximum), the exerted pressure required to create this buckling phenomenon is $220 + 4.76 = 224.76$ bar which is much higher than the maximum applied

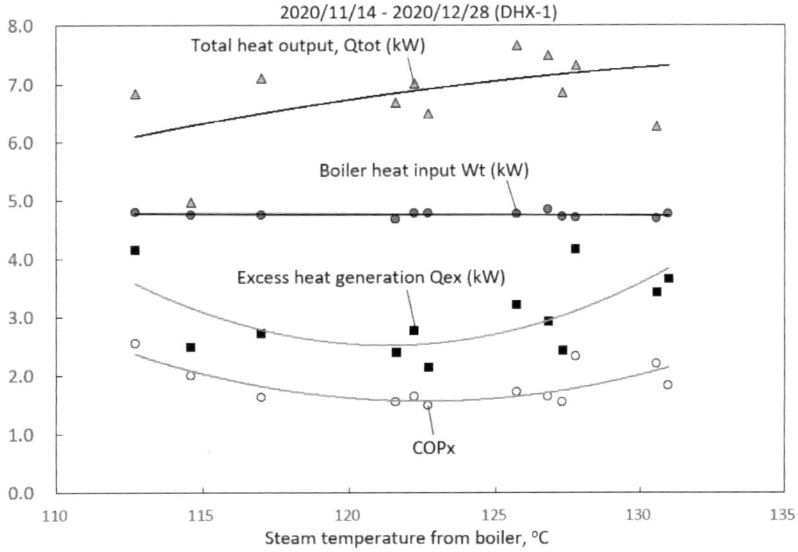


**Figure 15.** Excess heat generation in kilowatt levels in DHX-1 with $COP_x > 1.2$.

**Table 5.** Chemical element analysis of ruptured copper pipes of DHX-1.

| SEM results, wt% Mean (min/max) | Original Cu pipe | Ruptured Cu pipe (Specimen 1-1) | Ruptured Cu pipe (Specimen 1-10) |
|---|---|---|---|
| C | 5.78 (2.90/9.88) | 17.41 (12.21/23.18) | 16.16 (7.91/21.25) |
| O | 2.80 (0.67/9.84) | 19.37 (16.53/21.82) | 22.24 (20.4/23.25) |
| Cu | 91.1 (86.28/94.75) | 62.74 (57.71/67.13) | 60.83 (54.52/68.62) |
| Cl | 0.19 (0/0.79) | 0.31 (0.13/0.52) | 0.24 (0.08/0.45) |
| Fe | 0.12 (0.01/0.19) | 0.17 (0/0.72) | 0.53 (0.13/0.89) |

total pressure of water (21.2 bar) including the thermodynamic pressure 3.6 bar at 140°C (10°C lower than the steam temperature) and the pressurization by RO pump (around 17.6 bar). This implies that a huge force is induced from cavitation with a bubble size around 20 nm (according to bubble dynamics) or two-phase flow instabilities [22]. LENR could be the cause for this pipe buckling.

It is also noted that the outside surface of the ruptured inner pipe became shining black (Fig. 16). The surface contains 16∼17% carbon (C), 19∼22% oxygen (O) and 0.17∼0.53% Fe. The chemical element increases by 3 to 4 folds in C, 10 folds in O and 1.4 to 4.4 folds in Fe, as shown in Table 5. This might possibly have been caused by the phenomenon of nuclear transmutation. This may result from LENR. Little scale was found in the DHX since the water was purified using RO equipment.

Figure 17 shows the SEM images of a piece of the ruptured copper pipe (Specimen 1-10) (Fig. 18). Figure 19 shows the SEM image of the outside surface of the ruptured inner copper pipe in DHX-1. The surface looks like CuO nanowire made from thermal oxidation of copper in oxygen at 500-700°C [16]. If this is so, the buckling of inner copper pipe may be resulted from a high temperature around 500-700°C induced by LENR. High temperature causes softening of the copper pipe and creates deformation of the pipe by the outside pressure around 21.2 bar. The same buckling phenomenon of the inner pipe was also found in the DHX-1B when $COP_x > 2$, as shown in Fig. 20.

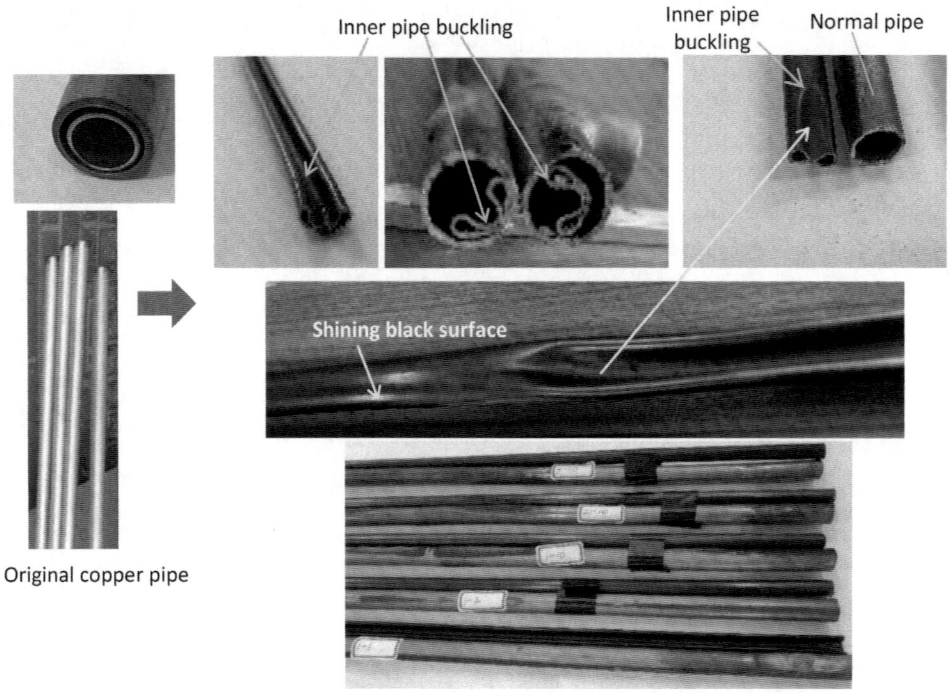


**Figure 16.** Buckling of inner copper pipe in DHX-1.

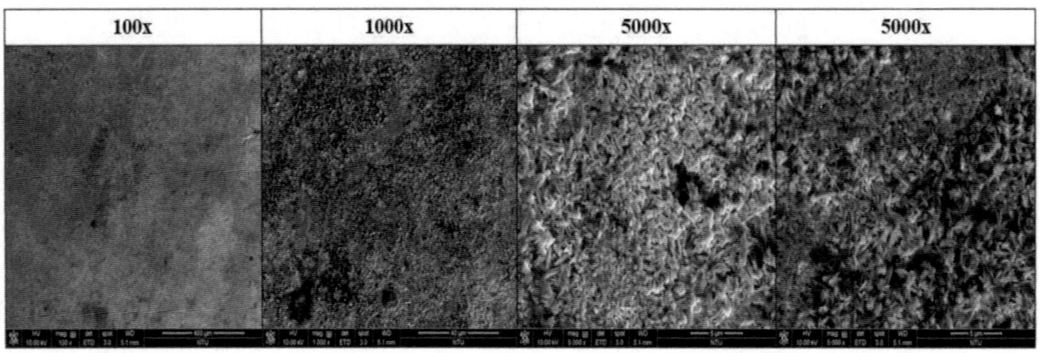


**Figure 17.** SEM image of ruptured copper pipe surface (Specimen 1-10).

## 4. Discussion and Conclusions

In the present study, we built two kinds of machines which can produce excess energy: a triple-pipe heat exchanger heated by a vapor compression system (VCS) and a double-pipe heat exchanger heated by a steam boiler (DHX). The excess energy index $COP_x$ is used to detect the existence of excess energy. Excess energy occurs if the measured

**Figure 18.** Pieces of ruptured copper pipe for SEM image inspection.

$COP_x$ is greater than 1 or 1.2 if taking into account the worst-case experimental error. The highest $COP_x$ is 2.39 in the VCS machine and 2.55 in the DHX machine, both beyond the worst-case experimental error.

The solid evidence for the occurrence of LENR comes from the buckling, shining black look, and crack and leakage of the copper pipes found in both VCS and DHX-1, when the measured $COP_x$ is greater than 1.2. Another side evidence is the fact that the chemical elements of C, O and Fe on copper pipe surface increases several folds. They may be attributed to a nuclear transmutation, or potential hydrogen and oxygen interactions during cavitation [17]. The dissolved oxygen or nitrogen in water may play a role since it may help create a CNO cycle reaction [18].

In the VCS machine, the $COP_x$ was found to increase with decreasing inlet water temperature, which implies higher dissolved O and N (from air) in the water at lower temperatures. This may also happen in the DHX machine since the RO water is prepared before the tests and it is open to air and allows dissolving of O and N from air into water. The lower inlet water temperature implies a larger temperature gradient created by the heat source (R22 in VCS and steam in DHX) which may result in two-phase flow instabilities [21] which facilitate the LENR reaction.

In the present study, we experimentally confirm that excess energy or LENR can be induced by a heat transfer process involving cavitation in heat-exchange system. We found that something in common in the two kinds of machines (VCS and DHX) built in the present study, including cavitation within tiny passages (0.8$\sim$2.3 mm gap), intense dynamic implosions from collapse of nanobubbles smaller than 100nm, and two-phase flow oscillation/resonance in heat transfer process.

We draw the following conclusions from the present study:

1. Excess energy in kilowatt levels can be induced from simple heat exchanging process involving cavitation under certain design and operating conditions.
2. The excess energy index ($COP_x$) is defined based on the first law of thermodynamics and can be used in the experimental data analysis to identify the occurrence of excess energy phenomena. The maximum $COP_x$ obtained in the present study is 2.39 in the VCS-2c machine and 2.55 in the DHX-1 machine.

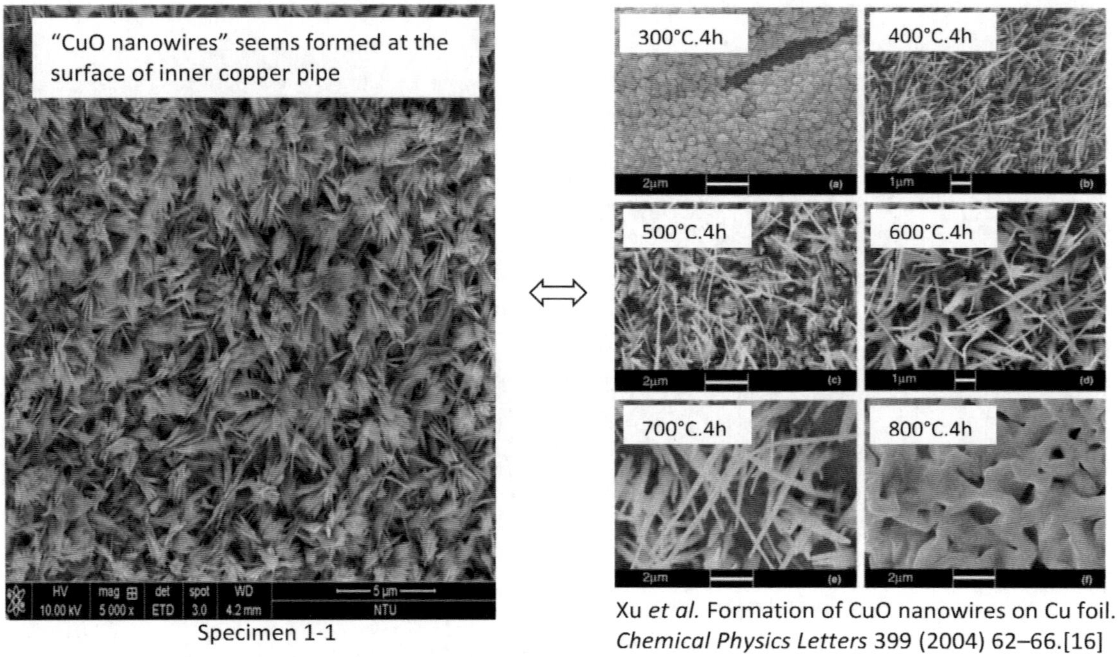


**Figure 19.** SEM images of outside surface of ruptured inner copper pipe in DHX-1.

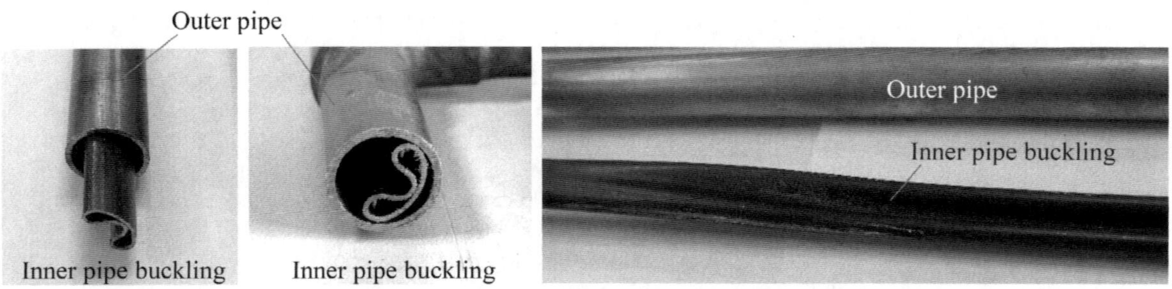


**Figure 20.** Buckling of inner copper pipe found in DHX-1B.

3. We experimentally confirmed that excess energy process is reproduceable and controllable.
4. Material problems will be eventually encountered for the machines with $COP_x > 1$, since extremely high pressure or high temperatures will be induced by LENR.
5. The phenomenon of $COP_x > 1$ is related to cavitation, nano-bubbles implosion, or the liberation of stored intermolecular bond energy in electrodynamic process [20]. The two-phase flow instabilities cavitation may play an important role in LENR [19]. More scientific research is needed.
6. As engineers, we are focusing on finding the optimal design and duplicating the best performance obtained during the studies to provide a reliable, simple and cheap machine to harvest energy from water.

## Acknowledgements

The present study was supported by Ministry of Science and Technology, Taiwan. The original design and manufacture of the machine VCS's and DHX's used in the present study were provided by CETWO AIR Co. (Taiwan) and Advanced Thermal Devices Inc. (Taiwan), respectively. The authors appreciate very much for many instructive opinions given by Mr. Bob Greenyer and participants of ICCF-22 and ICCF-23.

## References

[1] Alan J. Walton, Geo. T. Reynolds, Sonoluminescence, *Advances in Physics* **33(6)** (1984) 595-660

[2] Michael P. Brenner, Sascha Hilgenfeldt, Detlef Lohse, Single-bubble sonoluminescence, *Reviews of Modern Physics* **74(2)** (2002) 425-484

[3] Bob Greenyer. Affordable ultrasonic driven transmutation? Martin Fleischmann Memorial Project, Oct 6, 2019 http://www.quantumheat.org/index.php/en/home/mfmp-blog/552-ultr-affordable-ultrasonic-driven-transmutation

[4] Bob Greenyer. ULTR – Experiment. Martin Fleischmann Memorial Project, April 09, 2021. https://www.youtube.com/watch?v=By1xWRLE4pY

[5] Thomas N. Claytor, Roger S. Stringham, Malcolm M. Fowler, Cavitation effects on various metals in D2O, *J. Condensed Matter Nucl. Sci.* **29** (2019) 75–84

[6] M.E.M. Azouzi, C. Ramboz, J. Lenain, F. Caupin, A coherent picture of water at extreme negative pressure, *Nat. Phys.* **9 (1)** (2013) 38–41

[7] Q. Zheng, D.J. Durben, G.H. Wolf, C.A. Angell, Liquids at large negative pressures – water at the homogeneous nucleation limit, *Science* **254** (5033) (1991) 829–832.

[8] Max I. Fomitchev-Zamilov, Cavitation-induced fusion: proof of concept. Natural Philosophy Alliance Vol 9, 2012. https://www.youtube.com/watch?v=d5OVYKOhOGQ

[9] Cavitation heat generator. https://www.youtube.com/watch?v=wWDFX08rJvo. August 2016

[10] Scale-Free Heating - ShockWave Power Reactor. Hydro Dynamics Inc. (2020) Cavitation Reactor and Process Intensification Technology - Scale-Free Heating (hydrodynamics.com)

[11] Cavitation Energy Systems Technical Presentation (cavitation engine) https://www.youtube.com/watch?v=Ji0GAt-q2EM. April 2017

[12] V. Nikolsky, O. Oliynyk, ?. Lipeev, V. Ved, Examining a cavitation heat generator and the control method over the efficiency of its operation. *Eastern-European Journal of Enterprise Technologies* ISSN 1729-3774 4/8(**88**) (2017) 22-28

[13] Cavitation Heater – Overunity. https://www.youtube.com/watch?v=DjGSXKSLpfY

[14] Bin-Juine Huang, Ming-Li Tso, Ying-Hung Liu, Jong-Fu Yeh. Excess energy from a vapor compression system, ICCF-22, Assisi, Italy, Sept 8-13, 2019

[15] H. B. Ma, C. Wilson, Q. Yu, K. Park, U. S. Choi, Murli Tirumala, An experimental investigation of heat transport capability in a nanofluid oscillating heat pipe. *ASME Journal Heat Transfer* **128** (2006) 1213-1216

[16] C.H. X, C.H. Woo, S.Q. Shi. Formation of CuO nanowires on Cu foil. *Chemical Physics Letters* **399** (2004) 62–66.

[17] Bob Greenyer, Potential hydrogen and oxygen interactions during cavitation, A proposed explanation for Bin-Juine Huang's cavitation driven excess heat system - YouTube. https://www.youtube.com/watch?v=UH76XaTz4qE&list=PLZk-jCLezj9gZcm6_lSplPGR6By5j2EQJ&index=31&t=1s. Martin Fleischmann Memorial Project, June, 2021

[18] H. A. Bethe, Energy production in stars, Physical Review **55** (1939) 436-456

[19] Lewis Larsen, Lattice White Paper: LENRs enable green radiation-free nuclear power and propulsion Lattice Energy LLC, May 16 (2019) Chicago, IL USA

[20] Graneau, N., S. Verdoold, G. Oudakker, C. Yurteri and J. Marijnissen. Renewable energy liberation by nonthermal intermolecular bond dissociation in water and ethanol. *J Applied Physics* **109** (3) (2011) 034908

[21] Ruspini, L.C., Marcel, C.P., Clausse, A. Two-phase flow instabilities: A Review. *Int J Heat and Mass Transfer* **71** (2014) 521-548

[22] Bryan, G.H. Application of the energy test to the collapse of a long thin pipe under external pressure. *Cambridge Phil. Soc. Proc.* **6** (1888) 287-292

[23] Rothwell, Jed. Review of the calorimetry of Fleischmann and Pons. LENR-CANR.org August 2020, Revised October 2020

J. Condensed Matter Nucl. Sci. 36 (2022) 266–273

Research Article

# Conductivity and Molar Conductivity of LiOD Heavy Water Solution

Hui Zhao*

*State Key Laboratory of NBC protection for Civilian, Beijing 102205, China and also with Institute of Chemistry, CAS, Beijing 100190, China*

Wu-Shou Zhang†

*Institute of Chemistry, CAS, Beijing 100190, China*

Wu-Yun Xiao‡, Ye Chen, and Sheng Qi

*State Key Laboratory of NBC protection for Civilian, Beijing 102205, China*

**Abstract**

In cold fusion research, the LiOD heavy water solution has been used as electrolyte in Pd-$D_2O$ electrolytic systems for over 30 years. The conductivity of LiOD not only affects the electrical and thermal properties but also can be used to determine the LiOD concentration. In this paper, conductivities of LiOD heavy water solution ($0.01409 \sim 0.09721$ mol·$L^{-1}$) at $10 \sim 70$°C were measured. Changes of conductivity $\kappa$ in mS·$cm^{-1}$ with concentration $c$ in mol·$L^{-1}$ and temperature $T$ in °C can be expressed by the quadratic form: $\kappa = 60.56c - 14.25c^2 + 2.514cT - 0.5459c^2T$. This relationship can be simplified to $\kappa = 123.41c - 27.90c^2$ or $c = 8.1031 \times 10^{-3}\,\kappa + 1.484 \times 10^{-5}\,\kappa^2$ at 25°C.


*Keywords:* LiOD heavy water solution, Conductivity, Molar conductivity, Lithium, Cold fusion

## 1. Introduction

In 1989, Fleischmann and Pons reported anomalous excess heat in the electrolytic system of LiOD heavy water solution with a Pd cathode [1]. Since then, for more than 30 years, many researchers around the world have focused on this system [2]–[5]. Recently, our lab explored excess heat in a Pd-LiOD+$D_2O$ open system, similar to that used by Fleischmann-Pons [1] and Miles [2], with Seebeck envelope calorimetry. During open electrolysis, it was found that the LiOD concentration decreased, while the cell voltage increased with time in the constant current mode. To quickly determine the LiOD concentration, and to understand the electric and thermal behaviors of electrochemical

*E-mail: nicholas_chao05@163.com
†Corresponding author: wszhang@iccas.ac.cn
‡Corresponding author: xiaowuyun@sklnbcpc.cn

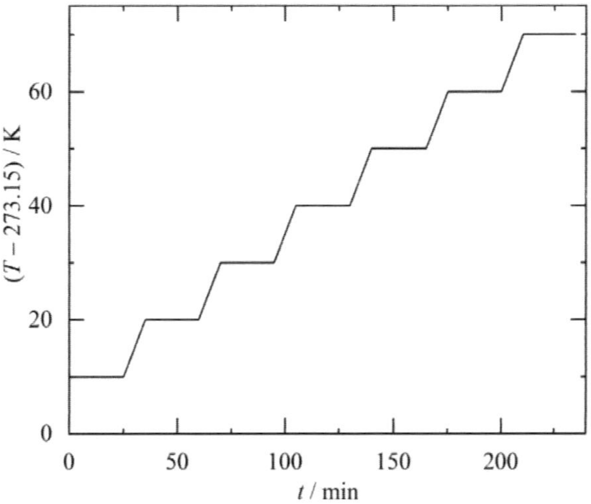


**Figure 1.** Heating curve of LiOD+$D_2O$ solution.

cells, the electrolyte conductivities of LiOD heavy water solutions under different concentrations and temperatures must be obtained first. However, there was only data of 0.1 mol·L$^1$ LiOD solution, from 1990 [7]. In this paper, the conductivities of LiOD heavy water solution at various concentrations and temperature are measured. At the same time, the molar conductivities are also given. We hope these data can support future works on the Pd-LiOD+$D_2O$ electrolytic system.

## 2. Experimental

The LiOD heavy water solution is made before conductivity measurements. The method is as follows. First, weigh 0.14524 g $Li_2O$ powder (Alfa, 99.5%, Order #41832; Lot #S23G011) with balance (Mettler Toledo XPR205R. max. 220 g, d = 0.01 mg) and put it into a 5 mL beaker, then add heavy water (J&K, 99.8 at.%D, Order #261750; Lot #L240U16) and ultrasonically mix for 10 minutes to ensure that $Li_2O$ is completely dissolved. Finally, transfer the solution from the beaker to a volumetric flask with capacity of 10 ± 0.02 mL at 20°C and add heavy water to the graduation line. After these procedures, we get 0.9721 mol·L$^{-1}$ LiOD heavy water solution. Solutions of 0.07279, 0.1126, 0.1215, 0.2411 and 0.4867 mol·L$^{-1}$ are obtained by diluting the 0.9721 mol·L$^{-1}$ LiOD solution. The solution of 0.1126 mol·L$^{-1}$ is further diluted to that of 0.01409, 0.02816 and 0.05630 mol·L$^{-1}$, respectively. In the end, nine different concentrations of LiOD heavy water solutions are obtained.

Each solution is placed in a centrifuge tube, which is immersed in a water bath (PolyScience AP15R-30-A12E) with temperature stability of ± 0.005°C. The conductivity of the solution is directly measured by Mettler Toledo S230 conductivity meter equipped with InLab®731-ISM using a 4-ring graphite electrode. Its measurement range is 0.01 to 1,000 mS·cm$^{-1}$ at temperature of 0 ∼ 100°C (The unit of conductance is S (Siemens), which is the inverse of Ω (Ohm), i.e., 1 S = 1 $\Omega^{-1}$).

The measurement is started at 10°C and heated at a rate of 1°C per minute, then the temperature is kept at a constant value for 25 minutes at 10, 20, 30, 40, 50, 60 and 70°C, respectively, as shown in Fig. 1. The conductivity meter is set to measure every 5 minutes and it automatically saves the data.

Before the formal measurement, the conductivity meter was calibrated with a conductivity standard solution of 111.3 mS·cm$^{-1}$ at 25°C (Shanghai Inesa Scientific Instrument Co., Ltd, Order #624101; Lot #9028N002102F00023).

**Table 1.** Conductivities ($\kappa$/mS·cm$^{-1}$) of LiOH aqueous solution at 25°C.

| References \ $c$/mol·L$^{-1}$ | 0.1 | 0.5 |
|---|---|---|
| Ritley *et al.* [7] | 13.5 | 82.6 |
| Darken *et al.* [8] | 20.51 | 89.72 |
| This work | 19.17 ± 0.07 | 86.27 ± 0.01 |

**Table 2.** Conductivities ($\kappa$/mS·cm$^{-1}$) and activation energies ($E_a$) of LiOD heavy water solution at different temperatures and concentrations.

| $c$/mol·L$^{-1}$ \ $T$/°C | 10 | 20 | 30 | 40 | 50 | 60 | 70 | $E_a$/ J·mol$^{-1}$ |
|---|---|---|---|---|---|---|---|---|
| 0.01409 | 1.364 | 1.714 | 2.103 | 2.506 | 2.920 | 3.343 | 3.758 | 13,647 |
| 0.02816 | 2.673 | 3.357 | 4.122 | 4.908 | 5.736 | 6.591 | 7.520 | 13,863 |
| 0.05630 | 5.273 | 6.594 | 8.080 | 9.615 | 11.213 | 12.80 | 14.460 | 13,574 |
| 0.07279 | 6.567 | 8.232 | 10.074 | 11.954 | 13.954 | 16.034 | 18.195 | 13,672 |
| 0.1126 | 9.938 | 12.437 | 15.195 | 18.034 | 20.954 | 23.851 | 26.885 | 13,369 |
| 0.1215 | 10.692 | 13.379 | 16.356 | 19.425 | 22.529 | 25.851 | 29.230 | 13,496 |
| 0.2411 | 20.218 | 25.276 | 30.839 | 36.483 | 42.264 | 48.149 | 54.391 | 13,262 |
| 0.4867 | 38.218 | 47.690 | 58.161 | 68.736 | 79.253 | 89.816 | 101.115 | 13,040 |
| 0.9721 | 67.6667 | 84.287 | 102.977 | 122.069 | 141.379 | 160.345 | 181.954 | 13,255 |

Then we measured another standard solution with conductivity of 12.88 mS·cm$^{-1}$ ± 1.5% at 25°C (Mettler Toledo InLab® Solutions, Order #51350094; Lot #1F325A). The result was 13.28 ± 0.01 mS·cm$^{-1}$ at 25°C and it is consistent with the standard value within 3.1%.

At the same time, we made a LiOH aqueous solution by mixing $Li_2O$ and $H_2O$ and measured conductivities at two different LiOH concentrations at 25°C as shown in Table 1. We found that our results are consistent with the interpolated values based on data in [8] within 6.5%. This means our measurement method is reliable and the sample preparation of LiOD heavy water solution using $Li_2O$ instead of lithium metal as in [2]–[7] is not a problem. However, conductivities of 0.1 and 0.5 mol·L$^{-1}$ LiOH aqueous solution measured by Ritley *et al.* [7] are 34% and 8%, respectively, less than the accepted values in [8]. Therefore, their results of the LiOD heavy water solution are only for reference here.

## 3. Results

### 3.1. Conductivity

The conductivities $\kappa$ of LiOD heavy water solution at different temperatures $T$ and concentrations $c$ are listed in Table 2 and shown in Fig. 2.

Although the linearity is not obvious on the logarithmic scale, it can be observed from Fig. 2(a) that the electrical conductivity increases linearly with increasing temperature at a constant concentration as described by:

$$\kappa = a_0 + a_1 T \tag{1}$$

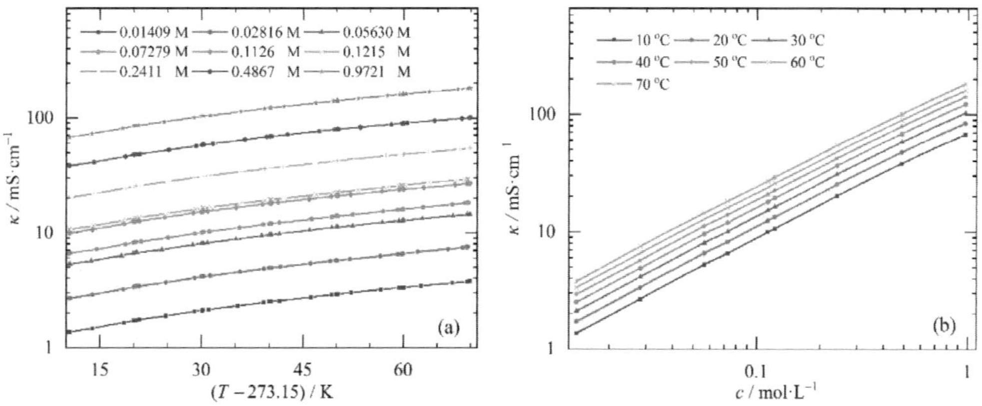


**Figure 2.** Conductivities of LiOD heavy water solution at different temperatures (a) and concentrations (b).

In addition, we find that a quadratic function can be used to express the relationship between conductivity and LiOD concentration at a constant temperature as shown in Fig. 2(b) and described by:

$$\kappa = b_1c + b_2c^2 \quad (2)$$

Considering the simultaneous influences of temperature and concentration, the relationship between these three parameters can be expressed as:

$$\kappa = d_1c + d_2c^2 + d_3cT + d_4c^2T \quad (3)$$

where $d_1$, $d_2$, $d_3$ and $d_4$ are constants. A program is used for fitting the data listed in Table 2 and we obtain the expression:

$$\kappa = 60.56c - 14.25c^2 + 2.514cT - 0.5459c^2T \quad (4)$$

with $T$ in °C, $R^2 = 0.9997$ and RMSE = 0.7193. The fitting results are shown in Fig. 3, where *X*, *Y* and *Z* axes refer to the LiOD concentration (mol·L$^{-1}$), temperature (°C) and conductivity (mS·cm$^{-1}$), respectively.

At 25°C, Eq. (4) can be simplified to:

$$\kappa = 123.41c - 27.90c^2 \quad (5)$$

The conductivity of 0.1 mol·L$^{-1}$ LiOD heavy water solution at 25°C is 12.08 mS·cm$^{-1}$. It is ∼ 40% lower than the conductivity of LiOH aqueous solution at the same alkali concentration as listed in Table 1. On the other hand, this value is 37% higher than 8.8 mS·cm$^{-1}$ in [7]. Because their conductivity at 0.1 mol·L$^{-1}$ LiOH aqueous solution is 34% less than the accepted value in [8], this means their results have prominent negative shifts both for LiOH+$H_2O$ and LiOD+$D_2O$ solutions.

The LiOD concentration can be determined from conductivity based on Eq. (5) at 25°C:

$$c = 8.1031\times10^{-3}\kappa + 1.484\times10^{-5}\kappa^2 \quad (6)$$

This equation can be used to calculate the LiOD concentration from conductivity data in experiments. In crude estimation, the first term is sufficient for dilute solutions.

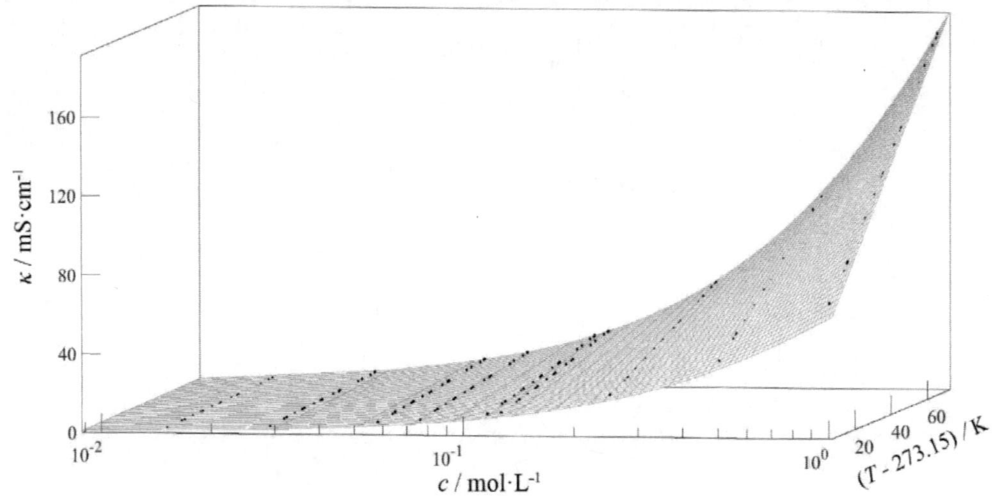


**Figure 3.** The fitting surface among conductivity, temperature and LiOD concentration.

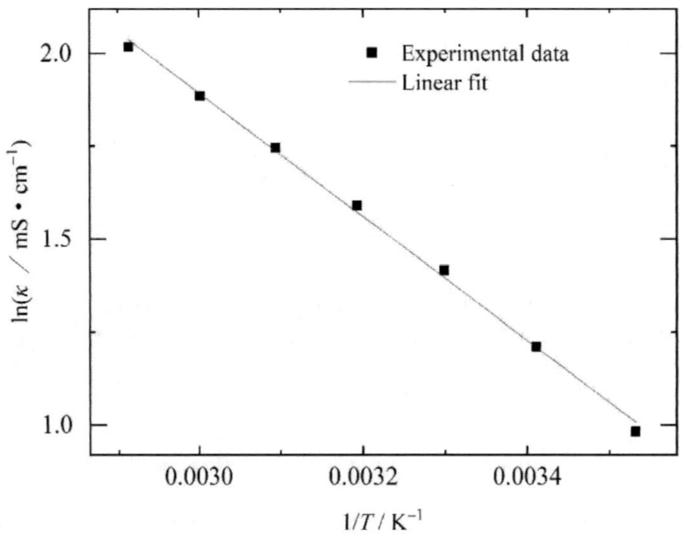


**Figure 4.** Arrhenius plot of conductivities of LiOD heavy water solution at 10 ~ 70°C ($c = 0.02816$ mol·L$^{-1}$).

On the other hand, the increase of electric conductivity of LiOD+$D_2O$ solution with temperature can be described by the Arrhenius law [9], whose logarithmic form is:

$$\ln\kappa = \ln\kappa_0 - \frac{E_a}{RT} \tag{7}$$

**Table 3.** Molar conductivity ($\Lambda$/S·cm$^2$·mol$^{-1}$) of LiOD+$D_2O$ solution at different temperatures and concentrations.

| $c$/mol·L$^{-1}$ \ $T$/°C | 10 | 20 | 30 | 40 | 50 | 60 | 70 |
|---|---|---|---|---|---|---|---|
| 0.01409 | 96.80625 | 121.64656 | 149.25479 | 177.85664 | 207.23918 | 237.26047 | 266.71398 |
| 0.02816 | 94.92188 | 119.21165 | 146.37784 | 174.28977 | 203.69318 | 234.0554 | 267.04545 |
| 0.05630 | 93.65897 | 117.12256 | 143.51687 | 170.78153 | 199.16519 | 227.35346 | 256.83837 |
| 0.07279 | 90.21844 | 113.09246 | 138.39813 | 164.22586 | 191.70216 | 220.27751 | 249.96565 |
| 0.1126 | 88.25933 | 110.45293 | 134.94671 | 160.15986 | 186.09236 | 211.8206 | 238.76554 |
| 0.1215 | 88 | 110.11523 | 134.61728 | 159.87654 | 185.42387 | 212.76543 | 240.57613 |
| 0.2411 | 83.85732 | 104.83617 | 127.90958 | 151.31895 | 175.29656 | 199.70552 | 225.59519 |
| 0.4867 | 78.52476 | 97.98644 | 119.50072 | 141.22868 | 162.83748 | 184.54078 | 207.75632 |
| 0.9721 | 69.60879 | 86.7061 | 105.93252 | 125.57247 | 145.43668 | 164.94702 | 187.17622 |

**Table 4.** The limiting molar conductivities ($\Lambda_0$) and slopes of fitting ($\beta$) at different temperatures.

| $T$ / °C | $\Lambda_0$ / S·cm$^2$·mol$^{-1}$ | $\beta$ / S·cm$^2$·L$^{1/2}$·mol$^{-3/2}$ |
|---|---|---|
| 10 | 99.62 | 30.87 |
| 20 | 125.07 | 39.56 |
| 30 | 153.35 | 49.20 |
| 40 | 182.54 | 59.56 |
| 50 | 213.06 | 71.38 |
| 60 | 244.56 | 84.41 |
| 70 | 276.59 | 95.52 |

where $\kappa_0$ is the pre-exponential factor, $E_a$, the activation energy, $R$, the universal gas constant and $T$, the absolute temperature in K. Fig. 4. shows the experimental data and the fit of Eq. (7) at 0.02816 mol·L$^{-1}$. The slope of line enables the determination of activation energy, 13.863 kJ·mol$^{-1}$. As listed in the last column of Table 2, $E_a = 13.2 \sim 13.6$ kJ·mol$^{-1}$ while $c = 0.01409 \sim 0.9721$ mol·L$^{-1}$, which are greater than the estimated value of 12.6 kJ·mol$^{-1}$ for LiOH aqueous solution in [9].

### 3.2. Molar Conductivity

Molar conductivity $\Lambda$ refers to the electrical conductivity of a solution of different concentrations converted to unit molar concentration:

$$\Lambda = \kappa/c \tag{8}$$

Table 3 lists the calculated molar conductivity at different temperatures and concentrations based on data in Table 2.

The limiting molar conductivity ($\Lambda_0$) is defined as the molar conductivity of a solution at infinite dilution. The molar conductivity of a strong electrolyte solution has a linear relationship with the square root of the concentration [10] is:

$$\Lambda = \Lambda_0 - \beta\sqrt{c} \tag{9}$$

where $\beta$ is constant in S·cm$^2$·L$^{1/2}$·mol$^{-3/2}$. $\Lambda_0$ can be obtained by extrapolation to infinite dilution as shown in Fig. 5. The limiting molar conductivities $\Lambda_0$ and slopes $k$ obtained from these curves are listed in Table 4.

It can be seen from Fig. 5 and Table 4 that with the increase of temperature, the limiting molar conductivity and the slope of the fitting increase at the same time.

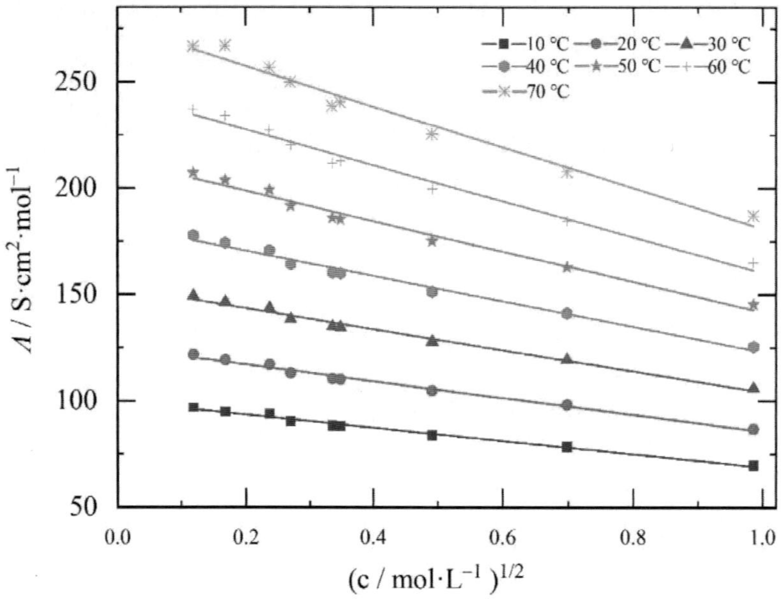


**Figure 5.** The fitted lines of molar conductivity against the square root of LiOD concentration at different temperatures.

## 4. Summary

We measured the conductivity of LiOD heavy water solution at temperatures of 10 $\sim$ 70°C and concentrations of 0.01409 $\sim$ 0.9721 mol·$L^{-1}$, and obtained a quadratic relationship among the conductivity, the temperature, and the concentration. The LiOD concentration can be calculated by equations we obtained using temperature and conductivity. This provides a method for understanding the properties of the Pd-LiOD+$D_2O$ solution in electrolysis experiments.

## 5. Acknowledgments

This work was supported by NSFC (21153003, 20973185) and 973 program of MOST in China (2009CB226113).

## References

[1] M. Fleischmann, S. Pons and M. Hawkins, Electrochemically induced nuclear fusion of deuterium, J. Electroanal. Chem. **261** (1989) 301−308 and errata in **263**, 187.

[2] M.H. Miles, R.A. Hollins, B.F. Bush, J.J. Lagowski, and R.E. Miles, Correlation of excess power and helium production during $D_2O$ and $H_2O$ electrolysis using palladium cathodes, J. Electroanal. Chem. **346** (1993) 99−117.

[3] M.C.H. McKubre, S. Crouch-Baker, A.M. Riley, S.I. Smedley and F.L. Tanzella, Excess power observations in electrochemical studies of the D/Pd system; the influence of loading. Proc. ICCF3. Nagoya, Japan, Oct. 21–25, 1992, pp. 5−20.

[4] E. Storms, C. Talcott. Electrolytic tritium production, Fusion Tech. **17** (1990) 680−695.

[5] F. Celani, A. Spallone, P. Tripodi and A. Nuvoli. Measurements of excess heat and tritium during self-biased pulsed electrolysis of Pd-$D_2O$. Proc. ICCF3. Nagoya, Japan, Oct. 21–25, 1992, pp. 93−105.

[6] M.H. Miles, H.D. Arman, J.D. Carrick, C.K. Gren, K.A. Haggerty, H.Y. Kim, A.G. Ky, J.E. Markham, C.F. Meeks and D.E. Noga. The elevation of boiling points in $H_2O$ and $D_2O$ electrolytes. Proc. ICCF9, Beijing, China, May 19−24, 2002, pp. 246−249.

[7] K.A. Ritley, P.M. Dull, M.H. Weber, M. Carroll, J.J. Hurst and K.G. Lynn. The behavior of electrochemical cell resistance; a possible application to cold fusion experiments, Fusion Tech. **17** (1990) 699−703.

[8] L.S. Darken and H.F. Meier, Conductances of aqueous solutions of the hydroxides of lithium, sodium and potassium at 25°, J. Am. Chem. Soc. **64** (1942) 621−623.

[9] D.N. Bennion, E.L. Littaue, Mathematical-model of a lithium-water electrochemical power cell, J. Electrochem. Soc. **123** (1976) 1462−1469.

[10] C. H. Hamann, A. Hamnett, W. Vielstich. Electrochemistry, 2nd, completely revised and updated edition(2007). John Wiley & Sons.

J. Condensed Matter Nucl. Sci. 36 (2022) 274–284

Research Article

# Optical Observation of Spontaneous Heat Burst Phenomena during Hydrogen Desorption from Nano-Sized Metal Composite

Takehiko Itoh*

*Research Center for Electron Photon Science, Tohoku University, 982-0826 Japan*
*CLEAN PLANET Inc., 105-0022 Japan*

Yoshinobu Shibasaki, Tomonori Takahashi, and Mari Saito

*CLEAN PLANET Inc., 105-0022 Japan*

Jirohta Kasagi and Yasuhiro Iwamura

*Research Center for Electron Photon Science, Tohoku University, 982-0826 Japan*

**Abstract**

We have constructed a photon measurement system to observe the radiant power emitted from an NiCu multilayer film for a wide energy range, from visible light to mid-infrared light. Using the new system, we attempted to measure the radiation from the sample when a heat burst occurred while anomalous excess heat was being continuously generated. About 10 big burst events were observed during the 5-day continuous measurement: it was found that the radiation intensity increased simultaneously as the heater temperature suddenly increased. We have made a comparison of the radiant spectrum of the burst event with the spectrum in normal times. There is no particular difference due to the burst; both spectra can be simply interpreted as grey body radiation emitted from a thermally equilibrated object. The correlation of radiant intensity between near-IR and visible light has been deduced; although it is expected there is a linear correlation between them, two correlation lines can be drawn instead of one. This indicates that the contribution of photon emission from the non-equilibrium stage may have an effect.


*Keywords:* Excess heat, Anomalous heat, Heat burst, Multilayer thin film, Nano-sized metal composite, Nano material, Hydrogen gas, Gas loading, Optical observation

## 1. Introduction

In recent years, in a system consisting of nanostructured metals (Ni, Cu, etc.) and hydrogen (deuterium or hydrogen), anomalous heat generation that produces huge thermal energy (at least one order of magnitude larger than ordinary

*Corresponding author: itoh@lns.tohoku.ac.jp

© 2022 ICCF. All rights reserved. ISSN 2227-3123

chemical reactions) without $CO_2$ emission has been reported [1]–[6]. If power generation technology utilizing this phenomenon can be put into practical use, clean, powerful and inexpensive energy will be obtained, and the social and economic impact will be immeasurable.

We have been studying on this subject since 2015. The following has become clear from our measurements, although systematic basic data are still required.

(1) This heat generation phenomenon largely depends on the substance and structure of the nanostructured metal material [3]–[6].

(2) Surface analysis of nanostructured metal materials suggests that the reaction occurs locally on the surface of the material [3], [7], [8].

In addition to the material issues, we often observe sudden heat generation in the thin film samples during the measurements. We call this a 'heat burst.' The heat burst consists of a simultaneous sudden rise of the temperature of both the heater and the film surface measured by the radiation thermometer. We have interpreted it as follows: there is sudden energy generation in the sample, part of which is dissipated quickly by radiation from the surface, and part of which is propagated to the inside, raising the heater temperature [6]–[8]. In the subsequent measurement of visible light, we observe a sudden increase in light intensity associated with the heat burst [9].

In the experiments performed so far, only the temperature and a small fraction of the heat radiation were measured (with a radiation thermometer), despite the rather high temperature of the sample. The entire radiation spectrum was not observed. We have now improved the experimental equipment to allow for a wide range of radiation spectrum measurements in addition to temperature measurements, as reported in another presentation at ICCF23 [Kasagi et al.]. The measurement has been extended to the visible light, near infrared, and mid-infrared regions so that the whole picture of the radiation can be drawn almost perfectly. Observation of the radiant spectrum as a whole is indispensable for explaining the heat burst phenomenon. For example, when a reaction that locally generates high energy (such as a nuclear reaction) occurs suddenly and the generated energy is transferred to several electrons, the electrons may move in a random direction in the sample and lose their kinetic energy. The small region may locally form a hot thermal equilibrium, the temperature much higher than that heated up only by the heater and, then, one may expect a radiation spectrum with two temperature components [10].

The aim of the present work is to (1) try to observe the heat burst phenomenon with a photon detection system newly developed for wide-region measurement of the radiation, and (2) to explore the mechanism of the burst generation by measuring the time correlation between temperature and radiation, as well as examining the radiant spectrum in detail.

## 2. Experimental Equipment and Procedure

The experiment was performed by improving the reaction chamber used in the optical measurement reported in [9]. The improvements allowed multiple radiation detectors to be attached. The experimental procedure and the sample tested in the present work are basically the same as previously reported in [6]–[9]. We describe only procedures that are important in the present experiment.

### 2.1. Sample - Heater Structure

The metal thin film sample, which is the key to the success or failure of AEH (anomalous excess heat) generation, is installed in a vacuum chamber into which hydrogen gas is introduced. Two sheets of nanostructure samples are set on both sides of the ceramic heater (SAKAGUCHI E.H VOC Corp.), as schematically shown in Fig. 1. The ceramic heater is made of alumina, the size is 25 mm × 25 mm × 0.25 mm, and an R-type thermocouple is placed in the center. As described in [3], [4], [6], [8], the nanostructure metal is formed by alternately depositing Cu and Ni on a Ni plate. The metal is covered and fixed by a holder with a 20 mm$\phi$ hole in the middle and facing the film-forming surface

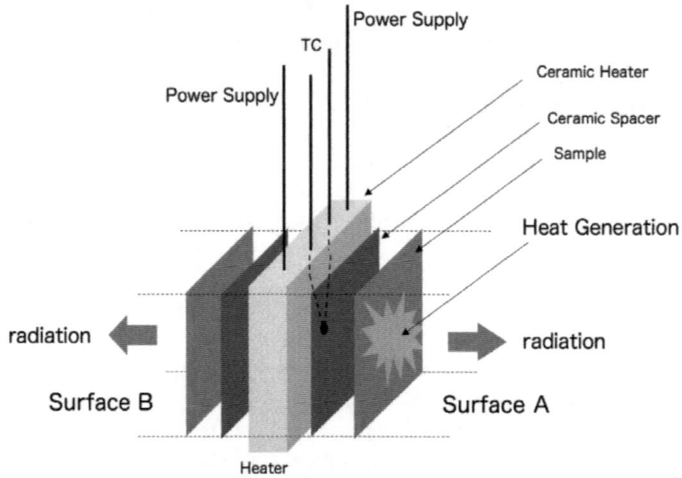


**Figure 1.** Schematic of components around nano-sized multilayer metal composite.

outward. The holder (not shown) is made of ceramic (Photoveel). A 1-mm-thick ceramic (Photoveel II k70) plate is sandwiched between the heater and the sample. In addition to the heater thermocouple for temperature measurement, we set K-type thermocouples on the ceramic holder and the chamber wall.

The heater input was supplied from a regulated DC power supply (PSW-360L80: TXIO TECHNOLOGY Corp.). The current and voltage values are measured with a digital multimeter (GDM-8342: TXIO TECHNOLOGY Corp.) in parallel with capturing the monitor value of the power supply into the data logger, and the input power is cross-checked. The temperature values measured by the thermometers were recorded with a data logger (GL820: Graphtec Corp.) every second.

## 2.2. Sample Preparation

The sample is a NiCu multilayer film which is formed on a Ni substrate plate (size: 25 mm × 25 mm × 0.1 mm, purity: 99% up, Furuuchi Chemical Co.), as follows.

(1) The Ni plate was washed with acetone, alcohol, and ultrapure water in order to remove the oil component.

(2) After that, in order to remove the processing history, annealing was performed at 900°C for 72 hours under a high vacuum ($<10^{-4}$ Pa) condition.

(3) After that, chemical etching was performed using nitric acid to remove impurities on the surface.

(3) Using a magnetron sputtering device, a nano-multilayer film in which 6 layers of Cu: 2 nm and Ni: 7 nm are alternately formed.

Figure 2 shows a STEM image of the cross section of the multilayer film.

## 2.3. Optical Measurement Apparatus

Figure 3 shows a top view of the arrangement of photon detection devices employed in the present work. Two sheets of samples put on both sides of the heater are placed in the center of the vacuum chamber. Photons emitted from the surfaces of the samples are detected outside the vacuum windows. A photodetector TMHK-CLE1350 (wavelength

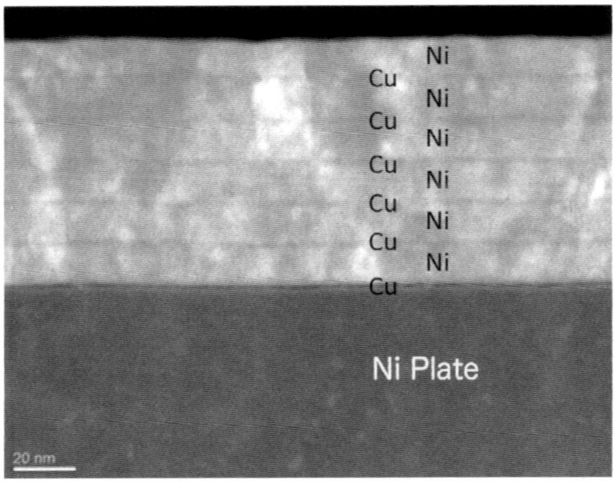


**Figure 2.** STEM (Scanning transmission electron microscope) image of Cu-Ni multilayer thin film.

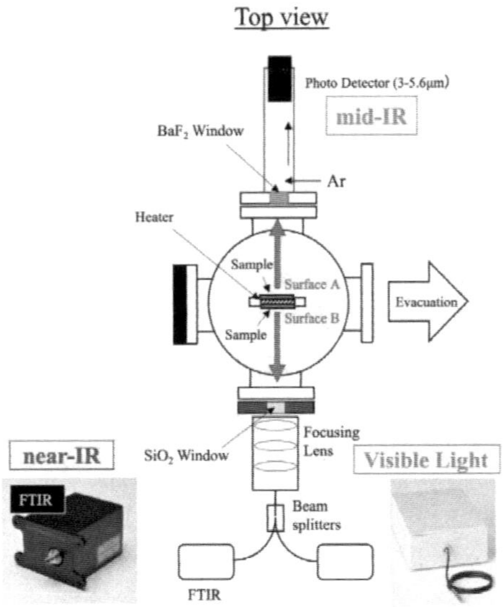


**Figure 3.** Schematic of experimental apparatus. Top view of apparatus and optical measurement device.

3-5.5 $\mu$m photon energy: 0.23-0.41 eV) for mid-infrared measurement is placed on one side of the chamber via a BaF2 window. On the other side through a quartz window two spectrometers are placed, one for near-infrared and the other for visible light; the light emitted from the sample is focused on a fiber by a focus lens and is split by a splitter. Employed are FTIR (Fourier Transform Infrared Spectrometer C15511-01: Hamamatsu Photonics K.K.) to measure 1.5-2.5 $\mu$m light (photon energy: 0.5-0.83 eV) and spectroscope (PMA-012 C10027: Hamamatsu Photonics

K.K.) to measure 0.3-0.9 $\mu$m light (photon energy: 1.4-4.1 eV). Each detector faces two samples by rotating the sample 180 degrees. In the present work, we report the results of long-term observation with the sample fixed without rotation. In the following description, the surface which faces the mid-infrared detector (MIR) will be referred to as Surface A, and the other surface facing to the near-infrared (FTIR) and visible light (Light) spectroscope will be referred to as Surface B.

### 2.4. Experimental Procedure

The NiCu multilayer film was used for the long-term observation. Before the measurement run, the sample was baked, and a measurement without hydrogen was performed as follows:

(1) Mount the two samples in the sample holder and place it in the chamber, which is evacuated to $1\text{x}10^{-7}$Pa or less.

(2) Turn on the heater and keep the heater temperature at 900°C. After about 3 days of baking, turn off the heater.

(3) Reference experiment without hydrogen: Power is applied to the heater without hydrogen absorption, and the heater temperature and radiation are measured.

The observation run with hydrogen was started by letting the sample absorb hydrogen gas. We set the sample temperature to 270-300°C, maintain it under the pressure of hydrogen gas 200 Pa for 15 hours. After that, the evacuation of the vacuum chamber is started at the same time as setting the heater input voltage to the value of the measurement condition.

The heater temperature Tc, radiation thermometer MIR, chamber wall temperature Tw, holder temperature Tpr, heater input voltage and current, and vacuum chamber pressure are constantly read every second and recorded by the logger. In addition to this, spectral measurements in the infrared and visible light region are recorded every 30 minutes. Data collection continues for at least a few days at the voltage once set, since $H_2$ desorption is expected to continue measurably.

## 3. Results & Discussion

Figure 4 shows an example of heat bursts observed through temperature Tc; the CuNi sample was heated to about 900°C, and the heater power is kept constant at about 33 W for 130 hours. In Fig. 4(a), the red line indicates Tc and the black line is the heater input (Pin). As shown, heat bursts with a sharp rise in Tc were frequently without any change in Pin. Figure 4(b) shows the excess power evaluated by a simple method that has been used so far [3], [4]. Since the evaluation is based only on the Tc values, errors of about 30% are expected, but it can be said that bumping on the order of 1 W occurs momentarily at intervals of 10 to 20 hours.

Results of photon observation are shown in Fig. 5. Figure 5 (a) shows the temperature of the heater (red line) and the radiation intensity measured by MIR (green line), for the A side of the sample. Vertical dotted lines with the alphabet above the graph indicate when large bursts occur (heater temperature rises by about 10°C). As expected, the mid-infrared radiation also increases in response at each time.

Figure 5 (b) shows the radiation intensity detected by FTIR and Light spectrometer for the B surface of the sample. The brown square in Fig. 5 (b) shows the near-infrared intensity (integrated intensity from 0.5 to 0.75 eV in the measured 0.3-0.9 eV spectrum) and the blue circle indicates the visible light intensity (integrated intensity from 1.5 to 1.8 eV in the measured 1.5-3 eV spectrum). It can be concluded that heat bursts are observed as a wide range of simultaneous radiation increases, from mid-infrared rays, which mainly contribute to heat radiation, to visible light, which is a high-energy component of heat radiation.

Figure 6 is an enlarged view of Fig. 5 between 60 to 80 h. Figure 6 (a) is for Tc (red line) and MIR (green line) and Fig. 6 (b) for FTIR (brown square) and Light (blue circle). It is shown that not only large bursts of Tc but also small

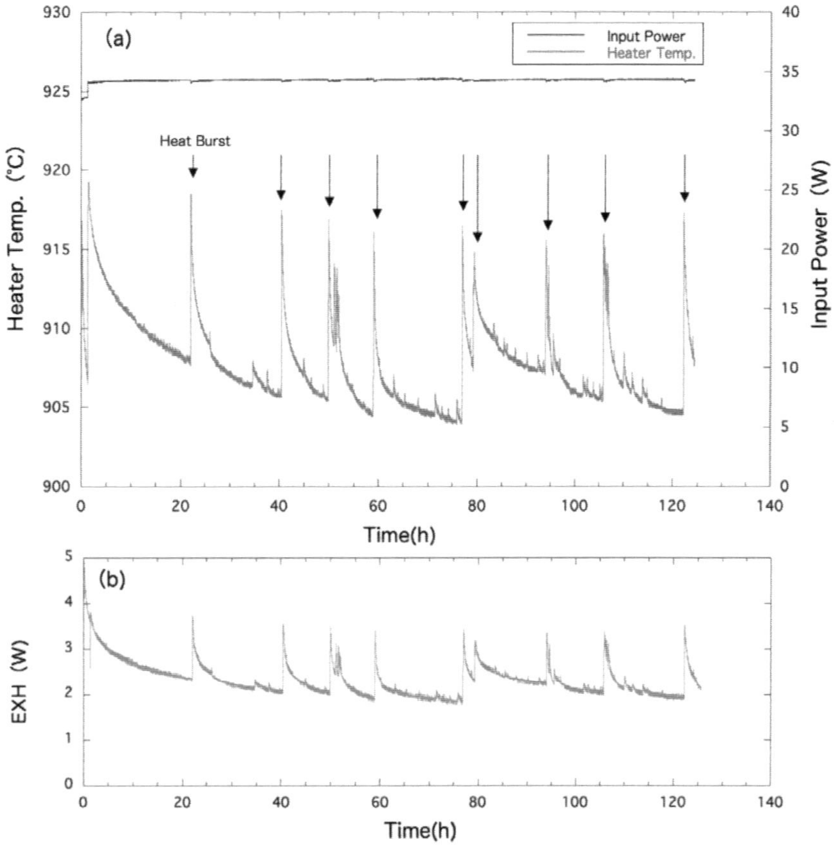


**Figure 4.** Observation of heat bursts: (a) time evolution of heater temperature and input power. (b) Excess heat evaluated from heater temperature.

ones cause an increase in radiant intensity of high-energy photons. The largest burst in this region is the one at 77.6 h. This burst is used to compare the spectrum of burst events with the spectrum in normal times. Such a comparison is made in Fig. 7. Figure 7 (a) shows the FTIR spectrum and Light at time 77.6 h and 77.1 h. The blue dots correspond to data at time 77.1 h, and the red dot 77.6 h. Figure 7 (b) and Fig. 7 (c) are enlarged views of the low energy side (0.55-0.75 eV) and the high energy side (1.6-1.8 eV), respectively. Both spectra can be simply interpreted as grey body radiation emitted from a thermally equilibrated object. We can deduce values of temperature to be 923.1 K for the burst and 915.5 K for the normal time.

Figure 8 shows the correlation between the near-infrared radiant intensity and the visible light intensity. Light intensity is integrated yield between 1.6 eV and 1.8 eV and near IR intensity between 0.5 eV and 0.75 eV. In the case of gray-body radiation, if the emissivity of the surface does not change, the radiant intensities in near-infrared and in visible light are linearly connected in a narrow temperature range as in this measurement. The correlation of radiant intensity shown in Fig. 8 roughly indicates that, but it is better to say that two correlation lines can be drawn instead of one, as shown by the red dotted lines in Fig. 8. This indicates that the contribution of photon emission from the non-equilibrium stage may have an effect.

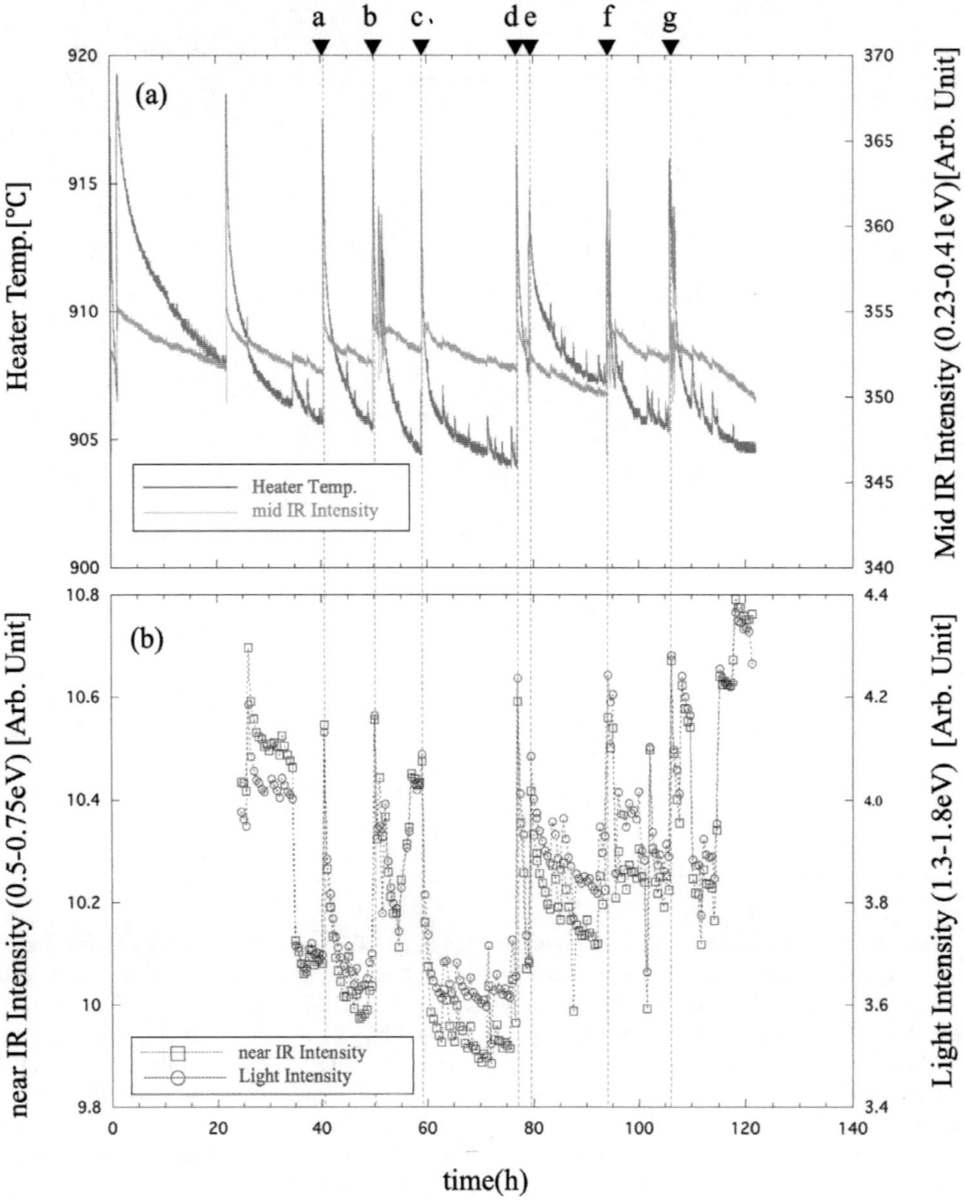


**Figure 5.** Simultaneously detection of heat burst by radiation of spectrometer and a thermocouple located in the center of heater. (a) Heater temperature and Mid-IR intensity, (b) Near IR intensity (integrated yield between 0.25-0.75 eV) and light intensity (integrated yield between 1.3-1.8 eV).

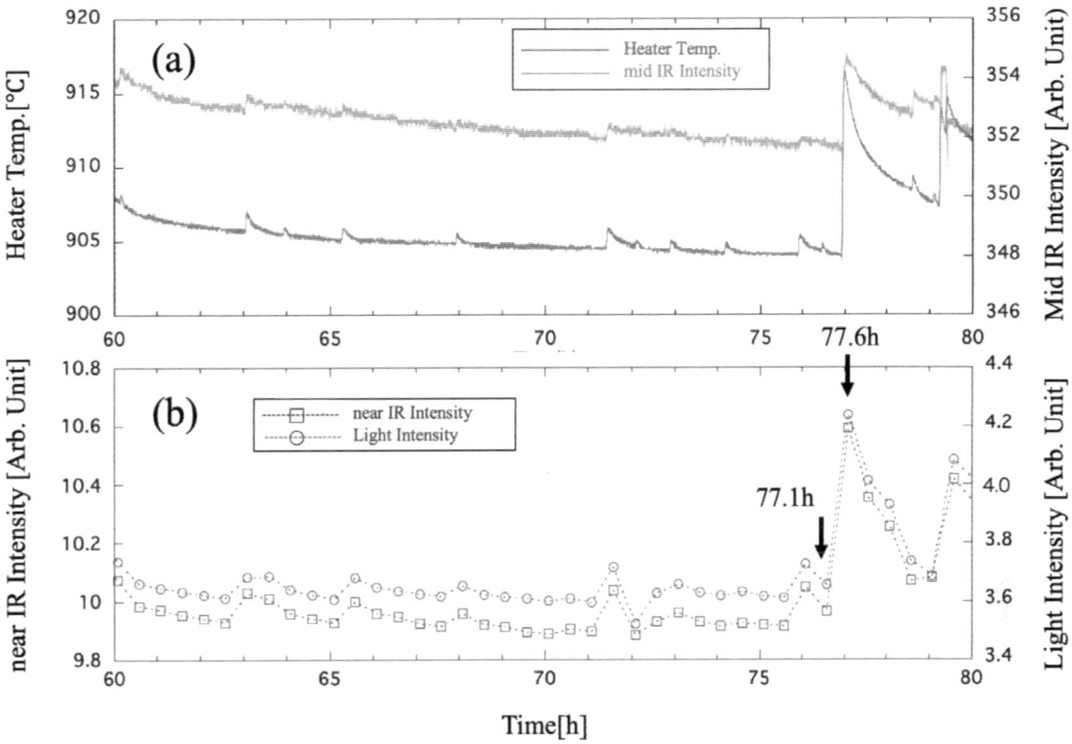


**Figure 6.** Simultaneous measurement of heat bursts from time 60 h to 80 h (a) Heater temperature and Mid-IR intensity, (b) Near IR intensity and light intensity.

The present observation shows that when a heat burst occurs on the sample surface, the radiation in the entire region from the mid-infrared to visible light increases immediately. It is considered that when the heat burst occurs locally on the surface of the sample, part of the power is dissipated immediately from the surface as the radiant power, and at the same time, the generated heat is transferred to the core of the sample to raise the heater temperature. The existence of pre-equilibrium photon emissions is one of the issues to be pursued in the future. Thus, it is highly desirable to investigate the heat burst phenomena in more detail, for example, to obtain precise time correlation between the bursts in heater temperature and in the radiant power, to analyze the shape of the spectrum, and so on. A detailed analysis is in progress.

## 4. Summary

We have constructed a photon measurement system to observe the radiant power emitted from NiCu multilayer film for a wide energy range, from visible light to mid-infrared light.

In the present work, we attempted to measure the radiation from the sample when a heat burst occurred while AEH is being continuously generated. About 10 big burst events were observed during the 5-day continuous measurement. It was found that the radiation intensity increased simultaneously for all detectors as the heater temperature suddenly increased. We have made a comparison of the radiant spectrum of the burst event with the one in normal times. There is

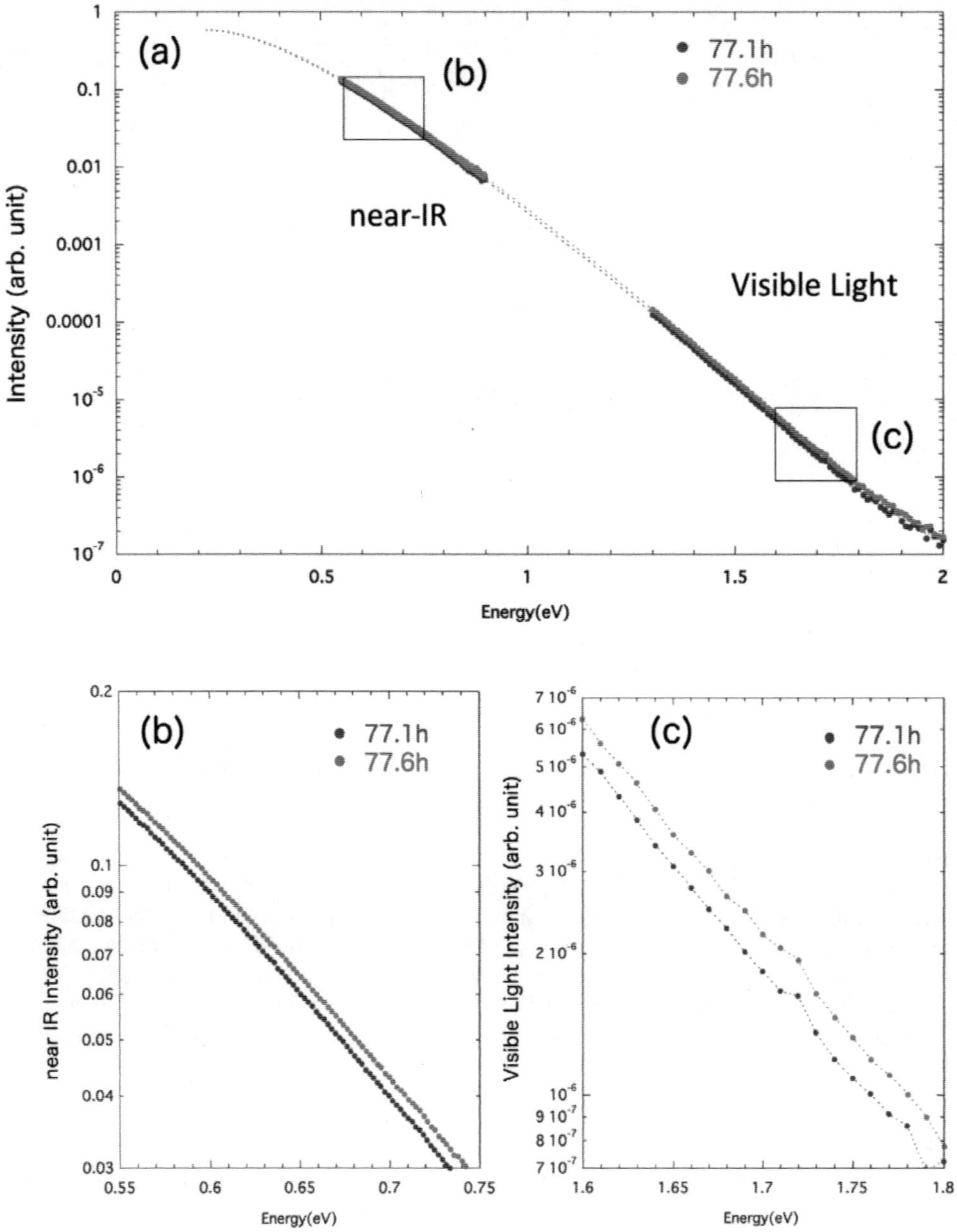


**Figure 7.** Visible light and near-infrared spectra at time 77.1 h (immediately before heat bursts) and 77.6 h (when heat bursts occur).

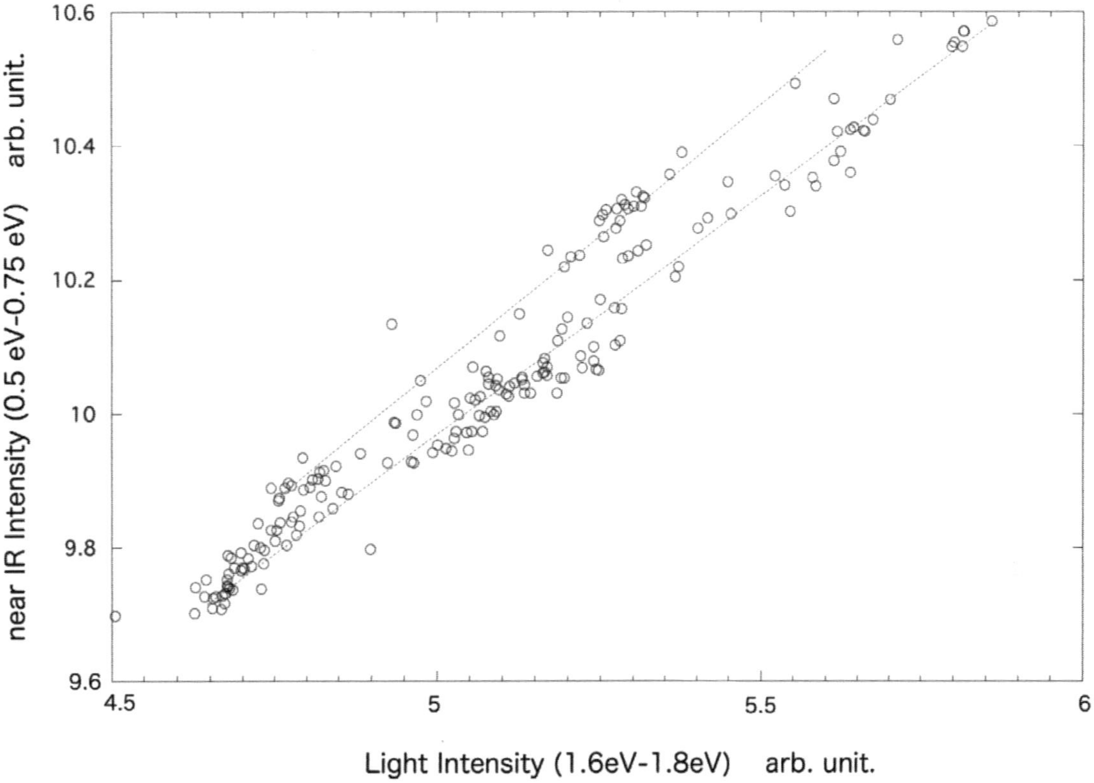


**Figure 8.** Correlation between light intensity and near-infrared intensity.

no particular difference due to the burst; both spectra can be simply interpreted as grey body radiation emitted from a thermally equilibrated object.

The correlation of radiant intensity between near-IR and visible light has been deduced; although it is expected a linear correlation between them, two correlation lines can be drawn instead of one. This indicates that the contribution of photon emission from the non-equilibrium stage may have an effect.

It is highly desirable to investigate the heat burst phenomena in more detail, for example, to obtain precise time correlation between the bursts in heater temperature and in the radiant power, to analyze the shape of the spectrum, and so on. A detailed analysis is in progress.

## 5. Acknowledgements

This work is partially supported by CLEAN PLANET Inc. The authors acknowledge also the members of CLEAN PLANET Inc., Mr. H. Yoshino, Mr. S. Hirano, and Mr. H. Shimokawa, for their significant assistance.

## References

[1] A. Kitamura, A. Takahashi, R. Seto, Y. Fujita, A. Taniike and Y. Furuyama, "Effect of Minority Atoms of Binary Ni-Based Nano-Composites on Anomalous Heat Evolution under Hydrogen Absorption", *J. Condensed Matter Nucl. Sci.*, **19** (2016) 1-10.

[2] A. Kitamura, A. Takahashi, K. Takahashi, R. Seto, T. Hatano, Y. Iwamura, T. Itoh, J. Kasagi, M. Nakamura, M. Uchimura, H. Takahashi, S. Sumitomo, T. Hioki, T. Motohiro, Y. Furuyama, M. Kishida, H. Matsune, "Excess heat evolution from nanocomposite samples under exposure to hydrogen isotope gases", *International Journal of Hydrogen Energy* **43** (2018) 16187-16200.

[3] Y. Iwamura, T. Itoh, J. Kasagi, A. Kitamura, A. Takahashi, K. Takahashi, R. Seto, T. Hatano, T. Hioki, T. Motohiro, M. Nakamura, M. Uchimura, H. Takahashi, S. Sumitomo, Y. Furuyama. M. Kishida and H. Matsune, Anomalous Heat Effects Induced by Metal Nano-composites and Hydrogen Gas, *J. Condensed Matter Nucl. Sci.* **29** (2019) 119-28.

[4] S. Murakami, T. Itoh, Y. Iwamura, M. Saito and J. Kasagi, "Excess Energy Generation Experiments using a Nano-sized Multilayer Metal Composite and Hydrogen Gas", *Proceedings of the 20th Meeting of Japan CF Research Society*, JCF20, December 13-14, 2019, Reference Hakata Eki-Higashi Rental Room, Fukuoka, Japan p.86-96.

[5] M. Saito, T. Itoh, Y. Iwamura, S. Murakami and J. Kasagi, "Elemental Analysis for Elucidation of the Anomalous Heat Generation Phenomena", *Proceedings of the 20th Meeting of Japan CF Research Society*, JCF20, December 13-14, 2019, Reference Hakata Eki-Higashi Rental Room, Fukuoka, Japan p.74-85.

[6] Y. Iwamura, T. Itoh, J. Kasagi, S. Murakami and M. Saito, "Excess Energy Generation using a Nano-sized Multilayer Metal Composite and Hydrogen Gas", *J. Condensed Matter Nucl. Sci.* **33** (2020) 1-13.

[7] Y. Iwamura, T. Itoh, M. Saito, S. Murakami and J. Kasagi, "Evidence for Surface Heat Release Reaction over Nano-sized Multilayer Metal Composite with Hydrogen Gas", *Proceedings of the 21st Meeting of Japan CF Research Society*, JCF21, December 11-12, 2020, Virtual Meeting, p.1-p14.

[8] Y. Iwamura, T. Itoh, J. Kasagi, M. Saito, T. Takahashi, Y. Shibasaki, and M. Murakami, "Progress in Energy Generation Research using Nano-Metal with Hydrogen/Deuterium Gas", *Proceedings of the 23rd Meeting of International Conference on Condensed Matter Nuclear Science, Xiamen, China*, to be published.

[9] Takehiko Itoh, Yoshinobu Shibasaki, Jirohta Kasagi, Shouichi Murakami, Mari Saito and Yasuhiro Iwamura, "Optical Observation on Anomalous Heat Generation from Nano-sized Metal Composite", *Proceedings of the 21st Meeting of Japan CF Research Society*, JCF21, December 11-12, 2020, Virtual Meeting, p.15-25.

[10] J. Kasagi, "Possible radiation from thin metallic multilayers with hydrogen-permeated anomalous excess heat: Can we observe hot spots or Bremsstrahlung?", *Abstract of ICCF22* 8-13 Sept 2019 Assisi Italy p.10

J. Condensed Matter Nucl. Sci. 36 (2022) 285–301

Research Article

# Progress in Energy Generation Research Using Nano-Metal With Hydrogen/Deuterium Gas

Yasuhiro Iwamura* and Jirohta Kasagi
*Research Center for Electron Photon Science, Tohoku University, 982-0826 Japan*

Takehiko Itoh
*Research Center for Electron Photon Science, Tohoku University, 982-0826 Japan*
*CLEAN PLANET Inc., 105-0022 Japan*

T. Takahashi, Mari Saito, Y. Shibasaki, and Shoichi Murakami
*CLEAN PLANET Inc., 105-0022 Japan*

**Abstract**

The field of condensed matter nuclear science originated in 1989 with Fleischmann and Pons' electrolysis of Pd with heavy water. However, anomalous heat generation and other related phenomena were observed not only by electrochemical methods, but also by loading of deuterium or hydrogen in metals such as Pd and Ni. In this paper, we will first give a short overview of the progress in these gas loading type experiments. After that, we will describe the recent progress of our research team. The authors have been studying energy generation using nano-sized multilayer metal composites with hydrogen gas. Two nano-sized metal multilayer composite samples, which were composed of Ni, Cu, and other thin films on bulk Ni (25 mm $\times$ 25 mm $\times$ 0.1 mm), were used. These samples were fabricated by the Ar ion beam method, or by the magnetron sputtering method. Heat bursts and excess energy generation were observed during the experiments under vacuum conditions ($<10^{-4}$ Pa) using nano-sized metal multilayer composites on Ni substrate and hydrogen gas. Released energy normalized to the total amount of absorbed hydrogen reached 16 keV/H or 1.5 GJ/H-mol. Sometimes spontaneous heat bursts were observed. This suggests that the burst heat release reactions occurred in the near surface region of the nano-sized multilayer metal composite. Furthermore, we have succeeded in intentionally inducing heat bursts, based on the observations of the spontaneous heat bursts. By measuring the optical spectrum emitted from the sample, we detected a simultaneous increase in mid infrared radiation flux when a heat burst occurred.


*Keywords:* Excess heat, Anomalous heat, Heat burst, Multilayer thin film, Nano-sized metal composite, Nano material, Hydrogen gas, Gas loading

*Corresponding author: iwamura@lns.tohoku.ac.jp

© 2022 ICCF. All rights reserved. ISSN 2227-3123

## 1. Introduction

The field of condensed matter nuclear science originated in 1989 with Fleischmann and Pons' electrolysis of Pd with heavy water [1]. However, some researchers started to do different experiments, especially gas loading experiments. Figure 1 shows an overview of the gas loading experiments. De Ninno, Scaramuzzi of ENEA tried to detect neutrons or protons from deuterated Ti [2]. According to the book "Cold Fusion; Advances in Condensed Matter Nuclear Science", Fralick of NASA first did excess heat experiments using Deuterium gas and Pd [3]. Li and Biberian also observed excess heat using Pd and deuterium gas [3]–[4]. Piantelli and Forcadi first did experiments using Ni and Hydrogen [5]. Francesco Celani has been making excess heat experiments with his original method of high voltage pulses to Pd and constantan wires [6]. Brillouin is attempting to commercialize a pulsing method called Q-pulse using Ni and light hydrogen [7].

These studies do not specifically consider the nanoscale. However, the following studies were based on nanoscale materials in addition to gas loading. Arata firstly demonstrated long-term excess heat data using Pd black and Deuterium [8]. Iwamura firstly showed transmutation reactions using deuterium gas with Pd nano-film [9]. Based on these studies, Mizuno, Takahashi, Kitamura, Iwamura and their colleagues made further progress. A Collaborative Research Project between six Japanese organizations funded by NEDO (New Energy and Industrial Technology Development Organization) on anomalous heat effects was done from Oct. 2015 to Oct. 2017 using Ni, Pd, Cu, Zr nanoparticles. Anomalous heat generation, which is too much to be explained by any known chemical process, was observed. Qualitative reproducibility was confirmed between Kobe University and Tohoku University [10]–[11]. The authors replicated the experiments using nano Pd/Ni fabricated by glow discharge with $D_2$ gas developed by Mizuno [12]. In these experiments, nano-sized particles and diffusion of hydrogen and deuterium were one of the key factors to produce the heat effects.

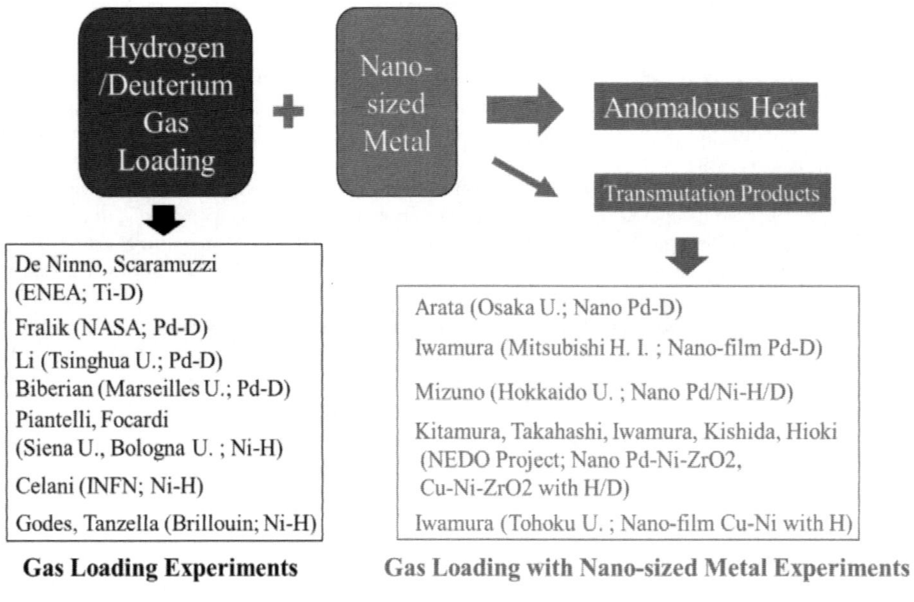


**Figure 1.** Overview of gas loading experiments and gas loading with nano-sized metal experiments in this field.

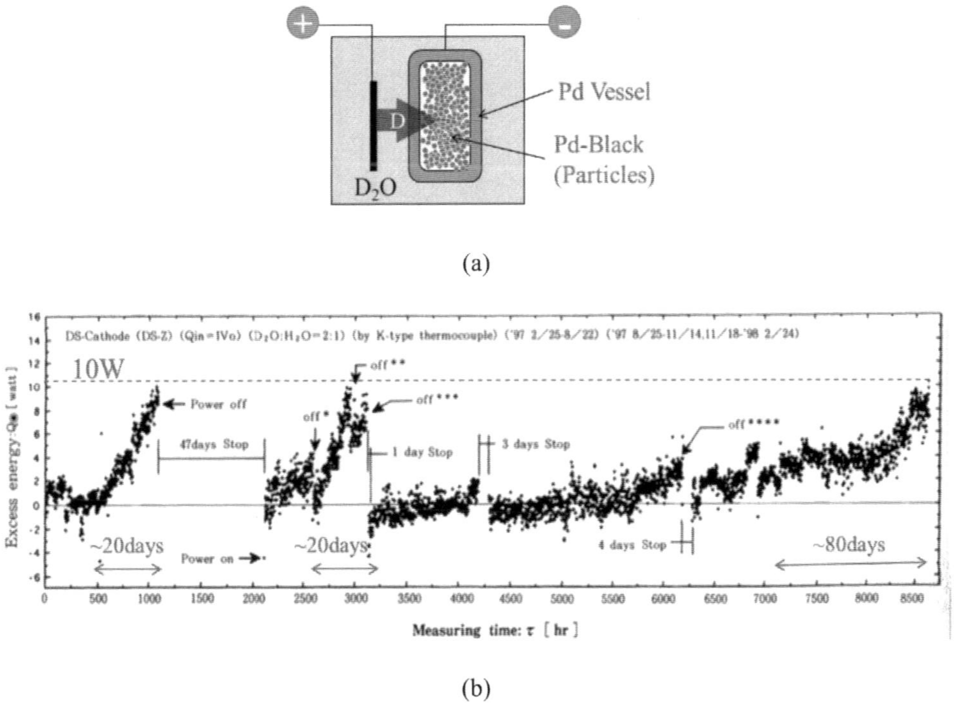


**Figure 2.** Arata experiment done with gas loading and Pd black particles [8].

Figure 2 shows the first experiment done with gas loading and Pd black particles. In this experiment, as shown in Fig. 2(a), Pd-black particles are packed in a Pd vessel and electrolysis is carried out in heavy water, which results in deuterium gas loading into Pd-black. In this process, heat generation is observed for long durations, as shown in Figure 2(b). This heat generation was evaluated by performing the same experiment with light water electrolysis at the same time, using the light water experiment as a control experiment.

Collaborative research between Technova Inc., Nissan Motor Co. Ltd., Kobe Univ., Kyushu Univ., Nagoya Univ. and Tohoku Univ. was done from Oct. 2015 to Oct. 2017 funded by NEDO (New Energy Development Organization) in Japan. An example of a NEDO experiment with gas loading and metal nanocomposites is shown in Fig. 3. Heat generation experiments using nano-sized metal composite supported by zirconia or by silica with $H_2$ or $D_2$ gas were done. Anomalous excess heat generation were observed for all the samples at elevated temperature (150°C-350°C), except for the Pd nanoparticles embedded in mesoporous $SiO_2$. The amount of anomalous heat generation per hydrogen atom ranged from 10 eV/H or D to 100 eV/H or D, which could not be explained by any known chemical process [10]. Excess heat experiments using the same material at Kobe and Tohoku Universities showed similar experimental results. Coincident burst-like increase events of the pressure of reaction chamber and gas temperature, which suggested sudden energy releases in the reaction chamber, were observed under specific conditions [11].

Next, we describe the recent results of our research team. Initially, one of us (Iwamura) observed transmutation reactions induced by deuterium gas permeation through nanosized Pd multilayers doped with Cs and other elements [9]. In this study, for example, we observed the transmutation of Cs into Pr, which has been reproduced by

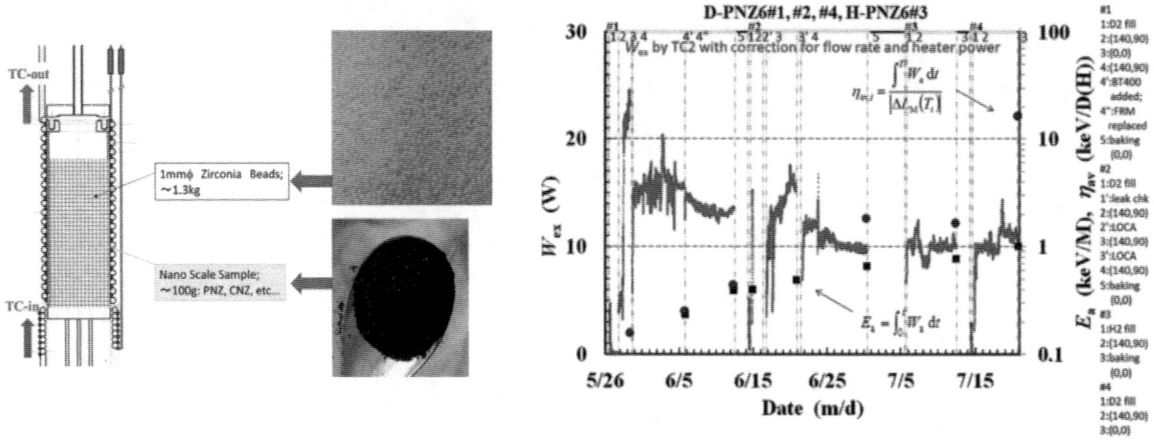


(a) (b)

**Figure 3.** NEDO experiment with gas loading and Metal Nanocomposites [8].

other independent research institutes. This study shows that the diffusion of deuterium and nano-sized multilayers are key factors in inducing the transmutation reaction [9], [12]–[13]. In addition, from the results of the NEDO project mentioned in Fig. 3, we learned that the anomalous phenomena occur not only with deuterium but also with light hydrogen, and that the nano-sized composite particle is one of the key factors to inducing the anomalous phenomena [10]–[11]. By combining the two methods, we have developed a new type of experimental procedure to induce an exothermic reaction by loading light hydrogen in a nickel-based composite material with nano-sized multilayers, and then diffusing the light hydrogen through the nano-multilayers by rapid heating. This new experimental method gave us larger excess energy per H atom [15]–[18].

Recently the author has decided to name the heat generation phenomenon in the system of nano-sized composite metallic materials and hydrogen as QHE (Quantum Hydrogen Energy). The reason for this is that condensed matter nuclear reactions are associated with a complicated and dangerous image that is difficult for the general public to accept.

QHE can be defined as “An exothermic reaction induced by quantum phenomena during the diffusion process in a nanoscale metal composite material with hydrogen”. The characteristics of QHE are as follows: 1) No $CO_2$ emission and more than 1000 times higher output energy than the combustion reaction of the same amount of hydrogen; 2) QHE does not emit harmful levels of radiation to the human body and has the potential to become a compact, high-power, $CO_2$-free energy source; 3) QHE could be used as an energy source for the production of electricity and heat.

Figure 6 summarizes the progress from the time of the discovery to the present. The most significant differences between Cold Fusion and QHE are in materials and operating temperatures. In the early stage, expensive materials such as Pd and deuterium were mainly used. However, now we are using nickel and ordinary hydrogen, which is more cost effective. In the beginning, the electrochemical method was mainly used. Electrochemistry is an excellent method for packing deuterium into Pd metal at high density. However, from the perspective of using the excess heat generated for practical purposes, a temperature increase of a few degrees in room temperature heavy water does not have much practical impact, although it was an epoch-making event from a scientific point of view. QHE enables us to obtain heat in the temperature range of several hundred degrees, which makes it possible to use high-temperature gas to turn turbines, which will have an impact on industrialization. Recently, there has been an increasingly strong

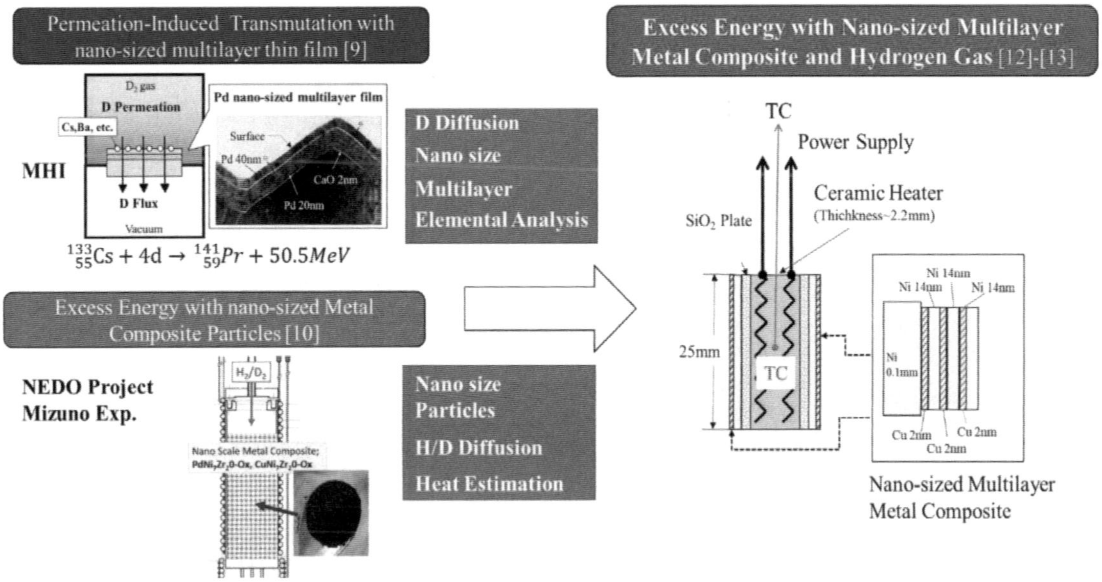


**Figure 4.** Background and motivation of the present experimental method.

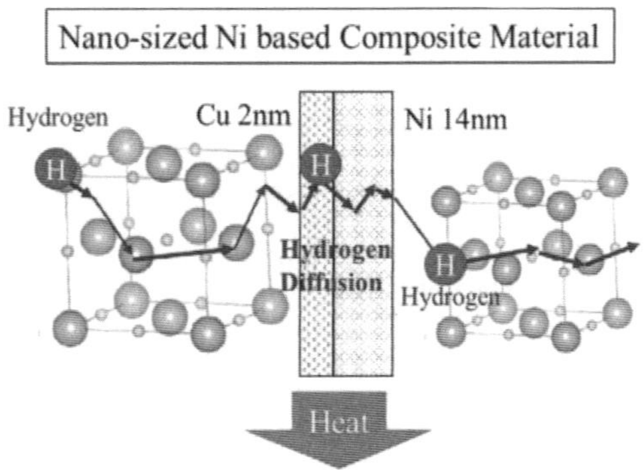


**Figure 5.** Quantum Hydrogen Energy (QHE).

demand around the world for energy sources that do not emit $CO_2$ to prevent global warming. QHE using nano-metal and hydrogen could be just the right technology to meet this global social demand.

## 2. Experimental

A schematic of our experimental apparatus is shown in Fig. 7(a). It is basically the same as in Ref. [17], with improvements in some points [15]–[16]. Two nano-sized metal multilayer composite samples, which were composed of Ni, Cu, CaO thin films on bulk Ni (25 mm × 25 mm × 0.1 mm), placed in the center of the chamber. $H_2$ gas and its

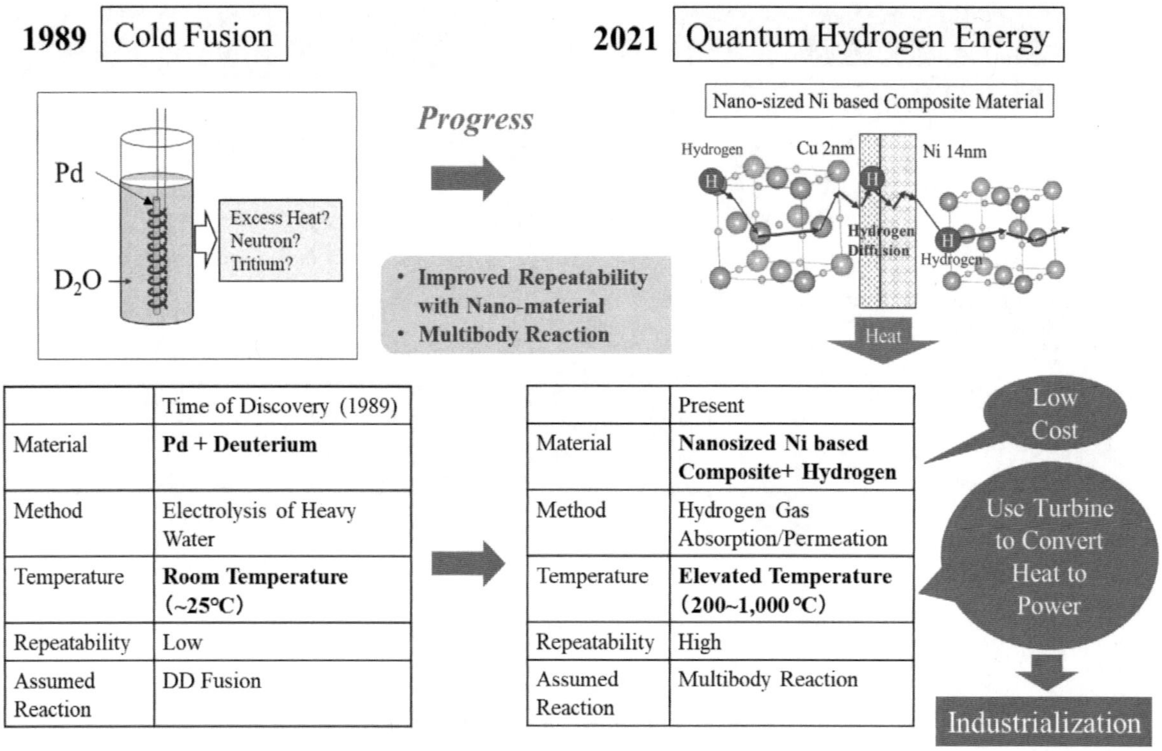


| | Time of Discovery (1989) |
|---|---|
| Material | **Pd + Deuterium** |
| Method | Electrolysis of Heavy Water |
| Temperature | **Room Temperature (~25°C)** |
| Repeatability | Low |
| Assumed Reaction | DD Fusion |

| | Present |
|---|---|
| Material | **Nanosized Ni based Composite+ Hydrogen** |
| Method | Hydrogen Gas Absorption/Permeation |
| Temperature | **Elevated Temperature (200~1,000 °C)** |
| Repeatability | High |
| Assumed Reaction | Multibody Reaction |

**Figure 6.** Progress from the time of discovery (1989) to the present.

pressure were monitored by a Pirani gauge. The chamber was evacuated by a turbo molecular pump. The multilayer samples were heated up by a ceramic heater (MS-1000R; Sakaguchi E. H Voc Corp.) in which a thermocouple (TC; Pt-PtRh13%) was embedded.

In the papers [15]–[16], the surface temperature of a sample was measured by an infrared radiation thermometer (IR-CAQ3CS; Chino Corp.). Now, we are able to measure surface temperatures for the two nano-sized metal multilayer composite samples by introducing two thermometer detectors. They were made of InGaAs and dual wavelength mode, 1.55 $\mu$m and 1.35 $\mu$m, were usually used. During the surface temperature measurement, it was possible to measure the emissivity of the sample surface by switching between single wavelength mode and dual wavelength mode. Heater input power was supplied by a DC power source in constant voltage mode. The input voltage and current were measured both by voltage and current monitors provided by the power supply and an independent voltmeter and amperemeter, respectively. Input power is calibrated using the voltmeter and the amperemeter readings. Gamma-rays and neutrons were monitored by a NaI (Tl) scintillation counter (TCS-1172; Hitachi, Ltd.) and He-3 counter (TPS-1451; Hitachi Ltd.) during all experiments for safety reasons.

A detailed drawing of the Ni based nano-sized metal multilayer composite is shown in Fig. 7(b). It was composed of a Ni Plate (25 mm square and 0.1 mm thick) and Cu-Ni multilayer thin film (20 mm diameter circle and about 100 nm thick). These samples were fabricated by the Ar ion beam method, or the magnetron sputtering method. Two nano-sized metal multilayer composite samples were heated by the ceramic heater (25 mm square and 2.2 mm

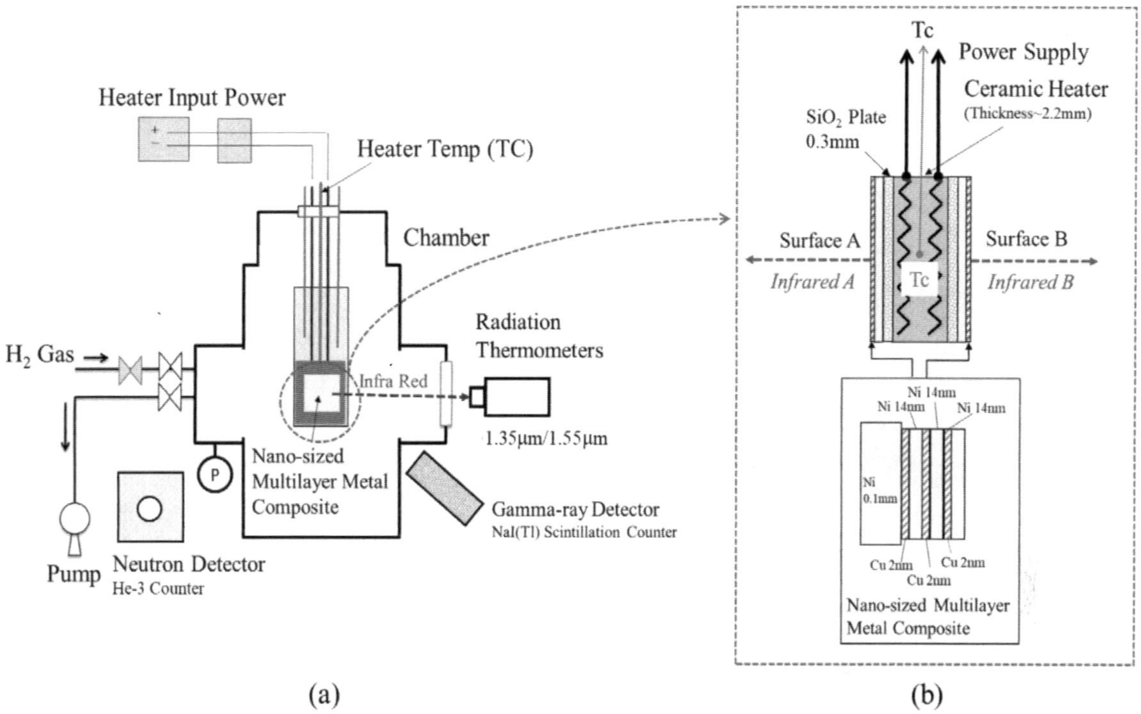


**Figure 7.** Experimental set-up; (a) Schematic of experimental apparatus, (b) Detail drawing around nano-sized multilayer metal composite.

thick) through $SiO_2$ plates (0.3 mm thick). If certain energy generation reactions occur on the surface of the samples, the temperature of the embedded thermocouple will rise. Simultaneously, infrared emission detected by the radiation thermometer, which corresponds to the surface temperature of the sample, would increase. Photos of the experimental set-ups and STEM image of Cu-Ni multilayer thin film are shown in Fig. 8.

The experimental procedure is as follows. Two nano-sized metal multilayer composites were placed in the chamber and baked for 2-3 days at heater temperature 900°C to remove $H_2O$ and other hydrocarbons from the surface under vacuum conditions.

According to general knowledge, Cu and Ni diffuse into each other during the baking process, forming a Cu-Ni alloy. However, in our multilayer film, Cu and Ni are not simply alloyed, especially when CaO, $Y_2O_3$, etc. are added to Ni. The cause of this phenomenon is still under investigation, but it might be due to the effect of oxygen formed on the surface during the sputtering process.

After the baking process, $H_2$ gas was introduced into the chamber up to about 200 Pa at 250°C. To change the hydrogen loading conditions, the pressure of $H_2$ gas was sometimes increased up to 30 kPa. $H_2$ gas was loaded for about 16 hours. Then, $H_2$ gas was evacuated by the turbo molecular pump and simultaneously the samples were heated up by the ceramic heater up to 500∼900°C. These processes trigger heat generation reactions and anomalous heat. Typically, after 8 hours, the heater input was turned down and the samples were cooled down to 250°C. These processes ($H_2$ loading, heating up and cooling down samples) were repeated several times, with different heating temperatures or $H_2$ loading pressure.

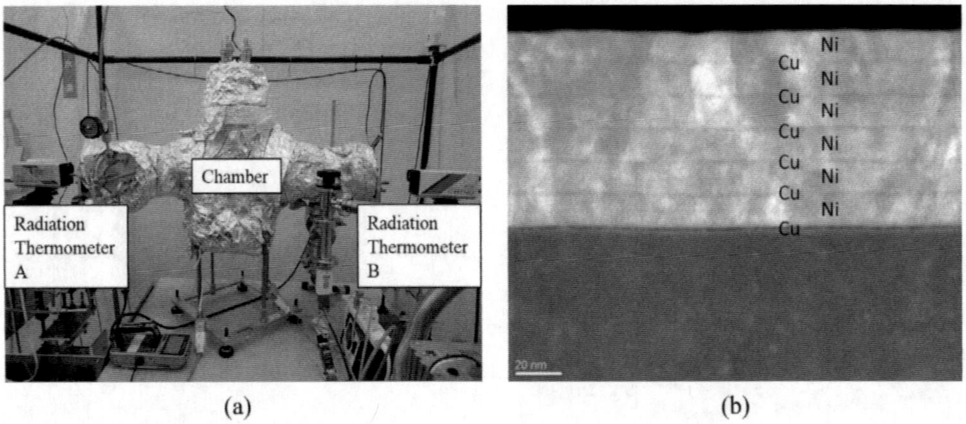


**Figure 8.** Photos of experimental apparatus and cross-sectional view of nano-sized metal multilayer composites; (a) Outer view of the present experimental Set-ups, (b) STEM (Scanning transmission electron microscope) image of Cu-Ni multilayer thin film.

During the above experimental procedure, hydrogen atoms are assumed to diffuse from the Ni plate through the nano-sized metal multilayer to the surface. The diffusion mechanism of hydrogen atoms is well known as "quantum diffusion" [19]. Hydrogen atoms are hopping from a site to another site in metal. We assume that hydrogen flux is one of the key factors to induce the anomalous heat generation phenomena, and the hydrogen flux is intentionally arranged by the present experimental method. Hydrogen flux $\boldsymbol{J}$ from the nano-sized metal multilayer composite to the chamber is caused by a gradient of hydrogen concentration and gradient of temperature as shown in eq. (1) [20].

$$\boldsymbol{J} = -\mathrm{nD}\left(\nabla \mathrm{c} + \frac{cQ^{*}\nabla T}{k_B T^2}\right) \tag{1}$$

where n is the number of lattice atoms per unit volume, c is hydrogen concentration defined as the hydrogen/host-metal atom ratio, D is diffusion coefficient and Q* is the heat of transport.

## 3. Results and Discussion

### 3.1. Excess Heat Evaluation

Excess Heat is evaluated based on the model described in Fig. 9(a) and the following equation (2).

$$k_{eff}\frac{T_C - T_W}{L_{eff}}A_{eff} + A_S\sigma\left\{\varepsilon_A\left(T_{SA}^4 - T_W^4\right) + \varepsilon_B\left(T_{SB}^4 - T_W^4\right)\right\} + A_{Rloss}\varepsilon_{Rloss}\sigma\left(T_{Rloss}^4 - T_W^4\right) = P_{in} + H_{ex}, \tag{2}$$

where $k_{eff}$ is equivalent thermal conductivity, $T_c$ is the thermocouple temperature embedded in the ceramic heater, $T_w$ is the wall temperature of the chamber, $L_{eff}$ and $A_{eff}$ are effective length and effective surface area between the sample holder and wall, respectively. $A_s$ is the surface area of the sample, $T_S$ is the surface temperature, $\varepsilon$ is the emissivity of the sample, $\sigma$ is the Stefan–Boltzmann constant. subscript A and B means surface A and B, respectively. $A_{Rloss}$, $\varepsilon_{Rloss}$ and $T_{Rloss}$ are effective surface area; effective emissivity and effective surface temperature for radiation loss except from the sample surface, which is mainly derived from the sample holder. $P_{in}$ is the electrical heater input and $H_{ex}$ is excess power. This equation is obtained under the following assumptions.

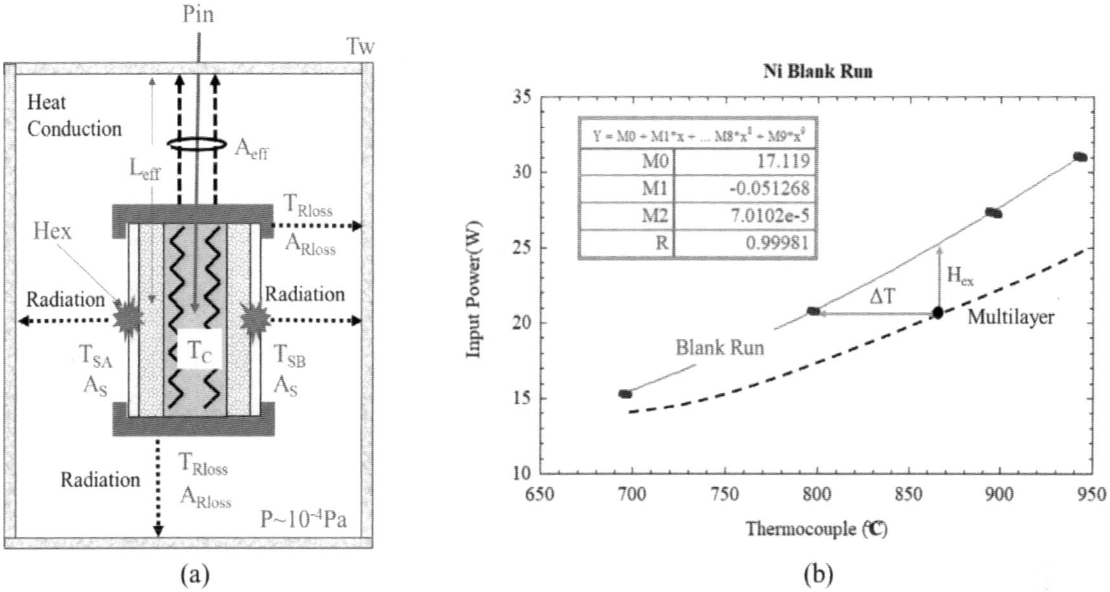


**Figure 9.** Excess heat evaluation; (a) Model of excess heat evaluation, (b) Relationship between input power (W) and thermocouple temperature (°C) for a blank run and a multilayer run.

1) Thermal conduction via $H_2$ gas is negligible because $H_2$ pressure is low.
2) Radiation from the chamber wall is negligible because $T_w$ is room temperature.
3) The electrical input power is constant.

A blank run, in which same sized Ni bulk samples without multilayer thin films were used, was performed with the same procedure described above. Figure 3(b) shows the relationship between input power given to the ceramic heater and heater temperature detected by the thermocouple. As the radiation loss term from the sample holder is the same for Ni bulk and multilayer samples for the same temperature, generated excess heat power is evaluated based on the blank run result as shown in Fig. 3(b). Equation (2) for Ni bulk (subscript "0") is written as:

$$k_{eff}\frac{T_{C0}-T_W}{L_{eff}}A_{eff}+A_S\sigma\left\{\varepsilon_{A0}\left(T_{SA0}^4-T_W^4\right)+\varepsilon_{B0}\left(T_{SB0}^4-T_W^4\right)\right\}$$
$$+A_{Rloss}\varepsilon_{Rloss}\sigma\left(T_{Rloss0}^4-T_W^4\right)=P_{in}. \qquad (3)$$

In the papers [15]–[16], excess heat analysis was done based on the assumption that $\varepsilon$ is constant for Ni based nano-sized metal multilayer composite and Ni bulk as a first step of data analysis. Emissivity $\varepsilon$ can be measured by switching between two wavelengths mode and single wavelength mode. Actual measured emissivity was in the range of 0.1-0.2 at surface temperature 700-750°C, depending on the condition of the sample such as oxidation of surface or surface roughness, the vacuum of the experimental apparatus. However, the difference in emissivity between Ni bulk and Ni multilayer composite samples is within 0.05 ($<5\%$), and the assumed condition is satisfied. Emissivity can be considered almost the same for the Ni bulk and multilayer composite samples.

Excess heat $H_{ex}$ is written based on the equations (2) for multilayer composite and Ni bulk (subscript "0") samples.

$$\begin{aligned} H_{ex} &= k_{eff}\frac{T_C - T_{C0}}{L_{eff}}A_{eff} + A_S\sigma\left\{\varepsilon_A T_{SA}^4 - \varepsilon_{A0}T_{SA0}^4 + \varepsilon_B T_{SB}^4 - \varepsilon_{B0}T_{SB0}^4 - T_W^4\left(\varepsilon_A - \varepsilon_{A0}\right)\right\} \\ &+ A_{Rloss}\varepsilon_{Rloss}\sigma\left(T_{Rloss}^4 - T_{Rloss0}^4\right) \\ &\approx k_{eff}\frac{T_C - T_{C0}}{L_{eff}}A_{eff} + A_S\sigma\left\{\varepsilon_A T_{SA}^4 - \varepsilon_{A0}T_{SA0}^4 + \varepsilon_B T_{SB}^4 - \varepsilon_{B0}T_{SB0}^4\right\} + A_{Rloss}\varepsilon_{Rloss}\sigma\left(T_{Rloss}^4 - T_{Ploss0}^4\right) \\ &\because T_{SA,B}^4 \gg T_W^4 \end{aligned} \tag{4}$$

Here, we assume the following relations based on our experimental data. $T_{SA}$ and $T_{SB}$ can be expressed as a linear function of $T_c$ within the experimental parameters.

$$\varepsilon_A \sim \varepsilon_{A0},\ \varepsilon_B \sim \varepsilon_{B0}, T_{SA} \sim \alpha_A T_C + \beta_A,\ T_{SB} \sim \alpha_B T_C + \beta_B,\ T_{Rloss} \sim \alpha_{Rloss}T_C + \beta_{Rloss}.$$

$\Delta T$ is defined as

$$\Delta T = T_c - T_{c0}. \tag{5}$$

Therefore, the following expression is obtained.

$$\begin{aligned} H_{ex} \approx \Delta T &\left\{\frac{k_{eff}}{L_{eff}}A_{eff} + A_S\varepsilon_{A0}\sigma\alpha_A\left(T_{SA} + T_{SA0}\right)\left(T_{SA}^2 + T_{SA0}^2\right)\right. \\ &\left. + A_S\varepsilon_{B0}\sigma\alpha_B\left(T_{SB} + T_{SB0}\right)\left(T_{SB}^2 + T_{SB0}^2\right) + A_{Rloss}\varepsilon_{Rloss}\sigma\left(T_{Rloss} + T_{Rloss0}\right)\left(T_{Rloss}^2 + T_{Rloss0}^2\right)\right\}. \end{aligned} \tag{6}$$

This equation shows that excess heat can be written as a function of $\Delta T$. Therefore, excess heat evaluation by the Ni bulk calibration curve shown in Fig. 3(b) is valid under the assumptions described above.

### 3.2. Excess Heat Generation

An Example of excess heat generation presented at ICCF22 is shown in Fig. 10 [15]. In this case, the sample which covered 6 CuNi layers (Cu = 2 nm and Ni = 14 nm) were used. Red and blue lines represent excess heat and pressure of the chamber, respectively. At the beginning of the experiment, hydrogen gas was introduced to the chamber and absorbed into the Ni based nano-sized multilayer metal composite at 250°C. The pressure for each experiment gradually decreased as shown in Fig. 10. The amount of hydrogen absorbed by each sample was estimated based on the pressure change and temperature of the chamber. After about 16 hours, $H_2$ gas was evacuated and simultaneously each sample was heated up by the ceramic heater. After that, excess heat more than input power was observed. The input power was stable during a one cycle; for example, 19 W was applied to the ceramic heater from about $6 \times 10^4$ sec to about $1 \times 10^5$ sec.

Released excess energy per hydrogen was evaluated based on the experimental result. The amount of absorbed hydrogen, total excess energy and excess energy per absorbed hydrogen are shown in Fig. 10. The amount of excess energy was calculated as 1.1 MJ by time integration of excess power. Although it seems highly unlikely that all the absorbed hydrogen atoms reacted, we can still estimate that average released energies per absorbed total hydrogen was 16 keV/H. Obviously, the released excess energy per hydrogen atom is too large to be explained by any known chemical reactions.

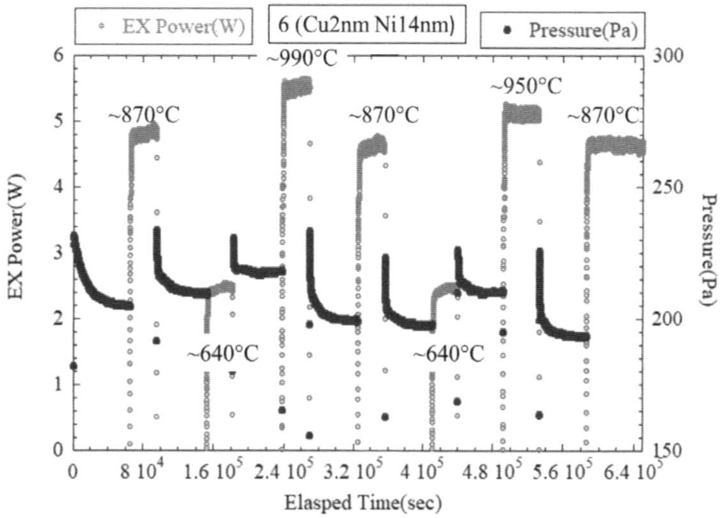


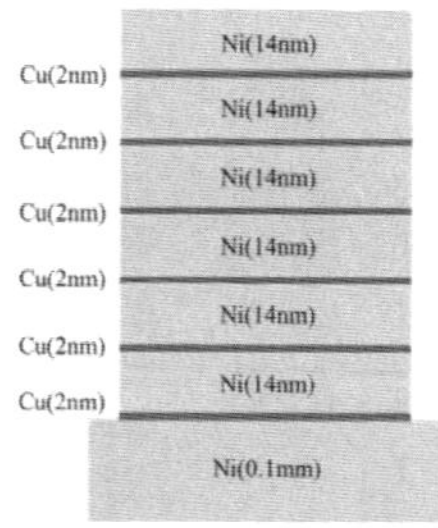


| | |
|---|---|
| Absorbed H [mol] | 7.3E-04 |
| Absorbed Number of H | 4.4E+20 |
| Total Excess Energy [J] | 1.1E+06 |
| Excess Energy per H [J/H-mol] | 1.5E+09 |
| Excess Energy per H [eV/H] | 1.6E+04 |

**Figure 10.** An example of excess heat generation [15].

### 3.3. Heat Burst Phenomena

In this section, we describe spontaneous heat bursts and intentionally induced heat bursts, which were based on the observations of the spontaneous heat bursts. The research implications of heat bursts are as follows. Usually, in the case of heat measurement as shown in this paper, the heat generation is evaluated using a calibration curve, and there are many cases where people question whether the heat dissipation characteristics are different between the time of calibration and the time of the heat generation experiment due to variations in the way the sample is mounted, etc. We are constantly checking various measurements to see if there are any changes in the heat dissipation characteristics from experiment to experiment. We conclude that the heat burst phenomena cannot be explained by the difference in the heat dissipation characteristics between a calibration and an experiment. The observation of the heat burst phenomena increases the reliability of the heat measurement experiment.

#### 3.3.1. *Spontaneous Heat Burst*

Figure 11 shows temperature time variations for the thermocouple surface A and B. It can be seen that the heat release reaction occurred at the surface A at first and afterwards at the surface B. There, heat bursts propagated to the thermocouple. Input power and room temperature were constant during these events. Therefore, these events were not caused by the change of electrical input power or heat coming in from the environment. The samples consisted of 6 layers of Cu (2 nm) and Ni (14 nm) fabricated by magnetron sputtering on Ni bulk, the same method as in [15]–[16].

The delay time from the burst-like heat release at surface A to the thermocouple is about 33 sec as shown in Fig. 11(a). Delay time from the heat burst at surface B is estimated to be about 40 sec, as an inflection point of $T_c$ can

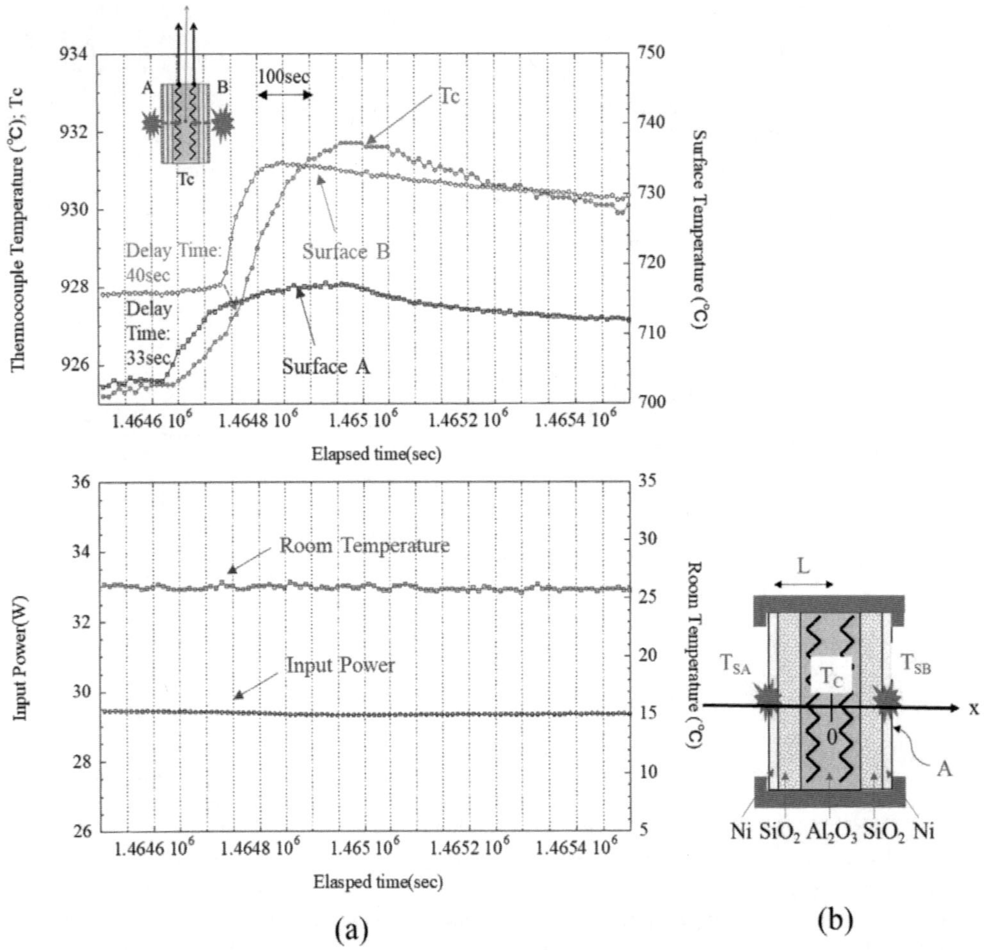


**Figure 11.** An example of a spontaneous heat burst; (a) Heat burst reactions occurred at the surface A at first and afterwards at the surface B. The heat bursts propagated to the thermocouple. Input power and room temperature were constant during these events, (b) Numerical model for one dimensional non-steady state heat conduction analysis [17].

be seen as shown in Fig. 11(a). Although a 3-dimentional heat analysis would be desirable, a rough estimation of the time constant for the observed burst phenomena can be made.

Figure 11(b) shows the numerical model for one dimensional non-steady state heat conduction analysis. The time constant $\tau$ is given as the following equation:

$$\tau = \rho C_p V \frac{L}{k_{Al_2O_3+SiO_2} A} \tag{7}$$

where $\rho$ is the density, $C_p$ is the specific heat, and V is the volume of the sample holder. And k is the thermal conductivity for the mixture of $Al_2O_3$, $SiO_2$, A is the surface area and L is the distance between the surface and center of the heater. As Ni is thin and thermal conductivity is larger than $Al_2O_3$ and $SiO_2$, the contribution of Ni for

$\tau$ is negligible. Based on the rough estimation, $\tau$ is calculated as 36 sec. This agrees with the experimental results, therefore the obtained delay times appear to be reasonable. This example gives us clear evidence that burst heat release reactions occurred in the near surface region of the nano-sized multilayer metal composite with hydrogen gas.

### 3.3.2. *Intentional Heat Burst*

When analyzing the phenomenon of spontaneous sudden heat bursts described above, it was observed that there were cases where the heat burst phenomenon occurred after small changes in input power caused by fluctuations in the outside temperature, etc. Therefore, we conducted an experiment to see what kind of heat burst would occur if we intentionally perturbed the input power. As a result, it became possible to intentionally induce heat burst phenomena depending on (1) the composition and state of the sample, (2) the surface temperature of the sample, and (3) the amount and time range of the input power perturbation.

An example of a heat burst phenomenon induced by an intentional perturbation of input electrical power is shown in Fig. 12(a). At the beginning, 25.8 W was input, and then decreased to 25.0 W. Then the $T_C$ and surface temperature A and B decreased gradually. When the input was returned to the original input of 25.8 W, the surface temperature A, B, and $T_c$ all increased and became larger than the values at the original input of 25.8 W. At maximum, $T_c$ increased by about 9°C, surface A by about 17°C, and surface B by about 25°C. The increases in surface A and B temperatures were larger and steeper than that of $T_c$ as shown in Fig. 12(a). It can be considered that heat burst energy generated at the surface A and B, which is induced by the intentional perturbation of input power, propagated to the thermocouple.

On the other hand, an example of no heat burst phenomenon induced by a similar perturbation is shown in Fig. 12(b). In this case, when the input power, which was initially 26.3 W, was reduced to 25.8 W, the thermocouple and surface temperatures A and B gradually decreased as in Fig. 12(a). Then, when the input power was returned to 26.3 W, the heat burst phenomenon did not occur, and the thermocouple and surface temperatures A and B returned to their original values.

The same experimental setup was used for these two examples, but the compositions of the samples were different. Figure 12(a) is a 6-layer CuNi sample, and Figure 12(b) is a 6-layer CuNi sample with CaO inserted into the Ni layer. The conditions under which this phenomenon occurs are being investigated in detail.

The experimental results related to this phenomenon were observed by the Focardi team [5]. They used Ni rods and hydrogen. They observed excess heat by once bringing the temperature of the Ni rods down to near room temperature and then abruptly returning it to the original input power level. Although the experimental methods are different, the same physicochemical mechanism related to the diffusion of hydrogen seems to be behind the experimental results.

### 3.3.3. *Observation of Heat Burst Phenomena by Thermocouple and Mid-IR*

In order to elucidate the mechanism of the anomalous heat phenomena, our research team has been conducting experiments using the same type of sample holder described in 3.3.1 and 3.3.2 to measure the light spectrum emitted from the sample surface during the generation of anomalous heat [18]. Here, we introduce the recently observed results [21].

Figure 13 shows the time variation between the integrated spectrum of infrared light in the mid-IR region (0.23-0.41 eV) and the thermocouple temperature ($T_c$) embedded in the center of the heater when multiple spontaneous heat bursts occurred. As shown in the figure, the mid-IR data is the infrared radiation emitted from surface A. In addition to this mid-IR, visible light emitted from surface B and near-IR (0.5-0.75 eV) were also measured in this experiment, but they are not shown here for simplicity. Many sudden spontaneous increases are clearly seen, even though the heater input power is constant. $T_c$ and mid-IR (0.23-0.41 eV) seem to be synchronized. This suggests that there was sudden energy generation near the surface of the nano-sized multilayer composite, part of it was dissipated quickly by radiation from the surface and was partly propagated to the inside, raising the heater temperature in the same way

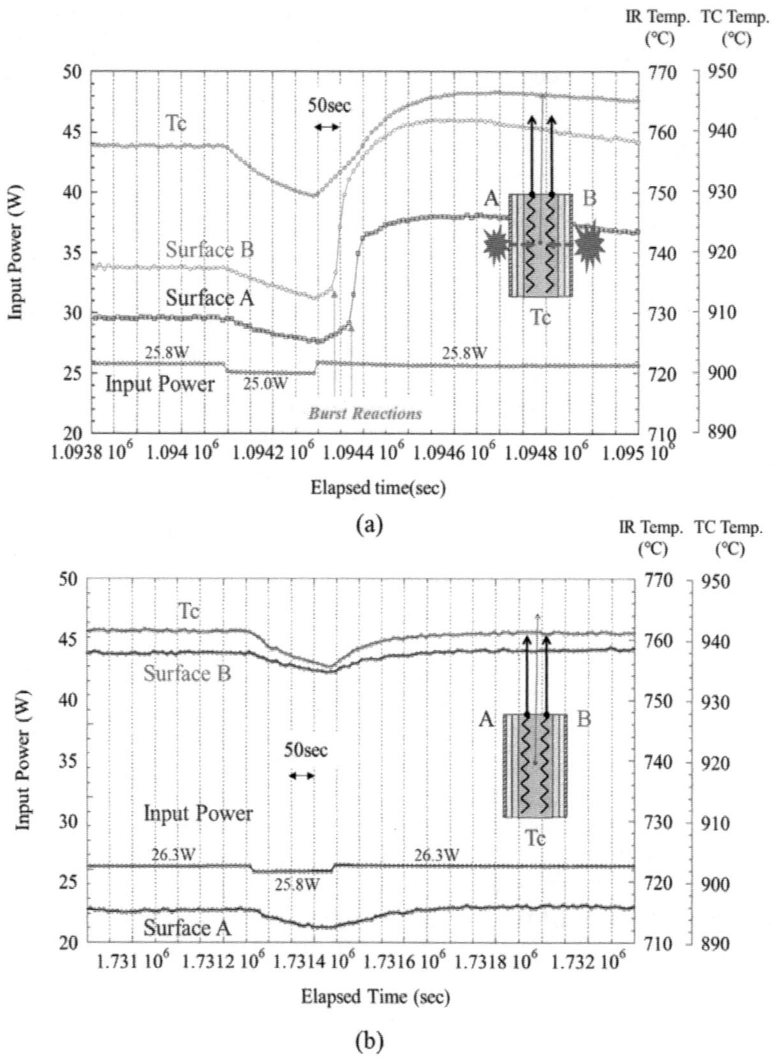


**Figure 12.** Different responses to intentional perturbations of input power: (a) An example of heat burst phenomenon induced by an intentional perturbation of input electrical power, (b) An example where no heat burst phenomenon was induced by a similar perturbation.

described in Fig. 11 and 12. The radiation thermometers used in Figs. 11 and 12 calculated the surface temperature from the infrared intensity at two wavelengths, 1.55 $\mu$m (0.80 eV) and 1.35 $\mu$m (0.92 eV). In other words, the surface temperature was calculated from the infrared intensities at 0.80 eV and 0.92 eV, and the heat burst was observed almost synchronously with $T_c$. It is important to note that not only 0.80 eV and 0.92 eV, but also 0.23-0.41 eV infrared, which is lower in energy and longer in wavelength, correlated with $T_c$.

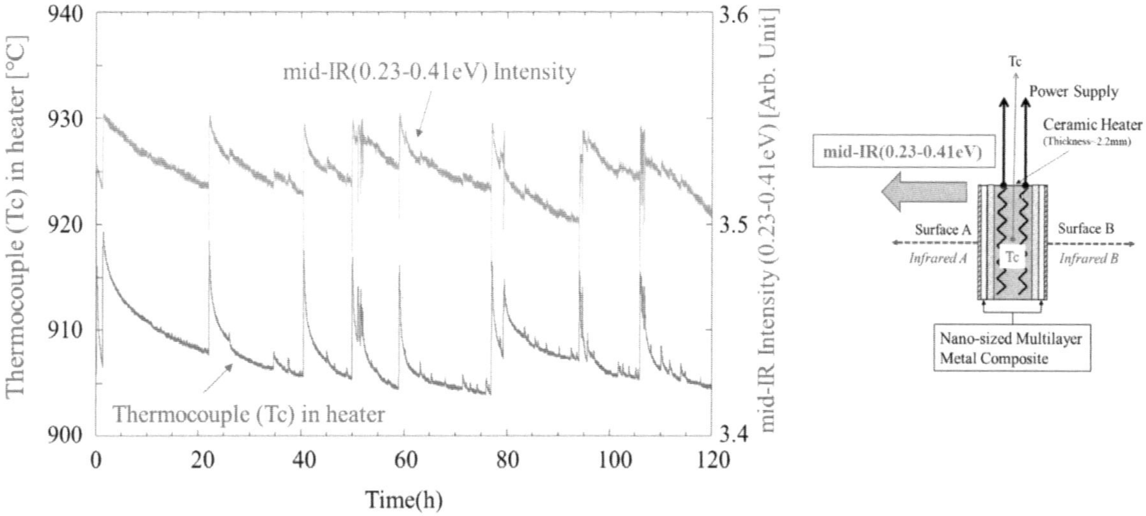


**Figure 13.** Observation of heat burst phenomena by thermocouple and mid-IR.

## 4. Future Scientific and Technological Challenges

Here, we briefly describe the future challenges for QHE. First of all, it is now known that the use of nano-sized multilayer composite metals and hydrogen can generate abnormal heat, which is several thousand times higher than the chemical reaction per unit of hydrogen fuel, but the mechanism of the phenomenon is still unknown. Therefore, the most important task is to identify the reaction responsible for this anomalous heat generation. It is also necessary to identify the reaction products, and to show that the heat generation and the products are consistent. This will require a very clean experimental environment and a synchrotron radiation facility such as SPring-8 for trace element analysis. It is also important to develop a theoretical model to explain the reaction.

Let us consider the development of an energy generator by QHE. There are many theoretical difficulties, but there are fewer inherently difficult engineering challenges in temperature range, materials, and fuels than tokamak plasma fusion. Therefore, in parallel with the theoretical clarification, it is urgent to develop a demonstration device that can convince people that it can be used as an energy source. At the same time, public acceptance is an extremely important issue for QHE, which should be solved in the future.

## 5. Concluding Remarks

Anomalous heat using nano-metal with hydrogen/deuterium gas could become a technology to prevent global warming, if it can be commercialized. An overview of the progress of gas loading type experiments, in which deuterium or hydrogen and metals such as Pd and Ni were utilized, was described. Among them, nano-sized Ni based composite material began to attract attention from industry because of its abundance on earth, lower cost, and its ability to react with hydrogen at several hundred degrees centigrade. Our team developed a new type of energy generation method using a nano-sized metal multilayer composite and hydrogen gas. We have now started to call the generated energy "Quantum Hydrogen Energy (QHE)". According to our experimental results, Exothermic reactions that cannot be explained by chemical reactions were observed in Cu/Ni and Cu/Ni multilayer metal composites. The amount of heat generation reached to 16 keV/H. Not only steady-state heat generation, but also heat burst phenomena were observed. Experimental data analysis showed that the exothermic reaction occurred near the surface. Recently, we have succeeded in intentionally inducing heat bursts. By measuring the optical spectrum emitted from the sample, we detected a simultaneous increase in mid infrared radiation flux when a heat burst occurred. This emerging technology is expected to be commercially feasible.

## Acknowledgements

The authors would like to thank President H. Yoshino and all the members of CLEAN PLANET Inc. for their significant assistance. The authors also thank Dr. S. Yamauchi, Mr. Y. Shibasaki and Prof. H. Kikunaga and of Tohoku University for their support. This work is supported by the Research Center for Electron Photon Science of Tohoku University, CLEAN PLANET Inc., The Thermal & Electric Energy Technology Foundation and the Tanaka Kikinzoku Memorial Foundation.

## References

[1] M. Fleishmann and S. Pons, Electrochemically Induced Nuclear Fusion of Deuterium, *J. Electroanol. Chem.*, **261** (1989) 301-308.

[2] A. De Ninno et. al, Evidence of Emission of Neutrons from a Titanium-Deuterium System., *Europhys. Lett.,* **9 (3)** (1989) 221-224.

[3] Cold Fusion: Advances in Condensed Matter Nuclear Science, Ed. Jean-Paul Biberian, Elsevier, Amsterdam, 2020.

[4] B. Liu, X. – Z. Li, Q. M. Wei, N. Mueller, P. Schoch and H. Oehre in Twelfth International Conference on Condensed Matter Nuclear Science, Yokahama, Japan, 2005, pp. 75.

[5] S. Focardi, V. Gabbani, V. Montalbano, F. Piantelli and S. Veronesi, Large excess heat production in Ni-H systems, *Il Nuovo Cimento A (1971–1996)* **111** (1998) 1233–1242.

[6] F. Celani et al., Deuteron Electromigration in Thin Pd Wires Coated with Nano-Particles: Evidence for Ultra-Fast Deuterium Loading and Anomalous, Large Thermal Effects. in ICCF-14 International Conference on Condensed Matter Nuclear Science. 2008. Washington, DC.

[7] R. Godes, R. George, F. Tanzella and M. McKubre, Controlled Electron Capture and the Path toward Commercialization, *J. Condensed Matter Nucl. Sci.* **13** (2014) 127–137.

[8] Y. Arata and Y-C. Zhang, Anomalous Difference between Reaction Energies Generated within $D_2O$-Cell and $H_2O$-Cell, *Jpn. J. Appl. Phys.* **37** (1998) 1274–1276.

[9] Y. Iwamura, M. Sakano and T. Itoh, Elemental Analysis of Pd Complexes: Effects of D2 Gas Permeation, Jpn. *J. Appl. Phys.*, 41 (2002) 4642-4650.

[10] A. Kitamura, A. Takahashi et.al, Excess heat evolution from nanocomposite samples under exposure to hydrogen isotope gases, *Int. J. of Hydrogen Energy* **43** (2018) 16187-16200.

[11] Y. Iwamura et.al, Anomalous Heat Effects Induced by Metal Nanocomposites and Hydrogen Gas, *J. Condensed Matter Nucl. Sci.* **29** (2019) 119–128.

[12] T. Mizuno, Method of Controlling a Chemically Induced Nuclear Reaction in Metal Nanoparticles, *J. Condensed Matter Nucl. Sci.* **13** (2014) 422–431.

[13] Y. Iwamura, T. Itoh and S. Tsuruga, Transmutation Reactions Induced by Deuterium Permeation through Nano-structured Pd Multilayer Thin Film, *Current Science*, **108** (2015) 628-632.

[14] Y. Iwamura, Review of permeation-induced nuclear transmutation reactions, Cold Fusion: Advances in Condensed Matter Nuclear Science, Ed. Jean-Paul Biberian, Elsevier, Amsterdam, (2020)191-204.

[15] Y. Iwamura, T. Itoh, J. Kasagi, S. Murakami and M. Saito, Excess Energy Generation using a Nano-sized Multilayer Metal Composite and Hydrogen Gas, *J. Condensed Matter Nucl. Sci.* **33** (2020) 1–13.

[16] Y. Iwamura, Heat generation experiments using nano-sized metal composite and hydrogen gas, Cold Fusion: Advances in Condensed Matter Nuclear Science, Ed. Jean-Paul Biberian, Elsevier, Amsterdam, (2020)157-165.

[17] Y. Iwamura, T. Itoh, M. Saito, S. Murakami and J. Kasagi, "Evidence for Surface Heat Release Reaction over Nano-sized Multilayer Metal Composite with Hydrogen Gas", Proceedings of the 21st Meeting of Japan CF Research Society, JCF21, December 11-12, 2020, Online Meeting, p.1-14.

[18] T. Itoh, S. Shibasaki, J. Kasagi, S. Murakami, M. Saito and Y. Iwamura, "Optical Observation on Anomalous Heat Generation from Nano-sized Metal Composite", Proceedings of the 21st Meeting of Japan CF Research Society, JCF21, December 11-12, 2020, Online Meeting, p.15-25.

[19] Y. Fukai, Evidence for Quantum Diffusion of Hydrogen in Ta: Quench-Recovery Experiments Revisited, *Japanese Journal of Applied Physics*, Vol.23, No.8 (1984) 596-598.

[20] H. Wipf, Electro- and Thermo-Transport of Hydrogen in Metals, *Hydrogen in Metals II*, Ed. by G. Alefeld and J. Völkl, Springer-Verlag, Berlin, Heidelberg (1978) 273-304.

[21] T. Itoh, Y. Shibasaki, J. Kasagi, T. Takahashi, M. Saito and Y. Iwamura, Optical Observation of Spontaneous Heat Burst Phenomena during Hydrogen Desorption from Nano-sized Metal composite, Proceedings of ICCF23, to be published.

J. Condensed Matter Nucl. Sci. 36 (2022) 302–304

Research Article

# Low Energy Nuclear Reactions with Emission of Two Photons

Pankaj Jain*, Ankit Kumar, K. Ramkumar, and K. P. Rajeev

*Physics Department, Indian Institute of Technology, Kanpur, 208016, India, E-mails: ankt@iitk.ac.in, ramkumar@iitk.ac.in, kpraj@iitk.ac.in*

Raj Pala

*Department of Chemical Engineering, Indian Institute of Technology, Kanpur, 208016, India, E-mail: rpala@iitk.ac.in*

**Abstract**

We propose a new mechanism for inducing low energy nuclear reactions (LENRs). The process is initiated by an electromagnetic perturbation. The initial two body nuclear state emits a photon and forms an intermediate state which makes a transition into the final nuclear state with the emission of another photon. We need to sum over all energies of the intermediate state. Since the energy of this state is unconstrained, we get contributions from very high energies for which the barrier penetration factor is not necessarily small. By considering fusion of $^1$H and $^2$H to form $^3$He, we determine the conditions under which this mechanism leads to fusion at observable rates. We show that this mechanism works only inside a medium and not in free space. We show that a clear experimental signature of this process is emission of two photons in coincidence whose total energy is related to the Q value of this process. Hence this process can be confirmed or ruled by the presence or absence of such photons in the final state.


*Keywords:* LENR, electromagnetic perturbation, fusion, two photons emission, coincidence

## 1. Introduction

In recent papers [1], [2] we have studied the possibility that nuclear reactions may take place at low energies within the framework of the second order perturbation theory. We investigated this by considering two reactions

$$^1\mathrm{H} + {}^2\mathrm{H} + \gamma(\omega_\mathrm{i}) \rightarrow {}^3\mathrm{He} + \gamma(\omega_\mathrm{f}), \tag{1}$$

and

$$^1\mathrm{H} + {}^2\mathrm{H} \rightarrow {}^3\mathrm{He} + \gamma(\omega_1) + \gamma(\omega_2) \tag{2}$$

Here a $^1$H nucleus undergoes fusion with $^2$H nucleus to form $^3$He. The first process is assisted by the presence of a photon in the initial state while the second proceeds by spontaneous emission of two photons. Let $\vec{r}_1$ and $\vec{r}_2$ be the

*Corresponding author: pkjain@iitk.ac.in

positions of the two initial state nuclei. The interaction Hamiltonian can be expressed as,

$$H_I(t) = -\frac{Z_1 e}{m_1 c}\vec{A}(\vec{r}_1, t)\cdot\vec{p}_1 + \frac{Z_2 e}{m_2 c}\vec{A}(\vec{r}_2, t)\cdot\vec{p}_2 + \frac{e^2 g_p}{2m_p c}\vec{\sigma}\cdot\vec{B} + \cdots \tag{3}$$

where $m_1$ and $m_2$ are the masses of the two particles, $Z_1$ and $Z_2$ their atomic numbers, $\vec{A}$ is the electromagnetic potential, $\vec{B}$ the magnetic field and $g_p$ the proton g factor. Here we have included the magnetic moment term due to proton and have set $\vec{p}_1 = -\vec{p}_2 = -\vec{p}$. The transition amplitude at second order can be expressed as,

$$\langle f \mid T(t, t_0) \mid i\rangle = \left(-\frac{i}{\hbar}\right)^2 \sum_n \int_{t_0}^{t} dt' e^{i(E_f - E_n)t'/\hbar} \langle f \mid H_I(t') \mid n\rangle \int_{t_0}^{t'} dt'' e^{i(E_n - E_i)t''/\hbar} \langle n \mid H_I(t'') \mid i\rangle \tag{4}$$

Here $|i>$, $|n>$ and $|f>$ are the initial, intermediate, and final states respectively and $E_i$, $E_n$ and $E_f$ their eigenvalues. Here we need to sum over all energy eigenvalues $E_n$. Hence, the sum over $E_n$ gets contributions from very high energy eigenstates and has the possibility of evading the Coulomb barrier. The applicability of this mechanism has been investigated in detail in [1], [2].

## 2. Results and Discussion

The mechanism described in section 1 requires a very high energy intermediate state, denoted by $|n>$ in Eq. (4). For large values of $r = |\vec{r}_2 - \vec{r}_1|$, the potential is nearly zero and this wave function behaves approximately as free particle wave function $\sin(k_n r + \phi)$, where $k_n$ is the wave number and $\phi$ the phase shift. In order for the matrix element $<n|H_I(t'')|i>$ to take significant values, we need another high energy particle. This is provided by the emitted photon, which for the current process may have energy of the order of 5 MeV. Hence, we may argue that the state $|n>$ balances the momentum of the emitted photon and in the sum over n we may simply include just this one state. This single high energy state then contributes to the second matrix element $<f|H_I(t')|n>$, leading to a substantially increased transition rate. However, the mechanism does not work in this simple manner. Explicit calculations [1], [2] show that the matrix element $<n|H_I(t'')|i>$ cannot be approximated by a delta function and we cannot just set the wave number $k_n$ equal to that of the photon. The main reason is that the wave function $|n>$ deviates very strongly from free particle (zero potential) wave function at small values of r. Hence, we obtain contributions from a wide range of energy eigenvalues $E_n$ (or $k_n$) of the intermediate state $|n>$. The matrix element shows rapid oscillations about 0 as a function of $E_n$ and we find that the sum (or integral) over $E_n$ leads to very small values if we use free space (absence of medium) eigenfunctions.

However, the situation can be very different in medium where the eigenfunctions may be greatly modified at large distances [2]. In [1], [2] we suggested several possibilities of medium modification of eigenfunctions which can lead to enhanced rates. Here we briefly describe the mechanism proposed in [2]. Let us first recall that, due to spherical symmetry, the free space eigenfunctions behave as $\sin(k_n r + \phi)$ at large values of r. However at large distances, it may not be reasonable to assume spherical symmetry in medium. Hence the wave function may be significantly modified. In [2] we assumed cylindrical symmetry along the z-axis. We ignore the center of mass motion, assuming it to be located at the origin of the coordinate system and focus on the relative motion. The relative coordinates are denoted by (z, $\rho$, $\theta$). We assume that the potential is negligible in the region $\rho_1 < \rho < \rho_2$. Here $\rho_1$ is of order 1 atomic unit and $\rho_2$ is determined by the lattice structure and is assumed to be of order 5 atomic units or larger. In this gap region, it is reasonable to assume that the eigenfunctions behave as linear superpositions of $\sin(k_z z)$ and $\cos(k_z z)$. With such wave functions, we find that the reaction rates can be much larger in comparison to what is obtained without a medium. Here we are using cylindrical coordinates and $k_z \approx k_n$ is the wave number which determines the z dependence of the wave function. Indeed, the rate was found to be large enough to lead to an observable signal. Hence, we conclude that low energy nuclear reactions are possible given suitable medium conditions. We point out that the assumed wave functions are very reasonable if the medium displays cylindrical symmetry. At short distances, we can expand such wave functions in spherical harmonics and only a few leading order terms dominate for the energy scale relevant to the problem. Furthermore, if the medium satisfies different symmetry conditions, we expect that a suitable generalization of our mechanism may still be applicable in some cases.

Our mechanism leads to a directly testable experimental prediction. It was found [1], [2] that the reaction involving emission of two photons (Eq. (2)) has much higher rate in comparison to the reaction shown in Eq. (1). In fact, the rate of the latter reaction is too small and not observable. The process in Eq. (2) leads to emission of two photons in coincidence with total energy roughly equal to the Q value of the process, which is approximately equal to 5.4 MeV. Here we need to account for a small energy loss due to nuclear recoil and due to propagation in medium. Due to the production of two photons in coincidence, the process provides a clean signal which can be tested in future experiments.

## References

[1] Jain, A. Kumar, R. Pala, and K. P. Rajeev, Photon induced low energy nuclear reactions, Pramana, J. Phys 96, 2022.

[2] P. Jain, A. Kumar, K. Ramkumar, R. Pala, and K. P. Rajeev, Low energy nuclear fusion with two photon emission, JCMNS 35, 1, 2022.

J. Condensed Matter Nucl. Sci. 36 (2022) 305–311

Research Article

# Decay-Instability of Transmuted Chemical Elements Obtained in LENR Experiment

A. Klimov*

*Power Engineering Technical University, Moscow, Russia*

**Abstract**

It is well known that there are chemical element transmutations and its isotope composition transformations in LENR experiments [1], [2]. Time evolution of the transmuted chemical elements and their decay-instability have been studied in this work. New transmuted chemical elements (such as Fe, Cu, Ca…) were obtained from the initial pure cathode material Ni (99.99%) by heterogeneous plasma formation in a plasma-vortex reactor [1], [2]. The chemical compositions of the initial cathode material and the transmuted chemical elements were measured by the EDS method and the ICP-MS method. It was revealed that there is considerable decay-instability of the obtained transmuted chemical elements at the time. This instability is accelerated dramatically by weakly ionized non-equilibrium plasma (WINP) action. It was confirmed that relative concentration of the initial Ni-atoms (cathode material erosion) is not changed, but the relative concentration of the transmuted atoms is dramatically decreased by WINP's action. The author supposes that the theoretical model of bi-nuclear atom [4] can explain the obtained experimental results. Two-step transmutation method of the new chemical element creation in LENR experiment is proposed in this work at the first time.


*Keywords:* LENR, ICP-MS, transmutation, time evolution, non equilibrium plasma

## 1. Introduction

There are many reports and much evidence connected with the observation of large bolide impacts on Earth's surface and large plasmoid interaction with mountain surfaces (for example, Tunguska's bolide, Chebarkulskii's bolide, Dalnogorskii's plasmoid (mount 611), [3], and others). It was revealed that there is a transmutation of the initial chemical elements and their transformation in the soil samples obtained in the mountain region burned by bolide's explosion (or plasmoid's action). It is interesting to note that these soil samples after the plasmoid's action have strong magnetic properties. For example, a pure silicon sample has strong magnetic properties [3]. Note that these transmuted elements are not stable ones. It is revealed that there is decay instability of these elements by external heating of these samples up to a temperature of $T_a > 2000°C$. The main task of this work is a study of decay-instability of the transmuted chemical elements obtained in *LENR plasma experiment* [1], [2].

*Corresponding author: klimov.anatoly@gmail.com

© 2022 ICCF. All rights reserved. ISSN 2227-3123

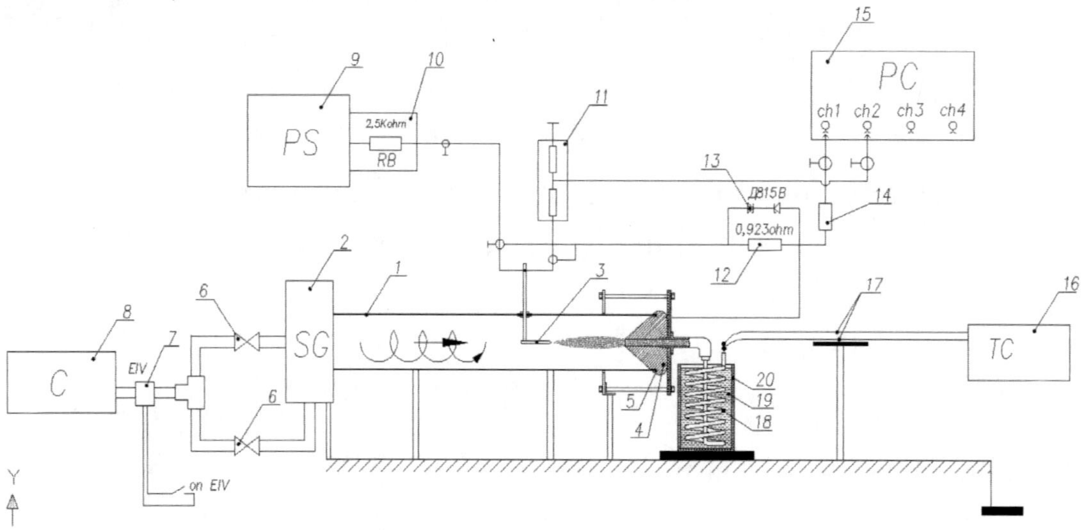


**Figure 1.** Schematic of the experimental PVR set up. 1 - quartz tube, 2 - swirl generator, 3,4 - electrodes, 6 - valve, 7 - gas flow meter, 8 - compressor, 9, 10 - power supply, 17 - thermocouple, 18-20 - water bath calorimeter.

## 2. Experimental Setup

Physical parameters and properties of a stable heterogeneous vortex plasmoid (plasma formation with eroded nanoclusters) created in high-speed swirl flow have been studied in our previous works [1], [2]. A schematic of the experimental plasma-vortex reactor (PVR) used in these works is shown in Fig. 1. This setup was described in our work [1], [2] in detail. The PVR's operation regime is shown in Fig. 2. One can see anode water steam injector (2), heterogeneous plasma (3) created by pulse-repetitive electric discharge and cathode (4) with of eroded metal nanocluster cloud in this picture.

The typical experimental conditions were the following:

| | |
|---|---|
| 1. Mean excess power output | 1 - 10 kW |
| 2. Mean power input | 0.1 - 1 kW |
| 3. Testing gas mixture | water steam + argon |
| 4. Gas mixture mass flow | <10 G/s |
| 5. Combined discharge | DC + high frequency discharge |
| 6. COP | 2 - 10. |

A schematic of the experimental setup to obtain a transmuted chemical element sediment in the LENR plasma experiment is shown in Fig. 3. One can see plasma vortex reactor PVR (1) and water bath calorimeter (2) in this picture. The eroded dusty particles (eroded cathode material) are captured by water in this calorimeter. Then these sediment particles are evacuated from the bath and analyzed by the EDS-method and ICP-MS method. Photos of typical particles are shown in Fig. 4.

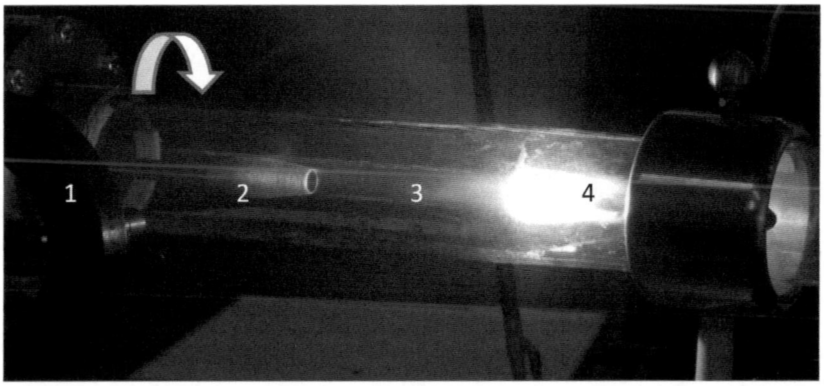


**Figure 2.** Plasma vortex reactor (PVR). Gas mixture Ar : $H_20 = 10 : 1$. Axial velocity $V_x$ is closed tangential velocity $V_t$: $V_x \sim V_t \sim 30$ m/s, $P_{st} \sim 1.5$ Bar. 1-swirl generator, 2 - anode-water steam injector, 3 - electric discharge plasma, 4 - cathode with eroded metal nano-clusters.

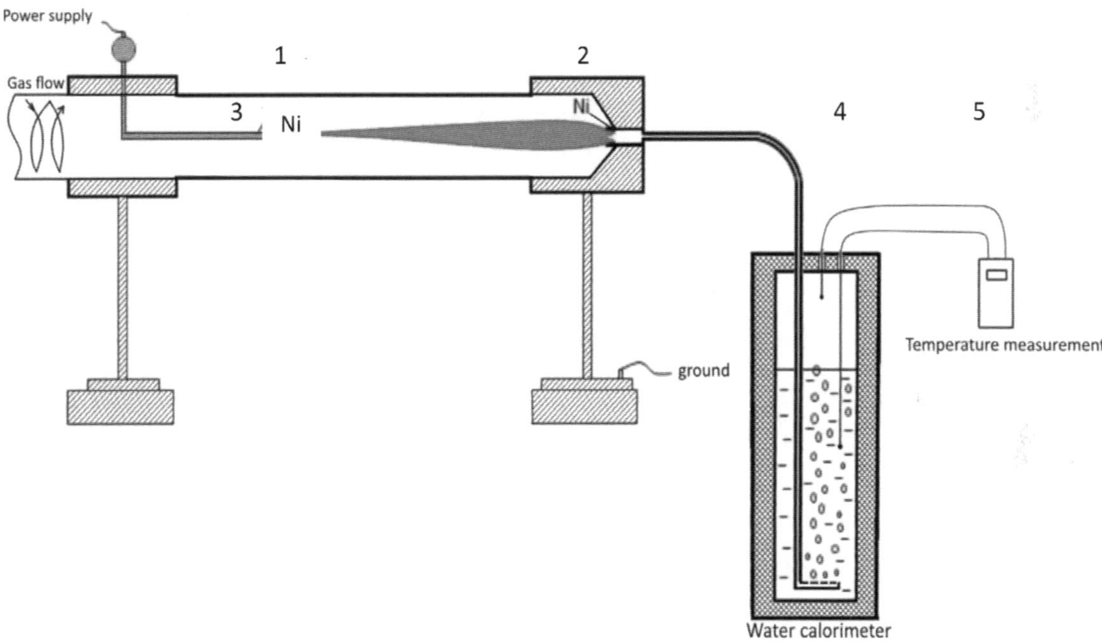


**Figure 3.** Schematic of the PVR with water calorimeter to obtain a sediment of eroded dusty particles in water. 1 - PVR, 2 - cathode (Ni), 3 - anode (Ni), 4 - water bath, 5 - thermocouple with recorder CENTER 306.

## 3. Main Experimental Results

Relative concentrations of new transmuted chemical elements inside eroded particles (sediment particles) are shown in Fig. 6 (right). All concentrations of these elements are normalized to their initial values. Note that pure Ni electrodes

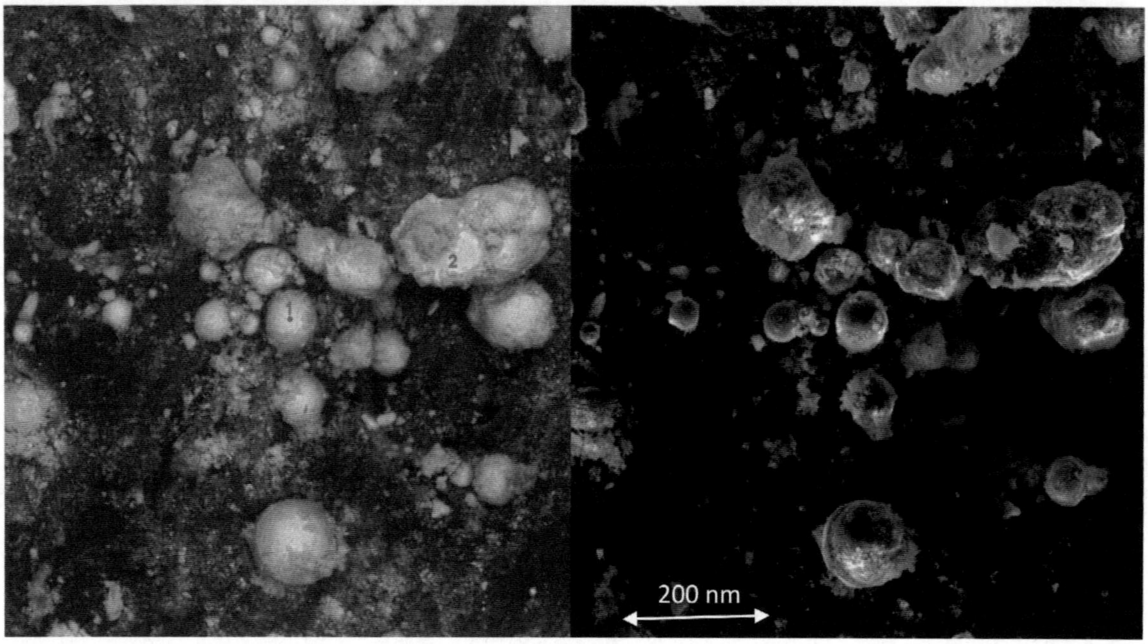


**Figure 4.** The typical eroded dusty particles obtained in the PVR with water bath calorimeter.

(99.99%) are used in the PVR. One can see that the relative concentrations of the following elements Fe, Cu, Zn are very high.

Then these sediment particles were acted by weakly ionized non-equilibrium plasma (WINP). A schematic of the experimental setup used in this experiment is shown in Fig. 5. An external magnetic field helps us to localize the exposed eroded particles with strong magnetic properties on the reactor's wall. This procedure increases plasma action on the eroded particle's surface. The magnetic field transforms a diffuse electric discharge to a constricted one. Argon injection (1) is used in this experiment to overcome possible oxidation of the tested eroded dusty particles at WINP's action.

Parameters of the power supply PS-1 were the following:

Mean discharge current $I_d \sim 64$ mA
Mean discharge voltage $U_d < 4.5$ kV.

Parameters of the plasma created by this PS were measured by optical spectroscopy method. These parameters obtained in this setup were the following:

Electron concentration $N_e \sim 10^9$ - $10^{11}$ cm$^3$
Electron temperature $T_e \sim 1$ eV
Gas temperature $T_g \sim 500$ K in argon flow
Ionization level $N_e/N_a \sim 10^{-8}$ - $10^{-6}$.

The chemical compositions of the transmuted chemical elements after WINP's action were measured by the EDS method, ICP-MS method and optical spectroscopy method, Figs. 6, 7. It was revealed that there is considerable decay-instability of the new transmuted chemical elements at the time (without plasma action). The chemical analyses

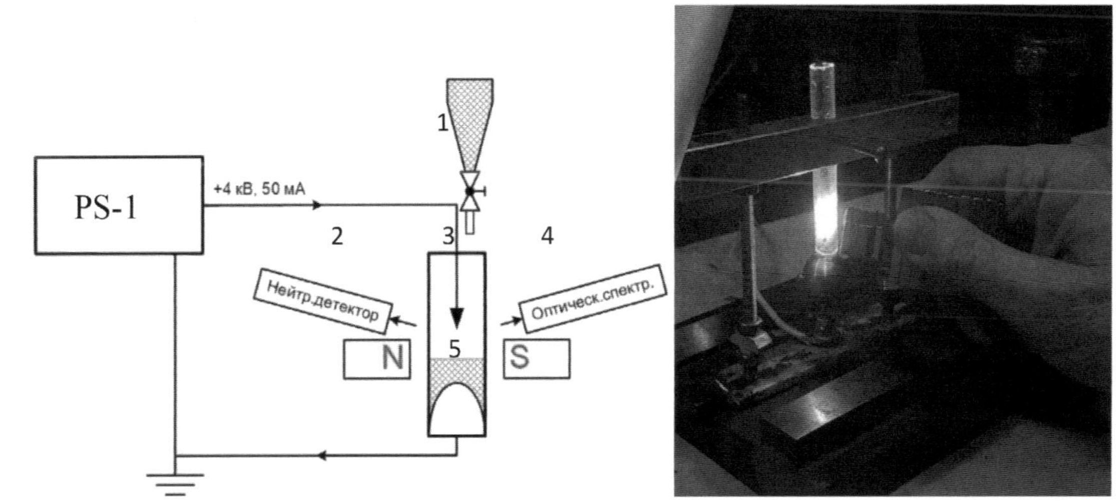


**Figure 5.** Schematic of the experimental setup to study WINP's action on the eroded dusty particles created by PVR (right). General view of experimental set up (left). 1 - argon injector, 2 - neutron detector, 3 - quartz tube reactor, 4 - optical spectrometer, 5 - tested sediment particles.

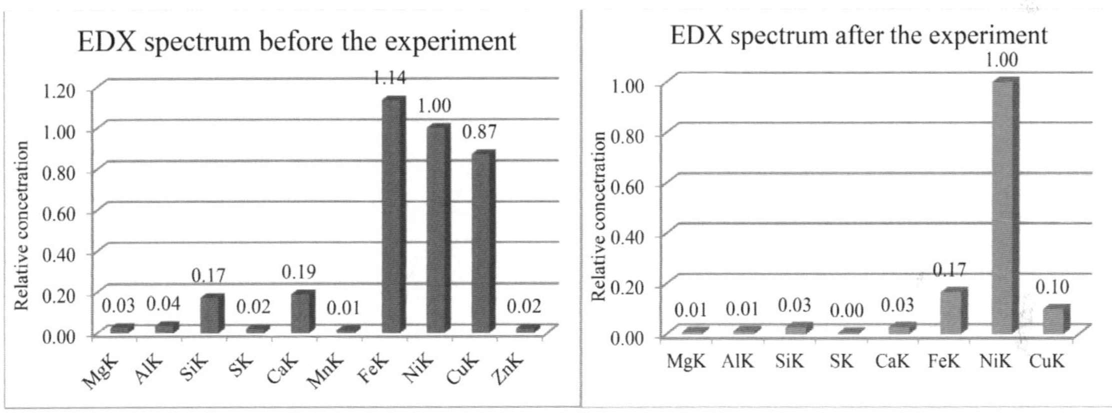


**Figure 6.** Relative concentrations of the transmuted chemical elements (eroded cathode particles) before WINP's action (left) and after its action (right).

were performed 1 week, 1 month, and 6 months after the experiment run. This instability is accelerated dramatically by factors of $10^3$ - $10^6$ by weakly ionized non-equilibrium plasma (WINP) action. It was confirmed that relative concentration of the initial Ni atoms (cathode material) is not changed in dusty eroded particles, but the relative concentrations of the transmuted atoms (Ca, Fe, Cu and others) are dramatically decreased, Fig. 6. An analogous result was obtained with Al-cathode (99.999%) exposed in the PVR with water steam, Fig. 7. One can see that the relative concentrations of transmuted chemical elements (such as Si, P, S, Cl, Ca, Fe, Zn) decreased dramatically from the WINP's action (Fig. 7, right).

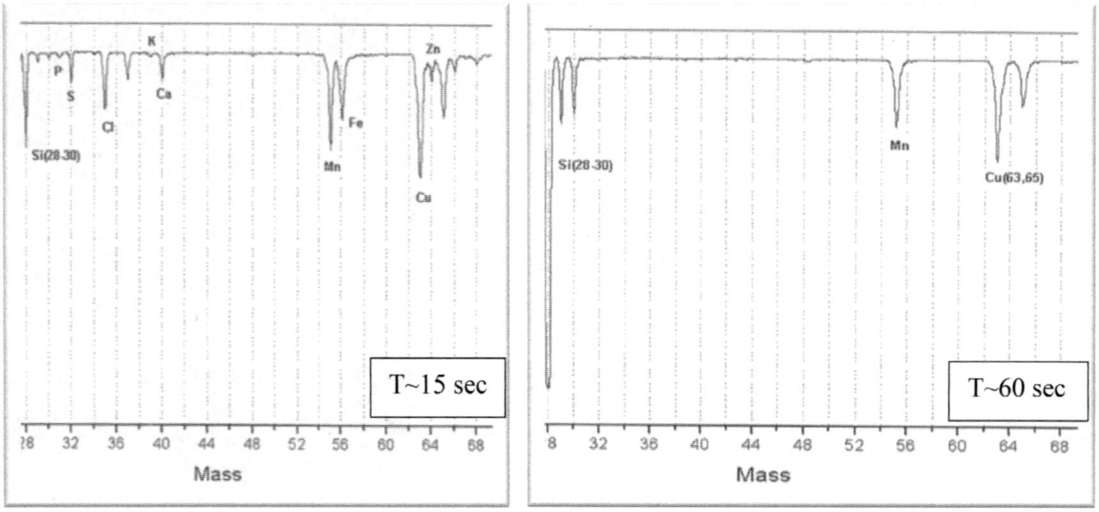


**Figure 7.** ISP MS method. Relative intensities of the transmuted element lines (rel. units). Al - electrode exposed in the PVR with water steam. After WINP's action on its surface. Left - after T∼15 sec of WINP's action, right - after T∼60 sec of WINP's action. Intensity of Al-line is very high and it is deleted from these spectra.

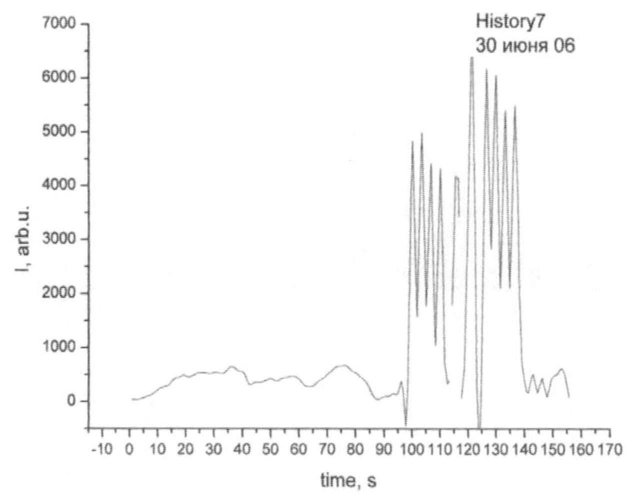


**Figure 8.** Optical Spectroscopy. Temporal evolution of the creation and decay of the transmuted lithium Li in the PVR. Ni-electrode (99.99%). Testing gas mixture: argon + water steam mixture. Optical $Li_\alpha$-line, $\lambda = 670$ nm.

Instability of the transmuted chemical elements was recorded by the optical spectroscopy method also [5]. The creation and decay of transmuted lithium is shown in Fig. 8. One can see that bright $Li_\alpha$-line ($\lambda = 670$ nm) appeared at the time T $>$ 95 sec and it disappeared at T $>$ 140 sec (where T – plasma operation time of the PVR).

Simultaneously the bright optical lines (duplet) of the potassium K appears in the optical spectrum at the time interval 600 sec > $\delta$T > 140 sec. Then K-lines disappeared and bright Na-lines (duplet) appeared at the time period $\delta$T > 600 sec. Thus, the new transmuted chemical elements are not stable.

The author supposes that the theoretical model of bi-nuclear atom [4] can explain the obtained experimental results.

## 4. Conclusions

1. The decay instability of the transmuted chemical elements obtained in the PVR setup was revealed in this work.
2. There is a considerable decay-instability of the new transmuted chemical elements at the time (without plasma action). The chemical analyses were performed 1 week, 1 month, and 6 months after the experiment run.
3. This instability is accelerated dramatically by a factor $10^3$ - $10^6$ by weakly ionized non-equilibrium plasma (WINP) action. Ionization level of this plasma was very small $N_e/N_a \sim 10^{-8}$ - $10^{-7}$. But the electron temperature $T_e$ of this plasma was very high, about of $T_e \sim 1 - 3$ eV.
4. This type of transmuted element instability is very similar to the one revealed in the soil samples after Dalnegorskii's plasmoid explosion [3].
5. Two-step transmutation method of the new chemical element creation in LENR experiment is proposed in this work at the first time.

## References

[1] A. Klimov, N. Evstigneev, et. al., High Energetic Nano-Cluster Plasmoid and Its Soft X-ray, Proc. ICCF-19, J. Condensed Matter Nucl. Sci., V.19, pp. 1-10, 2016

[2] A. Klimov, Calorimetric Measurements in Vortex Plasmoid, Proc. $12^{th}$ RCCNT and BL, Sochi, 2005, C.246

[3] Dalnogorsk. Altitude 611, 04.10.2016, http://mirtayn/ru/vysota-611

[4] V. Gurevich, et.al., Bi-Nuclear Atom, JTPh, V.79, pp. 1-5, 2009

[5] Klimov A.I., Grigorenko A., Efimov A., Evstigneev O., Ryabkov O., Sidorenko M., Soloviev A., Tolkunov B., Bychkov V. High-Energetic Nano-Cluster Plasmoid and Its Soft X-radiation. Optical and X-Radiation Spectra of Heterogeneous Plasmoid. Proc. AIS-2016, Kaliningrad, pp. 287-293

J. Condensed Matter Nucl. Sci. 36 (2022) 312–317

Research Article

# LENR- Experiment on Heterogeneous Hydrocarbon Plasma Jet Interaction with Ni-Foil-Target

A. Klimov*

*Power Engineering Technical University, Moscow, Russia*

A. Pashchina

*Joint Institute for High Temperature RAS, Moscow, Russia*

**Abstract**

Heterogeneous Hydrocarbon Plasma (HHP) was used in LENR experiments in the studies [1], [2] for the first time. This HHP was created by a pulsed erosive capillary plasma generator (PG). The working erosive substance PMMA (polymethyl-methacrylate, monomer-$C_5H_8O_2$) was used in this PG. This work is a continuation of the previous ones [1], [2]. A calorimetric experiment with PG and distant Ni-foil cathode has been carried out in argon atmosphere ($P_{st}$~1 Bar). It was determined that HHP consists of the carbon nano-clusters and hydrogen atoms + hydrogen ions. These species are connected with dissociation of the initial PMMA. The interaction of HHP with thin Ni-foil target (width 0.1 ÷ 1 mm) has been studied in this work. HHP- jet heats, melts and evaporates this thin Ni-foil target. In a result of this interaction a small hole with diameter 1–3 mm in the Ni-foil target was burned by HHP-jet. Parameters of this hole were measured. Ni-foil weight was measured before and after the experiment also. It was revealed that the value COP was about of 5 ÷ 6 in this experiment (where COP = $Q_T/Q_e$, $Q_T$ – thermal energy of heating, melting and evaporating of the Ni-foil-target, $Q_e$-electric energy input for HHP-jet creation). New transmuted chemical elements Li, Al, Ca,... were recorded in the HHP by the optical spectroscopy method, the EDS-method and the MS ICP method.


*Keywords:* LENR, heterogeneous hydrocarbon plasma, PMMA, Ni foil, transmutation

## 1. INTRODUCTION

The pulsed capillary erosive plasma generator (PG) used in this work was described in detail in our previous works [1], [2]. The PG consists of cathode (1), anode (2), dielectric working body (3) with a capillary discharge gap (Fig. 1). Dielectric working body was manufactured from PMMA. This working body was destructed and dissociated by plasma erosion. Hydrogen atoms (or ions) and carbon nano-clusters (final erosion products) are created by powerful pulsed electric discharge in the capillary gap. This PG was used in the gas calorimeter experiment in different tested gases [1].

*Corresponding author: klimov.anatoly@gmail.com

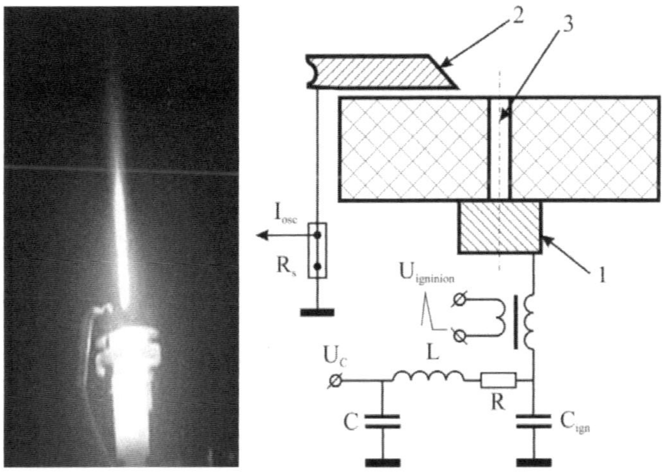


**Figure 1.** Electrical circuit of capillary erosive plasmotron (right). Operation regime of the PG (left). *1 – anode, 2 – cathode, 3 – capillary dielectric gap (PMMA).*

The hydrogen ion flux interacted with carbon nano-clusters and metal nano-clusters (products of electric discharge erosion) in the HHP's volume. An adiabatic gas calorimeter was used in this experiment. It was determined that the coefficient COP was about 1.4 $\div$ 4 in this experiment. The value COP depended on the test gas and PG's operation regime.

## 1.1. New Experimental Setup

The experimental setup used in this work is shown in Fig. 2. The testing section (1) 100 $\times$ 100 $\times$ 200 cm was manufactured from Nylon-6. Two optical windows (2) are arranged in this section. These windows help us to obtain a high-speed video and optical spectra from HHP. These optical spectra from HHP were obtained by an optical spectrometer (AvaSpec 2048).

This section is evacuated by vacuum pump (8) and then filled by argon (7) at the initial pressure $P_{st}$ ~0.25 – 1.5 Bar. The pulsed erosive capillary PG (3) with cathode (5) arranged inside test section (1) is used in this work. The external cathode (5) is manufactured from thin Ni- foil (width 0.1–1 mm). This Ni-target-cathode is grounded. The anode (6) is arranged behind a capillary gap of the PG (3). The anode is manufactured from nickel. Dielectric capillary gap is manufactured from PMMA. The pulsed PG creates an erosive HHP-jet (4). This heterogeneous plasma jet acts on the surface Ni-foil-target (5).

The scheme of this PG's power supply is shown in Fig. 1. The power supply (PS) has the following parameters:

Capacity storage, C – C = 470 $\mu$F,

Filter capacity $C_{ign}$ – $C_{ign}$ = 0,1 $\mu$F,

Ballast inductance L – L = 17 $\mu$Hn and $L_2$ = 225 $\mu$Hn.

PS's voltage- $U_c$ = 600–800 V.

Experimental conditions were the followings:

Cathode target - Ni-foil, Cu-foil, and others.

Cathode foil thickness - 0.1–1 mm.

Distance between cathode's target and PG – 5–20 mm.

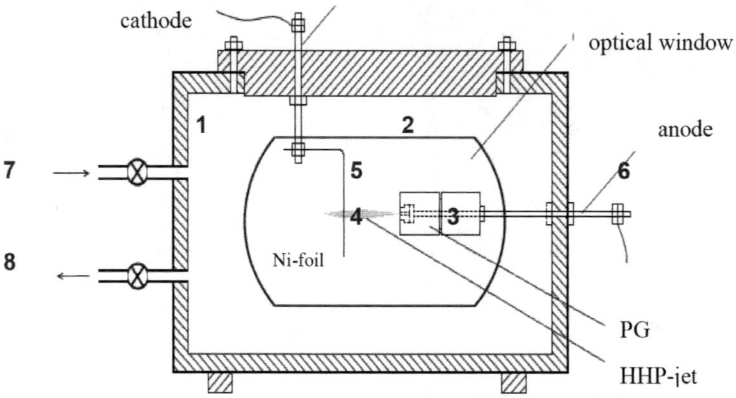


**Figure 2.** Schematic of experimental set up. 1- working section, 2- optical window, 3- erosive capillary plasma generator (PG), 4- HHP, 5- Ni-foil-target, 6- PG's cathode, 7- argon injection, 8- vacuum pump.

*PG's capillary discharge gap*:
Test substance #1 - PMMA ($C_5H_8O_2$), [H2] = 8%
Test substance #2 - C2F4, [H2] = 0%
Test substance #3 – $Al_2O_3$ ceramics, [H2] = 0%
Capillary diameter – 1mm, lengh-5 mm.
*Test gas – Argon*
Static pressure - $P_{st}$= 0.25–1.5 Bar.

## 2. MAIN EXPERIMENTAL RESULTS

1. An experimental study of interaction of a hydrogen ion flux with Ni-nano-clusters (or carbon clusters) has been carried out in argon atmosphere at the static pressure $P_{st}<1.5$ bar. These particles were created by an erosive PG with capillary pulsed electric discharge and Ni-foil-cathode, Fig. 1, (5). This interaction was studied by a high-speed camera. The typical high-speed video frames of interaction HHP-jet with Ni-foil-target is shown in Fig. 3. One can see that there is a hole creation in the Ni-foil-target by HHP-jet at the time delay $T_d>3$ ms after PG's start operation. Then HHP-jet penetrates through this foil target.
2. It was found that there is a hole in the Ni-foil-target created *by erosive plasma jet with hydrogen ions* (tested capillary substance-PMMA), Fig. 4 (left). This hole in the Ni-foil-target is absent when using of HHP-jet without hydrogen ions (tested capillary substance-C2F4 or $Al_20_3$ ceramics). Only a small crater in the metal target is created in this regime, Fig. 4 (right).
3. The hole's diameter and the metal weight decrease in the Ni-foil-target evaporated by HHP- jet were measured.
4. It was determined that there is a considerable extra energy release in this calorimetric experiment. The estimated value COP was about ~5 ÷ 6.

*The LENR power balance* was estimated by the following formulas and the following experimental results obtained in this experiment.

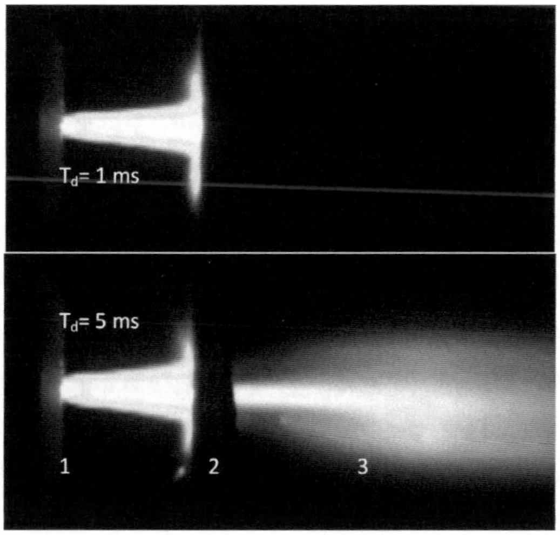


**Figure 3.** Interaction of a pulsed HHP-jet with the thin Ni- plate-target. $T_d$- delay time from PG's ignition. Hole in the Ni- target (down) with diameter 3–5 mm burned by HHP-jet. 1- PG, 2- Ni-foil-target, 3-HHP-jet. Time exposure- 0.2 $\mu$s. Frame frequency- 10 kHz.

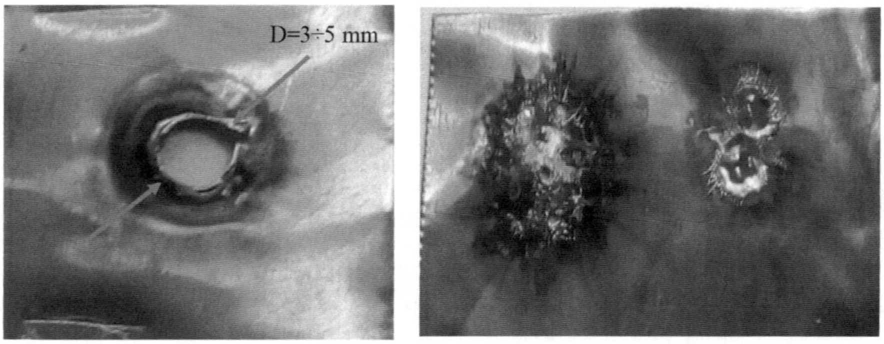


**Figure 4.** Interaction of a pulsed HHP-jet with the thin Ni- plate-target. Left: - small hole in the Ni- target with diameter 3–5 mm burned by HHP-jet. PMMA. Argon, $P_{st} = 0.5$ Bar. Right: - small crater in the Ni- target. C2F4. Argon, $P_{cT} = 0.5$ Bar.

- Capacity storage energy in the PG: $E_c = CU^2/2 = 85$ J.
- Pulse discharge time: $T_i \sim 10$ ms,
- Time duration of the hole's creation in Ni-foil-target: $T_{bn} \sim 3$–5 ms
- Energy input to erosive plasma (from V-Amp signals): $E_p \sim 40$ J,
- Electric energy used for hole's creation: $E_{bn} \sim 0{,}3\ E_p \sim 15$ J, at $T_{bn} \sim 3$ ms and $T_i \sim 10$ ms
- Evaporated metal mass wight drop of Ni-foil target: $\delta M_r \cong 11$ mG

One can estimate a total thermal energy $Q_t$ considering Ni-target's heating ($Q_h$), melting ($Q_m$) and its evaporation ($Q_v$) by HHP-jet:

$$Qt = \Sigma Qi = Q_h + Q_m + Q_v \sim 90J$$

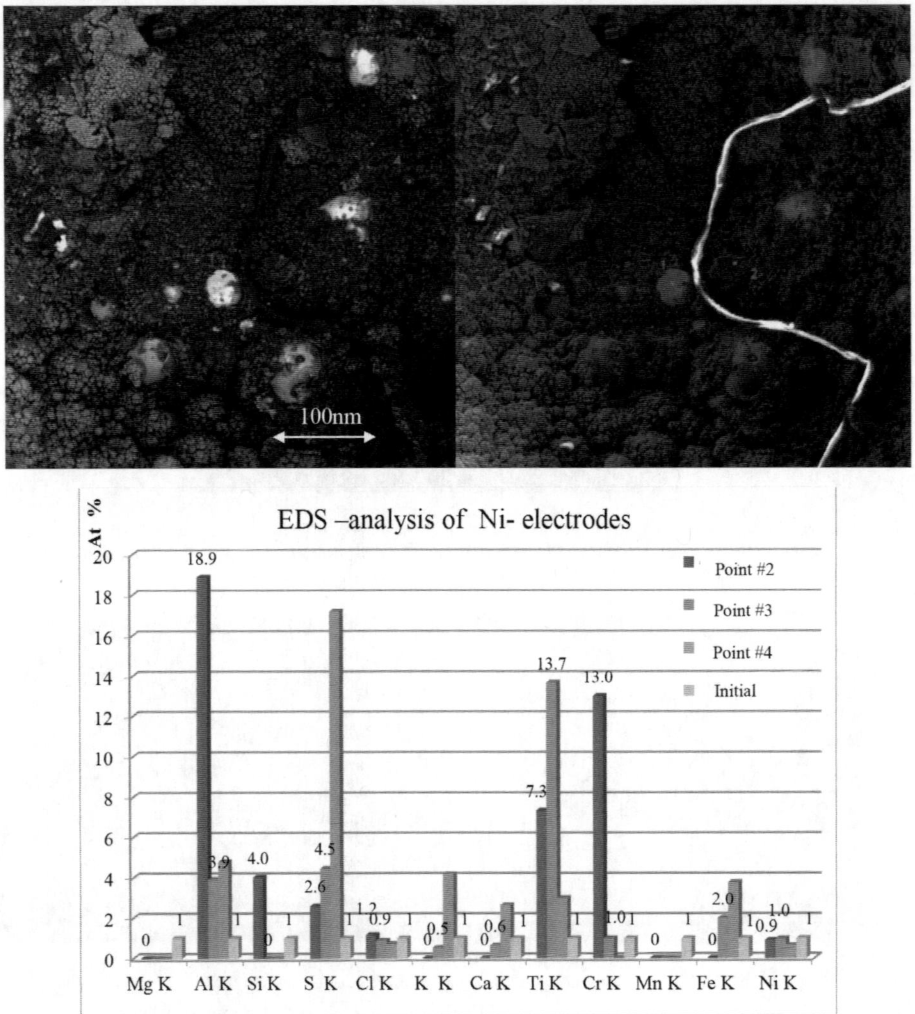


**Figure 5.** EDS analysis of erosive dusty particles. Photo of erosive dusty particles obtained by electron microscope (top). Relative concentrations of the new transmuted chemical elements on the dusty particle's surfaces (bottom).

So, the value COP $= Q_t/E_{bn} \sim$ **6.** Ionization energy of metal atoms and plasma radiation losses are ignored in this estimation of COP value.

A PMMA mass loss of $\delta M_T$ was measured in the PG. This value was about of $\delta M_T \cong$ 0,35 mG/pulse. So, the value of maximum hydrogen atom numbers/pulse created by the PG was about of $N_H \sim 10^{19}$ atoms/pulse.

One can estimate the specific energy q of hydrogen ions with Ni-foil target interaction:

$$q = Qt/N_H \sim 100\text{eV/atom},$$

taking into account $T_{bn} \sim 0.3\ T_i \sim$ 10 ms.

## 3. TRANSMUTATION OF INITIAL CHEMICAL ELEMENTS

The transmutation of initial chemical elements was revealed in this experiment by the EDS-method, MS ICP method and optical spectroscopy method. The typical dusty particles created on Ni-foil surface by HHP-jet are shown in Fig. 4. Relative concentrations of the new transmuted chemical elements (normalized by their initial ones) on a dusty particle's surface are shown in Fig. 5. One can see a considerable concentration jump of the new transmuted elements, such as Al, S, Ti, Fe, Cr and others. Note that pure Ni- electrodes (99.99%) are used in these experiments. These chemical elements are absent in the initial working substance PMMA. It is revealed that new transmuted elements are not stable at plasma action [5]. Optical spectra prove that new elements Al, Ca, Ti, Zn and others are created at the time period of HHP – Ni target interaction only. These elements are strong excited ones. Intensive optical lines of the Ca II, Ti II, Zn II, .... are recorded in this experiment. So, these results prove that there is real transmutation of initial chemical elements in our LENR experiment.

## 4. CONCLUSIONS

1. An experimental study of *hydrogen ion flux with Ni-nano-clusters (and carbon clusters) interaction* has been carried out in argon atmosphere at the static pressure $P_{st} < 1.5$ bar. These particles were created by an erosive plasma generator PG with capillary discharge and distant Ni-foil cathode.
2. Dynamics of Ni-foil-target evaporation by HHP-jet was studied by a high-speed video camera. It was revealed that there is hole creation (or crater creation) on an Ni-target surface.
3. The hole's diameter and the metal weight decrease in Ni-foil burned by HHP- jet were measured.
4. It was revealed that the burned hole in Ni-foil target is created *by erosive plasma jet with hydrogen ions* (working capillary substance-PMMA) *only*. This hole in the Ni-foil target is absent when using an erosive plasma jet without hydrogen ions (working capillary substance-$C_2F_4$ or $Al_2O_3$ ceramics).
5. It was determined that *there is a considerable extra energy release in hydrogen ion flux+ Ni nano-clusters interaction* in this work. The COP value is about ∼6 in this experiment (where COP = $Q_T/Q_e$, $Q_T$ – thermal energy of heating, melting and evaporating of the Ni-foil-target, $Q_e$-electric energy input for HHP-jet creation).
6. *The transmutation of the initial chemical elements* is revealed in this work by EDS-method, MS-ICP method, and optical spectroscopy method.
7. The authors suppose that *LENR* is responsible for extra energy release and chemical element transmutation in this experiment.

## References


[1] Ball Lightning in Laboratory, M. Chemistry, 1994, p. 265.

[2] Klimov A., Could We Use Hydrocarbon Fuel Correctly Today? 5 July 2019, https://regnum.ru

[3] Klimov A., Bityurin V., et.al., Plasma-Assisted Combustion of Hydrocarbon Fuel in High-Speed Airflow, AIAA Paper 0670-2006. p. 10.

[4] Klimov A., Bityurin V., Study of Plasma-Assisted Combustion of Hydrocarbon Fuel in High-Speed Airflow, Combustion Laws, Moscow, RAS, 2006, pp. 93–10.

[5] Klimov A., Decay-Instability of Transmuted Chemical Elements Obtained in LENR Experiment, Abstracts of ICCF-23, I-15, 202, http://ikkem.com/iccf-23.php

J. Condensed Matter Nucl. Sci. 36 (2022) 318–326

Research Article

# Temperature and Pressure Dependence of Anomalous Heat Generation Occurring in Hydrogen Gas Absorption by Metal Powder

Tomotaka Kobayashi*, Junsuke Shigemura, and Ken Naitoh
*Faculty of Science and Engineering, Waseda University, Tokyo, Japan*

Yutaka Mori, Reiko Seto, and Joji Hachisuka
*Technova Inc., Tokyo, Japan*

**Abstract**

As reported in previous papers, anomalous heat generation occurs when nickel or palladium powder is exposed to hydrogen or deuterium gas, while there is also a chemical effect that occurs when hydrogen (or deuterium) gas is absorbed by mixed oxides of Pd and Zr. Many experiments were previously conducted while increasing either initial temperature or pressure in the reaction chamber. In this report, we conducted fundamental experiments of hydrogen gas absorption by metal powder, while increasing both temperature and pressure simultaneously. The experiments were conducted in conditions up to 244°C and 0.5 MPa, using a small constant-volume reaction system developed by us. As a result, about 4 K of temperature rise was observed in the experiment using nickel powder, and about 10 K was observed in the experiment using Pd-Ni-Zr composite powder. In these experimental results, there are positive correlations between the temperature rise and both the initial temperature and loaded gas pressure. This result may show that the temperature rise is assisted by not only preheating but also gas absorption under higher pressure.


*Keywords:* Anomalous heat, Gas absorption, Temperature dependence, Pressure dependence

## 1. INTRODUCTION

Anomalous heat generation when nickel or palladium powder was exposed to hydrogen or deuterium gas was reported in previous papers [1], [2]. It was also reported that anomalous effects occurred when hydrogen (or deuterium) gas is absorbed by mixed oxides of Pd and Zr, which may also be related to a chemical reaction [3]. In many previous studies, experiments were conducted while increasing either initial temperature or pressure in a reaction chamber. A few studies were done at both high temperature and high pressure [4], whereas they are not related to power or heat generation. Systematic data gathering and scanning are necessary to go further.

*Corresponding author: winningshot1996@asagi.waseda.jp

© 2022 ICCF. All rights reserved. ISSN 2227-3123

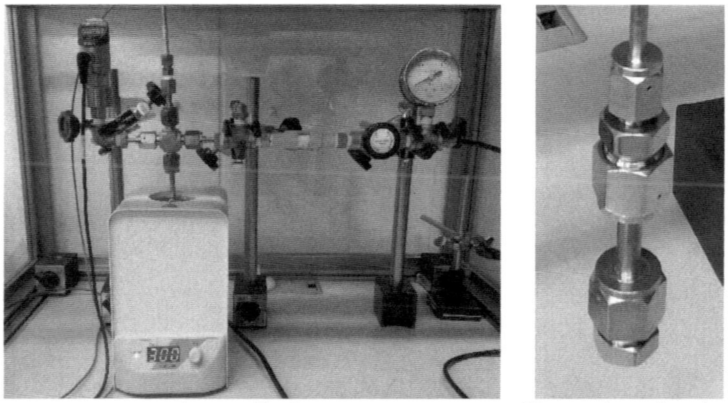

a) overall view b) reaction chamber

**Figure 1.** Reaction system.

In our previous report [5], we developed a small constant-volume reaction system (shown in Fig. 1) to validate excess heat generation, while increasing both temperature and pressure simultaneously. This is because a temperature increase alone may lead to a less powerful reaction due to the possibility that when hydrogen gas is heated, it degasses out of metal powder such as palladium or nickel. In this report, we conducted fundamental experiments of hydrogen gas absorption by metal powder, up to 244°C and 0.5 MPa.

## 2. EXPERIMENTAL DEVICE

Figure 2 shows the outline of the reaction system. The reaction system is constructed of a heater, reaction chamber, K-type thermocouple, two pressure gauges, two valves, gas supply device, and vacuum pump.

The reaction chamber is made of stainless steel (SUS316L). The volume of the reaction chamber is about 2 mL. The heater (GBSD-300, AS ONE CORPORATION) is used for preheating the reaction chamber. The maximum temperature to be set for the heater is 300°C. The K-type thermocouple (HTK0227, Hakko electric co., ltd.), whose tip is embedded in the sample, is used to measure the sample temperature. The gas supply device (LMX2-J-A7, FRONTO Co., Ltd) is used for providing gas while the experiments. The maximum pressure of gas allowed from the gas supply device is 0.7 MPa. The pressure of the gas provided in the reaction chamber is measured by the pressure gauges. The scroll pump (ISP-250C, ANEST IWATA Corporation) is used for evacuation. The ultimate pressure (the lowest pressure that can be reached within a realistic evacuation time) of the vacuum pump is 1.6 Pa.

To prevent damage to the reaction system, a safety valve (RAT2V-1000, IBS Inc.) is installed. It opens when the pressure in the reaction chamber exceeds 1.0 MPa.

Specifications of the reaction system are shown in Table 1.

## 3. EXPERIMENTAL PROCEDURE

The experimental procedure is shown in Fig. 3. First, the sample (nickel powder or Pd-Ni-Zr composite powder, designated PNZ10r [6]) is set in the reaction chamber. Then, the chamber is evacuated by the vacuum pump. After that, the chamber is preheated with the heater. After preheating, hydrogen gas (or nitrogen for the control experiment)

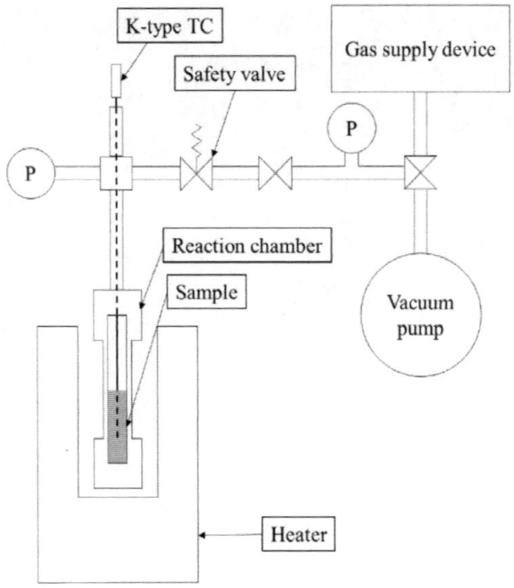


**Figure 2.** Schematic of the reaction system.

**Table 1.** Specifications of the reaction system.

| Material of reaction chamber | SUS316L |
|---|---|
| Volume of reaction chamber | 2 mL |
| Ultimate pressure of vacuum pump | 1.6 Pa |
| Maximum pressure of loading gas | 0.7 MPa |
| Maximum temperature of heater | 300°C |
| Cracking pressure of relief valve | 1.0 MPaG |

is loaded into the chamber. The temperature increase of the sample is measured by the K-type thermocouple. At the end of the experiment, the loaded gas is exhausted, and the chamber is evacuated again to remove all of the loaded gas.

## 4. EXPERIMENTAL RESULTS

Experimental conditions are shown in Table 2. In the experiments, 3.0 g of nickel powder (NFP201S, JFE MINERAL COMPANY, LTD.) or Pd-Ni-Zr composite powder [6] was used as a sample.

The experiments were conducted using hydrogen gas, and nitrogen gas as the control experiment. (The control experiment using nitrogen gas is done before using hydrogen gas.) The experiments were conducted in the range of loaded gas pressure 0.12–0.5 MPa, and initial sample temperatures 95–244°C.

Results and conditions of the other experiments are shown in Table 3.

A result of the experiment using nickel (run No. 4) is shown in Figs. 4 and 5. An orange line for nitrogen gas loaded and a blue line for hydrogen gas loaded in Fig. 4 show the time-histories of temperature for the nickel sample, while Fig. 5 shows the temperature rise from the control experiment for nitrogen gas loaded, i.e., the temperature difference between cases of hydrogen and nitrogen loaded. In the experiment, a temperature rise began about 400 seconds after

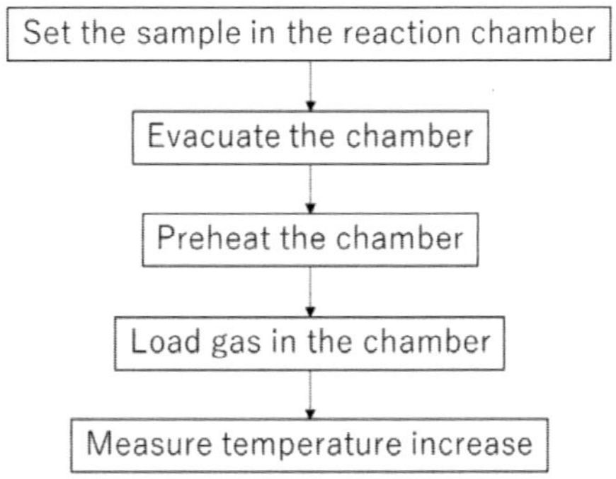


**Figure 3.** Experimental procedure.

**Table 2.** Experimental conditions.

| **Sample** | **Nickel powder or Pd-Ni-Zr composite powder** |
|---|---|
| Sample mass | 3.0 g |
| Kind of gas | Hydrogen or Nitrogen (control experiment) |
| Gas pressure | 0.12-0.5 MPa |
| Initial temperature | 95-244°C |

**Table 3.** Results of the experiments.

| **No.** | **Sample** | **Gas pressure** | **Initial temperature** | **Temperature rise** |
|---|---|---|---|---|
| 1 | Nickel 3.0 g | 0.12 MPa | 239°C | 2.2 K |
| 2 | Nickel 3.0 g | 0.5 MPa | 94°C | 1.2 K |
| 3 | Nickel 3.0 g | 0.5 MPa | 238°C | 4.2 K |
| 4 | Nickel 3.0 g | 0.5 MPa | 240°C | 4.4 K |
| 5 | Nickel 3.0 g | 0.5 MPa | 173°C | 2.1 K |
| 6 | Nickel 3.0 g | 0.5 MPa | 241°C | 3.9 K |
| 7 | Nickel 3.0 g | 0.5 MPa | 173°C | 1.9 K |
| 8 | Nickel 3.0 g | 0.3 MPa | 243°C | 4.0 K |
| 9 | Nickel 3.0 g | 0.3 MPa | 243°C | 3.9 K |
| 10 | Pd-Ni-Zr 3.0 g | 0.5 MPa | 238°C | 6.7 K |
| 11 | Pd-Ni-Zr 3.0 g | 0.12 MPa | 240°C | 1.5 K |
| 12 | Pd-Ni-Zr 3.0 g | 0.5 MPa | 244°C | 10.5 K |
| 13 | Pd-Ni-Zr 3.0 g | 0.3 MPa | 244°C | 4.0 K |

hydrogen gas loading. A temperature increase of about 4 K was observed in the experiment using nickel powder and hydrogen gas.

A result of the experiment using Pd-Ni-Zr (run No. 12) is shown in Figs. 6 and 7. Figure 6 shows the time-histories of temperature for the Pd-Ni-Zr sample, with an orange line for nitrogen gas loaded and a blue line for hydrogen gas, while Fig. 7 shows the temperature rise from the control experiment with nitrogen gas loaded, i.e., the temperature difference between cases of hydrogen and nitrogen loaded. In the experiment, temperature rise began immediately after hydrogen gas loading, and the temperature rise was sustained for about 200 seconds. A temperature rise of about 10 K was observed in the experiment using Pd-Ni-Zr composite powder and hydrogen gas.

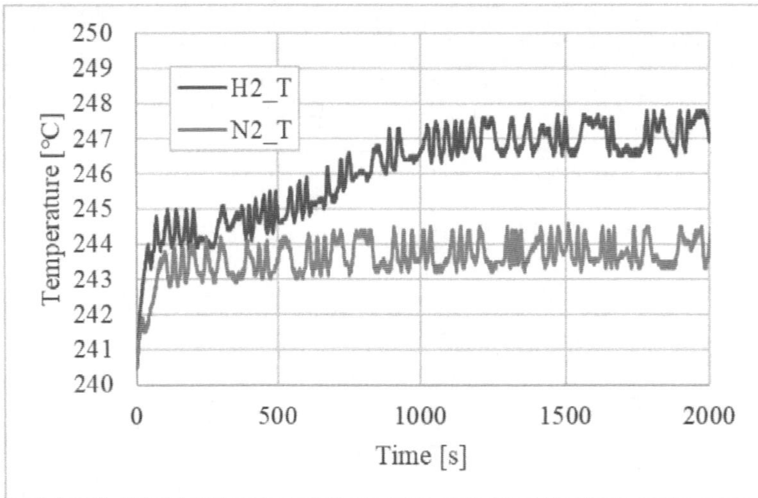


**Figure 4.** Result of the experiment (No. 4).

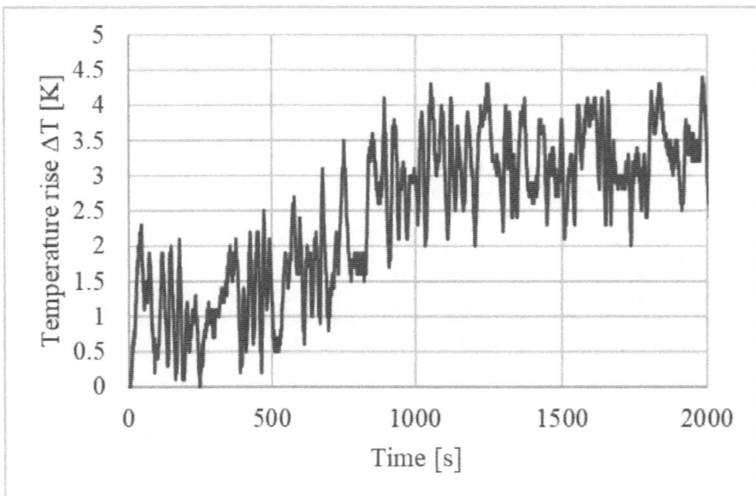


**Figure 5.** Temperature difference from the control experiment (No. 4).

## 5. TEMPERATURE AND PRESSURE DEPENDENCE OF TEMPERATURE RISE

Figure 8-10 shows the experimental temperature rise plotted against initial temperature or gas pressure. According to Fig. 8, there are positive correlations between initial temperature and temperature rise, where the correlation coefficient

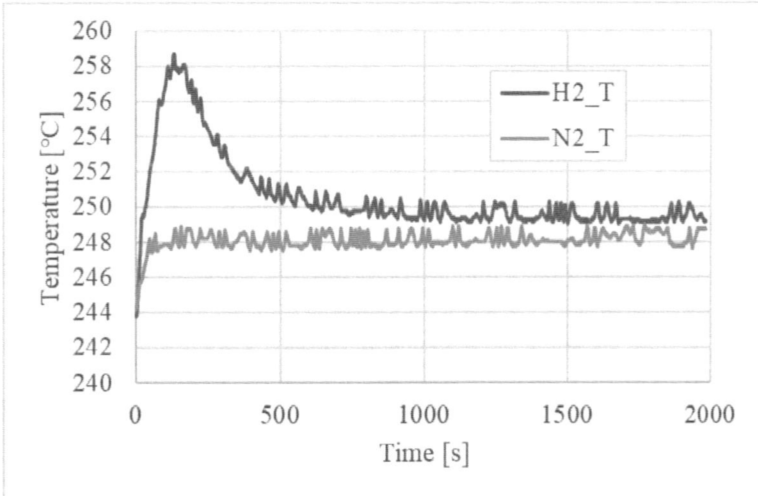


**Figure 6.** Result of the experiment (No. 12).

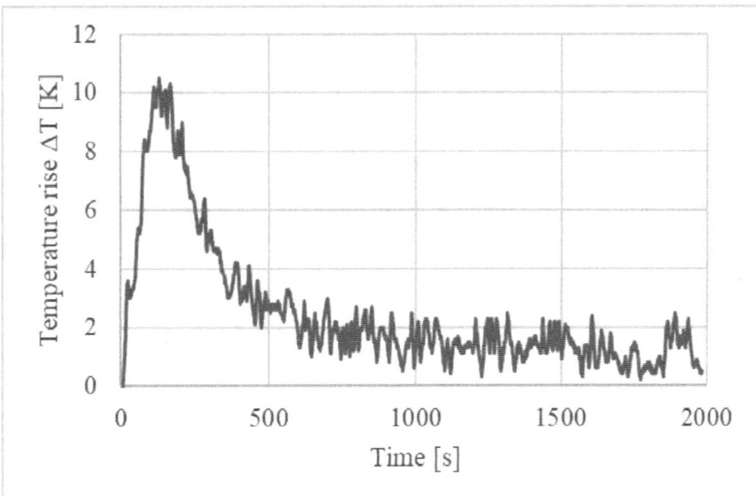


**Figure 7.** Temperature difference from the control experiment (No. 12).

between them is 0.95. In Figs. 9 and 10, there are positive correlations between loaded gas pressure and temperature rise, where the correlation coefficients between them are 0.84 and 0.91, respectively.

These results may show that temperature rise is assisted not only by preheating but also by gas absorption under higher pressure.

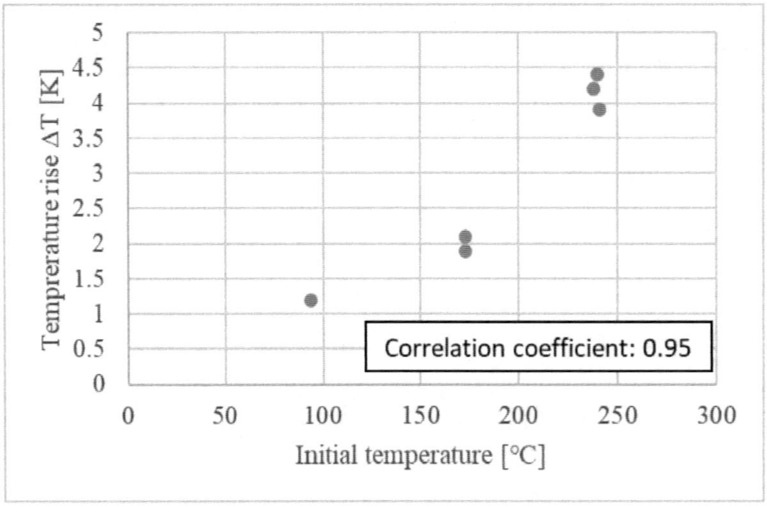


**Figure 8.** Result of the experiments using Ni at 0.5 MPa.

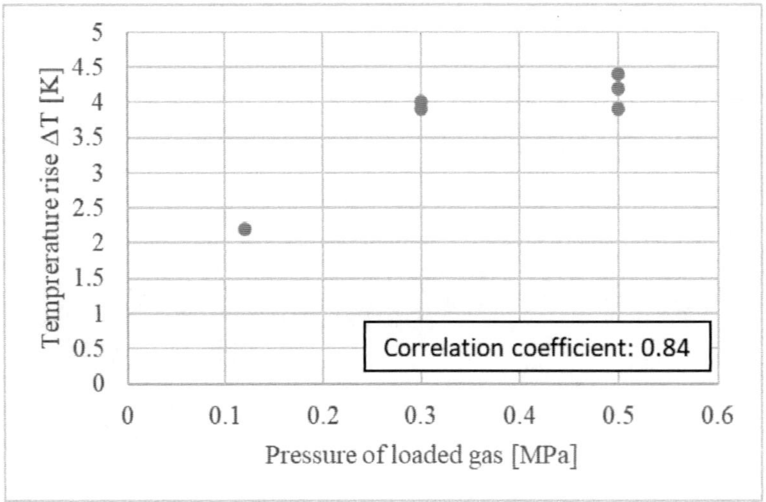


**Figure 9.** Result of the experiments using Ni at about 240°C.

## 6. CONCLUSION

We developed the small constant-volume reaction system and conducted fundamental experiments while increasing both temperature and pressure simultaneously.

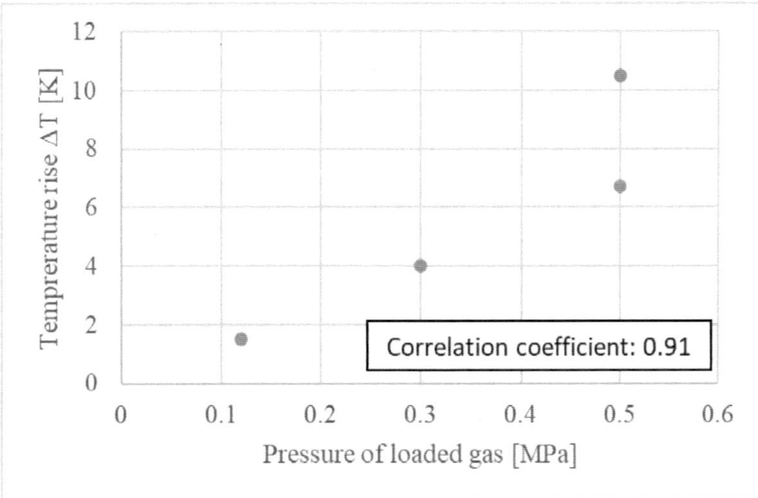


**Figure 10.** Result of the experiments using Pd-Ni-Zr at about 240°C.

As a result, a temperature rise of about 4 K was observed in the experiment conducted for nickel powder, continuing for about 1,000 seconds after the start of gas loading. We also conducted the experiments with Pd-Ni-Zr composite powder and about 10 K of temperature increase as maximum was observed in the experiments, until 200 seconds after the start of gas loading.

There are positive correlations between the temperature rise and both initial temperature and loaded gas pressure. This result may show that the temperature rise is assisted not only by preheating but also by gas absorption under higher pressure.

## 7. OUTLOOK

We developed a prototype of a high-power engine using the focusing compression due to supermulti-jets colliding with pulse [7]. This engine promotes chemical self-ignition of the fuel (hydrocarbon), because of both high temperature and high pressure around the center of the combustion chamber caused by colliding plural pulsated jets. We expect that the region of both high temperature and high pressure will also lead to anomalous heat generation if we put nickel (Pd-Ni-Zr) powder at the chamber center, while using collisions of pulsated jets of hydrogen gas. This is because the experimental results described in Sections 4 and 5, which are conducted under the condition of both high pressure and high temperature, show a possibility for accelerating the reaction in our prototype engine, which may lead to heat release larger than ordinary combustion. We are preparing to conduct some experiments using the prototype engine with hydrogen gas and nickel (or Pd-Ni-Zr composite) powder.


## ACKNOWLEDGMENT

This paper is part of the outcome of research performed under the grant of The Thermal & Electric Energy Technology Foundation (TEET). Sincere thanks are also due to the other members of the Naitoh Laboratory for their help.

## References


[1] A.G. Parkhomov, V.A. Zhigalov, S.N. Zabavin, A.G. Sobolev, T.R. Timerbulatov, "Nickel-hydrogen heat generator continuously working for 7 months," *22nd International Conference on Condensed Matter Nuclear Science (ICCF22) Book of Abstracts*, Assisi, p. 77, 2019.

[2] Y. Arata, Y. Zhang, "The Establishment of Solid Nuclear Fusion Reactor," *J. High Temp. Soc.*, vol. 34, pp. 85–93, 2008.

[3] Kitamura, A., Nohmi, T., Sasaki, Y., Taniike, A., Takahashi, A., Seto, R., & Fujita, Y. (2009). Anomalous effects in charging of Pd powders with high density hydrogen isotopes. *Physics Letters A, 373*(35), 3109–3112. https://doi.org/10.1016/j.physleta.2009.06.061

[4] Yu. N. Bazhutov, E. O. Belousova, V. P. Koretsky, A. G. Parkhomov, et al., "Investigation of Radiation Effects at Loading Ni, Be and $LaNi_5$ by Hydrogen," *J. Condensed Matter Nucl. Sci.*, vol. 13, pp. 19–28, 2014.

[5] T. Kobayashi, K. Naitoh, A. Takahashi, R. Seto, J. Hachisuka, et al. "Development of reaction system with small chamber for fundamental experiments measuring anomalous heat effect," *Proc. 20th Meeting of Japan CF Research Society JCF20*, Fukuoka, pp. 1–8, 2019.

[6] A. Takahashi, H. Ido, A. Hattori, R. Seto, et al., "Latest Progress in Research on AHE and Circumstantial Nuclear Evidence by Interaction of Nano-Metal and H(D)-Gas", *J. Condensed Matter Nucl. Sci.*, vol. 33, pp. 14–32, 2020.

[7] K. Naitoh, et al., "Fundamental Experimental Tests toward Future Cold Fusion Engine Based on Point-compression due to Supermulti-jets Colliding with Pulse (Fusine)," *J. Condensed Matter Nucl. Sci.*, vol. 24, pp. 236–243, 2017.

J. Condensed Matter Nucl. Sci. 36 (2022) 327–335

Research Article

# Upper Bound in the Fusion Products and Transmutation Enhancement in Alloys

Ankit Kumar, Pankaj Jain, K. P. Rajeev

*Department of Physics, Indian Institute of Technology Kanpur, India*

Raj Ganesh Pala

*Department of Chemical Engineering and Materials Science Programme, Indian Institute of Technology Kanpur, India*

**Abstract**

Significant elemental transmutations occur on cathodes comprised of either pure metals (Ni, Cu) or the alloys Kanthal ($Fe_{0.74}Cr_{0.21}Al_{0.05}$) and Ni-Fe ($Ni_{0.93}Fe_{0.07}$) in light water electrolysis highly biased (~30 V) against a pure graphite anode. The utilization of graphite as an anode is advantageous because it avoids anode-driven contamination and establishes a distinct upper bound in the fusion products (Cu), which is different from upper bounds in previous studies (Pb or Ag) with a Pt anode. Whenever isotopic shifts were measured following transmutation, they differed from the natural ratios. Pure Ni cathode predominantly transmutes to Fe and Cu, pure Cu cathode transmutes to Fe and Mg, Kanthal shows the appearance of Cu and Mg and the Ni-Fe alloy transmutes to Mg and Cu. While the extent of transmutation in pure cathodes is a few tens of percent after about twelve hours of Cathodic Hydrogen Loading (CHL), Kanthal and Ni-Fe alloy show comparable or larger amounts of transmutations in less than an hour of CHL.


*Keywords:* Low-energy-nuclear-reactions, Electrolysis, Alloys, Graphite, High-voltage

## 1. INTRODUCTION

In the past thirty years, multiple studies have shown nuclear signatures in various low-energy physical and chemical processes [1]–[3]. The history and motivation of most of these studies trace back to the work of Fleischmann and Pons, who reported that their electrochemical apparatus generated anomalous heat ("excess heat") beyond what can be attributed to chemical reactions [4]. In addition, small amounts of nuclear reaction byproducts (including neutrons and tritium) were also detected in their electrochemical experiments wherein Pd was the cathode. As the focus of the initial studies was mainly on excess heat obtained from H or D fusion, not much attention was paid to the nuclear transmutations of the cathode itself. In addition, to ascertain the conditions for reproducibility of nuclear events, the emphasis was on working with the purest forms of Pd "bulk" cathodes [5].

© 2022 ICCF. All rights reserved. ISSN 2227-3123

Low Energy Nuclear Reactions (LENR) experiments involving electrochemical conditions have been extensively explored, mostly with heavy water and less often with light water [6]. While the role of electrochemical potential in Cathodic Hydrogen Loading (CHL) is beyond debate, arguments have also been made that electronic density in the cathode plays a significant role in reducing the Coulomb barrier, thereby increasing the likelihood of a nuclear reaction at the cathode [7]. The importance of surface and micro/nanostructure has been documented by the codeposition experiments and those involving thin films and metal nanoparticles [8]–[11].

It has also been suggested that there is a crtical lower bound to cathodic hydrogen loading (CHL) below which excess heat effects are absent [12]. The viability of light water as opposed to heavy water-based electrolytes for LENR is yet to be established unambiguously. Circumstantial evidence for nuclear phenomena under electrochemical conditions has been documented via the detection of charged particles and gamma radiation [13]–[15]. Still, the precise conditions necessary for reproducible detection of nuclear products are unclear. However, a review of LENR electrochemical experiments suggests that nuclear transmutations of the cathodes might offer reproducible experimental evidence for nuclear phenomena under electrochemical conditions.

The primary objective of the present work is to explore the contrasting nuclear transmutation behavior of alloys, as opposed to pure metals, when they are utilized as cathodes in highly biased light water electrolysis with a graphite anode. We used Ni, Cu, Kanthal ($Fe_{0.74}Cr_{0.21}Al_{0.05}$) and Ni-Fe ($Ni_{0.93}Fe_{0.07}$) alloy as cathodes and pure graphite was the anode in all the experiments. Transmutation of the cathode is explored as a function of bias voltage (10–30 Volts), current (a few milliamps to 3 A) and time duration (1–24 hours) of CHL. Our experiments with graphite anodes establish a distinct upper bound for fusion products, Cu, in contrast to other fusion products observed in earlier electrolysis driven transmutation with Pt anodes. The spot analysis with Energy Dispersive Spectroscopy (EDS/EDX/EDAX) shows that part of the Ni cathode gets transmuted post-electrolysis to Cu, Mg, and Fe. We also observed that Cu cathode was partially transmuted to Fe and Mg. Transmutation rates are significantly higher in alloys compared to rates in pure metal cathodes. We found that a substantial fraction of Kanthal transmutes to Cu and Mg within one hour of electrolysis for CHL. Similarly, a significant percentage of the Ni-Fe alloy transmuted to Cu and Mg in one hour of CHL electrolysis. The isotopic ratios in Ni and Cu after electrolysis are found to be different from natural abundance ratios. Our experiments also show that the macroscopic control of electrochemical conditions (such as potential bias, time and current) does not adequately control the transmutation outcomes. That is, with the same macroscopic parameters, there are variations in certain features of nuclear transmutation. We are currently exploring upper bounds in fusion products in other cathodes and the underlying reasons for the effectiveness of alloys in enhanced nuclear transmutation.

## 2. EXPERIMENTAL PROCEDURE

Light water electrolysis is performed in a covered borosilicate glass beaker† with generated $H_2$ vented out. Various mechanical and chemical cleaning processes are employed to prepare the samples for the experiment. Cleaning procedures are crucial as the experiment is sensitive to inorganic and organic contaminants. The cathode (wire of diameter 0.5 mm) is first mechanically cleaned using fine emery paper. Next, we chemically clean the cathode by dipping the cathode into a dilute acid (sulphuric or hydrochloric acid) to remove oxides present on the metal. Any organic contaminants are washed off using acetone. The cathode is rinsed with deionized water after every chemical treatment. The cathodes are stored in a desiccator. A pure graphite (99.99% by EDS) cylindrical rod (radius 0.3 cm and total length 10 cm of which 1 cm is dipped in electrolyte) is used as the anode. The use of a graphite anode, instead of metallic anodes such as Pt, is an innovation we introduced. We will elaborate on the implications of graphite anodes for our experimental results later in the discussion section. In addition, using a graphite anode serves to react away the generated $O_2$, avoiding the dangerous and explosive combustion of $H_2$ and $O_2$, thereby significantly improving the safety of the experiment. Pure Ni, pure Cu and the alloys Kanthal and Ni-Fe in wire form (1 cm of cathode wire dipped in electrolyte in all cases) are utilized as cathodes in 1 molar potassium carbonate solution (99% minimum assay with

0.03% chloride, 0.01% nitrate, 0.02% sulphate, 0.001% iron, and 0.1% sodium as maximum impurity), made using deionized light water (pH 7), as the electrolyte. The typical current density range is from 15–22 A/cm$^2$. The high voltage bias generates a significant amount of gases at electrodes, resulting in a significant increase of the electrolyte's resistance, leading to an appreciable increase in temperature of the electrolyte. Some amount of electrolyte is added periodically to compensate for evaporation. EDS/EDX/EDAX, Wavelength Dispersive Spectroscopy [16] (WDS) are used for elemental analysis and Time of Flight Secondary Ion Mass Spectroscopy (ToF SIMS) [17] is employed for isotopic analysis. The instrument used for EDS and WDS is Electron Probe Micro-Analyser (EPMA), JXA-8230 from JEOL and the ToF-SIMS used is PHI TRIFT V NANO ToF from Physical Electronics.

## 3. RESULTS

The results of the elemental analyses are presented in figures 1 to 4. We used WDS and EDS to characterize cathodes and we present the more precise and accurate WDS data in those cases where they were available. We point out that EDS and WDS are surface analysis techniques, where the electron beam penetrates to a depth of about a micrometer and hence the compositional information extracted using these techniques would be applicable to the near-surface regions. We choose four or five locations (or spots), each of size less than a micrometer in diameter on the surface of the sample for spot analysis.

Figure 1 shows a comparison of the WDS of the pure and transmuted Ni cathode. The purity of the Ni electrode, from Alfa Aesar, is nominally 99.5%. Figure 1(a) shows that the Ni purity determined from WDS varies in the range 98.70% to 99.42% from four data points with an average of 99.20%. We also found trace amounts of Cu (0.50%), Fe (0.12%), Mg (0.10%), Cr (0.04%) and K (0.01%). These trace elements are not visible in Fig. 1(a) because of their tiny size. Figure 1(b) presents the elemental analysis after CHL electrolysis. We find the production of 10.97% Cu and 5.39% Mg on average at four spots of the cathode with 11 hours of electrolysis for CHL.

Figure 2 shows a comparison of the EDS of a pure and transmuted Cu cathode (1 M $K_2CO_3$ at 30 Volts, 3.5 A for 12 hours). Figure 2(a) shows the pre-sample analysis of the cathode, showing the presence of 100% Cu at all the five locations chosen on the sample. Post-electrolysis analysis Fig. 2(b) shows the presence of 6% Fe and about 2.2% Mg (on average) on the transmuted sample.

Figure 3 shows a comparison of the EDS of a pure and transmuted Kanthal (Fe-Cr-Al alloy) cathode (1 M $K_2CO_3$ at 30 Volts, 3.5 A for 1 hour.). Figure 3(a) provides the elemental composition of pure Kanthal, which closely matches the expected standard composition of Kanthal. Figure 3(b) shows the composition of the electrolyzed sample. In just one hour, we find Cu (20–53%) and Mg (6–11%) generated in large mass ratios due to transmutations.

Figure 4 shows a comparison of the WDS of a pure and transmuted Ni-Fe alloy (1 M $K_2CO_3$ at 25 Volts, 3.5 A for 1 hour). Figure 4(a) shows 6% Fe and 94% Ni in the cathode before electrolysis. Figure 4(b) shows that, on average, about 1.8% Cu and 5.9% Mg appear in the cathode within an hour of CHL electrolysis.

In Table 1, we present the isotopic abundance of Cu and Ni at three different locations on the cathode after electrolysis of the Ni-Fe alloy with that of their natural abundance obtained using ToF SIMS measurement. We note that the experimental ratios are quite different from their natural ratios. At all three locations, the abundance of $^{65}$Cu has reduced and that of $^{63}$Cu has increased significantly. Similarly, $^{58}$Ni, $^{60}$Ni and $^{62}$Ni abundance has reduced while that of $^{61}$Ni and $^{64}$Ni has increased in the reacted cathode. We note that the increase in the case of $^{64}$Ni is substantial.

The main observations of this work are:

1. Fusion products heavier than Cu are not generated in any of the highly biased light-water electrolysis conducted with four different cathodes, namely, either pure metals (Ni, Cu) or alloys Kanthal ($Fe_{0.74}Cr_{0.21}Al_{0.05}$) and Ni-Fe ($Ni_{0.93}Fe_{0.07}$) against graphite anode. This sharply contrasts the observations of other heavier fusion products when Pt is used as the anode [18]–[19].

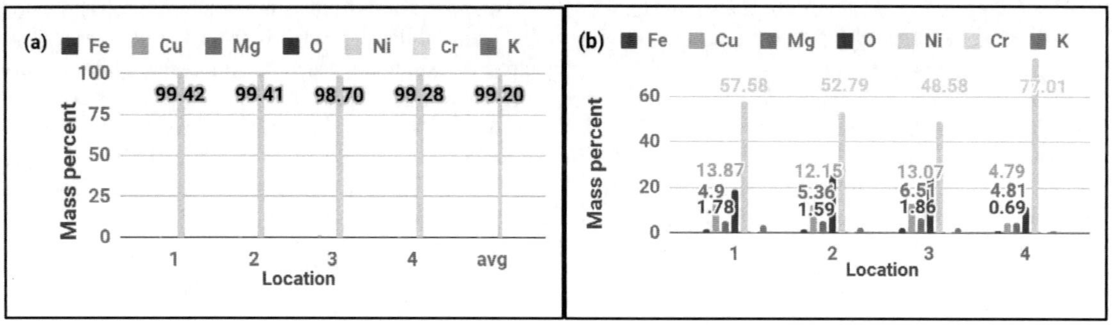


**Figure 1.** WDS of a Ni cathode before (a) and after (b) electrolysis in 1 M $K_2CO_3$ at 30 Volts, 2.5 A for 11 hours.

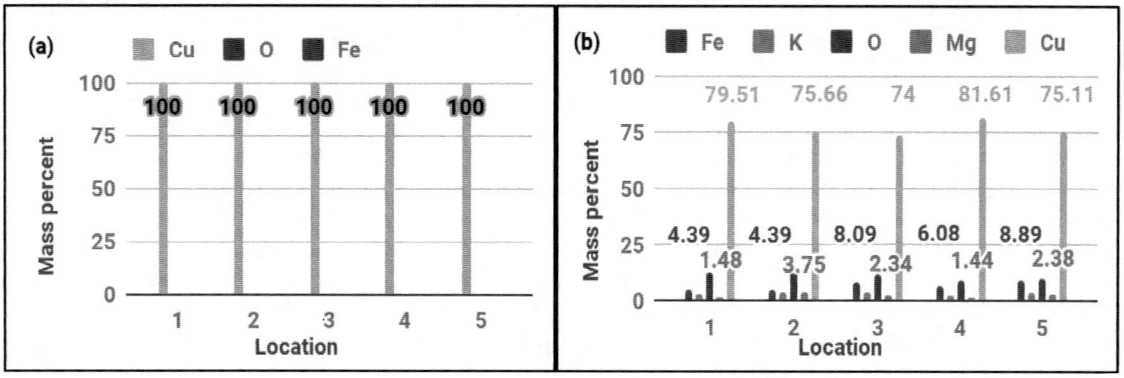


**Figure 2.** EDS of a Cu cathode before (a) and after (b) electrolysis in 1 M $K_2CO_3$ at 30 Volts, 3.5 A for 12 hours.

**Table 1.** Comparison of Cu and Ni isotopic ratios at three different cathode locations post electrolysis with that of natural ratios (ToF SIMS).

| Isotope | Natural ratios | Location 1 | Location 2 | Location 3 |
|---|---|---|---|---|
| $^{63}$Cu | 69.17 | 78.91 | 75.31 | 77.46 |
| $^{65}$Cu | 30.83 | 21.08 | 24.68 | 22.53 |
| $^{58}$Ni | 68.07 | 67.63 | 55.98 | 67.38 |
| $^{60}$Ni | 26.22 | 24.37 | 19.57 | 24.26 |
| $^{61}$Ni | 1.13 | 1.63 | 1.91 | 1.72 |
| $^{62}$Ni | 3.63 | 3.4 | 2.15 | 3.54 |
| $^{64}$Ni | 0.95 | 2.97 | 20.33 | 3.1 |

2. The utilization of graphite as an anode is beneficial not only in establishing the distinct fusion product (Cu) from the present highly biased electrolysis experiments, but also in avoiding other anode driven contamination.
3. With an Ni cathode, lower mass number Ni isotopes are depleted while there is a build-up of higher mass number Ni isotopes (Fig. 1, Table 1). In the case of Kanthal, composed of Fe, Cr and Al to begin with, the

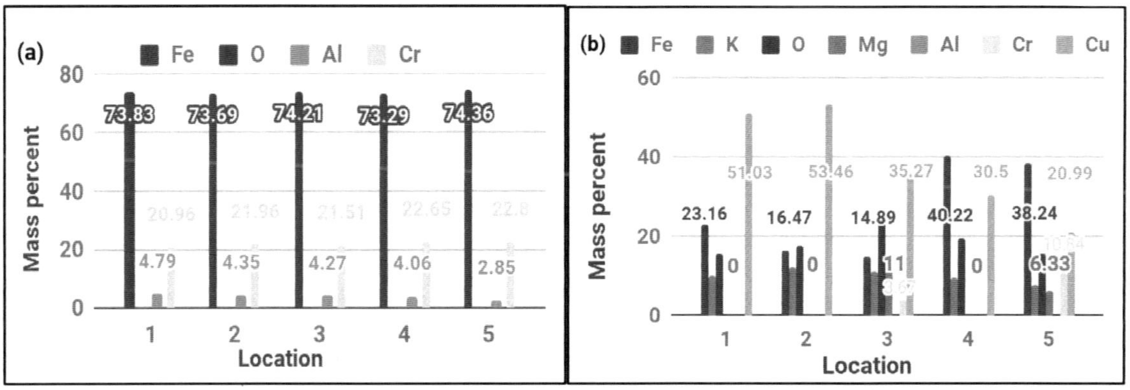


**Figure 3.** EDS of a pure and transmuted Kanthal cathode before (a) and after (b) electrolysis in 1 M $K_2CO_3$ at 30 Volts, 3.5 A for 1 hour.

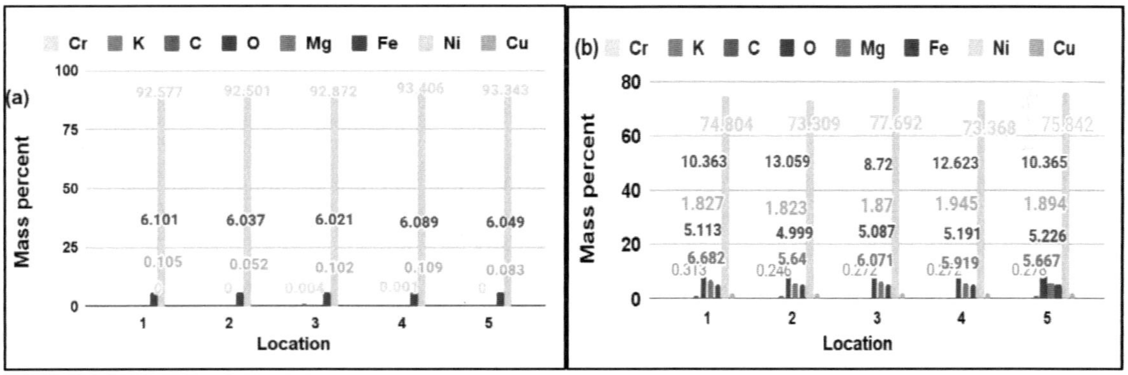


**Figure 4.** WDS of Ni-Fe cathode before (a) and after (b) electrolysis in 1 M $K_2CO_3$ at 25 Volts, 3.5 A for 1 hour.

heavier element Cu and also lighter elements such as Mg are generated. Nuclear transmutation into lower mass number elements such as Mg (all figures), Fe (Fig. 2) is also observed in the case of Cu cathode.

4. Observation of transmutation in pure Cu cathode (Fig. 2) goes against the general belief of the LENR community.
5. Alloys require less time for cathodic H loading than pure metals, as seen in the case of enhanced transmutations in Kanthal (Fig. 3) and Ni-Fe alloy (Fig. 4) in contrast to pure Ni (Fig. 1) and Cu (Fig. 2).

## 4. DISCUSSION

First, let us consider whether contamination can explain the presence of the new elements found in the cathodes. It is very difficult to conceive of a possibility that the new elements found on the cathodes are coming from the contaminants in electrolytes, electrodes or beakers. Even if one considers the possibility of contamination, the amount of contaminants should certainly be many times less than the observed amount of elements found in the cathode. Furthermore, the presence of contaminants would make it impossible to explain the isotopic shifts away from the natural ratios because the contaminants will tend to move these abundance values towards their natural values. Hence, we conclude that some novel nuclear reactions are definitely occurring in the cathodes in our electrolysis experiments, resulting in the transmutation products observed in the cathodes.

Now we begin our discussion of the main observations of this work listed at the end of the previous section.

### 4.1. Elemental Transmutations

Elemental transmutations were observed in all four experiments, whether we used pure metal or alloy cathodes. However, we point out that not all of these processes are easy to explain in terms of standard nuclear reactions, even if we ignore the problem of the Coulomb barrier.

In recent papers, the authors have proposed that a deuteron can undergo a nuclear reaction with a heavy nucleus in which the deuteron gets broken up, leading to neutron capture by the heavy nucleus [20], [21]. Hence, we may propose the reaction $Ni^{60} + d \rightarrow Ni^{61} + p$ to rationalize the increase in $^{61}Ni$ abundance over its natural abundance. But we note that such a mechanism does not explain how $^{58}Ni$ can get transmuted to $Ni^{60}$. We may conceive of the reaction $Ni^{58} + d \rightarrow Ni^{59} + p$ with subsequent weak decay of $Ni^{59}$ into $Co^{59}$, however, the ground-state of $Ni^{59}$ has a half-life of 76000 years! It may be possible that some excited states might be involved and such possibilities are to be explored in the future. Also, we do not find any experimental evidence for any nuclide with a mass number 59 in our ToF SIMS data. If we admit the possibility of deuteron capture, then we have $^{58}Ni + d \rightarrow {}^{60}Cu \rightarrow {}^{59}Ni + e^+$ with $^{60}Cu$ having a lifetime of only 24 minutes. We note that the availability of deuterons is not an issue because the abundance of deuterons in hydrogen is about 150 ppm and we expect this fraction of deuterons to be available in light water. However, it is not clear how deuterons can get captured by a heavy nucleus because of the large Coulomb repulsion between them.

We find transmutation products with mass numbers ranging both below and above the atomic mass number of the host metal in the cathode. To explain how elements with lower mass numbers can appear in the cathode, we have to invoke the possibility of nuclear fission. The presence of Mg probably can only be accounted for in this fashion.

The presence of K in the cathode in Kanthal is probably due to the $K^+$ ions in the electrolyte ($K_2CO_3$ in water) being driven into the cathode during the electrolysis process. The carbon and oxygen found on the cathodes could probably be accounted for as contaminants picked up from the atmosphere or during handling. We note that they could also be transmutation or fission products.

### 4.2. Shorter (Longer) Hydrogen Loading Time in Alloys (Pure Metals)

From the data presented in Figs. 1–4, we find that the pure metal cathodes require about ten times more time of CHL to produce roughly the same amount of transmutation products compared to alloy cathodes Kanthal and Ni-Fe. We found a conversion of about 40% of the Nickel from the nickel cathode in about 11 hours of electrolysis for CHL (Fig. 1), whereas in the case of Kanthal and Ni-Fe about 60% and 20% respectively of the original material gets transmuted to other elements in one hour of CHL (Figs. 3 & 4). We repeated this experiment many times and found that the alloys always take much less time for CHL than pure metals in showing transmutations. The reason for this behaviour of the alloys remains a mystery. Typically, alloys, in comparison to pure metals are better hydrogen evolution electrocatalysts [22]. Hence, greater amount of hydrogen at higher chemical potential will be available when alloys are utilized as cathodes. In addition, under the high bias in the light-water electrolysis, $K^+$ from the electrolyte will be cathodically reduced and alloyed with the cathode. Such an addition of electropositive metals to transition metals has been shown to enhance the hydrogen loading capacity of the alloys [23]. Please note that the CHL time corresponds to the time of electrolysis. While it is most likely that transmutation occurs during CHL/electrolysis, the possibility that transmutation occurs post-electrolsysis (say during EDS/WDS) cannot be neglected. Currently, our experiments are directed towards addressing this issue.

### 4.3. Copper as Cathode

We find transmutation products with pure Cu cathodes (Fig. 2). After 12 hours of CHL electrolysis, about 5% Fe and 2% Mg show up in the Cu electrode. The formation of these elements can probably be accounted for via a fission process. We point out that this may be the first time a LENR electrolysis experiment with a pure copper cathode has been conducted and the evidence of fission of copper nuclei presented. Most of the electrolysis experiments in LENR have been performed with Pd or Ni cathodes and the majority of the LENR community has been under the impression that nuclear reactions can occur only with these two elements as cathodes.

### 4.4. Graphite Anode Limits Contamination and Transmutation Artifacts at Cathode

The heaviest transmuted element found in all of our experiments with a graphite anode (irrespective of whether the cathode is Ni, Cu, Kanthal or Ni-Fe alloy) is Cu. In other electrolysis studies with Ni cathodes, and *Pt* anodes, the heaviest element found in the Ni cathode was Pb [11], [18]–[19]. Contrasting these studies suggests that the anode is not a passive participant in these electrochemical activated nuclear reactions. Based on the Pourbaix diagram of the anodes, we anticipate that under sufficient anodic potential, the anodes like Pt will get oxidized and dissolve in the electrolyte [24]. Subsequently, these ions can get reduced/deposited on the working cathode and may undergo transmutation and fission. Also, if the metal anode contains any impurities, they may get deposited on the cathode, which might lead to confounding experimental results due to the nuclear reactions of the impurities. Our data suggest that the graphite anode indirectly prevented the formation of heavier elements (beyond Cu) at the cathode. Overall, to avoid contamination from the anode and associated artifacts in transmutation products at the cathode, we suggest the utilization of a graphite anode.

### 4.5. Upper Bound on Fusion Products at the Cathode

In an effort to understand some general trends, we contrast the fusion products from our experiments in light of previous experiments. We observe two fusion reactions: 1. Ni to Cu and 2. Fe or Cr to Cu. The transmutation products from Pd/D co-electrodeposition on Au foil by Mosier-Boss et al. [25] yielded Cr, Fe and Ni only and hence, no fusion products were observed. A similar Pd/D co-electrodeposition experiment by Dash and Ambadkar [26] using Pd (anode) and Pd (cathode) in $H_2SO_4$-$D_2O$ electrolyte yielded Ag as a fusion product. The light water-electrolysis by Miley and Patterson with multilayer thin films of Pd and/or Ni cathode and Pt anode generated a large variety of fission/fusion transmutation products [19]. Four broad peaks occurring around Z =6–18 (Mg-Si), Z =22–30 (Fe-Zn), Z = 44–50 (Ag-Cd), and Z =75–85 (at Au) with Pb as the heaviest element found on the beads. The $D_2$ gas permeation over Cs, Sr, Ba, or other elements deposited over Pd/CaO by Iwamura [27] displayed transmutation of Cs into Pr, Sr into Mo, Ba into Sm, and W into Pt and the heaviest fusion product was Pb. Heavy water electrolysis by Mizuno et al. [28] yielded Pb as the heaviest fusion product.

First, in all transmutations observed via electrolysis, there is an upper bound of observed fusion products. Neither the model of Widom and Larsen [29] nor other models for transmutation [30]–[31] provide guidelines for rationalizing the upper bound for fusion products.

Second, as none of the above experiments were performed with graphite as anode, it will be insightful to repeat these experiments without a Pt anode to evaluate if the fusion products differ in the two scenarios.

## 5. CONCLUSION

We demonstrate nuclear transmutation in cathodes made of pure Ni, pure Cu, Ni-Fe alloy and Kanthal via light water electrolysis undertaken at a voltage bias of 25–30 V against graphite anodes at about 100°C and atmospheric pressure, for 1 to 12 hours of cathodic hydrogen loading. Cu is the heaviest fusion product on the cathode, and fission products such as Mg are observed. Cu used as a cathode with a graphite anode leads to induced fission processes in the Cu cathode. Graphite used as anode suppresses the formation of certain heavy elements on the cathode observed in earlier electrolysis transmutation experiments, implying that the anode actively participates in LENR processes occurring in the cathode. Results also demonstrate that alloys as cathodes are better than pure elements in inducing transmutations.

## ACKNOWLEDGMENT

The authors acknowledge the support extended by the Advanced Centre for Materials Science at the Indian Institute of Technology (IIT) Kanpur, and the Sophisticated Analytical Instrument Facility at IIT Bombay.

## ENDNOTES

† Borosilicate composition: we use beakers from Borosil™ brand the composition of which is, from their website, $SiO_2$ 81%, $B_2O_3$ 13%, $Na_2O$ / $K_2O$ 4%, $Al_2O_3$ 2% by weight).

## References

[1] J.P. Biberian, Ed. Cold Fusion: Advances in Condensed Matter Nuclear Science. Elsevier, 2020.
[2] E. Storms, Science Of Low Energy Nuclear Reaction, The: A Comprehensive Compilation of Evidence And Explanations About Cold Fusion. World Scientific, 2007.
[3] E. Storms, The explanation of low energy nuclear reaction: an examination of the relationship between observation and explanation. Infinite Energy, 2014.
[4] M. Fleischmann and S. Pons, “Electrochemically Induced Nuclear-Fusion of Deuterium” Journal of Electroanalytical Chemistry, vol. 261, no. 2a, pp. 301–308, Apr 10 1989.
[5] M. C. McKubre, “Electrochemical loading to produce the Fleischmann-Pons heat effect (FPHE)” in Cold Fusion, J.P. Biberian, Ed.: Elsevier, 2020, pp. 37–54.
[6] P. Mosier-Boss and L. Forsley, “Review of Pd/D co-deposition” in Cold Fusion, J.P. Biberian, Ed.: Elsevier, 2020, pp. 17–36.
[7] J. Kasagi, Y. Honda, and K. Fang, “Screening energy for low energy nuclear reactions in condensed matter” in Cold Fusion, J.P. Biberian, Ed.: Elsevier, 2020, pp. 167–187.
[8] T. Mizuno, “Method of controlling a chemically-induced nuclear reaction in metal nanoparticles” J. Condensed Matter Nucl. Sci, , J.P. Biberian, Ed.: vol. 13, pp. 422–431, 2014.
[9] T. Itoh, Y. Iwamura, J. Kasagi, and H. Shishido, “Anomalous excess heat generated by the interaction between nano-structured Pd/Ni surface and $D_2/H_2$ gas” J. Condensed Matter Nucl. Sci, J.P. Biberian, Ed.: vol. 24, pp. 179–190, 2017.
[10] S. Szpak, P. Mosier-Boss, C. Young, and F. Gordon, “The effect of an external electric field on surface morphology of co-deposited Pd/D films” Journal of Electroanalytical Chemistry, vol. 580, no. 2, pp. 284–290, 2005.
[11] M. Srinivasan and K. Rajeev, “Transmutations and Isotopic shifts in LENR Experiments” in Cold Fusion, J.P. Biberian, Ed.: Elsevier, 2020, pp. 233–262.
[12] F. Tanzella, S. Crouch-Baker, A. McKeown, M. McKubre, M. Williams, and S. Wing, “Parameters affecting the loading of hydrogen isotopes into palladium cathodes” in Sixth International Conference on Cold Fusion, Progress in New Hydrogen Energy, 1996, p. 171.
[13] A. Lipson, A. Roussetski, G. Miley, and C. Castano, “In-situ charged particles and X-ray detection in Pd thin film-cathodes during electrolysis in $Li_2SO_4/H_2O$” in ICCF9, Ninth International Conference on Cold Fusion, 2002.
[14] S. Focardi, V. Gabbani, V. Montalbano, F. Piantelli, S. Veronesi, and S. Veronesi, “Evidence of electromagnetic radiation from Ni-H systems” in Condensed Matter Nuclear Science, J.P. Biberian, Ed.: World Scientific, 2006, pp. 70-80.
[15] J. Dash and D. Chicea, “Changes In the radioactivity, topography, and surface composition of uranium after hydrogen loading by aqueous electrolysis” in Condensed Matter Nuclear Science, J.P. Biberian, Ed.: World Scientific, 2006, pp. 463–474.
[16] J. I. Goldstein, D. E. Newbury, J. R. Michael, N. W. Ritchie, J. H. J. Scott, and D. C. Joy, Scanning electron microscopy and X-ray microanalysis. Springer, 2017.

[17] A. M. Spool, The Practice of TOF-SIMS: Time of Flight Secondary Ion Mass Spectrometry. Momentum Press, 2016.
[18] K. P. Rajeev and D. Gaur, "Evidence for nuclear transmutations in Ni–H electrolysis" J. Condensed Matter Nucl. Sci, vol. 24, pp. 278–283, 2017.
[19] G. H. Miley and J. A. Patterson, "Nuclear transmutations in thin-film nickel coatings undergoing electrolysis" Journal of New Energy, vol. 1, no. 3, pp. 5–30, 1996.
[20] P. Jain, A. Kumar, R. G. Pala and K. P. Rajeev, "Photon-induced low-energy nuclear reactions" Pramana, pp. 1–14, 2022.
[21] P. Jain, A. Kumar, K. Ramkumar, K. P. Rajeev and R. G. Pala, "Low Energy Nuclear fusion with Two Photon Emission", Journal of Condensed Matter Nuclear Science, vol. 35, pp. 1–20, March 2022.
[22] J. Greeley, T. F. Jaramillo, J. Bonde, I. B. Chorkendorff, and J. K. Norskov, "Computational high-throughput screening of electrocatalytic materials for hydrogen evolution" Nature Materials, vol. 5, no. 11, pp. 909–913, Nov 2006.
[23] H. Mizoguchi, S. W. Park, and H. Hosono, "A View on Formation Gap in Transition Metal Hydrides and Its Collapse" Journal of the American Chemical Society, vol. 143, no. 30, pp. 11345–11348, Aug 4 2021.
[24] R. Chen et al., "Use of Platinum as the Counter Electrode to Study the Activity of Nonprecious Metal Catalysts for the Hydrogen Evolution Reaction" ACS Energy Letters, vol. 2, no. 5, pp. 1070–1075, May 2017.
[25] P. A. Mosier-Boss, S. Szpak, F. E. Gordon, and L. P. G. Forsley, "Triple tracks in CR-39 as the result of Pd-D Co-deposition: evidence of energetic neutrons" Naturwissenschaften, vol. 96, no. 1, pp. 135–142, Jan 2009.
[26] J. Dash and A. Ambadkar, "Co-deposition of palladium with hydrogen isotopes" in Condensed Matter Nuclear Science, J.P. Biberian, Ed.: World Scientific, 2006, pp. 477–484.
[27] Y. Iwamura, T. Itoh, and S. Tsuruga, "Transmutation reactions induced by deuterium permeation through nano-structured palladium multilayer thin film" Current Science, vol. 108, no. 4, pp. 628–632, Feb 25 2015.
[28] T. Mizuno et al., "Anomalous isotopic distribution of elements deposited on palladium induced by cathodic electrolysis" (in Japanese), Denki Kagaku, vol. 64, no. 11, pp. 1160–1165, Nov 1996.
[29] Y. N. Srivastava, A. Widom, and L. Larsen, "A primer for electroweak induced low-energy nuclear reactions" Pramana-Journal of Physics, vol. 75, no. 4, pp. 617–637, Oct 2010.
[30] P. Kalman and T. Keszthelyi, "Solid state internal conversion" Physical Review C, vol. 69, no. 3, Mar 2004.
[31] S. Bartalucci, V. I. Vysotskii, and M. V. Vysotskyy, "Correlated states and nuclear reactions: An experimental test with low energy beams" Physical Review Accelerators and Beams, vol. 22, no. 5, May 31 2019.

J. Condensed Matter Nucl. Sci. 36 (2022) 336–352

Research Article

# Composite Model(s) for Low Energy Nuclear Reactions in the Solid State: II

A. Meulenberg*
*Science for Humanity Trust, Inc., Tucker, GA, USA*

K. P. Sinha
*Department of Physics, Indian Institute of Science, (ret), Bangalore, 560012, India*

**Abstract**

Three partial models for Low-Energy Nuclear Fusion (LENR) of hydrogen, H, or deuterium, D, are brought together in the context of 30 years of searching for the answer to the source of nuclear fusion without the requisite kinetic energy to overcome a nuclear Coulomb barrier. The earliest of these D+D cold fusion (CF) models is Julian Schwinger's proposal (1990) to combine, in a single Hamiltonian, the attractive nuclear potential with the repulsive Coulomb potential to reach an excited state of $^4$He. The second was a pair of papers by Maly and Vavra (M&V) that, a few years later, provided a relativistic quantum mechanical model basis for deep electron orbits. A third was K.P. Sinha's 1999 model to use a natural electron pairing in a lattice to form charge-polarized $D_+$ $D_-$ pairs in a linear defect that is attractive rather than repulsive. Critical in Schwinger's model is the ability to dissipate nuclear energy before the actual fusion of the nuclei takes place. Critical to M&V's model is a deep-electron orbit predicted by relativistic quantum mechanics that, if occupied in deuterium, assured nuclear fusion. Critical in Sinha's model is the ability to free the hydrogen sub-lattice spacing in a linear defect from the encompassing crystal's lattice spacing. This present paper includes a connection with the more recently developed model (post-2015) of the deep-electron orbits and even more recent information (2019 and 2020) on new PdD phases and on models of properties of the hydrogen "chain". The importance and connection of each model's contributions with recent information are highlighted. We are not proposing a specific model supported with detailed calculations. We are suggesting that (aspects of) prior models can be combined to form a unified (composite) model that would be more convincing to the other members of the Condensed Matter Nuclear Science (CMNS) community (and others) than they would be individually.


*Keywords:* Early cold-fusion models, Linear defects, Nuclear potentials, Sub-lattice motion, Deep-orbit electrons

*Corresponding author: mules333@gmail.com

© 2022 ICCF. All rights reserved. ISSN 2227-3123

## 1. Introduction

Different models of the Low-Energy Nuclear Reaction have come and gone. Some were wrong and died a natural death, or are still being developed and 'pushed'. Some did not properly or fully represent the data available. Some were just premature and were not able to be appreciated at the time. We will address three of the last type and indicate how presentation and analysis of more recent models have brought them together. The present paper is an updated version of the unpublished text [1] of an oral presentation at ICCF-18. A poster presentation [2] in 2013 at that conference, provided a pictorial description of phonon activity in a linear array of D or H atoms that could be a common basis for the three models.

In 1990, Julian Schwinger, from his broad theoretical-physics background, presented a model for fusion of deuterium atoms into helium [3]. Breaking with the tradition of "non-interaction between nuclear and atomic forces", his model involved a coupling of the nuclear energy of D-D fusion into the PdD lattice, apparently even before the deuterons could be expected to access the non-Coulombic nuclear potential energy. Because of the low-energy conditions of the pre-fusion system and of the zero angular-momentum $^4$He states available, the energy was only slowly coupled into the PdD lattice phonons while the deuterons were still at large distances (many pico-meters) from each other and long before the expected post-fusion nuclear reactions, e.g., nucleon exchange and subsequent fragmentation, of the excited 4He* nucleus could take place.

In 1993 and 1995, Maly and Vavra showed that the Klein-Gordon and Dirac theories (both accepted relativistic quantum mechanical models) provided the basis for real deep-electron orbits with binding energies for hydrogen in excess of 500 keV [4], [5]. They gave details of such orbits and stated that, if occupied in deuterium, nuclear fusion was assured. More than 20 years later, the validity of this work has been confirmed and its value has become apparent. Its implications and applications have now been explored and greatly extended in the context of CF.

In 2000, K. P. Sinha, from his theoretical and solid-state physics backgrounds, presented a particular model for longitudinal-optical-phonon motion of multiple deuterium atoms in a linear defect [6]. The interaction of their bound electrons with the lattice and sub-lattice provided conditions that created high-effective-mass electrons in an independent deuterium sub-lattice within that defect.[1] These one-dimensional, atomic-orbital, electrons provided super-strong screening that allowed a tightly-bound linear molecule to form and to precipitate D-D fusion. This electron-lattice interaction model has been developed [7], [8] and extended [9], [10], [11] over the intervening years to address, not just the fusion of deuterium in the PdD lattice, but also issues of experimental conditions leading to enhanced LENR results [12], [13]. These extensions led to classical deep-electron transits, which were then supported by relativistic quantum mechanics in the form of deep-electron orbits (see Synthesis 3 below) in solutions of the Klein-Gordon and Dirac equations, e.g., [4]. The deep-electron models include a mechanism for: avoiding the fragmentation products

[1] For many years, the present co-author (AM) thought that Sinha was referring to the "effective" mass of an electron due only to Pd-lattice constraints. Since this solid-state physics concept is related to electron transport between lattice sites, he did not feel that it could be applied to atomic-electron levels. Only with the recognition that Sinha was referring to the linear hydrogen sub-lattice spacing (not that of the palladium lattice) and to the geometric (see ref. [33]) and material (field) constraints imposed by the lattice on the sub-lattice components, did he come to understand the full picture. Just as the effective mass of an electron is altered by the barriers between lattice sites ($m^* = F/a$), it is altered by its coupling with lattice electrons and by the potentials that can confine it to a reduced volume about its nucleus. Sinha's calculation (for his Eq. (21)), estimating the possible effective mass of the bound electron and based on available data, suggested it to be $m^* = 200\ m_e$. At the time, though it was known that strong internal and external fields and relativity could increase the energies and effective mass of bound electrons, this value appeared to be impossibly high. However, this is the range for effective mass of the highly-relativistic, deep-orbit, electrons. Was Sinha's high value; possible, entirely coincidental, a misinterpretation of the equations or parameters involved, or wishful thinking on his part? Relativity is not mentioned in Sinha's paper; but, it becomes very important in the 3rd Synthesis below. With this potential for misunderstanding in mind, these sentences were written hoping to clarify Sinha's intent in reference [6]. While the confinement $\Delta x$ of an electron is related to its momentum, $\Delta p$, (from the Heisenberg Uncertainty Relation, $\Delta p \cdot \Delta x \geq \hbar/2$), the relativistic effect, relating its mass to its energy, increases its effective mass ($m^* = \gamma m_o = E/c^2$) as well as decreasing its minimum size at velocities approaching the speed of light.

of 'hot' fusion [14], [15], [16], the means of coupling the nuclear energy of fusion into the lattice [17], [18]; the LENR results of the NiH system [19]; and the observations of transmutation in both the deuterium and light-hydrogen systems [20], [21].

Thus, by 1999, there was already a theoretical basis for a composite model of cold fusion. Sinha had provided a QM basis for bringing protons closer together in a (perhaps?) unique lattice-enabled solid-state ensemble. Schwinger had provided a nuclear "boost" to the attractive Coulomb potential drawing them closer. Maly and Va'vra had given the relativistic basis for electron stability in near-nuclear orbits. Sinha's lattice-based effective mass of the nuclear-bound electrons (effective at the lattice and atomic levels) was replaced by the relativistic effective mass of the deep-orbit electrons at the near-nuclear level. Sinha's and Schwinger's models provided possible (non-photonic?) mechanisms for populating the deep-orbits of hydrogen atoms; thereby greatly increasing their fusion probability, and that in a manner quite different from that of hot fusion.

Why were the pre-2000, fully-developed, theoretical-modeling papers ignored for so long? Why did it take so many years for these papers and their ideas to come together? Have they actually come together even now? In 1990, the field was new and Schwinger's model was presented to explain one of the major observations and anomalies at that time. A Nobel laureate and one of the best theoretical physicists at the time, Schwinger was publically ridiculed by some physicists for his non-traditional idea. On the experimental side, above-fragmentation deuterium fusion, a normal result of 100 million degree temperatures in the sun or multi-MeV accelerator beams in the laboratory, was known to produce neutrons and protons in equal numbers from d-d collisions. The 'cold' fusion results had already shown that the neutrons could be produced at a rate of less than $10^{-6}$ that of the protons (most often determined by the level of tritium resulting from the reaction). Thus, Schwinger was proposing a theoretical basis to provide a model to predict data that was being challenged and was already changing at the time. It was presented at a time that cold fusion was already beginning to be classified as "pseudo-science'. The 'moving' target of data, the novelty of his physics, and the changed perception of the field meant that his work appeared to have been misdirected and that, already, the small community of physicists capable of understanding and appreciating his work had nearly disappeared from this new field.

By 1993, when Maly and Vavra first published their work, there were few left in the CF field who could understand it, even if they had seen it and had time away from their own work to digest it (both are problems even today). In Sinha's case, the difficult description of a new concept in solid-state physics and the problems with combining terminology from three published versions (by 3 different authors) of a controversial development [22], [23], [24] in that field blurred the mathematical implications and obscured some of the descriptive statements that are important in this synthesis. The model that Sinha proposed is not expressed in modern physics. Therefore, from either the description or the mathematics, there is a problem of misinterpretation on the part of a reader because of a "language barrier". Sinha tried to express the problem mathematically by taking appropriate limits of a basic equation. However, in that activity and in those "words", there appeared to be a violation of accepted solid-state physics (the constancy of average crystal-lattice spacing). The co-author (AM) of this paper recognized this issue and he fought long and hard over it with Sinha in 2006. They worked together on many papers since then and did not revisit the issue until Schwinger's paper was found and Ed Storms expressed what appeared to be a similar thought in 2012 [25].

The key to the Sinha picture that differs from conventional solid-state models involves the average spacing between local atomic oscillators in a crystal. Since the physics models were developed in the context of a crystal with a fixed periodic structure, the Einstein oscillations [26] are considered to be about fixed points/spacing within the lattice. Sinha's insight came from the limit equations for a linear system of oscillators indicating that the system energy diminished with the lattice spacing. Since lower total energy means greater stability, the natural result would be a collapse of the spacing to zero. What is different about the Sinha system that allows it to do this when no other system does so? The difference is in the defects within the lattice. If the defects are "filled" with H or D, then the distance between these "hydrons" may not be tightly bound by the lattice spacing. If the hydrons form a sub-lattice, then the quantum mechanical description can be applied to it as a unique structure, influenced by the supporting lattice - but not defined by it.

This paper is not a review of the field of Low-Energy Nuclear Reaction (LENR) theory. It is more of a directed perspective. It provides the mathematical and physical basis that combines the concepts expressed in the Schwinger-Sinha (SS) and the relativistic QM models. It also describes how some other LENR models have contributed to this fuller picture that could perhaps become the theoretical foundation for explaining, among other things, the phenomenon of generating heat without energetic particles from the fusion of two deuterons.

The Schwinger model was specific to the PdD system. Nevertheless, the concept of combining the nuclear and Coulomb potentials in a common Hamiltonian to describe the low-energy, but complex, system applies to the NiH system (and others) as well. The Sinha model is generic. It applies equally to the deuterium and hydrogen (D and H) systems. Therefore, the use of H and D, or p and d for the atoms and ions under discussion throughout the paper, will be interchangeable (unless specifically noted).

## 2. Background

### 2.1. Overcoming the Coulomb barrier

One of the major problems with trying to achieve fusion of two nuclei is overcoming their mutual Coulomb barrier. When far apart, the neutral atoms have no repulsion. When the atoms are very close, their atomic-electron clouds, which cannot be readily shrunk beyond certain limits but can join to become a common molecular cloud, extend beyond both nuclei. As a result, the effective negative charge between the nuclei decreases with their reduced spacing. Kinetic energy of the electrons keeps them too far from the deuterons, on average, to counter even the screened nuclear-Coulomb barrier when the deuterons get too close together. This loss of compensating charge is the source of the transition from a zero to a strong repulsion with decreasing nuclear separation and is the reason that molecules have a nominally fixed spacing between atoms ($\sim$ 2A $= 0.2$ nm $= 200$ pm).

If two nuclei can be brought closer together (e.g., 1500 fm $=$ 1.5 pm) within the confines of a lattice and by their mutual oscillations, the electron distribution becomes less isotropic as the electrons are forced into alignment along the nuclear axis. For s-orbital electrons, it is not just a shift in average position of the electron cloud relative to its protons, altering a dipole moment. Lattice confinement can provide an alignment force shifting the axis of the instantaneously linear orbits. Any such alignment of the electron orbits along the axis between the nuclei further reduces the net distance between electrons and protons because it increases the average electron density between the protons.

Figures 1, from [27], indicate the effects of these alignments and proximity in the form of their electron probability distribution. The order-of-magnitude scale difference in the axes shows the strong compression of electron distribution into a cylindrical geometry and the concentration of the electrons into the nuclear regions. This greatly increases the percent of total time electrons spend between the nuclei. Nevertheless, the nuclear Coulomb repulsion has also greatly increased with proton proximity. Much of the time there is with no electron between the nuclei to provide screening of the repulsive barrier and the nuclei are kept apart.

In free space at normal molecular distances, the electron's kinetic energy prevents the electron cloud from being tightly 'packed' into the positive potential between the nuclei. The charge dipole-dipole polarization of the protons adds to some extent as an attractive force; but, once the wave functions of the two electrons overlap sufficiently, further 'self-compression' is not possible and the potential barrier remains. Thus, even this short-range $1/r^4$ attraction disappears for small r in normal atoms because, beyond a certain point, charge polarization is no longer as effective. This leaves the relatively large spacing (comparable to the average electron orbital radii) between nuclei in a molecule.

Akito Takahashi proposed a technique for overcoming this problem via a tetrahedral deuterium configuration [28] that counters the bound-electron's tendency to provide insufficient screening. The tight clustering of multiple atoms and their electrons means that the electron orbits overlap in the central region and are thus able to greatly compensate the

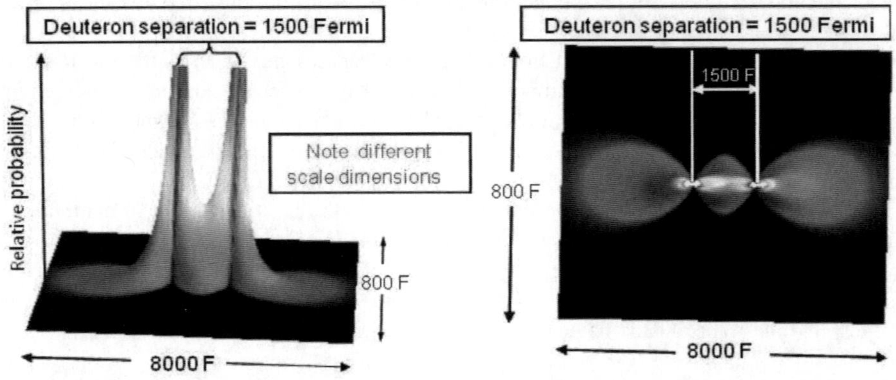


**Figure 1.** An example of the probability distribution of an electron bound to a pair of protons separated far below the normal molecular spacing [27].

accumulating positive nuclear charge when the atoms come together. In free space, and in most materials, the negative-charge distribution of nearly-isotropic, overlapping, electron clouds is not adequate to compensate the repulsive charge of the protons of the net-neutral atoms. Takahashi's concept of embedding the deuterium cluster within a lattice provided two benefits. By sharing their bound electrons with the lattice atoms, the deuterium atoms have less electron-cloud repulsion to overcome and the concentrated and oriented negative charge of the overlapping electron clouds is able to better screen the charge of the deuterons. By being in a sub-lattice, the deuterons can benefit from both the energy and, possibly, the coherent displacements of the collective modes of lattice oscillation (phonons) to help push the deuterium atoms together.

While Takahashi's model is a valid means of reducing or overcoming the nuclear Coulomb barrier, of overcoming the weaker, but significant, electron barrier, and of using the lattice phonon motions to provide sufficient energy to overcome the remaining barriers, a problem remains. The probability of bound nucleons in 4 separate sub-lattices achieving the synchronous motion necessary to simultaneously induce and take advantage of the lowered barrier seems low. Nevertheless, Takahashi has identified a means of overcoming the Coulomb barrier within the framework of a low-temperature environment. Remember, 1 eV is equivalent to $\sim$11,000°C and the compression alone of a hydrogen atom to a 1-D configuration, during the phonon-induced cycling, reduces the electron ground-state energy by over 40 eV. **This is an example of a geometric multiplier acting as a catalyst.**

K. P. Sinha, in his 2000 Infinite Energy paper [6], proposed the lochon model to produce $D^+$ $D^-$ pairs that are attractive rather than repulsive - at least for sufficient time for fusion to take place. This pairing concept goes back to a 1963 paper of his [29] (and to other authors also exploring the phonon-electron interaction at the time) to explain experimentally observed aspects of superconductivity. The pairing, in physical space (rather than in momentum space), is allowed by spin-spin coupling between the s-orbital electrons. Furthermore, the increased polarization of the local host lattice provides an increase in effective mass, m*, of the bound electrons [30]. The increased effective mass leads to a deepening of the electron orbits and thus to a reduction in the 'size' of the atom. This can equivalently be attributed to the alteration of the shape of the local potential well confining the atom and its electrons. This shrinkage of the atom size produces an energy component due to localization effects. The extra energy means that the electron energy level for the $D^-$ or $H^-$ ion in the lattice is lowered relative to the Fermi level or to the neutral-ion level and this negative state becomes more populated.

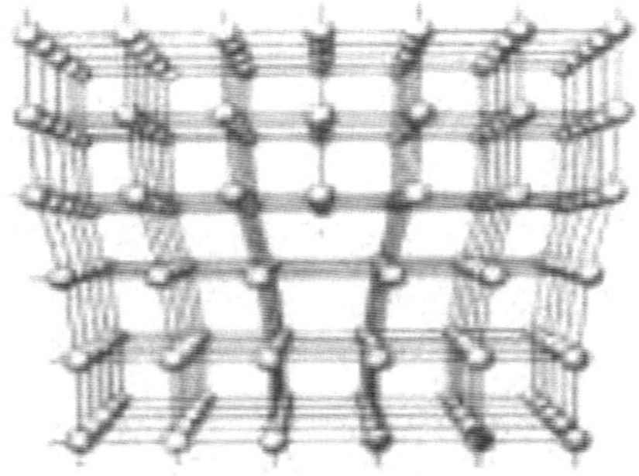

**Figure 2.** Example of a linear defect that could constrain a linear $H_n$ sub-lattice with less than the normal sub-lattice barriers, interposed lattice atoms, and lattice constraints.

Sinha's model depended on a reduction of the 'effective spacing' within the linear array of deuterium atoms to increase the effective mass of the electrons when treating the array as a linear lattice. Sinha's quantum mechanics appeared to provide the numbers supporting his claim as long as the spacing of the linear array, l, could decrease significantly in a regular manner (see Eq. (1)). This cannot happen within a PdD lattice. However, Sinha had considered that point and wrote that "When the atoms are located on regular crystal lattices, the shrinking in their size ... is constrained by the lattice constants." He continued by stating that, "... when the chains of deuterons lie in crevices, voids or empty channels of the matrix, the conditions are different. They will have their own spacing in the chain."

Is it possible to find such conditions within the Pd lattice? Fig. 2 shows one such possibility. However, a recent study [31] suggests that this type of dislocation defect might be more important as a source of vacancies in forming and leading to islands of delta $\delta$ and delta $\delta'$ phase in Pd that may be an even better fit to our requirements for a 1-D channel or "tube". This $\delta$-phase possibility is explored further and is proposed to be a site for the linear defect as needed for CF [32].

What Sinha apparently realized was that, even given an appropriate defect, the situation was extremely complicated and the system energy (Eq. (1), now modified to include a dependence on l as a variable) depended on multiple variables that were not independent. When he tried to establish the energy dependencies on the multiple parameters, he did not fully appreciate that not only were the two standard variables (x, the displacement of the nuclei from their equilibrium location, and r, the distance between the nuclei and the nearest electron) interelated, but all of the parameters in the equation were actually variable within different dimensional regimes and had a non-linear interdependence as well. The energy terms that depend on the proton-electron, the electron-electron, and the electron-lattice interactions, include a displaced oscillator mass, $M_d$, that produces a displaced oscillator energy, $E_d$, with an oscillator frequency, $\omega_d$, and displacement, x; the radius of influence of a bound charge (from polarization of the lattice), $r_i$; the distance between the electrons and their nearest proton, r; the effective dielectric constant of the local lattice region between the proton and electron, $\varepsilon^*$; the sub-lattice spacing, $\ell$; the displacement from this spacing, x; and the effective mass, m*, of a bare electron within the surrounding lattice. Thus,

$$E(x, \ell, r) = \hbar^2/2m^*r^2 - e^2/\varepsilon^*r - x\omega_d(2M_dE_d)^{1/2} + 1/2x^2\omega_d^2M_d(r_i/\ell)^3, \quad (1)$$

with $\hbar = h/2\pi$ (h* is used for $\hbar$ in Sinha's paper). The first two terms on the right are repectively the electron kinetic and potential energies of the hydrogen atom (perhaps extended by bond sharing) in the lattice. The kinetic energy is expressed in terms of the momentum related to the deBroglie wavelength. The last two terms are the 'displaced' oscillator kinetic and potential energies tying the hydrogen atom's kinetic and potential energy to the lattice. In this latter case, the ratio of the range of influence $r_i$ within the lattice to its spacing, $\ell$, within the sub-lattice is critical.

When the electron gets close to the nucleus, then $r_i \Rightarrow r$, the atomic radius, and this becomes a limiting factor in determining the sub-lattice spacing. The atomic radius in the hydrogen sub-lattice is not isotropic. It is compressed by the lattice defect walls and becomes 2-D [33] or 1-D like, even at low hydron concentrations within the sub-lattice. However, as more hydrogen atoms are 'crammed' into the defect by lattice loading, their high concentration further compresses and deepens the Coulomb potential well for the electrons. Thus, r can shrink toward 1-D values or even be reduced in all three dimensions (as in Fig. 1). This shrinkage and change from monopole- (of the seperated electrons and nuclei) to dipole-field distributions (of the closely-coupled pairs) alters the charge interaction with the lattice and thus the dielectric constant and effective mass of the electrons. So, the 'constants' of Eq. (1), including the displacement energy and oscillator mass and frequencies, have interdependences that are defect and loading dependent and can change at several levels of loading and at each of the several spatial regimes to be considered.

It was an almost impossible task to treat the equation without assuming that all of the 'constants' would be constant over the distance of interest from the host-lattice spacing to the near nuclear regime. But, with that assumption, strange effects were observed. Furthermore, this QM model does not include relativitic effects or even electron-spin coupling (which includes the Pauli exclusion principle). While this coupling leads to dimerization in the longitudinal-optical-phonon mode of the deuteron sub-lattice, it prevents significant shrinking of the sub-lattice spacing if the D-bound electrons remain in s-orbitals. However, recent modeling of the linear-H chain [34] has shown that for a 40 atom chain, the metal-insulator transition (MIT) occurs at an inter-proton distance of $R = \sim 1.7$ Bohr radii (or about 4/3 of the free-space $H_2$-molecule spacing). The transition is attributed to the "mixing" of the 1s, 2s and 2p states. The 2n states have a high enough density-of-states and radius that, if partially populated, will allow ready transfer of electrons along the chain without violating the Pauli principle. The paper suggests that significant non-thermal population of these 2p states begins at $R = \sim 1$ Bohr radii ($a_B$), which, at $\sim$53 pm, is much greater than the spacing chosen for Fig. 1. Thus, with the hydron sub-lattice spacing controlled by the Pd lattice ($d_{PdD} = \sim 405$ pm $> 7.5 a_B$) and the large energy difference between the ground state and excited levels, there should be little probability of significant bound-electron transport along the linear-D chain even with further considerations. It is this transport that could allow greater screening of the repulsive Coulomb forces between the sub-lattice nuclei.

The further considerations, in the present model, include a Pd lattice structure defect that confines the D-chain to linearity without "forcing" the sub-lattice spacing to that of the lattice. It is highly unlikely that this lateral confinement alone will reduce the sub-lattice spacing by the factor of 4 to 7 needed to fit the modeled 2p initial-filling condition for PdD. Nevertheless, it can confine the filling of 2p orbitals to only the ones along the D-chain axis. Since the transverse size of the 2p orbitals are calculated (for a linear chain without lateral confinement) to be about 5 $a_B$ [34], the aligned orbitals should be capable of forming sigma bonds between the hydrons (initially about 7.5 $a_B$ apart) as electrons begin to move into the 2p orbitals.

The sigma bonds (with both s and p electrons, if populated) will provide electron transport along the D chain to reduce the repulsive nuclear-charge effects and allow the chain to shrink together under the right circumstances (e.g., a high local Fermi level, a high hydron loading, and sufficiently weak coupling to the Pd lattice). The local fermi level is affected by both the sub-lattice spacing and the substrate itself (e.g., material, temperature, and doping). The higher the fermi level, the greater the electron population within the sub-lattice; then, the greater the screening of the protons' Coulomb barrier. In the reduced sub-lattice spacing as the hydrons move closer together in their phonon-driven oscillations, momentarily leaving gaps, more hydrons can be injected from the lattice if the hydrogen population is high enough. New modeling of a lattice-bound linear chain in a lattice defect would be required to determine the rate of filling the 2p orbital and its temperature dependence, since results depend on the selected boundary conditions.

### 2.2. Dissipation of nuclear energy without fragmentation

In addition to the problem related to lattice spacing, the present co-author was also unable to accept what he thought was Sinha's means of obtaining the high "effective-mass" electrons that would allow sufficient tunneling to achieve fusion by a muon-like induced-fusion mode. Nevertheless, he was attracted by the $D^+D^-$ model of tunneling through the nuclear-Coulomb barrier and, together, the two of us did the calculations necessary to show that this was possible and produced a result on the order observed in CF experiments [9]. However, like muon-induced fusion, this lochon model would produce fusion with fragmentation. Thus, an extended-lochon model was developed to get beneath the fragmentation levels [14], [16]. This latter effort found support in the decades-long pre-Cold-Fusion struggle over the deep-electron levels predicted by the relativistic Schrodinger equations of quantum mechanics and best expressed in [4].

Early in the history of Cold Fusion, a different approach was proposed (1990) to provide deuterium fusion without neutron emission. When Schwinger proposed a model for a phonon-induced transition between the D-D pair and the 1st excited state of $^4$He* [3], he stated that, unlike the standard 2-step process of tunneling to within the nuclear-potential well and then decay to the excited state(s), a single coherent process, rather than an event, could be considered for the transition. His idea did not gain much traction because it was very theoretical; it required a transfer of nuclear energy to the involved atoms prior to fusion. And, while it explained the dearth of neutrons in the CF process, it still provided fragmentation via the tritium plus proton (t+p) products. Soon thereafter, data started to indicate that the end product of the D-D CF process was almost entirely $^4$He and not tritium. This recognition of an important feature of CF was not integrated into a published model by Schwinger before his death in 1994. Nevertheless, such an extension is trivial.

Peter Hagelstein, for quite a while, logically advocated the decay of $^4$He* to below the fragmentation level, or to the ground state, via phonons from a recoiling nucleus to the lattice [35]. A problem with this model was the need to dispose of a very high level of energy (>20 MeV) in a very short time, via an extremely high number of coherent low-energy phonons, to be able to compete with the extremely-rapid fragmentation process. On the other hand, this model could almost fit with an extension of Schwinger's model that would permit phonon decay to the $^4$He ground state or to an arbitrary level beneath the fragmentation levels, via lattice-atom oscillation, before fusion actually occurred. Ed Storms also proposed such a pre-fusion decay process, but via nuclear photons[2] in a nuclear-active environment (NAE) of the lattice [25].

After ICCF-18, a theoretical effort of getting an electron close enough, for long enough, between a proton pair to draw them close enough for fusion to occur, was devoted to developing the relativistic quantum models that predicted near-nuclear electron orbits (see [36] and the references therein). Such an orbit could contain an electron with kinetic energies and effective mass in the 1 - 100 MeV range. These deep-orbit electrons would automatically lead to nuclear fusion and transmutations, along with the myriad other results observed in cold fusion. In the reviewing of these older models, a mechanism for populating the deep orbits from lattice confinement and distorted atomic- or molecular-electron orbits presented itself. This is further discussed below as Synthesis 3.

### 2.3. Lattice confinement effects

The reduction of the hydrogen sub-lattice volume (via confinement) due to localization in a linear lattice defect, is primarily in the direction of neighboring lattice-array atom(s). It forms the sub-lattice toward a 1- or 2-dimensional structure. This means that orientation of the bound sub-lattice electrons will be preferentially directed along the linear- or planar-array axes. As the population of hydrogen atoms in the 'gap' exceeds its first 'saturation' point, where the

[2]The present authors consider such nuclear photons to be unlikely since there are no known energy levels to provide them. However, the linear-H array would, as a molecule, have molecular levels that would provide a basis for photoemission as the H atoms move closer together on average.

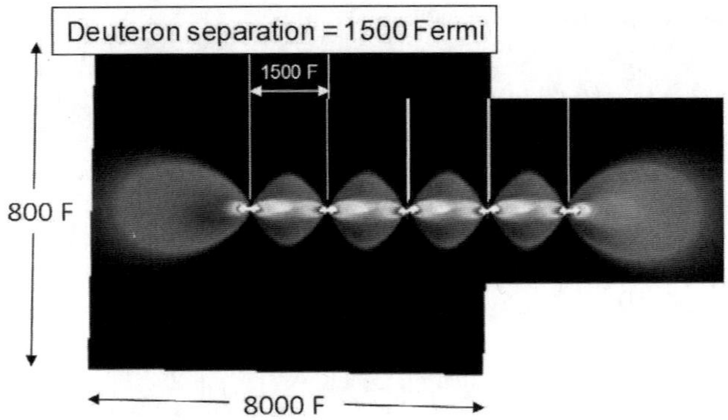


**Figure 3.** A linear-$H_n$ 'molecule'/sub-lattice with reduced spacing and increased linearity (based on an extension of Fig. 1).

number of lattice-provided potential minima are filled, the hydrogen electrons will begin to show a preference for bonding to other hydrogen atoms (dimerisation [37]), rather than to the host-lattice atoms. The molecular spacing will still be affected, but not defined, by the host lattice. Thus, oriented diatomic molecules may begin to form and interact within the confines of the gap. As more hydrogen is 'expressed' into the gap from the host during the H-loading step, the gap population becomes less and less dependent on the host-lattice spacing. A second 'saturation' point occurs when the lattice is filled with diatomic molecules oriented along the common axis (e.g., making the transition from Fig. 1 to Fig. 3). At this point, in a 1-D defect, a second phonon mode becomes dominant, in which pairs of atoms (the diatomic molecules) form a polymer (the linear array) and alternating pairs are cyclically squeezed together in a longitudinal-optical-phonon mode [38].

Ideally, each additional atom in a linear-H array, which is now an n-atom molecule not strongly bound by the lattice, reduces the array's average atomic spacing. The atoms at the end may be less tightly bound than average in this sub-lattice; but, at some point, the inner atoms can come together because there is always an electron between pairs of nuclei. The ability to maintain linearity is reduced by higher temperatures that increase intrusion of the lattice atoms into the linear-defect space and by the off-axis motion of the sub-lattice atoms. On the other hand, the amplitudes of linear-array phonon modes and the population of excited-electron orbitals increase with temperature. This means that at the phonon frequency, which depends on the sub-lattice modes, two or more hydrogen nuclei can come together close enough to greatly increase the probability of fusing.

This new mode, consisting of oriented molecules colliding in alternating pairs, has its own spacing conditions. They are not dependent on the host lattice and not even strongly dependent on the linear-lattice spacing of the hydrogen atoms in the gap. However, as pairs of pairs colliding in a line, its electron distribution has an unusual property. In the normal, molecular-hydrogen configuration (as in Fig. 1), the atoms and molecules are kept apart because the electron population density cannot be confined between the nuclei. It extends in all directions beyond that of the proton-pair spacing. Thus, the nuclear-Coulomb barrier rises as each nucleus moves inside the other's electron cloud and sees a reduced screening. In a linear chain of H atoms (e.g., the 5-atom chain of Fig. 3), when the atoms are close enough, the very electron overlap that prevented collapse of the molecule in the diatomic case, is now concentrated and aligned along the linear-chain axis. It therefore is able to provide additional charge compensation for the adjacent pair. As a consequence, **the inner pairs of nuclei in a linear array have a greatly reduced Coulomb barrier keeping them apart.** The level of approximation to linearity determines the extent of potential-well deepening [33], [39], the probability of electrons decaying to a near-nuclear deep-level [4], and then the probability of paired nuclei fusing.

Any residual barrier is a consequence of the deviation of the electron distribution from the 1-D ideal. However, as the growing phonon oscillations bring the hydrogen atoms closer together the resultant potential well of three (or more) nuclei (i.e., that dominated by the sub-lattice rather than by the lattice) becomes deeper and more one dimensional (Fig. 3). As the bound electrons move in this deepening potential well, they gain kinetic energy that comes from the nuclear Coulomb potential. As the electron orbits are changed by the confinement and motion, the atomic and molecular energy levels become distorted, spread out, and overlapped. Thus, as the electron orbitals begin to overlap, the electron concentration between the sub-lattice protons are free to increase despite the increased mutual repulsive potentials. The Pauli exclusion principle no longer precludes this increase since the extra electrons are not in the same state. In addition, when the nuclei get close enough during their phonon cycle, they experience a small portion of their nuclear attraction and this is a 'boost' in the direction of increased attraction. The extent of boost is dependent on the increase in number of phonons in the driving mode and this addition is distributed along the linear chain and into the lattice. Even if only the Coulomb attraction between nuclei and electrons is considered, this change in nuclear-Coulomb potential energy and fields will alter the average internal structure spacings (and thus mass) of the protons.[3]

The energy transfer from the nuclear potential to the linear chain initially is weaker than the Coulomb potential energy being transferred to the electrons. Nevertheless, both come from the same source, the sub-lattice nuclei. The energy gain to the sub-lattice from both the nuclear and the Coulomb forces increases with time. However, as soon as the nuclear potential begins to contribute significantly to the system's kinetic energy, the process becomes irreversible because that energy is not stored (e.g., in the electron's relativistic and kinetic energy). It is distributed into the lattice phonon field. We now have multiple answers (e.g., Synthesis 3 below) to the question about how atomic electrons can decay to the deep-Dirac levels without single-photon emission.

## 3. Bringing past models together

### 3.1. Synthesis 1: Atomic proximity effects and nuclear-to-lattice energy transfer

Common elements in much of present CMNS thinking on the subject are the longitudinal-optical-phonon motion of a linear-hydrogen sub-lattice and/or a deep-orbit- or multiple-electron presence to provide sufficient screening to aid tunneling/fusion. Sinha and Storms have specifically identified linear defects or voids within the lattice (or at the surface) as logical sites for a special action needed for CF. Sinha had specifically considered the $D^+D^-$ interaction that leads to D-D fusion [6]. Meulenberg identified the longitudinal-optical-phonon mode as a means of forming such pairs as resonant transients in the sub-lattice [2]. Storms considered a centrally located 'electron' as the 'attractor' that initiates a process of continuous photon emission ($\sim$24 MeV worth of IR radiation) leading to the $^4$He ground state. Staker identified negatively-charged vacancies within linear "tubes" of deuterium in the $\delta$ and $\delta'$ phases of PdD that might fit all these requirements [32].

Without identifying a specific structure, Schwinger calculated the probability of phonon coupling/emission adequate to make a transition from the D + D state directly to an excited $^4$He* nuclear state. This differs from Hagelstein's phonon-dissipation model of nuclear energies [35] in several ways:

- Schwinger's model does not provide details on a physical mechanism for the coupling of the nucleus to the lattice. It only provides the theoretical/mathematical connection.
- his energy transfer is a relatively 'slow' process; but, being pre-fusion, there is no competing decay process that would prescribe or proscribe the dissipation rate.

[3] Within the charged-quark model, polarization of the nucleus increases the average distance between the positive and negative quarks. This reduces the internal energy and thus the mass of the nucleons.

- it does not require tunneling through a Coulomb barrier (or a strong overlap of nuclear wave functions), to initiate the fusion process, until the nuclear energy levels are reduced below the neutron fragmentation level.
  - Tunneling is still possible. However, it has a very low probability until the nuclei get closer together later in the process. Early tunneling provides aspects of the known fusion-fragmentation results. Late tunneling gives the cold fusion results. In-between tunneling, would be the source of the greatly-reduced neutron fragmentation and the reduced proton fragmentation that has been observed in many CF experiments.

Therefore, with the above considerations, we can draw out a synthesis of the thoughts:

- Sinha's lochon model, leading to D-D fusion, provides a basis for fusion above the fragmentation levels and may account for the minor neutron emission observed in some of the CF experiments. It also provides the early stages leading to fragmentation-free LENR: the longitudinal optical-phonon mode bringing the lattice-bound hydrogen atoms closer together; the charge separation and increased effective electron mass resulting from coupling to the lattice; and the possibility of decreased spacing of the deuterium ions in a linear defect. These are probably valid and important concepts; but, the basic model must be extended to get below the $^4$He fragmentation levels that identify the conventional fusion process.
- The present authors have explored an approach to satisfying that requirement in the Extended Lochon Model [14], [15].
- The Schwinger, Takahashi, and Storms processes are alternative approaches to getting beneath the fragmentation level and are particularly appropriate to Sinha's defect-based, linear sub-lattice, model.
- Staker's exploration of the $\delta$ and $\delta'$ phases of PdD, perhaps, provides the best location for all of the models to come together.

### 3.2. Synthesis 2: Multi-atomic, linear sub-lattice, interactions

Variations on Storms' model of the central electron and Sinha's lochon model may reflect a real scenario whereby a linear constraint (the crevice, void, or defect) confines the hydrons (the hydrogen species under consideration, H or D). The defect 'walls' and the linear hydrogen structure further reduces the dimensionality of the electron paths/orbits about the defect nuclei. The electrons become even more confined to a 1-D path as the hydrogen atoms get closer together and the screened nuclear-Coulomb potentials overlap (e.g., Fig.1). The summed proton potentials build up a Coulomb barrier between the protons; but, at the same time, this becomes a more one-dimensional 'trap' for the bound electrons.

The bound electrons are still in near ground-state 1s orbitals (zero angular momentum). Thus, they follow near-linear paths that are now oriented by the 1-D confinement. How does this linear, multi-H, structure differ from that of a $H_2$ molecule? As a near 1-D molecule, it has much deeper ground-state energies [33], [39]. Therefore, the average spacing is less than that for a 3-D $H_2$ molecule. Since the longitudinal, optical-phonon modes include multiple atoms coming together in pairs, repeatedly, opportunity exists for additional H-atoms to be inserted into the defect to begin the formation of a higher H-atom-density sub-lattice. The linear sub-lattice could be a $H_3, \ldots, H_m$ molecule (e.g., $m = 5$ in Fig. 3) or a linear structure with alternating H and vacancies, e.g., [32]. The bound 2-s and 2-p electrons can be shared among the multiple paired nuclei and within the sub-lattice as a whole. However, since the 'shrinking' of the molecule increases the electrons' kinetic energies, some/many of them can shift to their higher n-levels (all of which have had their energy levels 'spread' from the compression and proximity) to form a chain of sigma bonds and, possibly, compressed pi bonds. This provides the basis for a localized 'metallic' hydrogen [34].

The fact that multiple hydrogen atoms can come together in a linear formation means that the shared electrons maintain their screening capability to much smaller spacing than is possible if only two atoms come together (as described in Figs. 1 and 3). Thus, as in Takahashi's tetrahedral-hydrogen cluster model [28], the Coulomb barrier between nuclei is greatly reduced in length and height. The electron concentration initially provides strong screening for the nuclei. As the average sub-lattice spacing decreases, the potential well seen by the electrons gets more one-dimensional and the average of one electron between each nucleus gives a net attraction until the nuclear-Coulomb barrier emerges to dominate at very close sub-lattice spacing ($\ell < 100$ fm?) and at even smaller inter-nuclear distances during oscillations. This potential well and barrier for the nuclei provides a basis for the oscillatory motion proposed by Schwinger and Sinha.

The radiation from this oscillation of charges (s-orbital electrons and nuclei in the linear molecules) may not be photonic in the normal sense, if none of the charges has sufficient angular momentum to form photons. Neither may the combined sub-lattice have this, if in a highly-linear array. Furthermore, the only electron orbits below the atomic ground states are those of the deep-electron orbits with binding energies in the 500 keV range (and the basis for Synthesis 3). Thus, energy transfer from the combined Coulomb and nuclear potentials would be primarily in the form of numerous low-energy phonons from the hydrogen atoms in molecular-vibrational modes. Nevertheless, the photon radiation that has been reported does have possible source modes. Double-photon emission can occur with a net zero change in angular momentum. It is highly forbidden ($\sim 10^{-8}$ times the normal dipole radiation probability, which puts it in the same range as the zero-to-zero angular momentum transition); but, in the slow, repetitive, phonon-emission environment, with no competition from normal single-photon radiation, larger energy steps could be taken occasionally by this double-photon process. Such emission, if aligned with the excited linear molecule, could stimulate laser-like pulses.

Other photo-emission mechanisms may also come into play. What happens when the stability of the electron orbits is strongly disturbed? Any photo emission will not now provide the clean spectral lines of the stable orbits (line broadening is characteristic). At the collision points of the defect oscillations described, many orbits, previously with energy levels separated by electron volts, now overlap. This allows electron-orbit transitions and a means of establishing a new energy and angular momentum balance without photo-emission. At the high-separation points of the sub-lattice atoms, the energy levels would have returned toward their atomic states and could then radiate the energy that they had collected earlier. This energy could be released at nearly any point in the orbit so that there would be a large energy and angular distribution of radiation (including radio-frequency that has been observed). Furthermore, since all of the H or D atoms in the sub-lattice are synchronized, if one emits a photon in the direction of the linear array then by stimulated emission many of the other atoms, which would have populated the same orbital(s), can emit in one direction or the other along that axis. This becomes a single-shot laser-like EM burst that could repeatedly occur at or near the sub-lattice oscillation frequency. Again, the EM frequency of the bursts may not be fixed between bursts. Nevertheless, the pumping action of a collapsing sub-lattice will constantly feed electrons into orbits from which they will radiate. While the radiation may be focused along the direction of electron motion by the coupled EM fields, the timing of emission and its ability to trigger a cascade is still random. Bidirectional, laser-like, optical bursts have also been observed in CF experiments.

As mentioned in a footnote above, the linear-H array is an excited molecule seeking a ground state. It can decay via photonic radiation. In the thermal environment of a lattice, "collisions" between lattice and sub-lattice atoms would assure sufficient angular momentum to provide that necessary for photonic emission.

At some point in the coming together of the hydrogen atoms in an oscillatory mode that radiates nuclear (and electron) potential energy in the form of phonons and photons, fusion is inevitable. If D-D fusion occurs below the fragmentation levels, both the deeply bound electrons and nuclei are in potential wells that have no internal resonances. The deuterons have formed an excited helium nucleus below the fragmentation or excited nuclear-energy levels. The electrons are much deeper than the atomic-electron ground-state energies of H, D, or $^4$He. The energy transfer from this excited nucleus down to the $^4$He nuclear-ground state is the same as was described in ICCF-17 and ICCF-18 papers [17], [18]. Once below the fragmentation level, Hagelstein's phonon transfer process from the nucleus to the lattice may no longer have a time-constraint problem (without a deep-orbit electron, there is no competing rapid-decay mechanism) and his QM calculations may be more appropriate to the system than the relative field-strength numbers provided in above references.

### 3.3. Synthesis 3: Deep-orbit electrons and their effects

The relativistic-electron deep-orbit (EDO) model [36] solves many problems associated with the general acceptance of cold fusion. However, it has its own problems with acceptance by mainstream physics. These include: the lack of experimental evidence for orbits below the atomic ground state; the apparent violation of the Heisenberg Uncertainty Relation (HUR); and the failure of electrons to automatically decay from atomic levels down to the empty deep levels. Each of these problems (and others) may have now been satisfactorily resolved. For example, the $\sim$100 MeV nuclear-bound electron of the EDO model (based on the accepted relativistic Quantum-Mechanics Klein-Gordon and Dirac equations) satisfies both the virial theorem (for stable orbits) and the HUR (for size and momentum constraints).

The linear-defect model has been proposed as a means of overcoming the EDO problems, but without providing details. Initially, the reason for non-decay from the atomic ground state of hydrogen to the deep orbits was the highly-forbidden photo transition of zero-to-zero angular momentum states. A two-level model of quantum mechanics predicts the relative probability of the deep-orbit population in hydrogen relative to that of the atomic levels at only $\sim$2 parts in $10^9$ [40], which is consistent with the forbidden nature of the transition. However, the relativistic Dirac equation does not have such $l = 0$ angular momentum deep-orbit levels, so other reasons for preventing photodecay had to be sufficient, e.g., [40]. They also had to be uniquely resolved in cold-fusion experiments.

The lack of photodecay of atomic electrons to an EDO greatly slows the process(es) for formation of the deep-orbit electrons. Nevertheless, in a two-state model, there would be an easily measured EDO population. In reality, we cannot consider EDO calculations for a two-state system since it appears that the decay process to the EDOs requires a medium and that, in turn, provides a rapid "sink" (via transmutation, e.g., [19], [20], & [21]) for the resulting femto-hydrogen or femto-helium atoms. Populating EDOs in higher-atomic number (e.g., $Z > 8$) nuclei, if possible, would be at a much lower rate. They would result in isotopes with Z-1 and with atomic mass almost indistinguishable from that of the original atoms (e.g., $<$2 keV out of $>$16 GeV).

The prior two Syntheses have provided a basis for the proximity between protons or deuterons necessary for fusion. Here we go one step further to look at details when they do get close. The resonant motion in linear and compressive-cluster defects periodically distort the atomic-electron energy levels so that they can overlap. This allows direct electron transfer to different orbitals, including those with angular momentum, and could remove the problem of the angular momentum requirement in photo-emission to the deep orbits. However, no appropriate energetic photons have been observed even in active cold-fusion experiments and there are reasons for that. Nevertheless, new opportunities for energy dissipation present themselves in this highly-compressed, oscillating, environment.

The virial theorem for a 1/r Coulomb potential requires dissipation of excess PE (energy conservation) to establish a stable orbit for an electron decaying from free space or from an excited orbit. Photoemission does this for atomic-electrons. However, there are other processes that will also work. In the highly unstable environment suggested for the above syntheses, photo-emission at characteristic lines cannot be expected. Other means of energy dissipation could include [42]: phonons (the electron between two protons is doing work on both in bringing them together); collision-induced absorption and emission; auger-like processes with nearby electrons; proximity-distortion of energy levels to allow forbidden photo-transitions, and even excitation of low-lying nuclear levels. Some such processes are reversible; most are irreversible. A common feature of the defect models is the periodic proximity effect within the sub-lattice. This characteristic frequency is dependent on the motion of the atoms, not of the much faster electrons as in atomic photo-emission. Does this mean that we should be looking for very-low-energy radiation (e.g. molecular spectra or radio frequencies [43])? Or, is the phonon field the most that we could expect to see?

We have introduced a term in a manner not usually mentioned with calculations of stable atomic energy levels: reversibility. In an s-orbital (with zero angular momentum), the electron passes through the nuclear region every orbit. This periodic change in kinetic and potential energies is reversible. It is reflected in the nuclear structure (and thus in its mass energy) as the velocity-related, bound-electromagnetic, field of the electron changes throughout each orbit. Because the EM energy radiates out from the electron before it is "returned" to the outer limits of its orbit, there is a time lag in the reversible process. This energy is stored as standing waves in the "lagtime" of this oscillating EM field. If conditions are right, this stored energy, along with the new fields associated with the changing KE and PE of the decaying orbit, form a photon that can be emitted from the atom in an irreversible action. Associated with this irreversible action is a reversible action within the nucleus.

The internal structure (and mass) of the nucleus is altered by the change in EM fields of the electron in its new orbit. If the electron is returned to its former orbit by another irreversible action (absorption of a photon), the nucleus can return to its former configuration through a reversible action. Reversibility and irreversibility are defined in this context mainly by timing. This change in nuclear energy is only a small perturbation. However, it is "quantized", not by any resonance within the nucleus, but by the stability of electron orbits outside. Thus, we need to look at magnitudes of changes in electron energy and field strengths, as well as in timing, to see the mechanisms of some of the effects we seek. This is just another form of relativity.

We generally look at photodecay as an instantaneous event. It is not. Even Feynman describes it in one example from his "Lectures" as being a continuous process that might take more than $10^7$ cycles per photon (in the decay from excited-electron orbits) to complete. The resulting photon may be considered to be a particle; but, it could be $10^7$ cycles long (e.g., 6m by 20ns in the lab frame). Is it a wave or a particle? This has been a silly argument for generations. Again, it is a matter of relativity. The inability to see these effects is an intellectual blindness that can be cured. Nuclear fusion is also considered to be an "instantaneous" event. Is it? Both Schwinger's phonon decay and Storm's "small-photon" decay processes are irreversible "pre-fusion" mechanisms leading to the fusion end-result. Both are encompassed in the above description based on phonon-generated sub-lattice compression-pumping of the bound-electron levels that provides a means of coupling the nuclear-Coulomb energies to the sub-lattice and beyond.

Just as the wave-particle argument has gone unresolved for many people, the "teaching" that nuclear forces are independent of Coulomb forces makes no sense. How can gamma-rays be emitted from excited nuclear levels if electromagnetic effects are divorced from the nuclear effects? Claiming that protons and neutrons are different states of the same particle has a similar problem. These are "stories" that have been embedded in the student's mind for multiple generations now. Experimental evidence alters and even disproves some of these beliefs. However, it appears easier (more efficient?) to accept/believe them than it is to take the time to examine new data and to think things through. After all, it may not be too difficult to assume that the proton is just an excited-state of a hydrogen atom (one with an infinite-orbit, i.e. free, electron). But, it may be even easier for one to accept that electrons cannot exist within a near-nuclear region (because of the Heisenberg Uncertainty Principle) than to look at relativistic effects that can overcome this excuse and open new areas to explore.[4]

As the average linear-H sub-lattice spacing decreases, it also alters the electrons' orbital radii, thus increasing their binding energies. This progressive energy-transfer process can be used to complete a transition from the atomic ground state to a deep-electron orbit for one or more of the bound electrons in the sub-lattice.[5] If this were to happen, the

[4]Nevertheless, in a following paper, one of the present authors (AM) will propose that the neutron might be a highly-relativistic excited-electron state of the hydrogen atom with sub-nuclear orbital radius. Thus, the neutron, proton, and hydrogen atom are all different states of the same atom (particle?).

[5]One reason that atomic electrons may not photonically decay to the deep-orbits is proposed to result from the deep-orbit electrons approaching an anapole configuration with no external EM fields from which to form a photon [41]. Phonon generation may avoid this requirement of external fields.

fusion process follows immediately as pairs of atoms in the affected portion(s) of the sub-lattice collapse together. It is also possible that the energy release of one fusion within the sub-lattice would initiate others (as in stimulated emission). Thus, we see how all of the models mentioned in this paper can participate in an observable cold-fusion result. Nevertheless, since the possible paths are so difficult to quantify over the large range of time, space, and energies, it is difficult to assign relative importance to each of them without more data specific to the proposed processes.

## 4. Conclusions

(1) Cold fusion could occur in or on the lattice, if a linear defect (or another mechanism) provides a means of multiple hydrogen atoms coming together simultaneously.

- Hydrons (hydrogen or deuterium atoms) in a host-bound, linear, sub-lattice (perhaps as in the $\delta$- and $\delta'$-phases of heavily-H-doped Pd [31], [32]) continually come close together as pairs (longitudinal optical-phonon mode).
- This cyclical pairing of nuclei allows the bound-electron screening of nuclear repulsion to be more effective, over time, than that possible when multiple hydrons come together randomly as in a quasi-free linear array [2] or in Takahashi's Tetrahedral Cluster array [28].

(2) Sinha's calculations of the increase in electron effective mass from lattice effects, as multiple hydron pairs repeatedly come together in a linear defect, gives a much greater depth to the attractive Coulomb potential well (electron/proton) before the repulsive Coulomb barrier (proton/proton) grows at close inter-nuclear distances. This strongly affects the shape of the average potential well experienced by the hydrogen nuclei, which, from Schwinger's paper, can include both the Coulomb contributions of the electrons and protons and the nuclear wells of the nuclei and their components.

(3) Schwinger's model for getting from the D-D pairs in a lattice to an excited $^4$He nuclear level without the normal neutron fragmentation (to n + $^3$He) can easily be extrapolated to the ground state. The fact that it could temporarily stabilize at or resonantly tunnel to the known excited states gives a mechanism for the t+p reaction observed in CF experiments where tritium or protons are measured. Nevertheless, this fragmentation result is still much reduced in CF results relative to the fragmentation of normal D-D fusion. Thus, tritium could be a natural result of the CF process, but its relative concentrations would depend on a number of parameters, details of which are yet to be determined. Schwinger's model might have been a first time for a great theoretical-physicist's mind to openly suggest/publish that nuclear forces and electromagnetic forces can couple.

(4) This process could also allow a normal fragmentation process (n + $^3$He) of conventional D-D fusion to occur at a very much reduced rate, since such fusion is now dependent on: time spent at tunneling distances, proximity of the bound electron(s) to the nuclei, and the actual tunneling lengths. Early in the process, the hydrons (H or D) are not much closer together at closest approach on average than the host-lattice spacing. The longer the process of sub-lattice shrinking goes on (before fusion), the more energy will have been dissipated to the lattice and the lower into the $^4$He nuclear potential well will tunneling take place. Thus, for a good CF process, early tunneling rates to the higher $^4$He nuclear states are highly reduced relative to later tunneling to deeper energies.

In summary, a number of the disparate models for cold fusion are converging to provide a coherent picture of at least one mode, but perhaps several, that seems to fit the data and does not violate any physics and chemistry principles involved. Just as the individual models of CF have evolved, it is expected that even these combined models will also evolve.

## Acknowledgement

This work has been supported in part by HiPi Consulting, Licking, MO, USA; by the Science for Humanity Trust, Inc, Tucker, GA, USA, and by the Indian National Science Academy.

## References

[1] A. Meulenberg, K.P. Sinha, "Composite model for LENR in linear defects of a lattice," ICCF-18, 18th Int. Conf. on Cond. Matter Nuclear Science, Columbia, Missouri, 25/07/2013, Abstract and presentation slides at **http://hdl.handle.net/10355/36818**, video at **https://www.youtube.com/watch?v=RcTSUJUCRHE**, Updated text (this paper) to be submitted to JCMNS in 2022 as "Composite Model(s) for Low Energy Nuclear Reactions in the solid state: II".

[2] A. Meulenberg, "Pictorial description for LENR in linear defects of a lattice," ICCF-18, 18th Int. Conf. on Cond. Matter Nuclear Science, Columbia, Missouri, 25/07/2013, J. Condensed Matter Nucl. Sci. **15** (2015), 117-124

[3] J. Schwinger, "Nuclear energy in an atomic lattice," in *The First Annual Conference on Cold Fusion*. 1990, Salt Lake City, Utah pp. 130-136 and Z. Phys. D 15 (1990) 221–225

[4] Maly J.A., Va'vra J., "Electron transitions on deep Dirac levels I," Fusion Science and Technol., **24**, N.3, pp.307-318, 1993, http://www.ans.org/pubs/journals/fst/a_30206

[5] Maly J.A., Va'vra J., "Electron transitions on deep Dirac levels II," Fusion Science and Technology, V.27, N.1, pp.59-70, 1995, http://www.ans.org/pubs/journals/fst/a_30350

[6] K.P. Sinha, "A Theoretical Model for Low-Energy Nuclear Reactions in a Solid Matrix," Infinite Energy 29, 54 (2000)

[7] K.P. Sinha, A. Meulenberg, "Lochon Catalyzed D-D Fusion in Deuterated Palladium in the Solid State," National Academy of Science (India) Letters, Vol.30, No. 7&8, 2007 arXiv:0705.0595v1.

[8] K.P. Sinha and A. Meulenberg, "Lochon-mediated Low-energy Nuclear Reactions," J. Condensed Matter Nucl. Sci. JCMNS-6 (2012) 55–63.

[9] K.P. Sinha and A. Meulenberg, "A Model for Enhanced Fusion Reaction in a Solid Matrix of Metal Deuterides," Proceedings of the 14th International Conference on Cold Fusion (ICCF-14) 10-15 August 2008 Washington DC p. 633, https://www.researchgate.net/publication/344772366_A_model_for_enhanced_fusion_reaction_in_a_solid_matrix_of_metal_deuterides_showFulltext=1&linkId=5f8ed899a6fdccfd7b6ec478

[10] A. Meulenberg and K.P. Sinha, "Lochon and Extended-Lochon Models for LENR in a Lattice," *Infinite Energy Magazine*, pp. 29-32, Issue 112, November/December 2013

[11] K.P. Sinha, "Model of low energy nuclear reactions in a solid matrix with defects," PDF "Special Section: Low Energy Nuclear Reactions," Section editors: Srinivasan, M. and Meulenberg, A., Current Science, Vol. 108, No. 4, p 516, 25 Feb. 2015, http://www.currentscience.ac.in/cs/php/feat.php?feature=Special%20Section:%20Low%20Energy%20Nuclear%20Reactions&featid=10094

[12] K.P. Sinha, A. Meulenberg, "Laser stimulation of low-energy nuclear reactions in deuterated palladium", Current Science Vol. 91, No. 7, pp. 907-912, 10/10/06 (03/06, Ref. No. O232) and included in ArXiv (cond. matter) http://arxiv.org/abs/cond-mat/0603213

[13] K.P. Sinha and A. Meulenberg, "Quantum-Correlated Fluctuations, Enhanced Tunneling and Low-Energy Nuclear Reactions in Condensed Matter," 16th International Conf. on Condensed Matter Nuclear Science, Chennai, Feb. 6-11, 2011, JCMNS-8 (2012) pp. 105-114.

[14] A. Meulenberg and K. P. Sinha, "Tunneling Beneath the $^4He*$ Fragmentation Energy," presented at 239th ACS National Meeting, San Francisco, CA March 2010, Journal of Condensed Matter Nucl. Sci., JCMNS, 4 (2011) 241–255.

[15] A. Meulenberg and K P Sinha, "Deep-electron orbits in Cold Fusion," 17th International Conference on Condensed Matter Nuclear Science, Daejeon, Korea, 12-17 August, 2012, J. Condensed Matter Nucl. Sci. **13** (2014), 368-377

[16] A. Meulenberg, "From the Naught Orbit to He4 Ground State" 16th International Conference on Condensed Matter Nuclear Science, Chennai, February 6 -11, 2011, JCMNS-**10** p 15 - 29 (2013),

[17] A. Meulenberg and K.P. Sinha, "Deep-Orbit-Electron Radiation Emission in Decay from 4He*# to 4He," 17th International Conference on Condensed Matter Nuclear Science, Daejeon, Korea, 12-17 August, 2012, J. Condensed Matter Nucl. Sci. **13** (2014), 357-368,

[18] A. Meulenberg, "Radiation Coupling: Nuclear Protons to Deep-Orbit-Electrons, then to the Lattice," ICCF-18, 18th Int. Conf. on Cond. Matter Nuclear Science, Columbia, Missouri, 25/07/2013, J. Condensed Matter Nucl. Sci. **15** (2015), 125 – 136

[19] A. Meulenberg, "Femto-atoms and Transmutation," 17th International Conference on Condensed Matter Nuclear Science, Daejeon, Korea, 12-17 August, 2012, J. Condensed Matter Nucl. Sci. **13** (2014), 346-357,

[20] A. Meulenberg, "Femto-Helium and PdD Transmutation," ICCF-18, 18th Int. Conf. on Cond. Matter Nuclear Sci., Columbia, Missouri, 25/07/2013, J. Cond. Matter Nucl. Sci. **15** (2015), 106-117,

[21] A. Meulenberg and J. L. Paillet, "Basis for femto-molecules and -ions created from femto-atoms," ICCF-19, 19th Int. Conf. on Cond. Matter Nuclear Science, Padua, Italy, 15/05/2015, J. Condensed Matter Nucl. Sci. **19**, pp. 202 – 209, (2016),

[22] A.S. Alexandrov, N.F. Mott, Polarons and Bipolarons, World Scientific, Singapore, 1995

[23] P.W. Anderson, "Model for electronic structures of amorphous semiconductors," Phys. Rev. Lett., 34, p. 953, 1975

[24] R.A.Street and N.F.Mott, "States in the gap in glassy semiconductors," Phys. Rev. Lett. 35, p.1293, 1975

[25] E. Storms, "The role of voids as the location of LENR," J. Condensed Matter Nucl. Sci. **11** (2013), 123-141, http://coldfusioncommunity.net/pdf/jcmns/v11/368_JCMNS-Vol11.pdf

[26] https://en.wikipedia.org/wiki/Einstein_solid

[27] T. Barnard, "Applying the Scientific Method to Understanding Anomalous Heat Effects: High Energy D2 Bond from Feynman's Integral Wave Equation," ICCF 18 July 21-27, 2013, University of Missouri Columbia, MO, USA, Abstract and annotated presentation slides at http://hdl.handle.net/10355/36839

[28] Akito Takahashi, "Background for Condensed Cluster Fusion," Proceedings of JCF15, Nov. 1 - 2, 2014, Sapporo, Japan, https://www.researchgate.net/publication/325807810

[29] A.P.B. Sinha and K.P. Sinha, "Electronic motion in some polar semiconductors," Ind. J. Pure and Appl. Phys. 1, 286 (1963)

[30] I.G. Lang and Yu. A. Firsov, "Kinetic theory of semiconductors with low mobility," Sov. Phys. JETP, 16, 1301 (1963)

[31] M.R. Staker, "Coupled calorimetry and resistivity measurements, in conjunction with an emended and more complete phase diagram of the palladium - isotopic hydrogen system," J. Cond. Matter Nucl. Sci. **29** (2019) 129-168.

[32] M.R. Staker, "A Model and simulation of lattice vibrations in a superabundant vacancy phase of palladium-deuterium," Mod. Sim. in Mat. Sci & Engr 28. (2020) https://doi.org/10.1088/1361-651X/ab9994

[33] S.H. Patil, "Hydrogen molecular ion and molecule in two dimensions," J. Chem. Phys. 118, 2197 (2003).

[34] M. Motta et al., (Simons Collaboration on the Many-Electron Problem), "Ground-State Properties of the Hydrogen Chain: Dimerisation, Insulator-to-Metal Transition, and Magnetic Phases," Physical Review X 10, 031058 (2020)

[35] P.L. Hagelstein, "Bird's eye view of phonon models for excess heat in the Fleischmann–Pons experiment," *J. Cond. Mat. Nucl. Sci.***6** (2011) 169

[36] J.L. Paillet, A. Meulenberg, "A study on electron deep orbits by quantum relativistic methods," pp. 301-331, in Cold Fusion,: Advances in Condensed Matter Nuclear Science, Edited by Jean-Paul Biberian, 2020 Elsevier Inc. and J.L. Paillet, A. Meulenberg, "Special relativity: the source of electron deep orbits," Foundations of Physics, Feb. 2017, Vol. 47, Issue 2, pp 256–264.

[37] https://en.wikipedia.org/wiki/Dimer_(chemistry)#Covalent_dimers

[38] https://en.wikipedia.org/wiki/Phonon#Acoustic_and_optical_phonons

[39] K. P. Sinha and A. Meulenberg, "A model for enhanced fusion reaction in a solid matrix of metal deuterides," Slide 8 of an annotated copy of presentation at ICCF-14, Aug. 2008, Wash., DC, https://www.researchgate.net/publication/344772366_A_model_for_enhanced_fusion_reaction_in_a_solid_matrix_of_metal_deuterides_showFulltext=1&linkId=5f8ed899a6fdccfd7b6ec478

[40] R. A. Rice, Y. E. Kim, and M. Rabinowitz, "Comment on 'Electron Transitions on Deep Dirac Levels I,"' *Fusion Technol*, 26, 110 (1994).

[41] V. Savinov, "Light emission by accelerated electric, toroidal, and anapole dipolar sources," Phys. Rev. A **97**, 063834 (2018).

[42] see examples in the references of S. X. Hu *et al.*, "Interspecies radiative transition in warm and superdense plasma mixtures, "Nature Communications | (2020) 11:1989 | https://doi.org/10.1038/s41467-020-15916-3 | www.nature.com/naturecommunications.

[43] M.R. Swartz, "Active LANR Systems Emit a 327.37 MHz Maser Line," Journal of Condensed Matter Nuclear Science 33 (2020) 81–110.

J. Condensed Matter Nucl. Sci. 36 (2022) 353–361

Research Article

# Non - Accelerator Measurement of the Long - Range Quark - Lepton Interaction in Solids

V. G. Plekhanov*

*Fonoriton Sci. Lab., Garon Ltd, Tallinn, 11413, Estonia*

**Abstract**

The measurements of the dependence of the energy of strong nuclear interaction on the distance between nucleons in nucleus were performed for the first time. By the adding one neutron to hydrogen nucleus we enable the strong interaction. The initial components of mixed crystals $LiH_xD_{1-x}$ are LiH (no strong interaction in the hydrogen nucleus) and LiD (with strong interaction in the deuterium nucleus) crystals are dielectrics with cubic symmetry. The change of the concentration of deuterium in mixed crystals $LiH_xD_{1-x}$ determines the change in the distance between nucleons, which is directly manifested in the position of the phononless free exciton emission line in the low - temperature luminescence spectrum. A series of such measurements made it possible to construct a function of the dependence of the energy of strong nuclear interaction on the distance between nucleons in the deuterium nucleus.



*Keywords:* Strong Interaction, Quarks, Gluons, Excitons, Phonons, Quantum electrodynamics, Quantum chromodynamics

## 1. Introduction

The hydrogen atom has a central position in the history of the 20 th - century physics. Simple atoms, like hydrogen, being essentially quantum electrodynamics (QED) systems, allow highly accurate theoretical predictions. QED is the first successful and still the most successful quantum field theory. After the discovery of the neutron in 1932 by Chadwick [1], there was no longer doubt that the building block of nuclei are proton and neutron (collectively called nucleons). The discovery of the neutron may be viewed as the birth of the strong nuclear interaction [2], [3]: it indicated that the nuclei consists of protons and neutrons and hence the presence of a force that holds them together, strong enough to counteract the electromagnetic repulsion. Theory strong nuclear interaction is the heart of quantum chromodynamics (QCD) which is part of the Standard Model (SM) [4], therefore the base exchange is the gluon which mediates the forces between quarks. Nucleus is a bound system of strongly interacting protons and neutrons (more generally baryons) holds atomic nuclei together and, in another context, binds quarks within hadrons. The baryons are bound states of the three quarks and mesons are composed of quark and antiquark (see, e.g. [5]).

*Corresponding author: vgplekhanov@gmail.com

The phenomenological basis for strong nuclear interaction was laid down by Yukawa's paper [6]. Yukawa's potential has the form

$$V = ge^{-kr}/r^{n}, \tag{1}$$

for some n. Here the strength of the force is measured by the constant k (see, also [2]). The most important feature Yukawa's forces is that they have a small range ($\sim 10^{-15}$ m). The central dogma of atomic physics after Yukawa's paper that proton - electron attraction could be explained in terms of classical electrostatic theory, while the strong force effects were essentially new and inexplicable. The best theoretical guess is the Yukawa potential, but it is a static potential not dependent on velocities of the nucleons. A static force is not a complete one because it cannot explain the propagation of the nuclear interaction. Moreover, as was indicated above, a phenomenological Yukawa potential cannot be directly verified experimentally. We should note that nowadays in text books and elsewhere the separation of electromagnetic and strong interaction tacitly assumed. It is very strange up to present time we do not even know the strong force very well. And what is more we have some contradiction taking into account that the forces between quarks must be long - range, because the gluons have zero mass. But as was mentioned above the force between colorless hadrons is short - range, when the distance between hadrons is more than nuclear size [5]. We can see that the border of the nuclear size transforms long - range interaction in the short - range one. It is very old question which up to present time has not any theoretical explanation.

Recently we have shown that the origin of the isotope is due the strong nuclear interaction in solids [7] and presented it in a series of international conferences on high energy physics. The present paper describes the non - accelerator measurement of the long - range quark - lepton strong interaction and their dependence on the distance between nucleons in the deuterium nucleus. Using our experimental results we found the maximum value of the strong interaction coupling constant equal 2.680.

## 2. Experimental Results

The main experimental results were obtained on devices used already in the earlier investigations [9]–[12]. For clarity, we should mention here that the experimental method involves a home-made helium cryostat and two identical double-prism monochromators as well as a photoelectrical recording system with signal accumulation in the memory of a personal computer. For measuring the luminescence spectra the crystals were excited by the light from mercury (250 W) and deuterium (400 W) lamps.

In our experiments we investigated two kinds of crystals (LiH and LiD) which are differ by a term of one neutron. Lithium hydride and lithium deuteride are ionic insulating crystals with simple electronic structure, four electrons per unit cell, both fairly well – described structurally (neutron diffraction) and dynamically (second – order Raman spectroscopy) and through ab initio electronic structure simulation. Among other arguments, LiH and LiD are very interesting systems due to their extremely simple electronic and energy structure and to the large isotopic effects when the hydrogen ions are replaced by the deuterium ones.

The single crystals of LiH and LiD were grown from the melt by the modified method of Bridgeman - Stockbarger (see [11]–[13] and references quoted therein). The crystals were synthesized from $^{7}$Li metal and hydrogen of 99.7% purity and deuterium of 99.5% purity. Virgin crystals had a slightly blue – grey color, which can be attributed to nonstoichiometric excess of lithium present during the grown cycle. On annealing for several days (up to 20) at 500°C under $\sim$ 3 atm of hydrogen or deuterium, this color could be almost completely eliminated. Because of the high reactivity and high hygroscope of investigated crystals an efficient protection against the atmosphere was necessing. Taking into account this circumstance, we have developed special equipment which allows us to prepare samples with a clean surface, cleaving them in the bath of a helium cryostat with normal or superfluid liquid helium [9]. The samples with such surfaces allow us to perform measurements lasting 15 hours. An improved of LiH (LiD) as well as mixed

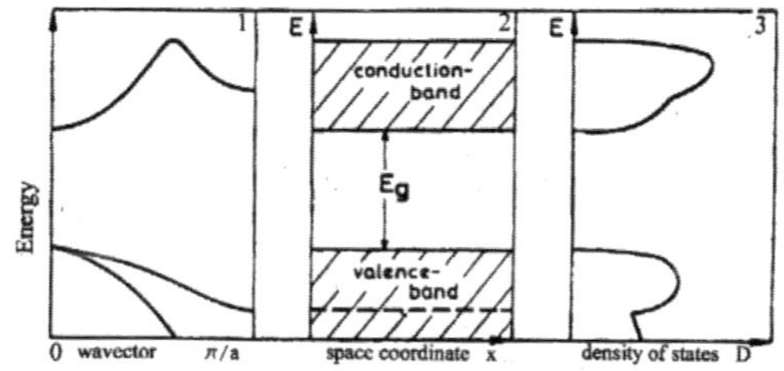


**Figure 1.** Various possibilities to present the band - structure of homogeneous undoped insulator (semiconductor). 1 - the dispersion relation, i.e. the energy E as a function of the wave vector k, 2 - the energy regions of allowed and forbidden states as function of a space coordinate x and 3 - the density of states (all curves are schematic ones).

crystals $LiH_xD_{1-x}$ were used in the present study. In spite of the identical structure of all free - exciton luminescence spectra, it is necessary to note a rather big variation of the luminescence intensity of the crystals from the different batches observed in experiment.

Before beginning a general discussion of long - range strong nuclear interaction, it is helpful to have an idea of electronic structure of bulk materials. As is well - known the nature of materials is determined by the interaction of their valence electrons with their charged nuclei and core electrons. This determines how elements react with each other, what structure the solid prefers, its optoelectronic properties and all other aspects of the material. The modern view of solid state physics is based on the presentation of elementary excitations, having mass, quasi-impulse, electrical charge and so on [14]. According to this presentation the elementary excitations of the non - metallic materials are electrons (holes), excitons (polaritons) and phonons. The theory discussed in modern text books (see, e.g. [15]) forms the basis for the modern theory of electrons in solid. It arises from the consideration of the periodicity of the crystal structure. This periodicity leads to the formation of energy bands (see Fig. 1). The importance of the electronic theory of solids as embodied in band theory is that it provides us with clear means of understanding how solids may be insulators, semiconductors, or metals. This depends upon whether or not it is Fermi surface. The existence of a Fermi surface produces metallic behavior, whereas at 0K, if the filled electron levels (bands in solids - see Fig. 1) are separated from vacant ones, we have insulating properties. The difference between a good conductor and a good insulator is striking. The electrical resistivity of a pure metal may be as low as $10^{10}$ ohm-cm at a temperature of 1 K, apart from the possibility of superconductivity. The resistivity of good insulator may as high as $10^{22}$ ohm-cm. The understand the difference between insulators and conductors, we shall use the band – gap picture (Fig. 1). The possibility of band gap is the most important property of solids.

Free exciton luminescence is observed when LiH (LiD) crystals are excited in the midst of the fundamental absorption. The spectrum of free exciton photoluminescence of LiH crystals cleaved in superfluid helium consists of a narrow (in the best crystals, its half - width is $\Delta E \leq 10$ meV) phononless emission line and its broader phonon repetitions, which arise due to radiated annihilation of free excitons with the production of one to five longitudinal optical (LO) phonons (see Fig. 1 in Reference [9]).

The phononless emission line coincides in an almost resonant way with the reflection line of the exciton ground state which is indication of the direct electron transition $X_1 - X_4$ of the first Brillouin zone [1]. The lines of phonon replicas form an equidistant series biased toward lower energies from the resonance emission line of excitons. The energy difference between these lines in LiH crystals is about 140 meV, which is very close to the calculated energy of the LO phonon in the middle of the Brillouin zone.

The isotopic shift of the zero - phonon emission line of LiH crystals equals 103 meV. As we can see from Fig. 1 (Reference 9) the photoluminescence spectrum of LiD crystals is largely similar to the spectrum of intrinsic

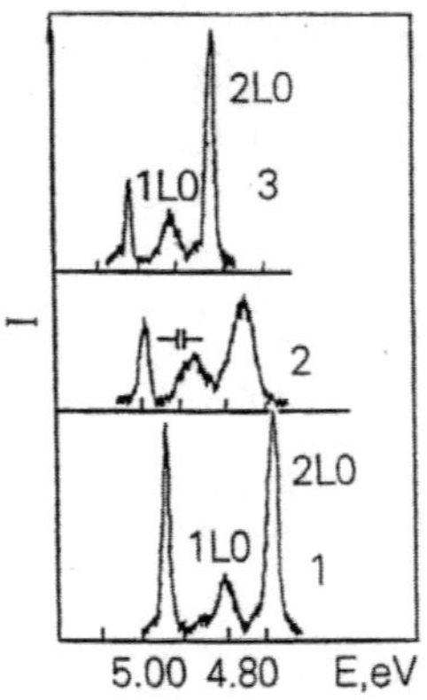


**Figure 2.** Photoluminiscence spectra of free excitons in LiH (1), $LiH_xD_{1-x}$ (2) and LiD (3) crystals cleaved in superfluid helium at 2 K. Spectrometer resolution is shown.

luminescence of LiH crystals (see, also Fig. 2). There are, however, some distinctions one is related. Firstly the zero - phonon emission line of free excitons in LiD crystals shifts to the short - wavelength side on 103 meV. These results directly show the violation of the strong conclusion that the strong force does not act on leptons. The second difference concludes in less value of the LO phonon energy, which is equal to 104 meV. The simplest approximation, in which crystals of mixed isotopic composition are treated as crystals of identical atoms having the average isotopic mass, is referred to as virtual crystal approximation (VCA) [15].

When light is excited by photons in a region of fundamental absorption in mixed $LiH_xD_{1-x}$ crystals at low temperature, line luminescence is observed (Fig. 2), like in the pure LiH and LiD crystals. As before [12], [15], the luminescence spectrum of crystals cleaved in superfluid liquid helium consists of the relatively zero - phonon line and its wide LO replicas. For the sake of convenience, and without scarficing generality, Fig. 2 shows the lines of two replicas. Usually up to five LO repetitions are observed in the luminescence spectrum as described in detail in [1]. In Fig. 2 we see immediately that the structure of all three spectra is the same. The difference is in the distance between the observed lines, as well as in the energy at which the luminescence spectrum begins, and in the half - width of the lines.

The measurements of the low temperature of reflection [15] and luminescence spectra of the whole series of mixed crystals is permitted to obtain the dependence of the interband transition (the long - range force dependence of strong nuclear interaction on the distance between nucleons in deuterium nucleus) energy on the deuterium concentration (Fig. 3). This dependence has a nonlinear character. As can be seen from Fig. 3, VCA method (the straight dashed line) cannot describe observed experimental results.

Comparison the experimental results on the luminescence (reflection) and light scattering [19] in the crystals which differ by a term of one neutron only we may conclude:

1. At the adding one neutron (using LiD crystals instead LiH ones) is involved the increase exciton energy up to 103 meV.
2. At the addition one neutron the energy of LO phonons is decreased on the 36 meV that is direct seen from luminescence and scattering spectra.
3. In the isotope effect, the energy of acoustic phonons does not depend on the replacement of H by D [18], which is proved by the identical structure of the light scattering spectra [19]. Along with this, a small change in the

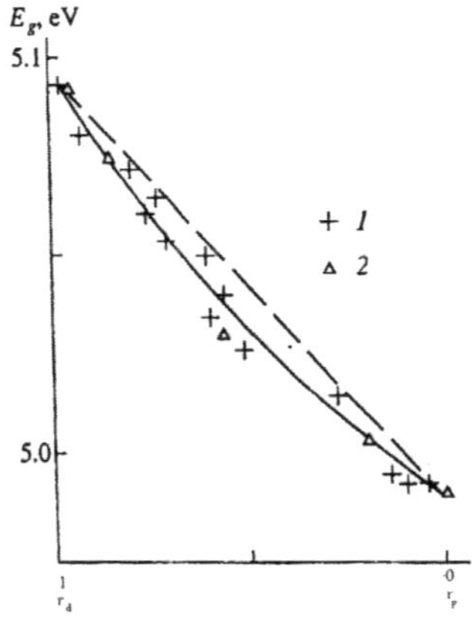


**Figure 3.** The force dependence of strong nuclear interaction on the distance between nucleons in deuterium nucleus. The straight dashed line is the linear dependence of the force dependence $F_s$=f(x) in the virtual model. The solid line corresponds to calculation using the polynom of second degree $F_s = f(r_d)$. Points derived from the reflection spectra indicated by crosses, and those from luminescence spectra by triangles.

energy of optical (LO) phonons (36 meV) observed in the luminescence and light scattering spectra indicates a non - electron - phonon renormalization mechanism of the energy of zero - phonon emission line of free excitons in LiD crystals.

## 3. Discussion

Traditionally nuclear - electron interaction in our case neutron - electron interaction taking into account the solving of Schrödinger equation using a model of Born - Oppenheimer (adiabatic) approximation [16]. Since electrons are much faster and lighter than the nuclei by a factor nearly 2000 the electron charge can quickly rearrange itself in response to the slower motion of the nuclei, and this is the essence of the Born - Oppenheimer approximation. This approximation results the omission of certain small terms which result from the transformation. As was shown in [18] the eigenvalue (energy) of the electronic Schrödinger equation (equation 6 in [17]) depends on the nuclear charges through the Coulomb potential, but it doesn't include any references to nuclear mass and it is the same for the different isotopes (see also [16], [18]). The independent of the potential energy (the eigenvalue of the Schrödinger equation) is the essence adiabatic approximation. However, we must repeat, that the Born - Oppenheimer approximation is the standard ansatz to the description of the interaction between electrons and nuclei in solids (see, however below). Now we should take into account a small contribution to isotopic shift through reduced electron mass $\mu = ((m_e M_{Nucl})/(m_e + M_{Nucl}))$ so far as $M_{Nucl}$ is different for the hydrogen and deuterium nucleus. In this case the contribution equals $\Delta E \simeq 6$ meV. Contribution to isotope shift of the zero - phonon line in luminescence spectra of LiD crystals as well as Lamb shift and specific Coulomb potential approximately equal 1, and 1 meV, respectively. This value is less than isotope shift in our experiments, by two orders or more. The last result is forcing us to search for new models and mechanisms of nuclear - electron interaction including results of subatomic physics, e.g. hadron - lepton interaction.

The formulation of the model explaining the non - accelerator experimental results on the mass isotope effect requires a more careful analysis of the origin of the four fundamental forces. It is reasonable to recall that more than three decades ago, Barut [20] hypothesized that all fundamental forces have an electromagnetic nature (see, also [21]). Out of four known interactions, three are described by SM - the electromagnetic, weak and strong ones [4, 5, 22]. The first two of them have a common electroweak gauge interaction behind them. The strong interaction in SM is described by the QCD, a theory with the gauge group SU(3)C. The effective coupling constant of this theory grows when the energy decreased. As a result, particles which feel this interaction cannot exist as free states and appear only in the form of bound states called hadrons (see, also [22]). Most of modern methods of quantum field theory work at small values of coupling constant, $\alpha_s$, [23], that is, for QCD, at high energies. Quarks and leptons, the so - called SM matter fields, are described by fermionic fields. Quarks take part in strong interactions and compose observable bound state hadrons. Both quarks and leptons participate in the electroweak interaction. The matter fields constitute three generation: particles from different generation interact identically but have different masses (see, e.g. [5], [22]). The gauge bosons, which are carriers of interactions are massless for unbroken gauge groups $U(1)_{EM}$ (electromagnetism - photons) and SU(3)C (QCD - gluons), masses of $W^{\pm}$ and $Z^0$ bosons are determined by the mechanism of electroweak symmetry breaking. It should note that the forces between the quarks must be long rang, because the gluons have zero mass. This does not imply that forces between hadrons are also long range, because hadrons have zero color charges overall. The forces between the colorless hadrons are the residues of the strong forces between their quark constituents, and cancel when the hadrons are far apart.

The diagram of Fig. 4 is drawn for the case of single – gluon exchange, but taking into account that strong constant coupling $\alpha_s \sim 1$, multiple exchanges, involving terms $\alpha_s^2$, ${\alpha_s}^2, \ldots$ must also be very probable. However, it turns out that for violent 'close' collisions at very high $q^2$, $\alpha_s \ll 1$ and single - gluon exchange is a good approximation, while at low $q^2$ or, equivalently, larger distances, the coupling becomes very large and the theory is incalculable (see also [2]–[21]). This large - distance behavior is presumably linked with of confinement of quarks and gluons inside hadrons. Since gluons are massless, one might expect the static QCD potential to have a similar 1/r form to that in QED. In fact the quark - antiquark potential is often taken to be of the form [5]

$$V_s = -(4/3)((\alpha_s)/r) + kr \tag{2}$$

where the first term, dominating at small r, arises from single - gluon exchange. It is similar to the Coulomb potential between two unit charges of

$$V_{em} = -(\alpha/r) \tag{3}$$

The factor 4/3 in (2) is plausible in view of the fact that there eight color gluon states, to be averaged over three quark colors, giving factor 8/3 as compared with QED (one 'color' of photon and one fermion charge). This has to divide by 2 because, for historical reasons [22], a factor 2 enters into definition of $\alpha_s$ in terms of the square of the strong color charge. However, at larger distances the second term in (2) is dominant and is responsible for quark confinement. The value of k is about 0.86 $GeVfm^{-1}$. In equation (2) the attractive force at large r is dV/dr = k = 0.85 $GeVfm^{-1}$ [4]. The estimation of the value of k = 13.9 tons weight — a great deal for those tiny quarks, each weighing less than $10^{-24}$ g. Because gluons carry a color charge (unlike photons which are uncharged) there is a strong gluon - gluon interaction.

Returning to our non - accelerator experimental results, we should underline that in this paper we measure the low - temperature luminescence spectrum of $LiH_xD_{1-x}$ mixed crystals. We should repeat that LiH crystals are without

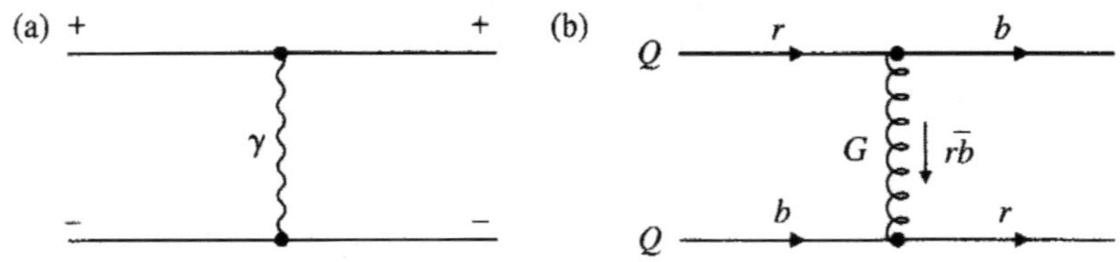


**Figure 4.** Comparison of the Feynman diagrams of electromagnetic (a) and strong (b) interactions. The electromagnetic interaction in (a) involves two types of electric charge, (+) and (−) is mediated by an uncharged photon. In (b) the strong interquark force involves six types of color charge. The diagram depicts the interaction of a red quark with a blue quark via the exchange of a red - antiblue gluon.

strong interaction in hydrogen nucleus and LiD crystals are with strong interaction in deuterium nucleus. When we add one neutron in the hydrogen nucleus, we artificial activated of the strong interaction. As far as the gravitation, electromagnetic and weak interactions are the same in both former crystals (LiH and LiD), it only emerges the strong interaction (see, also [9], [21]). Therefore a logical and historical [25], [26] conclusion is made in paper [8], [12] that the renormalization of the energy of electromagnetic excitations (isotopic shift equals 0.103 eV [15]) is carried out by strong nuclear interaction. The short range character of the strong interaction of nucleons in SM does not possess direct mechanism of the elementary excitation (electrons, excitons, phonons) energy renormalization, which was observed in our low temperature experiments. Second reason that the interpretation of our experimental results is very difficult task because they are first demonstration of the violation of the strong conclusion in nuclear and particles physics that the strong nuclear force does not act on the colorless leptons (see, e.g. [22], [23]). Moreover in paper [27] we read "We do not know the origin of the strong nuclear force very well". It should be added that before present work there were no measurements of the dependence of the energy (force) strong nuclear interaction on the distance between nucleons in the nucleus. In our tentative interpretation (the electromagnetic origin) we should suppose that the potential of strong interaction has the next form:

$$\mathrm{V(r)} = -((\mathrm{Z}\alpha)/(\mathrm{r})) + (1 + \mathrm{e}^{\mathrm{m'r}}\chi^2) \equiv \mathrm{V_{Coulomb}} + \delta\mathrm{V(r)}, \quad (4)$$

where Z is the charge of the massive central particle which causes the potential and $\alpha$ is the fine structure constant. Here is the second term term is Yukawa's short - range potential, and the first term which has a long - range character of interaction and depends on the distance between nucleons in deuterium nucleus. The most intensively studied atomic scale is at a few values of the Bohr radius, $a_0 \simeq 0.53 \times 10^{-10}$ m. For the Yukawa radius equal to $a_0$, the related mass m' of the intermediate new kind particle (boson) is 3.5 keV which interact with hadron (neutron) and lepton (electron) (for details see [21] and references therein). The origin of the first term in equation (4) is due the magnetic - like long - range interaction (possibly not point - like neutron and electron [28], [29]). This assumption is in accordance with very large value magnetic moment of neutrons (see, also [2], [3], [30]). Moreover, as we can see from results of Fig. 3 the force dependence of strong nuclear interaction on the distance between nucleons in deuterium nucleus has a nonlinear character. With increase the distance between nucleons the force of strong nuclear interaction is increased. We should stress that this is the first observation in experiments the force dependence of strong nuclear interaction on the distance between nucleons in nucleus. A possible interpretation is to assume that in addition to the 8 gluons predicted by QCD SU(3)C group there is a ninth gluon color singlet [22] as photon and may interacts with colorless hadrons and leptons.

$$\mathrm{g_9} = (1/(\sqrt{3}))(\mathrm{r\underline{r}} + \mathrm{g\underline{g}} + \mathrm{b\underline{b}}), \quad (5)$$

where $\underline{\mathrm{r}}$, $\underline{\mathrm{g}}$, $\underline{\mathrm{b}}$ are antiquarks. This massless photon like gluon may be strongly interacts between nucleons (neutrons) and leptons (electrons) [22]. So, the first point to realize is that despite the numerous textbook statements, all leptons are feeling the strong nuclear interactions. This strong conclusion is a direct consequence of our non - accelerator experimental results - the origin of the mechanism of interaction between hadrons and leptons is a task for future theoretical and experimental investigations.

## 4. Conclusion

The main feature of the present work is that measurements of the dependence of the energy of strong nuclear interaction on the distance between nucleons in a nucleus were performed for the first time, and these results allow us to find the maximum possible value of $\alpha_s = 2.4680$. Our non - accelerator results opens new avenue in the investigation of the propagation the force of strong nuclear interaction in the wide value range by means the condensed matter alike traditional accelerator methods. These results may shed light on a number of fundamental puzzles in modern physics, particularly on the unification of forces taken into account long - range strong nuclear force. Experimental observation of the renormalization of the elementary excitation energy of solids by the strong nuclear interaction stimulates its count in the process of description of the elementary excitation dynamics in quantum electrodynamics. Besides, we should highlight that such important information has been obtained via rather simple and inexpensive experimental physics equipment.

## Acknowledgements

Many thanks to Dr. K. Walker for reading my manuscript, G. Tumar and G.A. Plehhanov for improving my English. Special thanks to Jed Rothwell for his contribution to my article.

## References

[1] J. Chadwick, The existence of a neutron, Proc. Roy. Soc. (London) **136,** 692 - 708 (1932).
[2] E.M. Henley, A. Garcia, Subatomic Physics (World Scientific Publishing CO., Singapore, 2007).
[3] B.R. Martin, Nuclear and Particle Physics (Wiley and Sons Ltd., - VCH, Weinheim, 2006).
[4] W.N. Cottingham, D.A. Greenwood, Introduction to Standard Model of Particle Physics (Cambridge University Press, Cambridge, 2007).
[5] D.H. Perkins, Introduction to High Energy Physics (Cambridge University Press, Cambridge, 2000).
[6] H. Yukawa, On the interaction of elementary particles, Proc. Phys. Math. Soc. (Japan) **17,** 48 - 57 (1935).
[7] V.G. Plekhanov, A possible signature of neutron quarks - leptons via gluon interaction in solids, J. Cond. Mat. Nucl. Sci. **29**, 503 - 511 (2019).
[8] a) V.G. Plekhanov, Isotope effect renormalization of energy of electrons by strong nuclear interaction, Deposit. in VINITI, N202 - B2012, 1 - 13 (in Russian);
b) V.G. Plekhanov, Necessity addition, Phys. - Uspekhi (Moscow) **61**, 449 - 450 (2019).
[9] V.G. Plekhanov, Modern View of the Origin of Isotope Effect (LAMBERT Academic Publishing, Saarbrucken, Germany, 2018).
[10] V.G. Plekhanov, Isotope - induced energy - spectrum renormalization of the Wannier - Mott exciton in LiH crystals, Phys. Rev. B **54**, 3869 - 3877 (1996).
[11] V.G. Plekhanov, Direct observation of the strong nuclear interaction in the optical spectra of solids, in, Proc. $3^{rd}$ Int. Conf. High Energy Physics, Rome, Italy, 2017, p. 46.
[12] V.G. Plekhanov, Experimental study of the long - range quark - lepton strong interactions in solids, Chapter 1 in Understanding Quarks (Nova Science Publishers, Inc, New York, 2021).
[13] O.I. Tyutyunnik, V.I. Tyutyunnik, F.F. Gavrilov, and G.I. Pilipenko, Lithium hydride growth by Bridgmen - Stockbarger method, J. Crystal Growth 68, 741 - 749 (1984).
[14] D. Pines, Elementary Excitations in Solids (W.A. Benjamin Inc., New York, 1963).
[15] V.G. Plekhanov, Giant Isotope Effect in Solids (Stefan University Press, La Jola, CA, 2004).
[16] M. Born and H. Kun, Dynamical Theory of Crystal lattices (Clarendon Press, Oxford, 1954).
[17] V.G. Plekhanov, Measurements of the wide value range of strong interaction coupling constant, SSRG - IJAP **6**, 32 - 37, (2019).
[18] M. Wolfsberg, W.A. van Hook, P. Paneth, Isotope Effects in Chemical, Geological and Bio Sciences (Springer, Heidelberg, 2010).
[19] V.G. Plekhanov, Isotope effects in lattice dynamics, Phys. - Uspekhi **46**, 689 - 715 (2003).
[20] F.O. Barut, Unification based on electromagnetism, Annalen Phys. (Leipzig) **498** , 83 - 92 (1986).
[21] V.G. Plekhanov,Isotope Effect - a macroscoplically manifestation of a strong nuclear long - range interaction, Atomic Energy **131**, 123 - 128 (2021)(in Russian).
[22] G. Griffiths, Introduction to Elementary Particles (Wiley - VCH, Weiheim, 2008).
[23] A. Deur, S.J. Brodsky, G.F. Teramond, The QCD running coling constant, Prog. Part.. Nucl. Phys. **90**, 1 - 74 (2016).
[24] J.F. Donoghue, E. Golowich, B.R. Holstein, Dynamics of Standard Model, (CUP, Cambridge, 2008).

[25] E.U. Condon, Note on electron - neutron interaction, Phys. Rev. **49**, 459 - 461 (1936).
[26] L.L. Foldy, Neutron - electron interaction, Rev. Mod. Phys. **30**, 471 - 481 (1958).
[27] Ling Jun Wang, Unification of Gravitational and Electromagnetic Forces (Scholar's Press, Marietta, GA, USA, 2019).
[28] A.H. Compton, The size and shape of the electron, Phys. Rev. **14**, 27 - 259 (1919).
[29] L.D. Landau, About fundamental problems, in, Selected Papers, Vol. 2 (Science, Moscow, 1969) 421 - 424 (in Russian).
[30] V.G. Plekhanov, J.G. Buitrago, Evidence of residual strong interaction in nuclear atomic level via isotopic shift in LiH - LiD crystals, Prog. Phys. **15**, 68 - 71 (2019).

J. Condensed Matter Nucl. Sci. 36 (2022) 362–376

Research Article

# Investigation of LENR Processes Near Incandescent Metals

A.G. Parkhomov*

*KIT Research and Development Laboratory, Moscow, Russia*

R.V. Karabanov

*Scientific Research Center SINTEZTECH, Russia, Sochi*

E.O. Belousova

*Institute of Experimental Mineralogy, Russian Academy of Sciences*

**Abstract**

This paper describes experiments conducted to test the assumption that the source of the agent that causes nuclear transmutations in LENR can be metals heated to a sufficiently high temperature. Tungsten filaments in incandescent lamps, hot iron rods and tungsten powder were used as the heated metal. Heat releases much greater than the electricity used to heat the material, and the formation of many initially absent chemical elements was found. Based on the experimental results obtained by the authors and other researchers, a generalization of the properties of LENR is made, which can form the basis of a theory of this phenomenon.


*Keywords:* Hot metals, Nuclear transmutations, LENR, Incandescent lamps, LED, Calorimetry

## 1. INTRODUCTION

LENR studies have shown a wide variety of manifestations of this phenomenon. It appears in metals with hydrogen dissolved in them [1], [2], in plasma [3], [4], in gas discharge [5], in electrolysis [6] and in biological systems [7], [8]. In addition to energy releases, which far exceed the upper limits of chemical reactions, LENR is characterized by a huge variety of emerging chemical elements. For example, in a nickel-hydrogen LENR reactor created in our laboratory, which worked for 7 months continuously, Ca, V, Ti, Mn, Fe, Co, Cu, Zn, Ga, Ba, Sr, Yb, Hf were found [2]. This proves the presence of nuclear transmutations in the LENR process. The appearance of new elements is found not only in the "fuel" but also in the surrounding matter. In [9], [10], it was suggested that the most effective source of the agent that causes nuclear transmutations in LENR processes is metals heated to a sufficiently high temperature. This paper describes a number of experiments specifically carried out to test this assumption. Let us consider several

*Corresponding author: alexparh@mail.ru

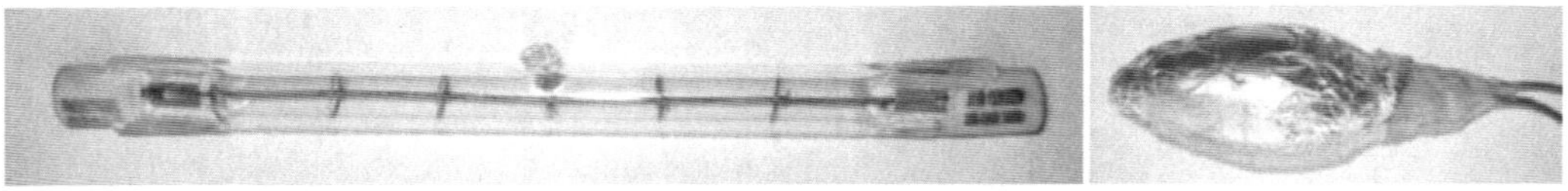

**Figure 1.** Halogen incandescent lamp and 40W incandescent lamp wrapped in aluminum foil.

confirmatory experiments reported at ICCF-23 and the 26th Russian Conference on Cold Nuclei Transmutations and Ball Lightning [11].

## 2. MEASUREMENTS OF EXCESS HEAT GENERATION IN EXPERIMENTS WITH HOT METAL

In the experiments described below, a tungsten filament was used as hot metal in incandescent lamps, in particular, in halogen lamps with a tubular quartz shell, with the nominal power of 150 or 300 W (Fig. 1), as well as in a conventional 40 W incandescent lamp. To reduce the power consumption, at which a sufficiently high temperature of the filament is achieved, you can use a reflective coating such as aluminum foil wrapped around the lamp. Aluminum foil was used with the 40-watt lamp.

To avoid problems measuring the electrical power consumed by the reactors, we use conventional alternating current with a frequency of 50 Hz. We use conventional transformers with a switchable or continuously adjustable secondary winding. A typical diagram of the power supply and measuring equipment can be seen in [2]. Digital multimers were used to measure current and voltage. In addition, energy meters were used, which allowed measuring voltage, current, power, power factor and the amount of electricity consumed with an error of no more than 1%.

Dependence of the specific resistance of tungsten on temperature is well known [12]. Therefore, temperature of the tungsten filament is easy to determine by measuring its resistance at room temperature $R_{20}$, as well as the voltage U and current I in the operating mode. Knowing the voltage and current, we determine the resistance $R = U/I$, and then the temperature by the formula $t(°C) = 197.6(R/R_{20}) - 1.57(R/R_{20})^2 - 176$. It should be noted that the described method of determining the temperature gives an average value, since the filament has colder (at the ends and near the supports) and hotter areas.

Using the same data, you can determine the power consumed by the lamp $P = U \cdot I$.

### 2.1. Experiment 1. Measurement of Heat Output Power Using an Air Flow Calorimeter

The halogen incandescent lamp was placed inside a cylindrical stainless steel container with two walls, the space between which can be filled with various substances (Fig. 2). To measure heat output power, air flow calorimeter was used (Fig. 3). The object under study was placed in a thermally insulated cylinder with a diameter of 20 cm and a length of 100 cm. Heat output power was determined by measuring the temperature of the air flowing over the object. The temperature difference between the outlet and inlet air was measured with a differential thermocouple. To create a stable air flow, a fan connected to a stable power source was used. Calibration measurements showed that the measurement error of this calorimeter at a heat output power of 100 to 2000 W does not exceed 3%.

Figure 4 shows the dependence of the thermal coefficient COP (the ratio of the heat output power to the electrical power consumed), and the temperature of the filament in the halogen lamp. The measurements were made with an empty container and with a container filled with lithium tetraborate $Li_2B_4O_7$ (10 g). Excess heat generation at temperatures above 2200°C is observed even in the case of an empty container, but a container filled with lithium tetraborate gives a stronger effect. With electrical power consumption of 292 W, the average temperature of the tungsten filament reached 2390°C. Heat output power measured by flow-through air calorimeter, was 428W. Thus, release of power in excess of the power used to heat the filament is 136 W. Further temperature rise leads to an increase

**Figure 2.** Halogen incandescent lamp inside a cylindrical container.

**Figure 3.** Air flow calorimeter.

in excess power. However, as can be seen in Fig. 4, the thermal coefficient decreases. This is due to the fact that at high temperatures, the power required for heating grows much faster ($\sim T^4$) than the excess heat release.

Note that this reactor does not contain hydrogen, yet despite this it reliably produces excess heat. This indicates the need to reconsider the established view that hydrogen is necessary for the LENR reaction. This conclusion is confirmed by nuclear transmutations found in the lead-tin alloy that does not contain hydrogen (see section "*Experiment 6. Halogen incandescent lamp and tin-lead alloy*").

## 2.2. Experiment 2. Measurement of Heat Generation by the Rate of Temperature Increase in Water

In this experiment, incandescent lamps were immersed in water (450 ml) poured into a glass Dewar vessel (Fig. 5). The heat output was determined by the rate of increase in the water temperature. Temperature was measured with an electronic thermometer, with a range up to 300°C and accuracy of 0.1°C. Calibration experiments have shown that by varying the measurement time, this calorimeter can measure the heat output power in the range of 10-500 W with an error of no more than 1%. To speed up the establishment of thermal equilibrium, a manual stirrer was used.

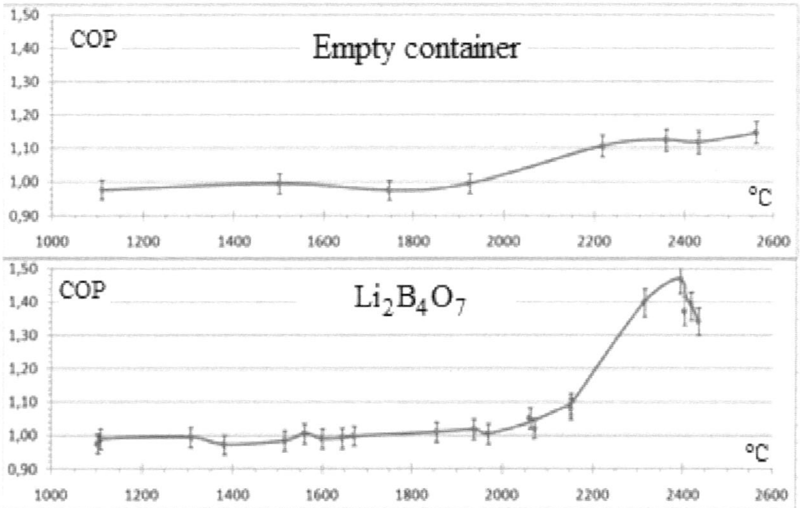


**Figure 4.** Thermal coefficient as a function of the filament temperature in halogen lamp, measured with an empty container and a container filled with lithium tetraborate.

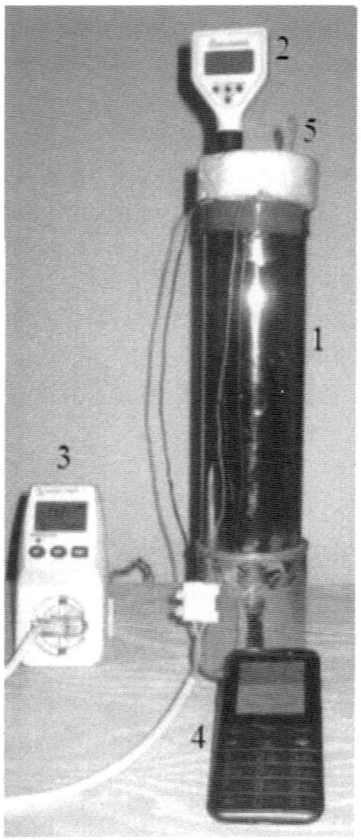


**Figure 5.** Water calorimeter. 1 – Dewar vessel with water in which an incandescent lamp is immersed, 2 – thermometer, 3 – meter of current, voltage and power consumption, 4 – stopwatch, 5 – stirrer.

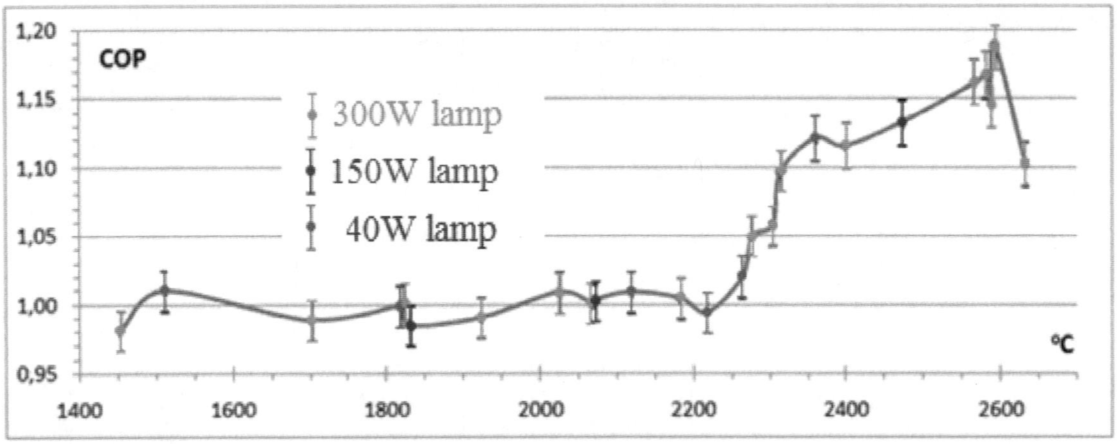


**Figure 6.** Dependence of the thermal coefficient COP on the filament temperature. Set of data obtained in different experiments.

Several different types of incandescent lamps have been tested. In all experiments, excess heat was detected when the average temperature of the filament exceeded 2200°C. Figure 6 shows a set of data obtained in experiments with an incandescent lamp with a nominal power of 40 W and halogen lamps of a tubular design (Fig. 1) with nominal power of 150 and 300 W. Due to the stronger cooling of the lamp shells with water compared to air, 60-70% higher than the rated power is required to achieve a sufficiently high filament temperature.

Figure 6 shows that for all the tested lamps, which are very different in power and design, a noticeable increase in the thermal coefficient is observed at temperatures above 2200°C. At a filament temperature of about 2500°C, the thermal coefficient reaches a value of 1.18, but at a higher temperature it decreases. This is due to the fact that at high temperatures, the power required for heating grows much faster than excess heat generation. This effect was also observed in type 1 experiment (see Fig. 4).

Excess heat generation can occur in lamp cylinders containing $SiO_2$ and in water. To make sure that new chemical elements actually appear in the substance near the incandescent lamps, special rather long-term experiments 5,6 and 7 were carried out, which convincingly confirmed the appearance of nuclides that were initially absent.

### 2.3. Experiment 3. Measurement of Excess Heat Generation in the Installation with Hot Iron (Reactor C3)

The disadvantage of experiments with incandescent lamps is that the mass of the incandescent tungsten filament is very small (about 100 mg). In the C3 reactor (Fig. 7), iron rods with a total weight of 60 g were heated by a tungsten spiral wound on a sapphire tube. The "hot zone" was surrounded by thermal insulation made of porous quartz, wrapped in nickel mesh. Between the grids there was 15 g of nickel powder saturated with hydrogen gas at a temperature of about 300°C.The saturation procedure is similar to the one described in [2]. The outer shell was a quartz tube filled with a mixture of hydrogen and argon. To measure the heat output power, an air flow calorimeter was used (see Fig. 3).

Figure 8 shows dependence of the excess heat output power and the thermal coefficient of the C3 reactor on the temperature. A noticeable excess heat release is observed already at the iron core temperature of 800°C and continuously increases with increasing temperature. The thermal coefficient at a temperature of about 1000°C reaches a value of about 1.3. Increase in temperature does not lead to an increase in COP due to the rapid increase in the power consumed by the electric heater.

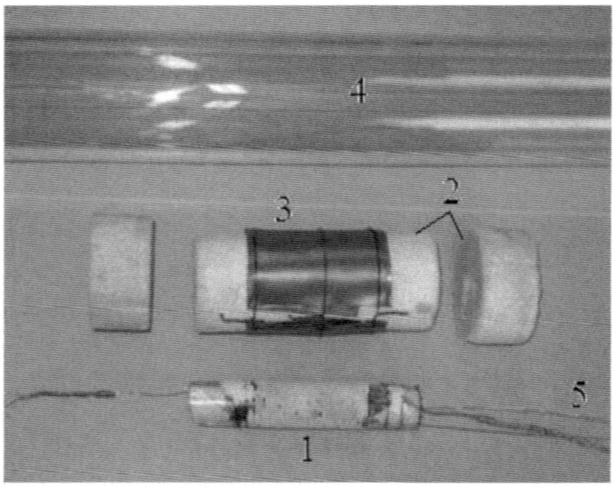


**Figure 7.** Details of the C3 reactor. 1 - iron rods in a sapphire tube, on which a tungsten heater is wound; 2-thermal insulation; 3-nickel mesh; 4-quartz tube; 5- W-Re thermocouple.

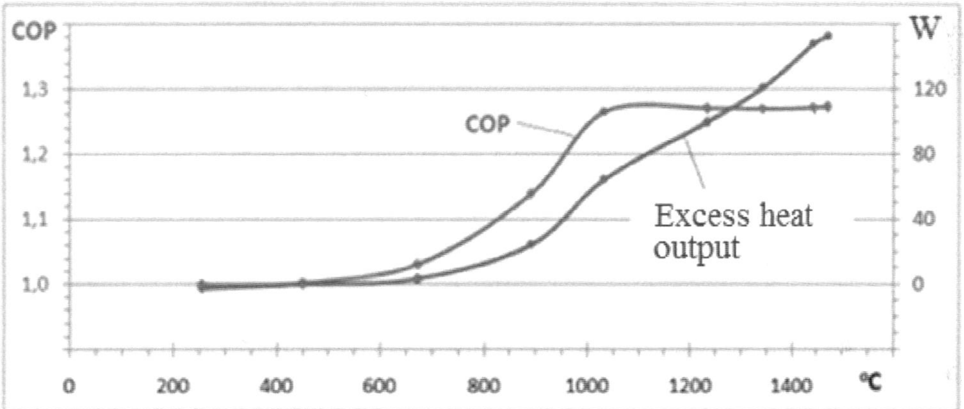


**Figure 8.** Excess heat output power and the thermal coefficient of the C3 reactor as a function of temperature.

### 2.4. Experiment 4. Measurement of Excess Heat Generation in the Installation with Hot Tungsten (Reactor W1)

The W1 reactor uses a tubular silicon carbide heater and a tungsten core, which allows for a higher temperature than the C3 reactor, which has an iron core. The design of the reactor is shown in Fig. 9. Inside the silicon carbide heater there is tungsten powder weighing 3.1 g. The heater is surrounded by thermal insulation made of porous ceramics. Between the thermal insulation and the outer quartz pipe is hydrogen-rich nickel mesh ("fuel"). A view of the reactor in operation is shown in Fig. 10.

Figure 11 shows the dependence of the excess heat output power and the thermal coefficient of the C3 reactor on the temperature. This reactor produced up to 1,000 W of excess power. A noticeable excess heat generation appears

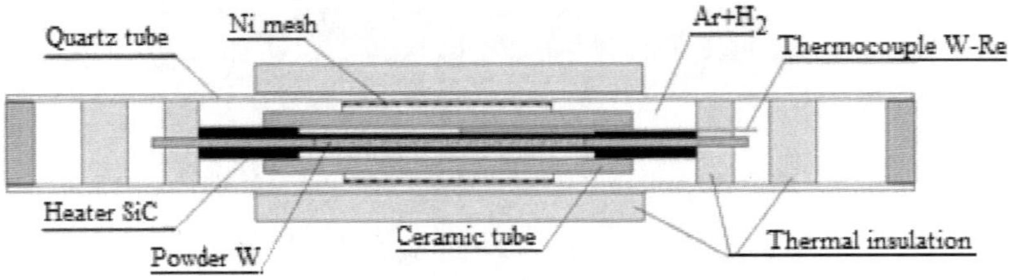


**Figure 9.** Design of the W1 reactor.

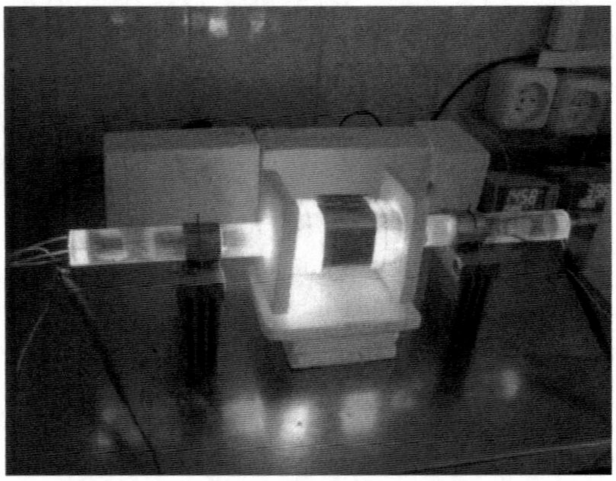

**Figure 10.** Reactor W1 and the process of operation.

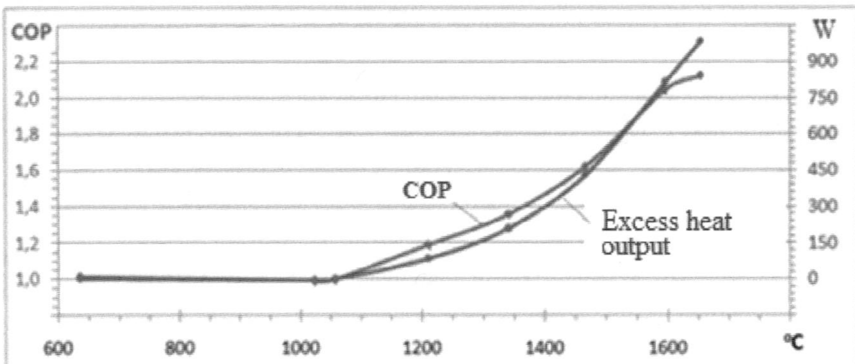


**Figure 11.** Excess heat output power and thermal coefficient of the reactor W1 as a function of temperature.

at a temperature of 1100°C and increases with increasing temperature. Thermal coefficient also increases, reaching a value of 2.2 at a temperature of about 1600°C. At higher temperatures, growth slows down.

Thus, the use of massive metal cores can significantly reduce the temperature at which excess energy is released.

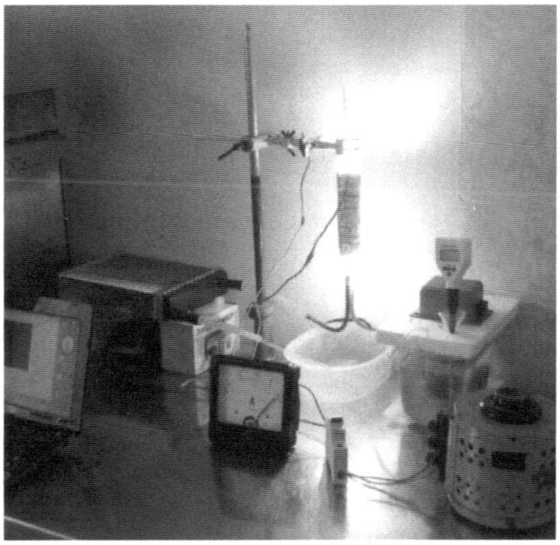

**Figure 12.** Experimental setup with a circulating $KNO_3$ solution and a halogen incandescent lamp.

## 3. INVESTIGATION OF ELEMENTAL AND ISOTOPIC CHANGES IN MATTER NEAR HOT METALS

In order to be sure that the LENR process is really occurring near the hot metals, it is important to make sure not only that a lot of heat is released in the substance surrounding a hot metal, but also that new chemical elements appear. The experiments described above showed the presence of excess heat release at a sufficiently high metal temperature, but they were not long enough to accumulate a noticeable amount of new elements. This problem was solved in further experiments.

### 3.1. Experiment 5. Halogen Incandescent Lamp and Circulating Solution

The halogen incandescent lamp (220V, 300 W) was located in a quartz tube, through which a 10% aqueous solution of $KNO_3$ flowed. The circulating solution was cooled by passing through a heat exchanger. The reactor operated for 20 hours at power consumption of 450 W. Heat output power, determined by the heating rate of the solution, is about 500 W. Average temperature of the tungsten filament is about 2400°C.

Samples of the solution taken before and after exposure to an incandescent lamp, after evaporation, were transferred for analysis of the elemental composition to the SYNTHESTECH Research Center. Two methods were used: X - ray fluorescence (XRF) and mass spectral (ICP MS). The results of the analyses are presented in Table 1.

It can be seen that the content of many elements after treatment of the solution increased by tens or even hundreds of times.

The calcium content increased the most (appearance of calcium is also typical for many other LENR experiments [1]). In this regard, we recall the research of Louis Kervran, who found that a chicken will continue to lay eggs, the shell of which contains a lot of calcium, even if it is completely deprived of its sources of calcium, replacing calcium with potassium [7]. He suggested that calcium occurs as a result of the nuclear reaction $^{39}K_{19}+^{1}H_{1}\rightarrow^{40}Ca_{20}+8.337$ MeV, which caused the ridicule of physicists: potassium and hydrogen cannot combine

**Table 1.** Results of analysis of the samples by XRF and ICP MS methods before and after processing with an incandescent lamp in $KNO_3$ solution (mass %).

| | before | | after | |
|---|---|---|---|---|
| | XRF | ICP MS | XRF | ICP MS |
| Li | | 0,007 | | 0,12 |
| B | | 0,017 | | 0,08 |
| Mg | | 0,050 | | 0,55 |
| Al | | 0,010 | | 3,30 |
| K | 99,91 | | 86,4 | |
| Ca | <0,01 | | 10,2 | |
| Fe | <0,01 | | 0,66 | |
| Cu | 0,018 | | 2,04 | |
| Cu | | 0,010 | | 1,10 |
| Zn | <0,01 | | 0,58 | |
| Zn | | 0,010 | | 0,13 |
| Ni | <0,01 | | 0,048 | |

because of the "Coulomb barrier", and if this somehow happened, the huge energy release would incinerate the chicken. This experiment shows the reality of the formation of calcium in a medium containing potassium by LENR.

### 3.2. Experiment 6. Halogen Incandescent Lamp and Tin-Lead Alloy

A 300-watt halogen incandescent light bulb was wrapped in lead-tin alloy tape and placed in a container of water. To avoid overheating and boiling, the water was cooled by pumping through a fan-cooled coil (Fig. 13). The lamp power consumption was 480 W. Heat output power of about 550 W was determined by the water heating rate. The temperature of the tungsten filament was 2400-2450°C. Working time was 40 hours.

Samples of the lead-tin alloy before and after processing in the described installation were transferred for analysis of the elemental composition by XRF and ICP MS methods to the SYNTHESTECH Research Center. The results of the analyses are presented in Table 2. It can be seen that the content of many elements after processing has increased many times. Especially strongly increased the content of lithium, sodium, aluminum, potassium, calcium, iron, cobalt, silver, cadmium, tungsten, bismuth.

Note that the XRF analysis provides information about elemental composition in the thin surface layer of the test sample. Therefore, it is quite possible to participate in transmutations in this layer of hydrogen, which is part of the water that washes the sample during the experiment. In contrast to XRF, ICP MS analysis provides thickness-averaged information. Hydrogen cannot penetrate deep into the lead-tin alloy. Therefore, if the participation of hydrogen in transmutations is necessary, the analysis of the composition of the samples during the ICP MS study would reveal significantly smaller changes than changes in the surface layer by the XRF method. However, strong changes were detected by both methods, which confirms that the participation of hydrogen in nuclear transmutations is optional. This is also indicated by the calorimetric experiment with an incandescent lamp and lithium tetraborate, described above (experiment 1).

**Table 2.** Content of some elements in the lead-tin alloy before and after processing (mass %).

| | before | | after | |
|---|---|---|---|---|
| | XRF | ICP MS | XRF | ICP MS |
| Li | | 0,0001 | | 0,0053 |
| B | | 0,0012 | | 0,012 |
| Na | | 0,13 | | 1,16 |
| Al | | 0,001 | | 0,024 |
| K | | 0,056 | | 0,75 |
| Ca | | 0,018 | | 0,34 |
| Fe | <0,01 | | 0,27 | |
| Fe | | 0,014 | | 0,13 |
| Co | | 0,0002 | | 0,014 |
| Ni | <0,01 | | 0,073 | |
| Ni | | 0,0006 | | 0,018 |
| Cu | | 0,012 | | 0,041 |
| Zn | | 0,0036 | | 0,040 |
| Pd | | 0,0002 | | 0,0005 |
| Ag | | 0,006 | | 0,024 |
| Cd | | 0,0005 | | 0,0011 |
| Sn | 45,7 | 43,6 | 40,3 | 46,00 |
| W | <0,01 | | 1,51 | |
| W | | 0,00003 | | 0,105 |
| Pb | 54,2 | 44,5 | 57,9 | 31,4 |
| Bi | | 0,0005 | | 0,057 |

### 3.3. Experiment 7. Halogen Incandescent Lamp and Boiling Solution

In a glass vessel with 900 ml of distilled water and 14 g of sodium bismuthate ($NaBiO_3$), a halogen incandescent lamp with a rated power of 150 W was immersed (Fig. 10). The duration of the experiment was 20 hours with power consumption of 270 W. Temperature of the tungsten filament was 2200-2300°C. Excess heat generation of about 25W was determined by the evaporation rate of water. Since sodium bismuthate is insoluble in water, it was suspended during the operation of the setup.

After the end of the experiment, XRF analyses of the sediment and the evaporated solution, as well as the initial $NaBiO_3$ powder, were performed at the SYNTHESTECH Research Center. The results are presented in Table 3.

In starting material, except for bismuth, only platinum was found. The presence of a number of other elements was revealed in the formed sediment and in the solution. In particular, as in experiment 6, a lot of tungsten appeared.

### 3.4. Experiments 8. Hot Metal or Light?

In 2013, at the Kurchatov Institute, Yu.N.Bazhutov and his colleagues conducted a series of experiments with solutions of LiOH, NaOH, and $Na_2CO_3$ illuminated by a laser or LED with a wavelength of 625-650 nm [13]. It was found that a tritium appears in the solutions (1 tritium atom per $10^{13}$-$10^{14}$ radiated photons). No noticeable excess heat generation was detected.

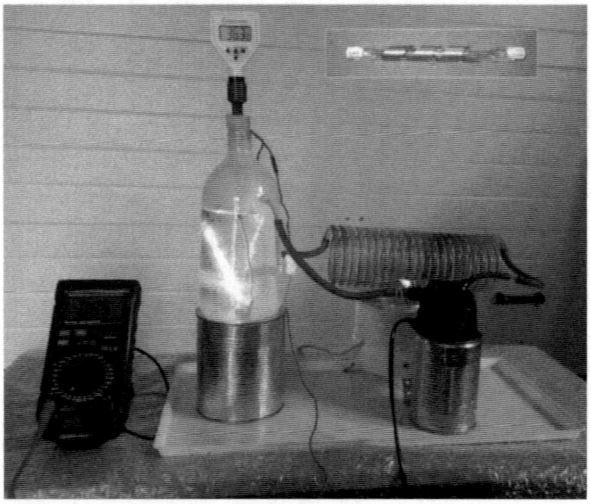

**Figure 13.** Halogen incandescent lamp wrapped with a tin-lead alloy tape in a vessel with water. The water is pumped through a fan-cooled coil.

**Figure 14.** Halogen incandescent lamp in boiling water with $NaBiO_3$ suspension.

Ubaldo Mastromatteo discovered the appearance of C, O, Na, Si, Al, Mg, S, Cl, K, Ca, and Cu in the atmosphere of hydrogen and deuterium as a result of two weeks of laser irradiation with 633 and 405 nm wavelengths of palladium film [14]. This experiment was recently reproduced by Jean-Paul Biberian [15]. After three months of irradiation in the atmosphere of hydrogen or deuterium of a palladium film with a 5 mW laser with a wavelength of 650 nm, initially absent N, O, Na, S, Al, Ca, Fe, Ni, Zn, Mo were detected. The presence of excess heat release was not controlled.

Thus, light exposure causes nuclear transmutations. Our experiments described in this paper are accompanied by very intense light radiation. Perhaps it is due to the action of light that the appearance of new elements and excess heat generation is associated?

**Table 3.** Results of the analysis of the content of chemical elements before and after the experiment (mass %).

| | starting powder | treated | |
|---|---|---|---|
| | | sediment | solution |
| S | <0.01 | <0.01 | 6.167 |
| Ca | <0.01 | 0.231 | <0.01 |
| Fe | <0.01 | 0.092 | <0.01 |
| Cu | <0.01 | <0.01 | 0.396 |
| Dy | <0.01 | <0.01 | 0.451 |
| Ta | <0.01 | 0.246 | <0.01 |
| W | <0.01 | 0.289 | 88.371 |
| Pt | 0.562 | 0.496 | <0.01 |
| Bi | 99.498 | 98.646 | 4.615 |

**Figure 15.** LEDs used in the calorimetric experiment.

LED lamps give very intense light without high temperatures, with low power consumption. If the cause of LENR is light, then the use of LED should provide much stronger effects compared to incandescent lamps.To test this possibility, a calorimetric experiment was conducted with a chain of 15 LEDs extracted from a household LED lamp (Fig. 15). The same calorimeter with water in the Dewar vessel was used, which was used in experiments with incandescent lamps (see Fig. 5).

The measurements showed that LEDs immersed in water do not give a noticeable excess heat release (COP = 1.00±0.01). The LED block, surrounded by aluminum foil or quartz, also gave no noticeable excess heat generation.

It can be concluded that, although light causes nuclear transmutations, they can only be noticed with very long-term illumination with powerful light. The efficiency of the reactors described in this paper is precisely correlated to the presence of hot metal.

## 4. IMPORTANT FEATURES OF LENR

Experimental results obtained by the authors and other researchers allow us to highlight a number of important features of LENR:

a) a huge variety of nuclides appears not only in the "fuel", but also in the surrounding matter, and, in general, nuclides that are not radioactive appear;
b) energy release is far beyond the upper limits of chemical reactions;
c) neutrons and gamma radiation are emitted during the LENR process. However, the radiation intensity is many orders of magnitude lower than in "normal" nuclear reactions;
d) sufficiently dense medium (solid, liquid, dense plasma) is required;
e) presence of hydrogen or deuterium is not necessary;
f) unusual tracks appear near the reactors.

## 5. ATTEMPTS TO EXPLAIN THE LENR PHENOMENON

Many attempts have been made to explain the LENR phenomenon. Numerous theories can be divided into several groups:

- proton (deuteron) overcomes the "Coulomb barrier" and merges with the nucleus;
- proton turns into a neutron, for which there is no "Coulomb barrier." This neutron generates a chain of nuclear transformations;
- LENR processes involve a catalyst;
- the atom goes into a compact state with the release of high energy. In addition, the atom in the compact state more easily overcomes the "Coulomb barrier".

But so far, it has not been possible to explain all the features of LENR, or at least some of them, without contradicting the rest. Basically, the search went in the direction of finding ways to overcome the "Coulomb barrier", which prevents a sufficiently close approach of pairs or groups of nuclei. An example of this theory is the well-known Widom-Larsen theory [16], which allows the possibility of converting protons into neutrons. The resulting neutrons react with the nuclei of the surrounding matter, forming new chemical elements. However, in reality, experiments did not detect the intense radiation of neutrons and hard gamma rays, which inevitably occur when neutrons are captured by nuclei. For example, the capture of a neutron by a proton is accompanied by the emission of a gamma quantum with an energy of 2.23 MeV. When a neutron is captured by nickel, gamma quanta with an energy of up to 9 MeV arise [17]. This fact alone refutes this type of hypotheses. Similar inconsistencies are typical for other approaches. It should be emphasized that none of the above approaches provides an explanation for the huge variety of nuclides that arise.

The problem of the appearance of a huge variety of emerging nuclides can be solved by the assumption that they appear as a result of energy-efficient rearrangements of nucleons located in two or more atoms [18], [19]. To do this, they must be exposed to action that covers several atoms at the same time. A possible way to solve this problem is described in [9], [10], [20].

## 6. CONCLUSIONS

Various experiments have shown that in a substance near metals, when they are heated to a sufficiently high temperature, there is indeed the release of excess heat and the appearance of initially absent chemical elements, which testifies to the course of LENR.

Our experiments, combined with numerous experiments by other researchers, give indications of a number of erroneous directions in the search for an explanation of the LENR phenomenon .In a number of experiments, LENR was detected in the complete absence of hydrogen. This indicates the need to reconsider the traditional approaches to explain LENR, in which hydrogen or deuterium are necessary. Also, theoretical explanations involving neutrons do not stand up to comparison with experimental results. Experiments also show that there is no need for a crystal lattice for LENR, since LENR occurs not only in solids, but also in liquids and dense plasmas.

## ACKNOWLEDGMENT

The authors would like to thank T.Timerbulatov, S.Zabavin and I.Stepanovfor supporting the research and helping with equipment; I.Kamputin for high-quality analysis; N.Samsonenko, L.Urutskoev, A.Chistolinov and V.Vysotskii for helpful discussions and valuable criticism; I.Savvatimova, A.Klimov, V.Zatelepin, R.Grinyer V.Zhigalov, G.Levi, A.Prosvirnov, S.Godin, V.Panchelyuga, A.Kovacs and many other people in Russia and abroad for their support and valuable advice.

## References

[1] A.G.Parkhomov, K.A.Alabin, S.N. Andreev, etc., Nickel-hydrogen reactors: heat release, isotopic and elemental composition of fuel,*RENSIT*, **9**(1)(2017) 74–93. Available from http://en.rensit.ru/vypuski/article/200/9(1)74-93e.pdf

[2] A.G.Parkhomov, V.A.Zhigalov, S.N.Zabavin, etc., Nickel-hydrogen heat generator, continuously worked for 7 months, *International Journal of Unconventional Science*, 23–24(7) (2019) 57–63 (in Russ.) Available from http://www.unconv-science.org/pdf/23/parkhomov2.pdf English translation: https://drive.google.com/file/d/10tPXOEGHfB95YQeOoePUftdByo7BVnvg/view

[3] A.V.Vachaev, N.I.Ivanov, A.N.Ivanov, G.A.Pavlova, Method of obtaining elements and device for its implementation, *Patent of the Russian Federation* No. 2096846, MKI G 21 G 1/00, H 05 H 1/24 (1994).

[4] Yu.N.Bazhutov, A.I.Gerasimova, V.P.Koretsky, A.G.Parkhomov. Features of electricity consumption, heat release and radiation in the process of plasma electrolysis. Proceedings of the 21st Russian Conference on Cold Nuclear Transmutations and Ball Lightning. Sochi, 2015: 122-131 (in Russ.).

[5] I.B.Savvatimova,Transmutation of Elements in Low-energy Glow Discharge and the Associated Processes, *J. Condensed Matter Nucl. Sci.* **8** (2011) 1–19.

[6] M. Fleischmann, S. Pons. Electrochemically induced nuclear fusion of deuterium. *Journal of Electroanalytical Chemistry and Interfacial Electrochemistry*. **261** (2A) (1989) 301–308.

[7] L.Kervran, *Biological Transmutations*. Happiness Press, USA, Magalia, California, 1998.

[8] A.A Kornilova, V.I Vysotskii. Synthesis and transmutation of stable and radioactive isotopes in biological systems. *RENSIT*, **9(1)**(2017) 52–64.

[9] A.G.Parkhomov. Weak Interactions as Essence of LENR. *International Journal of Unconventional Science*, **E4** (2019) 3–5. http://www.unconv-science.org/pdf/23/parkhomov1-en.pdf

[10] A.G. Parkhomov, R.V.Karabanov, LENR as a manifestation of weak nuclear interactions. New approach to creating LENR reactors, *RENSIT*, **13**(1) (2021) 45–58. Available from http://en.rensit.ru/vypuski/article/372/13(1)45-58e.pdf

[11] A.G. Parkhomov. New approach to the creation of LENR reactors, A.G. Parkhomov, R.V. Karabanov. Investigation of elemental and isotopic changes in matter near incandescent lamps. *Proc. of the 26th Russian Conf. on cold nuclear transmutations and ball lightning*. Moscow, 2020, p. 43-63 (in Russ.) https://www.ozon.ru/product/problemy-holodnoy-transmutatsii-yader-himicheskih-elementov-i-sharovoy-molnii-230215320

[12] I.S. Grigoriev, E.Z.Meilikhov (ed). Physical quantity. Moscow: Energoatomizdat, 1991. p. 438 (in Russ.)

[13] Yu.N.Bazhutov, A.I.Gerasimova, V.V.Evmenenko, V.P.Koretskiy, A.G.Parkhomov and Yu.A.Sapozhnikov. Calorimetric and Radiation Diagnostics of Water Solutions Under Intense Light Radiation. *J. Condensed Matter Nucl. Sci*, **19** (2016) 10–17.

[14] U.Mastromatteo, LENR Anomalies in Pd–H2 Systems Submitted to Laser Stimulation. *J. Condensed Matter Nucl. Sci.* **19** (2016) 173–182.

[15] J.-P.Biberian, Transmutation induced by laser irradiation. *Le colloque RNBE 2020*.
[16] A.Widom, L.Larsen, Ultra Low Momentum Neutron Catalyzed Nuclear Reactions on Metallic Hydride Surfaces. *arXiv:cond-mat/0505026.*
[17] T.S.Belanova, A.V Ignatyuk., A.B. Pashchenko. Radiation capture of neutrons. - Moscow: Energoatomizdat, 1986, 248 p.(in Russ.)
[18] V.Kuznetsov, G.Mishinsky,F.Penkov, V.Arbuzov,V. Zhemenik, *Annales de la Fondation Louis de Broglie*, **28**(2) (2003) 173–213.
[19] D.V.Filippov, L.I.Urutskoev, On the possibility of nuclear transformation in low-temperature plasma from the viewpoint of conservation laws, *Annales de la Fondation Louis de Broglie*, **29** (Hors série 3) (2004) 1187–1206.
[20] A.G. Parkhomov, E.O. Belousova, Huge Variety of Nuclides that Arise in the LENR Processes. Attempt at Explanation. *Journal of Modern Physics*, **13**(3) (2022) 274-284 https://www.scirp.org/journal/paperinformation.aspx?paperid=115710

J. Condensed Matter Nucl. Sci. 36 (2022) 377–394

Research Article

# The Nature of the D+D Fusion Reaction in Palladium and Nickel

Edmund Storms*

**Abstract**

The LENR reaction can be made to occur in Pd and Ni having different physical forms. The behaviors of solid Pd, sintered Pd powder, and sintered Ni powder are discussed in this paper. The required deuterium ions can be made available to the LENR process by using electrolysis or exposure to $D_2$ gas. Initiation of the nuclear reaction sometimes requires the creation of a large D/Pd ratio, but not always. In many cases, very little deuterium is required to produce detectable heating power, which is the thermal energy expressed in watts. In every case, the amount of power is not affected by the D/Pd ratio after the nuclear process starts. Also, increased temperature causes the amount of power to increase exponentially, with the activation energy being related to the source of deuterons available to the nuclear process. The reaction involving deuterium emits part of the nuclear energy as energetic ions having the characteristics of a hydrogen isotope, but not helium [1]. Very little photon radiation is detected outside the walls of the container in which the source is located. The behavior is consistent with the nuclear active environment (NAE) being in physical gaps having a critical width located outside the crystal structure, not in vacancies located within the lattice structure. Successful production of LENR involves the formation of these sites in high concentration and with reliability. This paper describes an effort to meet this challenge.


*Keywords:* LENR, D+D fusion, sintered Ni powders, sintered pd powders, energetics ions

## 1. Introduction

Effective study of LENR (Low Energy Nuclear Reaction) [2] requires the application of a model, theory, or perhaps even something as simple as an expectation. The choice of which idea to use is made difficult by so many being suggested, perhaps too many. I summarized and evaluated many of the published suggestions, including my own, in my book [3]. This paper will focus on the model I first suggested in 2003 and have expanded over the years [4], [5].

This model has two parts. One applies to the physical-chemical environment required for the nuclear process to occur and the second part applies to the nuclear reaction itself. I will address only the first part in this paper.

The model proposes that a particular condition must form in the material consisting of gaps having a critical width. This condition is called the NAE (Nuclear Active Environment). The hydrogen isotopes are assumed to accumulate within these sites by a conventional chemical process. For reasons still unknown, the assembly of hydrogen nuclei can then experience collective fusion with the production of energy and nuclear products including helium and tritium.

*Corresponding author: storms2@ix.netcom.com

© 2022 ICCF. All rights reserved. ISSN 2227-3123

Radiation is also produced, with most of it being unable to pass through the walls of the surrounding apparatus. The greater the number of the gaps that are populated by a hydrogen isotope, the larger is the resulting thermal power. In contrast, the complete absence of the critical gaps prevents LENR from occurring regardless of how the material is treated. Therefore, the concentration of suitable gaps, as determined by the nature of the material, is one of the critical variables. The other critical variable is temperature. Both are described in this paper.

It is important to realize that other mechanisms, besides LENR, can be caused to operate when extra energy is applied in the form of energetic deuterons [6]–[9] or photons [10]. These reactions are related to what is called hot fusion. Consequently, the behavior must not be confused with the mechanism causing LENR, because entirely different nuclear products are produced.

The necessary gaps can be created in many different ways. The method described here involves the use of powder to which physical pressure is applied with sufficient to force to increase the contact between the particles. This method has been used by other studies [11]–[13] but without the creation of gaps as the intended goal or without suitable methods of activation. As result, perhaps the treatments were not as effective as they might have been if the critical variables had been correctly identified. In the case of the method used here, extra energy is produced without fail after proper pre-treatment of the material based on consideration of the variables described here.

Based on my model, force-pressing a powder into a low-density structure would be expected to create a useful material because a large collection of gaps would form, some with a suitable dimension. In addition, a large fraction of the material would be directly accessible by the gas used to activate the sites. The need to achieve a large D/Pd ratio for activation, as found when solid Pd is used [14]–[16], would not be expected and was not found to be necessary.

The first step is to create a sufficient number of active gaps to produce enough power to make studies possible. That goal has been achieved in this study. The next step is to learn how to increase the number of active sites in order to increase the amount of excess power. That effort is underway.

## 2. Experimental

The information presented here comes from measurement of excess thermal power production using a sensitive Seebeck calorimeter [17]. The calorimeter consists of a water-cooled box held at a constant temperature of 10°C. The walls of the box are lined with 54 thermoelectric converters (Custom Thermoelectric, 12711-5P31-12CW) connected in series. The converters generate a voltage that is proportional to the average amount of thermal power being conducted through their thickness, which generates a Seebeck voltage proportional to the heat energy being lost from anywhere in the box. The relationship between the voltage and the heat energy is obtained from a suitable calibration.

Different cell designs can be placed in the box to study electrolysis, gas loading, or gas discharge without the need for modification. Magnetic fields and/or laser light can be easily applied and radiation detectors (Geiger-Mueller (GM) and other types) can be placed in the box near the sample. The calorimeter was tested for accuracy by measuring the enthalpy of fusion of tin at 212°C and the enthalpy of formation of $PdH_x$ [18] as a function of H content. Both measurements agree with the accepted values.

This work used either an electrolytic cell containing an electrolyte consisting of $D_2O$ plus LiOD or a cell in which the sample is exposed to fixed amount of low-pressure $D_2$ or $H_2$ gas. The complete history of the chemical reaction in the cell, including all energy entering and leaving the cell, is recorded electronically every minute using LabView.

The cells are heated by resistance wire wrapped around their outside, with the electrolytic cell reaching 90°C by applying 26 watts and the gas cell reaching 300°C by using 13 watts. The gas cell reaches a higher temperature because it is not limited by the boiling point of $D_2O$ and it is better insulated than the electrolytic cell. Temperature is measured using several RTDs (Pt resistance temperature detector) to an accuracy of $\pm 1$°C. The temperature is not used to measure the thermal energy and the measurements made by the calorimeter are independent of the sample temperature.

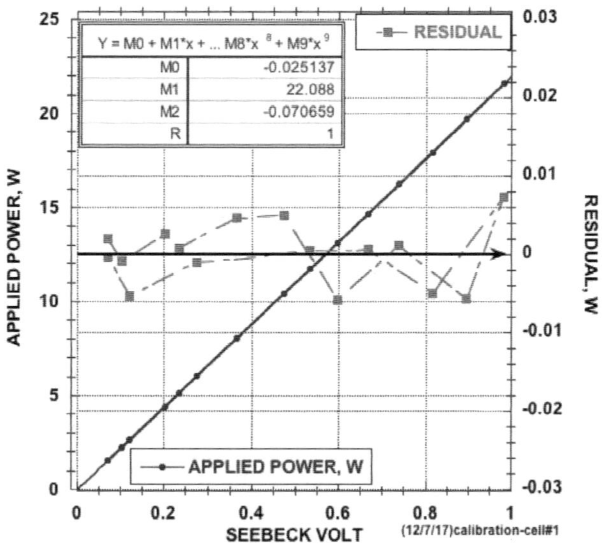


**Figure 1.** Typical calibration showing the relationship between Seebeck voltage and applied power. The points are taken first by increasing and then by decreasing the temperature in similar steps. A delay of 90 minutes between each measurement is used to allow the calorimeter to come to a constant temperature. The residual is the difference between the value of each point and the value obtained using the fitted equation. This value shows the amount of random variation in each measurement. The Standard Deviation of all the residual values is near ±0.005 W.

The calorimeter is calibrated frequently by applying power to the heating wire surrounding the cell being used without a sample present. A typical calibration is shown in Fig. 1. Each cell type is calibrated separately over the full range of power used to achieve the highest temperature. Excess power is identified when P in the following equation has a positive value:

$P = (M0) + (M1)*S + (M2)*S^2 - AP$ where the constants M0, M1, and M2 are listed in Fig. 1, S is the measured Seebeck voltage, P is the power measured by the calorimeter, and AP is the electrical power applied to all items in the box, including a fan.

The open calorimeter box containing an electrolytic cell is shown in Fig. 2. The balance with the oil container shown on the right is used to measure the amount of orphaned oxygen released when D reacts with Pd in the electrolytic cell.

The open Pyrex electrolytic cell is shown in Fig. 3. Both the anode and reference electrode are Pt and the recombiner is a commercial fuel-cell catalyst made by depositing nano-Pd on carbon cloth (FuelCellsEtc, HLGDE-W1S1010). The electrolyte is 30 ml of $D_2O$ to which about 0.1 g of Li metal was reacted. All gas entering or leaving the cell is measured as a weight change of displaced oil.

The assembly used to measure the temperature of the recombiner is shown in Fig. 4. All gas leaving the cell must pass next to the catalyst where the temperature increase caused by the $D_2 + O_2$ reaction can be measured by a nearby RTD. Any remaining $O_2$ (orphaned oxygen) is carried through a small plastic tube to the oil reservoir shown in Fig. 2.

Figure 5 shows the Seebeck calorimeter containing the gas-loading cell. This gas-cell has a volume of 17 ml. Two RTDs, one on the inside and one on the outside of the Pyrex wall are used to measure the temperature every minute. Resistance wire wrapped on the outside is used to heat the cell up to near 300° C. The cell is sealed using spring-loaded screws to hold the glass against a rubber seal. The cell is evacuated to a pressure of less than 5 x $10^{-6}$ torr before

**Figure 2.** The open Seebeck calorimeter containing the electrolytic cell, with the apparatus for measuring orphaned oxygen on the right. The amount of released $O_2$ is measured using the weight of displaced mineral oil from a reservoir.

being loaded with $D_2$ at a known pressure, generally between 200 and 700 torr. Because a small volume with a small initial pressure is used, the amount of D is limited to less than 0.0004 mol. The temperature and prevailing equilibrium pressure determine the amount of this D that reacts with the Pd. When fully reacted, the typical D/Pd ratio would be less than 0.02. Consequently, studies involving gas loading involve only the alpha phase in which the D atoms occupy a few random interstitial positions between the Pd atoms in the face-centered-cubic (fcc) structure.

When electrolysis is used, the D/Pd ratio is measured in three different ways. Examples are compared using unpublished data from a previous study in Fig. 6.

The orphaned oxygen method (OO) measures the amount of oxygen left behind when the $D_2O$ reacts with the Pd, as described above.

The temperature of the catalyst (RT) used to recombine any excess $D_2$ and $O_2$ allows the D/Pd ratio to be calculated as follows:

The number of moles of deuterium being made available for reaction with Pd by the electrolytic current can be determined using the equation:

$$\text{Moles H or D} = \mathrm{M} = \text{seconds} * \text{current}/98485. \tag{1}$$

Because the time interval is 60 seconds and the current is 0.101 A for this study, the total number of moles of deuterium generated in the cell during the measurement interval is:

$$\mathrm{M}_D = 60 * 0.101/98485 = 6.15 \times 10^{-5} \text{ moles/min} \tag{2}$$

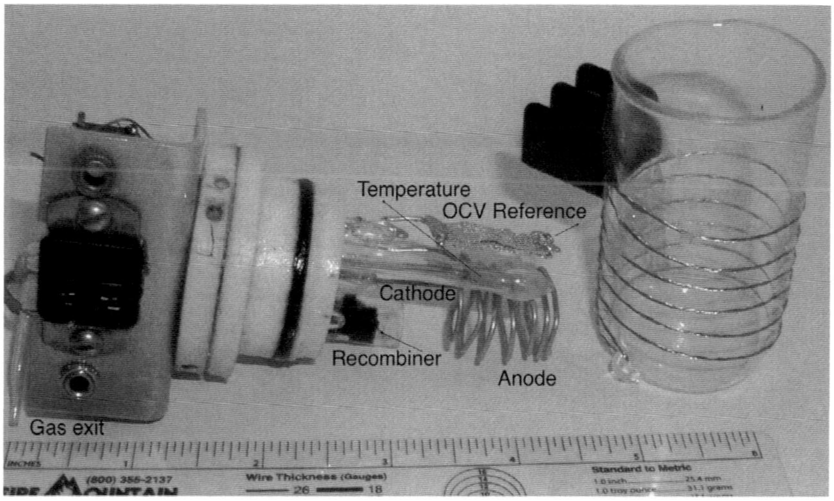


**Figure 3.** Components of the Pyrex electrolytic cell. The top is Teflon. All glass exposed to the electrolyte is Pyrex and all exposed metal is platinum, except the cathode, which is attached by a plastic clamp that allows rapid removal.

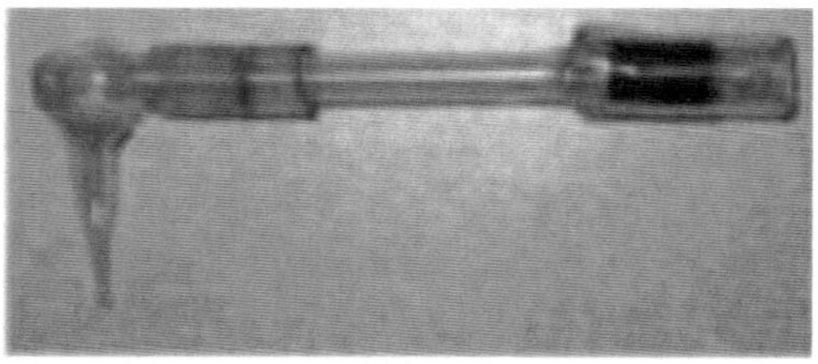

**Figure 4.** Assembly used to measure catalyst temperature after being removed from the cell shown in Fig. 3. Gas containing a mixture of $D_2$ and $O_2$ will form $D_2O$ at the catalyst, seen as the black region in the photograph. The resulting temperature increase is detected by an RTD contained in a sealed Pyrex tube located next to the catalyst. Excess $O_2$ can leave the cell through a gap between the inner and outer tubes.

Deuterium that reacts with the Pd is not available to combine with $O_2$ gas at the recombiner, hence does not contribute to an increase in recombiner temperature. Therefore, the increase in temperature can be used to determine how much D has reacted with the metal and the rate of this reaction.

Finally, in a few tests, the sample was weighed at the end of the loading process to determine the final amount of D in the Pd as follows:

Because D is lost from the Pd when the electrolytic current is stopped, a plot of deuterium content vs square root of time is extrapolated to zero time, as shown in Fig. 7, to determine the D content at the time the current was stopped. This method can only be applied to solid Pd because the porous material retains an unknown amount of fluid after it is removed from the electrolyte.

This loss of deuterium also occurs when the electrolytic current is stopped while the sample is in the electrolytic cell, as is occasionally done to determine the effect of the D/Pd ratio on thermal power production. Figure 8 shows how the D/Pd changes over time, as the final D/Pd approaches the equilibrium value created by the activity of $D_2$ present

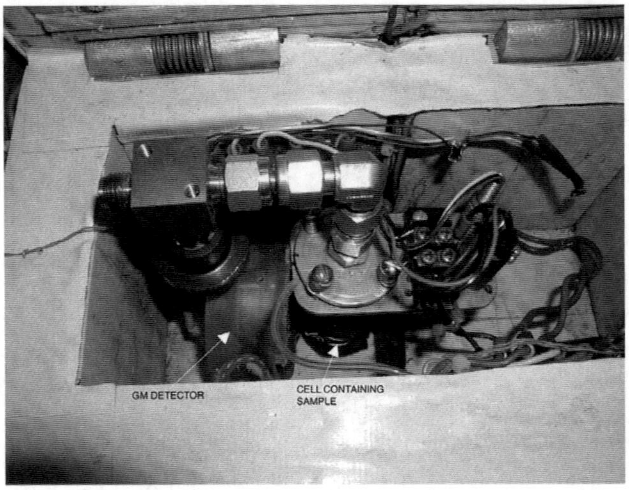


**Figure 5.** Seebeck calorimeter containing the gas-loading cell used to react the metal with $D_2$ or $H_2$ gas. A Geiger-Muller detector is shown next to the cell containing the sample. No radiation has been detected. The fan in the top of the box is not shown.

in the cell. Surprisingly, neither a large number of flaws nor the cracks in the surface of the pressed material seem to have much influence on this behavior.

When gas loading is used, the maximum D/Pd ratio at room temperature is calculated from the known volume of the cell combined with the known pressure of $D_2$ added with the sample, assuming all D had reacted. The amount of D combined with the Pd decreases as the sample is heated. Consequently, this method involves a change in the amount of D in the Pd as the temperature is changed. The initial amount of D is kept small because if enough $D_2$ were added to cause the beta phase to form, the pressure could become much too large for the cell to contain. Therefore, an amount of $D_2$ is used such that the pressure does not increase above 2 atm at 300°C. Therefore, the beta phase is not present during these studies.

Figure 9 shows the nature of the reaction when oxidized Pd reacts with a supply of $D_2$ gas sufficient to form some beta phase. The initial reaction is fast enough to heat the sample. The D/Pd ratio is calculated from the pressure change of a fixed volume of gas exterior to the cell. This behavior demonstrates just how reactive the Pd sample is to $D_2$ gas.

## 3. Results

### 3.1. Sample Preparation

The samples were prepared by pressing the powders in a steel die to form a disk shape about 12 mm in diameter and about 2 mm thick. This disk was then subjected to various treatments including heating in a vacuum at 900–1000°C and/or heating in air at various temperatures and times. The amount of physical pressure used to compact the powder was not explored but is expected to affect the amount of generated power. The formation of an oxide layer and the addition of various inert materials in the form of a fine powder are also found to affect success. Even though the conditions were not optimized, active samples could be made with reliability. Although many variables and treatments were explored during this study, this paper is focused only on a few behaviors produced by a few samples to show a few examples of typical behavior.

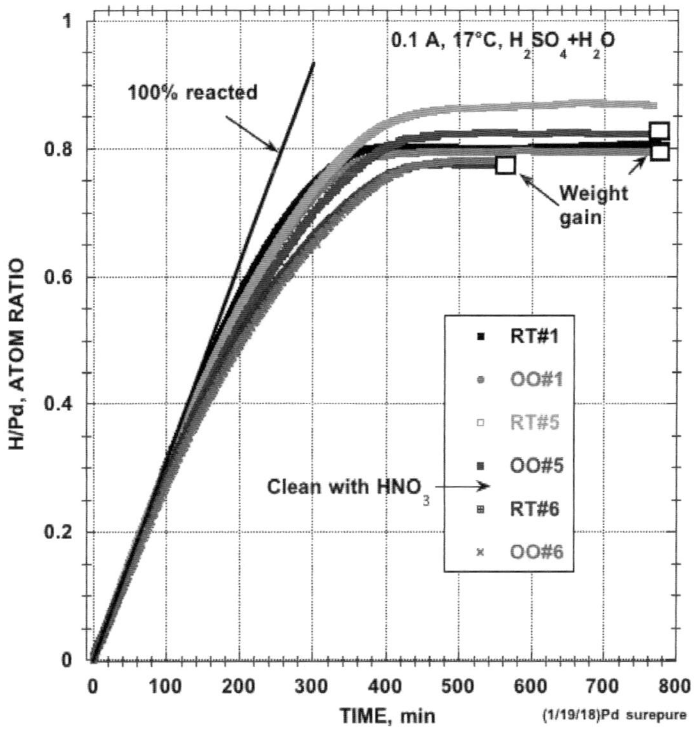


**Figure 6.** Comparison between the D/Pd ratio measured by orphaned oxygen (OO), catalyst temperature (RT), and weight gain after loading. Agreement between the different methods gives confidence in the accuracy of the measurement.

The pressed material has a surface structure shown in Fig. 10. The amount of free volume is variable among the samples but is between 28% and 50%. The samples of Pd generally weighed about 1.5 g. Their dimensions were measured and used to determine the free volume. The dimensions changed as a result of loading and heating in ways that revealed important information about the structure of the material. These behaviors will be discussed in a later paper.

A sample of pressed Pd powder (1.7 g) containing an additive was reacted at 1000°C in a vacuum for 19 hr. Afterward, it showed the loading behavior plotted in Fig. 11 when reacted with D using electrolysis. Despite the many ways the D could escape, this sample loaded to D/Pd = 0.9. Nevertheless, no excess power was produced when the sample was heated to 91° C in the electrolytic cell.

This sample was then heated in air to 800°C and cooled in the furnace, thereby causing a weight increase of 0.01442g consisting of oxygen as PdO that was formed throughout the open structure. The sample was again reacted with D. This time the apparent D/Pd increased to 0.95 as a result of the reduction of the extra PdO. Excess thermal power was made with and without electrolytic current being applied to the sample when heated to 91°C. The sample was then removed from the electrolytic cell and placed in the gas cell with about one atm of $D_2$. The cell was heated in steps to 215°C. Once again, excess power was made with the amount being consistent with the behavior in the electrolytic cell. The power produced with and without electrolytic current is compared in Fig. 12 to the power made when the sample is heated in $D_2$ gas. This agreement is not an artifact because 26 W of applied power was used to

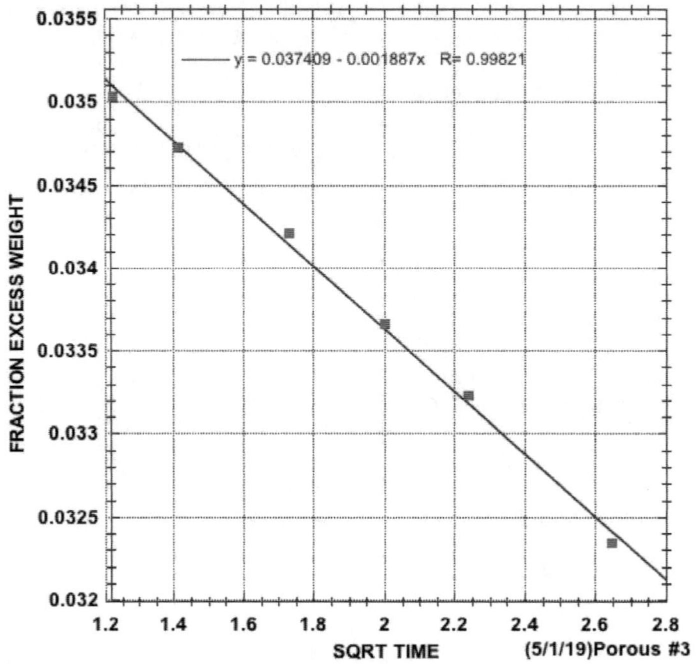


**Figure 7.** Typical weight loss of $D_2$ plotted as the square root of time (SQRT) when Pd is allowed to lose $D_2$ in air. The straight line is extrapolated to zero time to determine the D/Pd ratio at the time the sample was removed from the calorimeter.

produce a temperature of 93°C in the electrolytic cell while this same temperature was reached by applying only 7 W to the gas cell. In both cases, the applied power is within the range of the calibration used at the time. Also, this type of calorimeter is not sensitive to the temperatures in the cells.

The pure Pd powder does not produce LENR immediately after pressing or after sintering in a vacuum. Heating in the presence of $O_2$ or $H_2O$ gas is required to activate the material. Because a large surface is accessible to the oxidizing gas, an amount of reacted oxygen is sufficient to cause a detectable weight increase of several milligrams. The amount of power did not correlate with the amount of oxygen reacted, although too much or too little oxide formation would prevent the LENR reaction.

Somewhat different methods were used to activate nickel powder. This treatment will be described elsewhere.

### 3.2. Effect of Temperature

Temperature is one of the more important variables affecting the LENR reaction. Increased temperature can make an apparently dead material appear alive when the amount of power finally exceeds the sensitivity limit of the calorimeter. Indeed, I suspect people would have had much greater success in detecting LENR if they had simply increased the temperature. In addition, the activation energy calculated using temperature reveals important information about the mechanism.

Dong *et al.* [20] (Figure 13) provided a summary of published values for the relationship between Log EP vs 1/T, including the result of their own measurements. They used the resulting straight lines and slopes to support a resonant

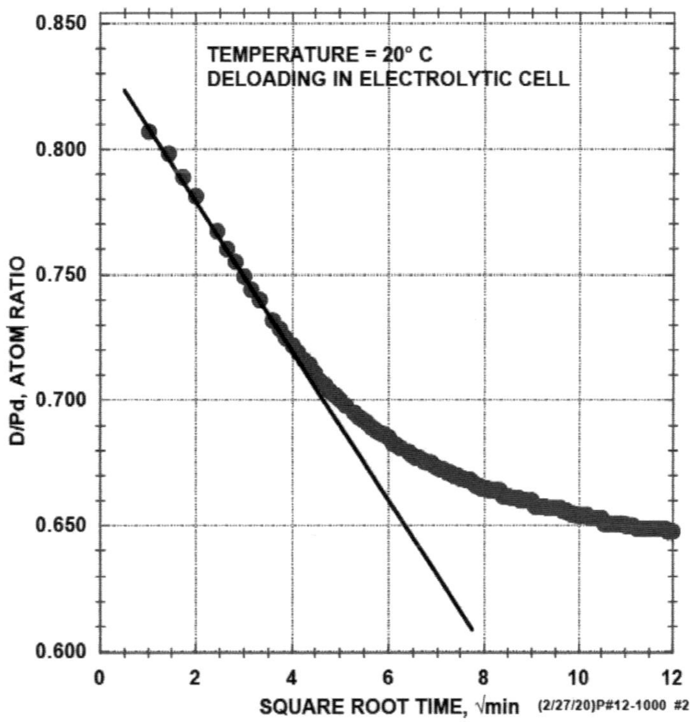


**Figure 8.** Example of deuterium loss when the electrolytic current is stopped at 20° C while a sintered sample is in the electrolyte. The loss rate increases when the temperature is increased. The D/Pd ratio is calculated using the orphaned oxygen method.

surface capture model [21]. In addition, they propose that the diffusion of D through various materials in the material is related to the effect of temperature.

Figure 14 compares several of my previous studies using solid Pd to the powdered Pd described in this paper. The data labeled B, which is a sample using powdered Pd, was heated in the electrolytic cell, first with 0.1 A applied to produce a measured D/Pd ratio of 0.85. Then current was turned off while the cell was again heated. At the end of the first run the D/Pd ratio has decreased from 0.85 to 0.56. At the end of the second run, the ratio had decreased to 0.51 and then to 0.46 at the end of the third run. In each case, the values all showed the same behavior. In other words, the D/Pd ratio had no effect on the amount of power. Also notice that a break in slope occurs, suggesting a change in the mechanism.

Another powdered sample shown by the red points labeled A produced less power but showed the same effect of temperature.

The points shown in green and labeled C were produced by a solid piece of Pd published by Storms [19], [22] to which 0.1 A was applied. Notice it has the same slope as the powder samples when studied in the electrolytic cell.

A very early study of solid Pd (1994) [23], labeled D, was unknowingly made in the transition region, which caused the slope to fall between the two effects. This demonstrates the importance of making measurements over a wide temperature range.

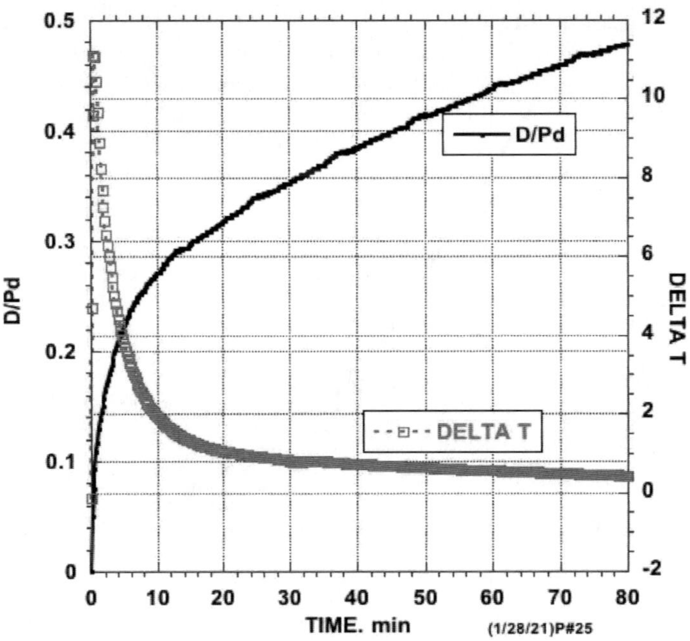


**Figure 9.** Change in sample temperature (degrees Celsius) compared to the resulting D/Pd ratio as 1.5 g of oxidized Pd reacts with $D_2$ gas initially at room temperature. The delta T is the temperature difference between the inside and outside of the cell. Most of the initial temperature increase results from the reaction of PdO with $D_2$.

The relationship between log EP and 1/T is shown in Fig. 15 using the values plotted in Fig 12 for a powdered sample. A similar effect of temperature is revealed in the low temperature region when electrolysis and the gas loading are used. In this case, the transition temperature is too high to reveal the behavior of the high temperature region when electrolysis is used. In contrast, this region can be easily identified when gas loading is used. It is important to note that the behavior in the low temperature region is the same as that caused by electrolysis.

The behavior of an activated sample of Ni powder is shown in Fig. 16. Once again, the behavior is the same as that produced by pressed Pd, but with a larger transition temperature.

## 4. Discussion

When pressed into a compact form, powdered Pd and powdered Ni can produce energy by LENR after suitable activation. Energy production can be initiated using electrolysis or by direct exposure to $D_2$ or $H_2$ gas. Both methods and both materials produce similar behavior, suggesting the thermal power results from the same mechanism. In both cases, the amount of energy is not influenced by the D content.

What do the observed behaviors mean in terms of the model being applied here?

The model identifies three processes that are required to operate before LENR can occur. First, sites must form into which D or H can assemble. For the sake of this discussion, only the role of D will be described. These sites are created by the proper initial treatment of the material. Next, a collection of D atoms must assemble at these sites. This process is controlled by the Gibbs energy difference between the D in the collective and where the D is located

**Figure 10.** SEM picture of a typical pressed sample.

in the surrounding material. Both the D/Pd ratio[18], and the temperature, with temperature having the greater effect, determine the Gibbs energy. An increased number of sites would be populated as both the temperature and D/Pd ratio are increased. This process is proposed to cause the temperature effect in the low-temperature region.

After the sites have become populated and the fusion process starts, the rate of thermal power would be determined by how fast the D replaces the atoms as they are converted to He. This replacement requires transport of the atoms though the material by diffusion. This process is proposed to cause the temperature effect in the high-temperature region. An obvious transition between these two mechanisms would be expected, as seen in Figs. 14, 15, and 16, because the mechanisms affecting power production had changed.

When suitable sites are not present, the LENR effect will not happen regardless of additional treatment. The lack of reproducibility and the difficulty in producing large amounts of power are proposed to result from a variable number of suitable sites being available, with the formation of a large number being especially rare.

Therefore, the model predicts two different effects for both temperature and D content, with a transition occurring between the two mechanisms at a characteristic temperature. How does this prediction compare to the behavior described here?

Three kinds of material were studied. Solid Pd (Fig. 14) was measured using electrolysis. Pressed Pd powder (Fig. 15) was measured using both electrolysis and exposure to low-pressure $D_2$ gas. The pressed Ni powder (Fig. 16) was exposed only to low-pressure $D_2$ gas. In each case, regions were observed in which temperature had a different effect, with each showing a transition temperature between the regions. The slope of Log Power vs 1/T can

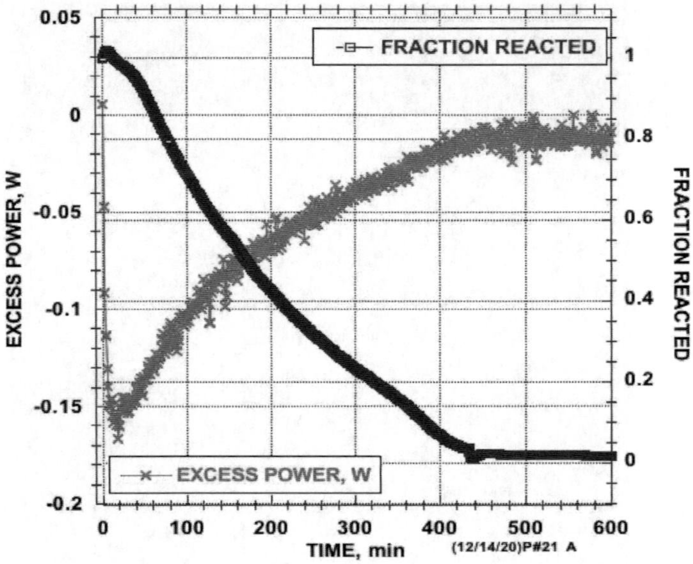


**Figure 11.** Relationship between the amount of excess energy and the reaction rate when 0.1 A is used to react a Pd sample with $D_2O$. Palladium, in both the solid and pressed-powder form, shows very similar behavior when loaded in the electrolytic cell [18], [19]. The excess energy is negative because the amount of energy used to decompose the $D_2O$ is greater than the energy released when $D_2$ reacts with Pd.

**Table 1.** Summary of values for Activation Energy

| MATERIAL | SLOPE, log/T | kJ/mol | eV/atom |
|---|---|---|---|
| Solid Pd | 1526 | 29.22 | 0.303 |
| Powder Pd | 678.5 | 12.99 | 0.135 |
| Powder Ni | 676.7 | 12.96 | 0.134 |
| Diffusion in $PdD_{0.85}$ | 1201 | 23.00 | 0.230 |
| Diffusion in Pd [25] | | 25.1 | 0.260 |

be used to calculate the activation energy for the process that controls the rate of power production, with the results summarized in Table 1.

If the diffusion of D limits the rate at which power could be produced, the activation energy for diffusion should match the activation energy for power production. Both solid Pd (Fig. 14) and powdered Pd produced activation energy of 29.2 kJ/mol (0.303 eV/atom) for the higher temperature region when studied in the electrolytic cell. The activation energy for the diffusion of D in $PdD_{0.85}$ can be calculated from data published by Majorowski and Baranowski [26] as plotted in Fig. 17. The indicated values in the figure are plotted in Fig. 18 to give activation energy of 23.0 kJ/mol. While this amount of D might be present as an average throughout the sample, the local D content would be expected to be much less as the local supply near the NAE was gradually reduced. Values for nearly pure Pd can be obtained from Lewis *et al.* [25]. They combined a large number of measurements made over a large temperature range to obtain activation energy of 25.1 kJ/mole for diffusion of H through Pd.

The agreement between the activation values suggests replacement of the D in the active sites by diffusion might be the variable limiting power production in the solid material.

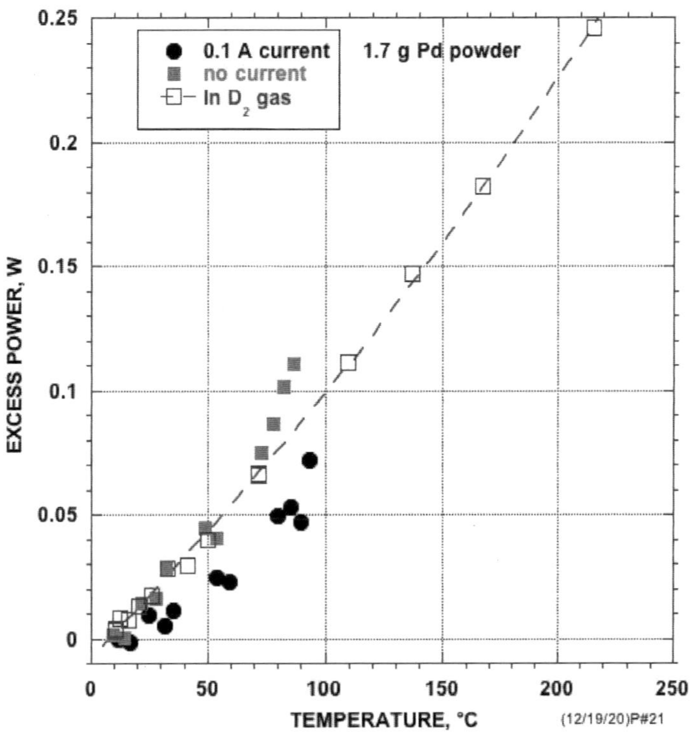


**Figure 12.** Comparison between power production when the measurement is done using an electrolytic cell and when the same sample of Pd is exposed to $D_2$ gas.

When studied in $D_2$ gas, the pressed powder also showed regions where the temperature had two different effects. However, the activation energy in the higher temperature region is much smaller than that found when electrolysis is used. Also, a higher transition temperature between the two mechanisms is measured.

This behavior can be explained as follows. The very open nature of the pressed material allows the active sites to be surrounded by $D_2$ gas. This access makes the D more easily available to the fusion process compared to the situation when electrolysis causes the supply of D to come from the surrounding solid structure. Consequently, the smaller activation energy results because the D can be obtained more directly from the gas, rather than having to diffusion through a solid structure.

Notice that the activation energies for powdered Pd and powdered Ni are almost identical when studied in $D_2$ gas. Perhaps this similar behavior results because both pressed materials obtain their D directly from the $D_2$ gas as a result of having the same physical structure.

The reported need to achieve a large D/Pd ratio in the low-temperature region [14], [27] before energy can be produced is proposed to apply only to solid Pd in which the stress produced by reaction with D is needed to form the required gaps. A large D/Pd ratio is not required to activate or increase power when the pressed powder is used because the gaps are already available.

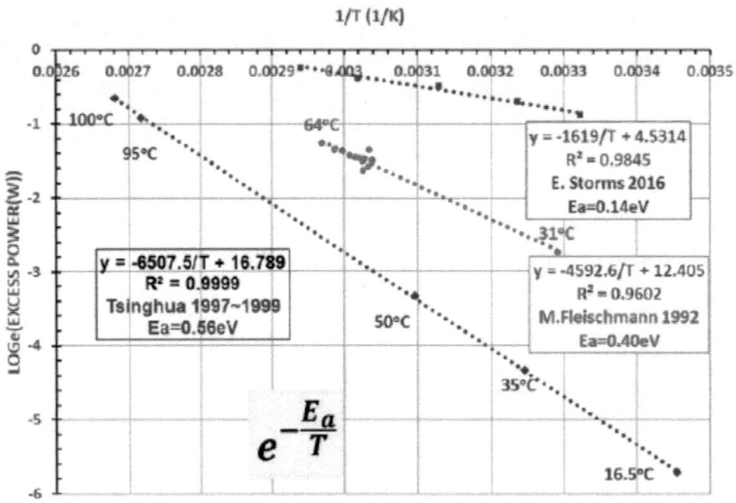


**Figure 13.** Summary of published values of log EP vs 1/T provided by Dong *et al.* [20].

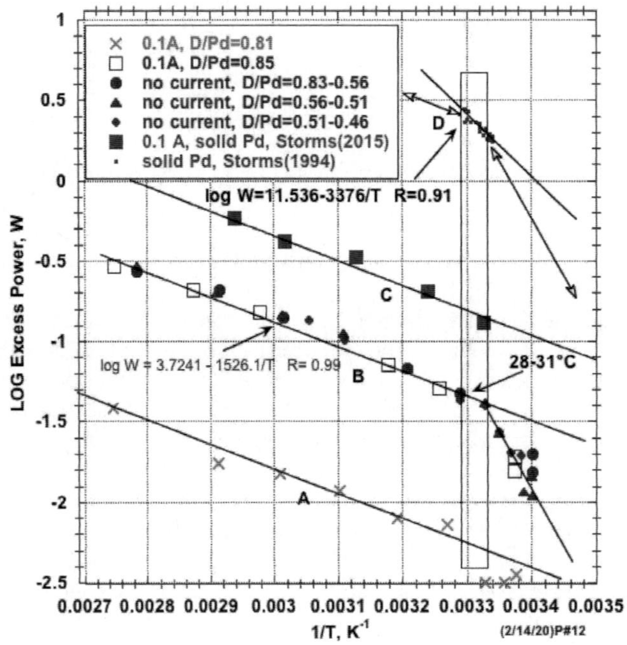


**Figure 14.** Comparison between the behaviors of different samples of solid and pressed Pd when heated in an electrolytic cell. The designations A, B, C, and D apply to independent measurements that are compared to show that the effect of temperature is independent of the amount of LENR power being produced and the form of the Pd. Also, the behavior is reproducible [19], [24] between measurements made years apart.

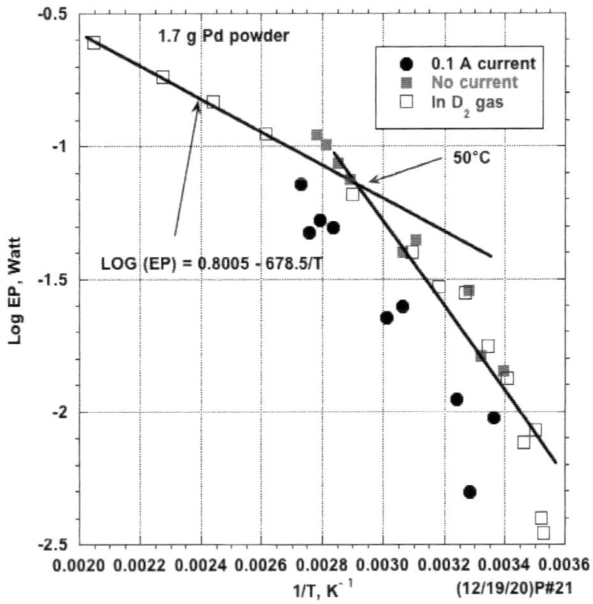


**Figure 15.** Relationship between log EP and 1/T for the sample of pressed Pd powder shown in Fig. 12.

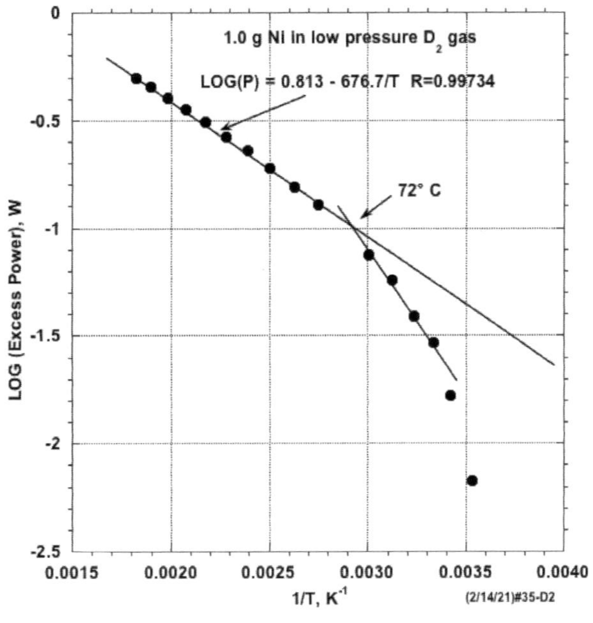


**Figure 16.** Relationship between Log excess power (P) and 1/T for 1 g of activated Ni powder when heated in low-pressure $D_2$ gas.

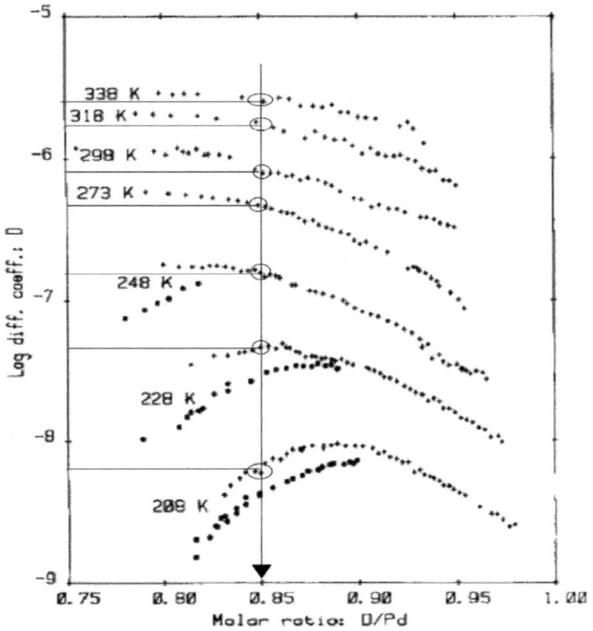


**Figure 17.** log Fitch diffusion constant as a function of D/Pd ratio at various temperatures. [26]

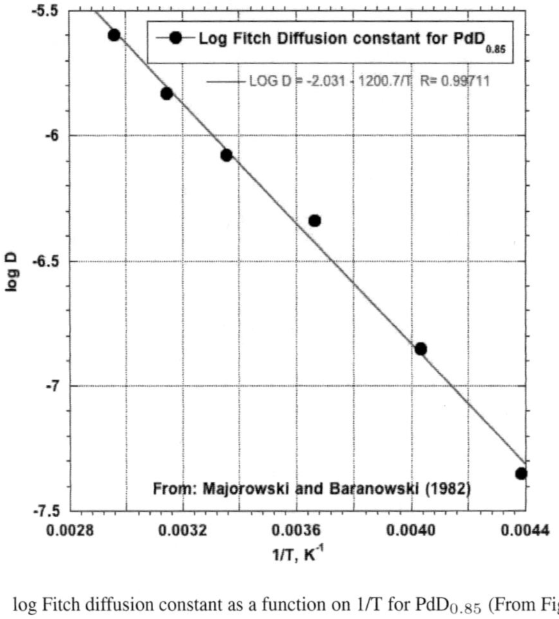


**Figure 18.** log Fitch diffusion constant as a function on 1/T for $PdD_{0.85}$ (From Fig. 17 [26]).

These repeated and consistent patterns of behavior demonstrate once again that the energy claimed to result from LENR is real and not the result of prosaic mechanisms or experimental error. Also, the resulting behavior is consistent with the Storms model of LENR [28].

## 5. Summary

The relationship between the part of the model related to this study and the observed behavior can be summarized as follows:

1. Pd can be caused to produce LENR equally well using electrolysis or gas loading. Therefore, the conditions uniquely created by the electrolytic action are not required.
2. The temperature has two different effects, with a clear transition between the regions.
3. The LENR process is sensitive to the D/Pd ratio only in the low-temperature region. In the high-temperature region, the LENR process will function at D/Pd ratios as low as 0.01.
3. The activation energy for the LENR process in solid Pd is nearly identical to the activation energy for the diffusion of D in PdD, suggesting the amount of power produced in the high-temperature region is limited by how fast the D can repopulate the active sites by diffusion through the PdD lattice. In contrast, the activation energy of pressed Pd and Ni in $D_2$ gas is much smaller, indicating that the D in pressed material has much easier access to the NAE because it has a more direct contact with the source of D.
4. The LENR effect is easy to produce at low levels of energy production in pressed powders of Pd or Ni after suitable activation, with an increased temperature being required to obtain significant power.

The model used to understand the behavior can be described as follows:

1. Sites must form in which clusters of D or H can accumulate. These sites are called the nuclear active environment (NAE). These sites are proposed to be physical gaps of a critical dimension.
2. The sites fill with D or H as result of Gibbs energy being released, as is characteristic of a normal chemical process. This process is sensitive to the temperature and (D+H)/Pd ratio.
3. Once a sufficient number of atoms have accumulated at a site, they rapidly fuse by an unknown mechanism.
4. The observed power is produced by the sum of the energy produced by each site operating independently, which is determined by how fast the D and H, on average, can be replaced. This process is sensitive only to local temperature.

### Acknowledgements

This work could not have been done without the financial support and inspiration provided by Brian Scanlan. The author is grateful to Robin Carter, Thomas Grimshaw, Thomas Claytor, Jed Rothwell, and Dieter Britz for their efforts in making the paper easier to understand.

### References

[1] E.K. Storms, B. Scanlan, Detection of radiation from LENR, in: D.L. Nagel, M.E. Melich (Eds), 14th International Conference on Condensed Matter Nuclear Science, www.LENR.org, Washington, DC, 2008, pp. 263-87.

[2] M. Fleischmann, S. Pons, M. Hawkins, Electrochemically induced nuclear fusion of deuterium, J. Electroanal. Chem. 261 (1989) 301-8 and errata in Vol. 263, 187-8.

[3] E.K. Storms, The explanation of low energy nuclear reaction, Infinite Energy Press, Concord, NH, 2014.

[4] E.K. Storms, What conditions are required to initiate the LENR effect?, in: P.L. Hagelstein, S.R. Chubb (Eds), Tenth International Conference on Cold Fusion, World Scientific Publishing Co., Cambridge, MA, 2003, pp. 285.

[5] E. Storms, How the explanation of LENR can be made consistent with observed behavior and natural laws, Current Science 108 (2015) 531-4.
[6] K. Czerski, New evidence of the cold nuclear fusion – accelerator experiments at very low energies, New Advances on the Fleischmann-Pons Effect, European Parliament, Brussels, 2013.
[7] R.J. Beuhler, G. Friedlander, L. Friedman, Cluster-impact fusion, Phys. Rev. Lett. 63 (1989) 1292.
[8] J. Kasagi, Y. Honda, Screening Energy of the d+d Reaction in an Electron Plasma Deduced from Cooperative Colliding Reaction, J. Cond. Matter. Nucl. Sci. 19 (2016) 127-34.
[9] E.N. Tsyganov, Cold Nuclear Fusion*, Phys. At. Nucl. 75 (2012) 153-9.
[10] T.L. Benyo, L.P. Forsley, L. Dudzinski, M.J. Forsbacka, NASA GRC Hosts Lattice Confinement Fusion Virtual Workshop, 2020.
[11] A. Takahashi, T. Yokose, Y. Mori, A. Taniike, Y. Furuyama, H. Ido, A. Hattori, R. Seto, A. Kamei, J. Hachisuka, Latest Progress in Research on AHE and Circumstantial Nuclear Evidence by Interaction of Nano-Metal and H(D)-Gas, ICCF-22, 2019.
[12] Y. Arata, Y.-C. Zhang, Solid-state plasma fusion ('cold fusion'), J. High Temp. Soc. 23 (special volume) (1997) 1-56.
[13] Y. Arata, Y.-C. Zhang, Sono implantation of hydrogen and deuterium from water into metallic fine powders, in: F. Scaramuzzi (Ed) 8th International Conference on Cold Fusion, Italian Physical Society, Bologna, Italy, Lerici (La Spezia), Italy, 2000, pp. 293.
[14] M.C.H. McKubre, S. Crouch-Baker, A.M. Riley, S.I. Smedley, F.L. Tanzella, Excess power observations in electrochemical studies of the D/Pd system; the influence of loading, in: H. Ikegami (Ed) Third International Conference on Cold Fusion, "Frontiers of Cold Fusion", Published by: Universal Academy Press, Inc., Tokyo, Japan, Held at: Nagoya Japan, 1992, pp. 5.
[15] T. Mizuno, T. Akimoto, K. Azumi, M. Enyo, Cold fusion reaction products and behavior of deuterium absorption in Pd electrode, in: H. Ikegami (Ed) Third International Conference on Cold Fusion, "Frontiers of Cold Fusion", Universal Academy Press, Inc., Tokyo, Japan, Nagoya Japan, 1992, pp. 373.
[16] H. Kamimura, T. Senjuh, S. Miyashita, N. Asami, Excess heat in fuel cell type cells from pure Pd cathodes annealed at high temperatures, in: M. Okamoto (Ed) Sixth International Conference on Cold Fusion,Progress in New Hydrogen Energy, New Energy and Industrial Technology Development Organization, Tokyo Institute of Technology, Tokyo, Japan, Lake Toya, Hokkaido, Japan, 1996, pp. 45.
[17] E. Storms, The Enthalpy of Formation of PdH as a Function of H/Pd Atom Ratio, ICCF-21, Fort Collins, CO, 2018.
[18] E. Storms, The Enthalpy of Formation of PdH as a Function of H/Pd Atom Ratio, J. Cond. Matter. Nucl. Sci. 29 (2019) 275-85.
[19] E. Storms, Anomalous Energy Produced by PdD, J. Cond. Matter. Nucl. Sci. 20 (2016) 81-99.
[20] Z.M. Dong, C.L. Liang , X.Z. Li, S.X. Zheng, Temperature Dependence of Excess Power in Both Electrolysis and Gas-loading Experiments, J. Cond. Matter. Nucl. Sci. 29 (2019) 85-94.
[21] A.F. AbuTaha, Cold fusion - the heat mechanism, J. Fusion Energy 9 (1990) 345.
[22] E.K. Storms, A study of those properties of palladium that influence excess energy production by the Pons-Fleischmann effect, Infinite Energy 2 (1996) 50.
[23] E.K. Storms, Some characteristics of heat production using the "cold fusion" effect, Trans. Fusion Technol. 26 (1994) 96.
[24] E.K. Storms, Measurements of excess heat from a Pons-Fleischmann-type electrolytic cell using palladium sheet, Fusion Technol. 23 (1993) 230.
[25] F.A. Lewis, K. Kandasamy, X.Q. Tong, Platinum and Palladium-Hydrogen, Soild State Phenomena 73-75 (2000) 207-69.
[26] S. Majorowski, B. Baranowski, Diffusion coefficients of hydrogen and deuterium in highly concentrated palladium hydride and deuteride phases, J. Phys. Chem. Solid. 43 (1982) 1119.
[27] A. Kubota, H. Akita, Y. Tsuchido, T. Saito, N. Haseqawa, N. Imai, N. Hayakawa, K. Kunimatsu, Hydrogen and deuterium absorption by Pd cathode in a fuel-cell type closed cell, in: H. Ikegami (Ed) Third International Conference on Cold Fusion, "Frontiers of Cold Fusion", Universal Academy Press, Inc., Tokyo, Japan, Nagoya Japan, 1992, pp. 565.
[28] E. Storms, How Basic Behavior of LENR can Guide A Search for an Explanation, JCMNS 20 (2016) 100-38.

J. Condensed Matter Nucl. Sci. 36 (2022) 395–407

Research Article

# The Presumption of Pseudoscience. In Defense of Electrochemists Martin Fleischmann and Stanley Pons, and Cold Fusion, of Course!

Sergei A. Tcvetkov*
*Kurchatov str. 4-13, Zarechny, Sverdlovsk region, 624250, Russia*

**Abstract**

In defence of electrochemists Martin Fleischmann and Stanley Pons, and cold fusion. With the help of science eliminated prejudices and errors, and achieved: protection produced truth, freedom for further development, the common good and internal well-being. On motives of the publications in Russian mass media. Even if cold fusion did not exist, it would still be worth inventing. The farther from the beginning of the cold nuclear fusion, the more imagination, transitions into the category of myth-making, violent perversion of facts, forgeries and unhealthy mockery of the authorities in the media and on the Internet. About the purely Russian invention in science - about the COMMISSION ON COMBATING PSEUDOSCIENCE AND FALSIFICATION OF SCIENTIFIC RESEARCH. Evaluation of the advantages of nuclear fusion over other methods of energy production. One gram of deuterium can become a continuous heat source with a capacity of 3 kW for a whole year! There is a strong suspicion that, having mastered the method of obtaining energy from cold fusion, we will only have one step left to completely release the energy of matter according to the famous formula $E = m \times c^2$! Cold fusion or LENR? The content of the scientific paper by M. Fleischmann and S. Pons in Journal of Electroanalytical Chemistry with specific results. About scientific ethics, the violation of which is blamed on M. Fleischmann and S. Pons. Intense heat release, and therefore the effect itself begins only on the 66th day ($\sim 5.65 \times 10^6$ seconds) CONTINUOUS operation of the electrolytic cell and continues for 5 days. What kind of non-reproducibility can we talk about here? There is not a single scientific paper in peer-reviewed journals that scientifically justifies the impossibility of cold nuclear fusion. Verification of devices for measuring physical quantities must be carried out with a device that has an accuracy class higher than the device being verified. Therefore, the tests on heat in MIT and Caltech, which are often referred to on the issue of the failure of cold fusion, are not really any tests. The "experts" are lying about neutron irradiation. The Coulomb barrier, thermonuclear fusion, plasma, sounding from the lips of plasma physicists have nothing to do with cold nuclear fusion. We must understand that cold nuclear fusion is a natural process that created and synthesized the entire world around us, and this process takes place both in the bowels of the Sun and inside the Earth. It can't be any other way. Cold nuclear fusion is not pseudoscience; it is only assumed to be pseudoscience. Long live cold fusion-the discovery of M. Fleischmann and S. Pons! And Stanley Pons should be awarded the Nobel Prize.


*Keywords:* Defence, Cold fusion, Purely Russian invention in science, LENR; Gram of deuterium, 66 day ($\sim 5.65 \times 10^6$ seconds), Neutron irradiation, Not pseudoscience, Discovery of M. Fleischmann and S. Pons.

*Corresponding author: E-mail: rossomahi@gmail.com

© 2022 ICCF. All rights reserved. ISSN 2227-3123

## 1. Introduction

As an epigraph I would like to give the verse and the words of two authentic Russians in literature and science:

> So the false wisdom shimmers and smolders
> Before the sun of the immortal mind.
> Long live the sun; yes, let darkness hide!
> Alexander Pushkin, (Bacchus song: "What silenced the joyful voice?" 1825)

> ... how much more must be learned, and how with the help of science, without violence, lovingly, but prejudices and errors are firmly eliminated, and the following is achieved: the protection of the acquired truth, the freedom of further development, the common good and inner well-being.
> Dmitry I. Mendeleev (from the dedication to his mother in his work "Investigation of aqueous solutions by specific gravity". 1887).

The history of the beginning and continuation of cold nuclear fusion is in its own way tragic and instructive;- every history is unique - like no other. History is our experience for future generations.

Their attitude toward cold fusion I would formulate so: Even if cold fusion did not exist, it would still be worth inventing.

Acting as a direct participant in many of these events, I have to say that the farther from the beginning of cold nuclear fusion, the more imagination we have seen: and the more transitions into the category of myth-making, violent perversion of facts, forgeries, and unhealthy mockery of the authorities in the media and on the Internet.

Sometimes it comes to outright lies. We need to fix this! I am for historical justice and truth. Because, is not the search for and preservation of truth really the primary task of science? There is always the availability of freedom degrees and multiple views on the event – this is my insider's view.

## 2. Russian Mass Media About of Cold Fusion

Here are some examples of publications and statements on TV in Russia for the past few years. (Below in cursive is a lie, a bold font - an explicit lie.)

In "Popular mechanic" №8, 2011 in the paper "Cold fusion: myth and reality"[1] Alexey Levin writes that "Massachusetts Institute of Technology's Staffers tried to reproduce the experiments of M. Fleischmann and S. Pons, *but again to no avail.*

It is therefore not surprising that *the demand for the great discovery had undergone a crushing defeat at the conference of the American Physical Society (APS), which met in Baltimore 1 May of that year.*"

Stanley Pons (left) and Martin Fleischmann (right).

**Creeping** cold fusion.[2]

Author: Evgeny Tsygankov 08.12.2016.

Cold fusion? A short look at the history.

*The date of birth of cold fusion can be considered in 1989. Then, in the English-language press was published* information about a report by Martin Fleischmann and Stanley Pons, in which it *was stated that nuclear fusion* was carried out in the following configuration: a current passes *through palladium electrodes* dipped in heavy water (with two deuterium atoms instead of hydrogen, $D_2O$), **as a result of which one of the electrodes melts**. Fleischmann and Pons *give this interpretation of what is happening*: the electrode melts as a result of the release of too much energy, the source of which is the fusion reaction of deuterium nuclei. **Nuclear fusion, therefore, supposedly takes place at room temperature**. Journalists called the phenomenon "cold fusion", in the Russian version, cold fusion became for some reason "**cold thermonuclear**", although the phrase contains a clear internal contradiction. If in some media the newfound cold fusion could be met with warmth, then in the scientific community the statement of Fleischmann and Pons *was treated very coolly*. **At an international meeting less than a month later**, to which Martin Fleischmann was also invited, *the statement was critically reviewed.*

*The simplest considerations pointed to the impossibility of nuclear fusion occurring in such an experiment.* For example, in the case of the $d + d \to {}^3He + n$ reaction, for the power discussed in the Pons and Fleischmann setup, **there would be a neutron flux that would give the experimenter a lethal dose of radiation within an hour**. The presence of Martin Fleischmann himself at the meeting directly pointed to the falsification of the results. Nevertheless, *in many laboratories, similar experiments were conducted,* according to the results of which **no products of nuclear fusion reactions were found.**

However, *this did not prevent one sensation from giving rise to an entire community of cold fusion adherents, which functions according to its own rules to this day.*

In the program "Meanwhile" with Alexander Arkhangelsky "Pseudoscience: the endless struggle?" on the channel Culture of Russian television at the end of October 2016[3], it was said that:

"The Presidium of the Russian Academy of Sciences has approved the new composition of the Commission on Combating Pseudoscience and Falsification of Scientific Research. Now it consists of 59 scientists, including physicists, biologists, astronomers, mathematicians, chemists, representatives of humanitarian specialties, and specialists in agriculture. When Academician Vitaly Ginzburg initiated the creation of the commission in 1998, pseudoscientific concepts particularly annoyed physicists and engineers. Back then, fantasies about new sources of energy and overcoming basic physical laws were popular. The Commission *has consistently defeated* the teachings on torsion fields, *cold nuclear fusion*, and anti-gravity. The most high-profile case was the exposure in 2010 of Viktor Petrik's invention of nanofilters for cleaning radioactive water..."

"*Huge damage* to thermonuclear fusion is caused by constantly appearing reports **about so-called** *cold nuclear fusion*, that is, a synthesis that takes place, not at millions of degrees, but, say, *at room temperature* on a laboratory table. *The message* from 1989 that *new elements* had been produced during electrolysis *on palladium catalysts*, those *hydrogen atoms had merged into helium atoms* - it was like a kind of information explosion.

Yes, the discovery, *in quotation marks, the "discovery"* of these scientists **were not confirmed by anything**. This damage the reputation of thermonuclear fusion also because the business easily responds to these strange scandalous requests, hoping for a quick easy profit, *it subsidizes start-ups dedicated to cold fusion.* **None of them were confirmed. This is an absolute pseudoscience, but, unfortunately, it is very harmful to the development of real thermonuclear fusion**. "These are the words of an expert, Doctor of Chemical Sciences, Professor Alexey Kapustin in the TV program of the NTV channel "We and science, science and we. Controlled thermonuclear reaction" from 26.11.2016."[4]

Thermonuclear Fusion: **the miracle** *that happens*.[5]

9 January 2017 Denis Strigun.

Cold nuclear fusion.

No matter how tiny it was, the chance to hit the jackpot in the *"thermonuclear (plasma fusion)" lottery* excited everyone, not just physicists. In March 1989, two fairly well-known *chemists*, the American Stanley Pons and the British Martin Fleischmann *gathered* journalists to show the world "*cold*" nuclear fusion. *It worked like this. A palladium electrode was placed in a solution with deuterium and lithium and a direct current was passed through it. Deuterium* and **lithium were absorbed** *by palladium* and, when they *collided, sometimes "linked" into tritium* and **helium-4**, *suddenly heating the solution sharply. And this is at room temperature and normal atmospheric pressure.*

First, details of the experiment appeared in The Journal of Electroanalytical Chemistry and Interfacial Electrochemistry only in April, *a month after* the press conference.

*It was contrary to scientific etiquette.*

Secondly, the nuclear physicists had *many questions* for Fleischmann and Pons. For example, *why in their reactor does the collision of two deuterons give tritium and* **helium-4**, when it *should give tritium and a proton or a neutron and helium-3*? Moreover, it was easy to verify this: provided that nuclear fusion took place in the palladium electrode, *neutrons with pre-known kinetic energy would* "**fly off**" *from the isotopes. But neither the neutron sensors* **nor the reproduction** *of the experiment by other scientists led to such results.* And **due to a lack of data**, **in May, these sensational chemists were recognized as "quacks".**

## 3. Classification of Lies

So, let us formulate the claims on which the refusal of the public is based on the recognition of the discovery of the phenomenon of cold nuclear fusion by M. Fleischmann and S. Pons. And note, it is the claims, and not the arguments and evidence against this phenomenon, based on the doubts of the so-called "experts", who have never been engaged in the repetition and verification of this phenomenon.

First: a press conference before the publication of a paper in a scientific journal - ay-ay-ay! not good: violation of scientific ethics!

Second: What are you doing? This can't be happening! We've been struggling with thermonuclear fusion for decades and we can't get any excess heat at hundreds of millions of degrees in the plasma, and you're telling us about room temperature and megajoules of heat in excess of the energy input? Nonsense!

Third: Even if this did happen, you all should have died a long time ago!

Fourth: Experiments by Lewis at Caltech (California Institute of Technology) and MIT (Massachusetts Institute of Technology) did not work. You're all lying!

Fifth: They also want to ask for money to continue these works? And from whom will this money be taken away?

Sixth: This will not happen while we are alive! Drive S. Pons out of the university and the USA!

I must say that the same scenario was tried in the early 2000s with a professor at Purdue University, Ruzi Taleyarkhan, for his bubble "thermonuclear" reactions, but the case went to court, and the professor was reinstated in his rights and positions.

## 4. Purely Russian Invention in Science

A few words about a purely Russian invention in science, the COMMISSION ON COMBATING PSEUDOSCIENCE AND FALSIFICATION OF SCIENTIFIC RESEARCH under the Presidium of the Russian Academy of Sciences, as they call themselves. The Commission on Pseudoscience has already hung up a medal for itself: "... for the consistent destruction of torsion fields, cold nuclear fusion, and anti-gravity." The "guard dogs of budget money" have completely dissolved; there is no one to flog them.

They believe that the cries about not giving money to these ignoramuses and adventurers in the Conferences and Symposiums section of the journal Advances in Physical Sciences volume 169 No. 6 for 1999[6], is the defeat of cold nuclear fusion? This is a strange way of conducting a scientific discussion, especially after instructions were sent out to the editorial offices of Russian scientific journals forbidding the publication of scientific papers that even once mention the words cold nuclear fusion.

The author has a sad experience of trying to publish the results of his research in at least two Russian academic journals. Let's hope that the leadership of the Russian Academy of Sciences will finally collect the last remnants of the brains leaking from the country and reconsider its attitude to science as a development and not a degradation of society.

It will finally close the shameful page of Russian science in the form of the Commission on Pseudoscience and stop wasting time and money on bans and obstacles to science, made by idlers.

## 5. The Price of the Question

Before we deal with these claims, we will try to evaluate the advantages of nuclear fusion over other methods of generating energy presently known. Take the amount of released energy per gram of reacting substance (energy density). It is in the reacting substance and not the material in which these reactions occur.

To begin with, let's look at the table of the amount of energy released per gram of the reacting substance for different methods of energy production and perform simple arithmetic, comparing these amounts of energy.

This data can be obtained from[7] and presented as a table:

| The way to get energy | kWh/kg | J/g | Times greater than previous row energy |
|---|---|---|---|
| **Burning oil (coal)** | 11.6 | 42 kJ/g(25 kJ/g) | 1 |
| **In the fission of uranium-235** | $22.9 \times 10^6$ | 82.4 GJ/g | 1,974,138 |
| **In the fusion of hydrogen nuclei** | $117.5 \times 10^6$ | 423 GJ/g | 5 |
| **The energy of a substance according to the formula $E = mc^2$** | $25 \times 10^9$ | 90 TJ/g | 213 |

It turns out that when burning oil or high-quality coal, you can get 42 kJ/g of thermal energy. During the fission of uranium 235, 82.4 GJ/g of heat is released; during the fusion of hydrogen nuclei, 423 GJ/g is be released; and according to the theory, if the entire 1 gram of any substance is annihilated in a reaction, that produces 104.4 TJ/g of energy (where k is a kilo = $10^3$, G-Giga = $10^9$, T-Tera = $10^{12}$).

It follows that one gram of deuterium can produce a continuous heat source of slightly more than 3 kW for a whole year!

And immediately the question: "Is it necessary to engage in the extraction of energy from water by fusion?" disappears, for any sane person. There is a strong suspicion that having mastered the method of obtaining energy in the fusion of hydrogen nuclei, we will only have one step left to completely release the energy of matter according to the famous formula $E = m \times c^2$!

Andrea Rossi during the tests of the device for the low-energy nuclear reaction.

Italian Andrea Rossi showed that for cold nuclear fusion, you can use simple hydrogen, which is available in inexhaustible quantities on the planet Earth, and in space.

Jules Verne.

This opens up even more opportunities for energy, and the words of the French science fiction writer Jules Verne in his work "The Mysterious Island", published in 1874, become prophetic: "...I think that water will one day be used as fuel and that the hydrogen and oxygen that are part of it will be used together or separately and will be an inexhaustible source of light and heat, much more intense than coal. ...I think that when the coal deposits are depleted, humanity will be heated and warmed by water. Water is the coal of the future." I put three exclamation marks to the great science fiction writer!!!

It is worth noting that by extracting hydrogen for cold nuclear fusion from water, humanity will receive oxygen necessary for breathing as a bonus.

## 6. Cold Nuclear Fusion or Low Energy Nuclear Reactions (LENR)?

At the end of the 90s, the defeated remnants of scientists who, out of curiosity, secretly continued to repeat the experiments of M. Fleischmann and S. Pons, decided to hide from the furious attacks of the "tokomafia" and the Commission established in Russia to Combat Pseudoscience at the Russian Academy of Sciences.

Renaming cold fusion as low energy nuclear reactions is obviously a weakness. This is an attempt to hide, so as not to be destroyed, this is a manifestation of the instinct of self-preservation. All this shows the severity of the threat not only to the profession but also to the very life of such researchers.

Italian Andrea Rossi understands that his activities to promote his Energy Catalyst (E-cat) pose a threat to his life. Therefore, his actions seem illogical to many. But that is how he defends himself. I saw for the first time, and perhaps for the only time, in Zurich in 2012, how a person who is engaged in the development and implementation of a new energy technology entered a meeting of scientists and engineers accompanied by a bodyguard wearing a bulletproof vest.

The pressure from academic groups in science is so strong and aggressive that only completely independent people, such as pensioners, can now engage in cold fusion. The rest of those interested are simply squeezed out of laboratories and universities. This trend was visible in the world of science until recently and only last year the European Commission finally began timidly funding two projects on cold fusion, calling it "Clean Hydrogen-Metal Energy"[8]. In 2019, the journal Nature, after 30 years of silence, published a controversial paper with two editorial comments. Although no outstanding results were obtained in the two-year studies, this paper brought the term cold fusion back to the pages of scientific journals and gave an impetus to a new beginning of publications on the topic of cold fusion in peer-reviewed journals.

## 7. Details of the Opening of the Field

Let us get back to our electrochemists. I would like to briefly recall the content of a scientific paper by M. Fleischmann and S. Pons[9] in the peer-reviewed journal with specific results. This information is taken from the abstract journal of the All-Union Institute of Scientific and Technical Information (RZH VINITI) of the USSR Academy of Sciences, published since 1952 – a periodical scientific and informational publication that publishes abstracts, annotations, and bibliographic descriptions of Russian domestic and foreign publications in the field of natural, mathematical and technical sciences, economics and medicine. Specifically, this is RJ 18B Nuclear Physics. -1989. -6. - ref. 6B1[10].

Electrochemically induced nuclear fusion of deuterium. / Fleischmann Martin, Pons Stanley / / J. of Elecroanal. Chem. -1989. - Vol. 261. - No. 2a. - pp. 301÷308.

At the University of Utah (USA), an experiment was performed aimed at detecting the fact of nuclear reactions

$$d + d \rightarrow^3 He(0,82 M эB) + n(2,45 M эB),$$
$$d + d \rightarrow T(1,01 M эB) + p(3,02 M эB)$$

under conditions where deuterium is embedded in the metal lattice of palladium, which means "an effective increase in the pressure that brings the deuterons closer together due to chemical forces", which contributes to an increase in the probability of quantum-mechanical tunneling of deuterons through the Coulomb barrier of the DD pair in the internodes of the palladium lattice. The electrolyte is a solution of 0.1 mol of LiOD in water of the composition 99.5% $D_2O$ + 0.5% $H_2O$. Palladium (Pd) rods with a diameter of 1 ÷ 8 mm and a length of 10 cm, wrapped with platinum wire (Pt-anode), was used as the cathode. The current density was varied in the range of 0.001÷1 A/cm$^2$ at the voltage at the electrodes of 12 V. Neutrons in the experiment was recorded in two ways. First, a scintillation detector, including a dosimeter with boron $BF_3$ counters (efficiency $2\times10^{-4}$ for neutrons of 2.5 MeV energy). Secondly, by the method of registering gamma-quanta, which are formed when a neutron is captured by a hydrogen nucleus of ordinary water surrounding an electrolytic cell, by the reaction:

$$p + n \rightarrow d + \gamma(2,23 M эB).$$

The detector was the NaI (Tl) crystal, and the logger was an ND-6 multichannel amplitude analyzer. Background correction was performed by subtracting the spectrum obtained at a distance of 10 m from the water bath.

Tritons (T) were extracted from the electrolyte using a special type of absorber (Parafilm's film), and then their $\beta$-decay was recorded on a Beckman scintillation counter (45% efficiency). The best results were achieved on a Pd cathode with a diameter of 4 mm and a length of 10 cm at a current density through the electrolyzer of 0.064 A/cm$^2$. Neutron radiation with the intensity of $4\times10^4$ neutrons/s, 3 times higher than the background, was recorded. The gamma-ray spectrum was found to have a maximum in the energy range of 2.2 MeV, and the gamma-ray counting rate was $2.1\times10^4$ s$^{-1}$. The presence of tritium with a rate of formation of $2\times10^4$ atoms/s was detected. In the process of electrolysis, a fourfold excess of the released energy over the total spent (electrical and chemical) energy was recorded. It reached 4 MJ/cm$^3$ of the cathode in 120 h of the experiment. In the case of a $1\times1\times1$ cm bulk Pd cathode, its partial melting was observed ($T_{melts}$=1554°C). Based on experimental data on tritium nuclei and gamma-quanta, the authors found the probability of a synthesis reaction to be $10^{-19}$ s$^{-1}$ per DD pair. At the same time, the authors note that if the main reason for the increased energy yield is considered to be nuclear reactions involving deuterons, then the neutron yield would be significantly higher (by 11–14 orders of magnitude). According to the authors, in the case of electrolysis of the $D_2O$+DTO+$T_2O$ solution, the heat release can increase to 10 kW/cm$^3$ of the cathode.

A few words about scientific ethics, the violation of which is blamed on M. Fleischmann and S. Pons. As is clear from the original paper, it was received by the editorial board of the journal on March 13, 1989, accepted for publication on March 22, 1989, and published on April 10, 1989. And where is the violation of ethics, and most importantly by whom?

From this description, it is unambiguously clear that an incredibly huge amount of excess heat has been obtained, several times greater than the energy spent on electrolysis and the possible chemical energy that can be released by the simple chemical decomposition of water into individual atoms. The detected tritium and neutrons indicate the process of nuclear fusion. Moreover, neutrons are registered in two independent ways by different devices.

In 1990, the following paper[11] was published in the same journal, specifically related to the heat release in these studies, from which it can be seen from Figure 1A that the intense heat release, and therefore the effect itself, begins only on the 66th day (~$5.65 \times 10^6$ sec) CONTINUOUS operation of the electrolytic cell and continues for 5 days. That is, to get the result and fix it, it is necessary to spend SEVENTY-ONE DAYS on measurements, not counting the time for preparing and manufacturing the experimental setup. For example, it took us all of April to make the first installation, launch it, and perform various calibrations, and it was only in mid-May 1989 that we received the first results.

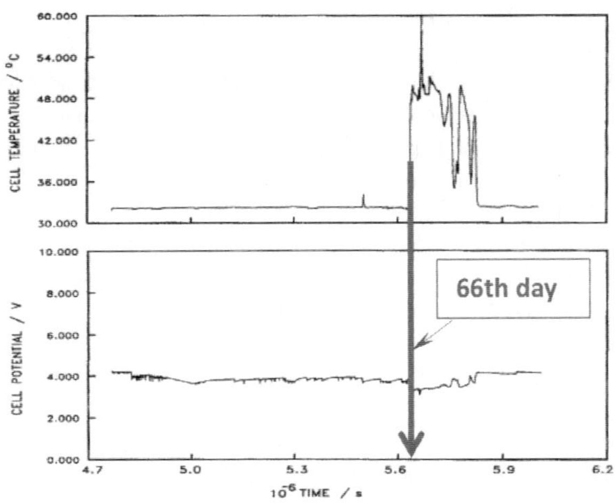


**Figure 1A.** Cell temperature vs. time (upper) and cell potential vs. time (lower) plots for a $0.4 \times 1.25$ cm Pd rod electrode in 0.1 M LiOD solution. Current density 64 mA·cm$^{-2}$, bath temperature 29.87°C.

The beginning of the process of heat release during electrolysis with a large delay was subsequently confirmed by D. Gozzi, et al.[12]. The beginning of a noticeable release of excess heat is registered here after 210 hours; this corresponds to 8.75 days and lasted for more than 30 days.

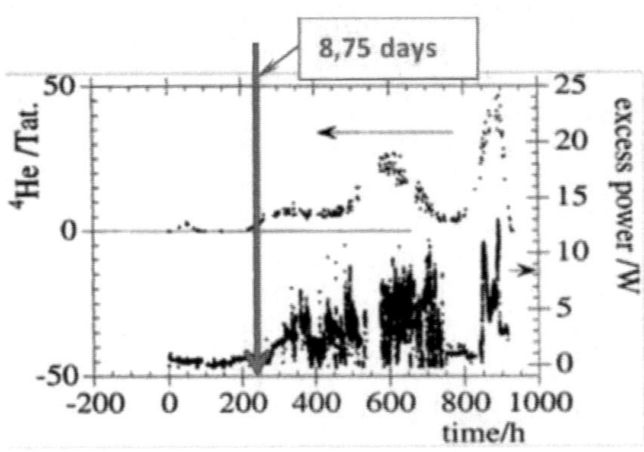

As well, Michael C. H. McKubre (Director of the Energy Research Center of SRI, USA), who presented his results at the 10th International Conference on Cold Fusion (ICCF-10) on August 25, 2003. The beginning of the release of excess heat was in 520 hours, which corresponds to 21.67 days.

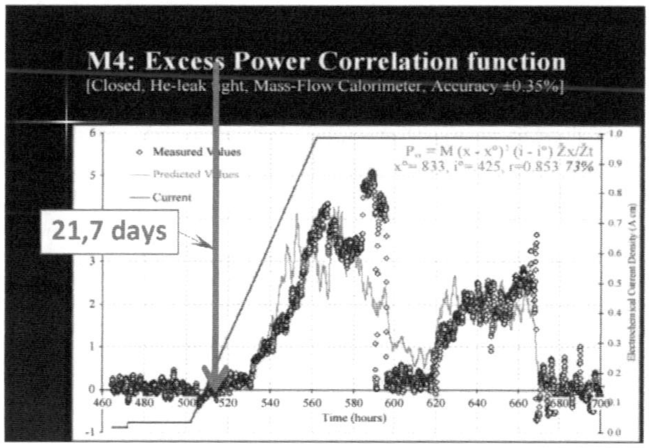


In their work in 1996, reported at the 6th International Conference on Cold Fusion (ICCF-6)[13] Stanley Pons demonstrated two things. The first and perhaps most important thing is that having moved from the United States in 1992 to the south of France in a new place after a considerable period, in another country, he was able not only to reproduce the experiment conducted in 1989 in Salt Lake City but also to get great results on heat! What kind of non-reproducibility are we talking about here? See:

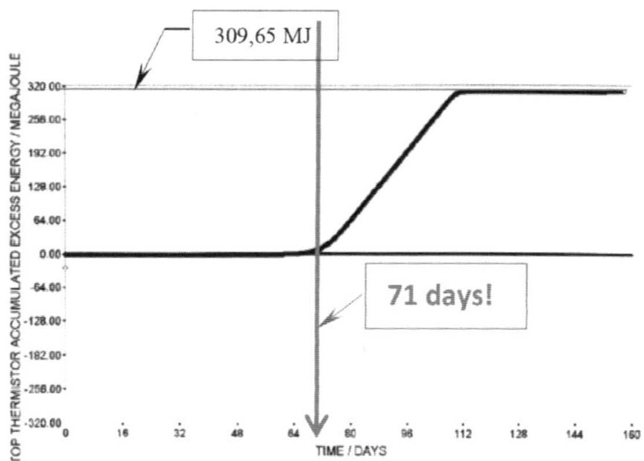


**Figure 2.** Accumulated excess energy as a function of time: top thermistor set.

According to these data, noticeable heat release begins on the 71st day of electrolysis! The change in heat release continues for more than 40 days and then constantly at the level of 310 MJ up to 160 days!

Therefore, how can we talk about the non-reproducibility of the experiments of M. Fleischmann and S. Pons by pointing to experiments in one single laboratory, which conducted the test only a month after the announcement of cold fusion, without involving and consulting the authors and never even published the results in a scientific paper? Self-serving motives and fear of the possibility of responsibility for unsuccessful experiments with plasma fusion are visible. With this statement in May 1989, the American Physical Society (APS), it turns out, put itself in an unpleasant position, replacing science with ordinary business, and for many years closed official research in the field of cold nuclear fusion. The members of this society, first of all, behaved contrary to all scientific ethics in the sense of refuting the results of scientific work with publication in a scientific journal, and instead relying on the New York Times, which in May 1989 published a devastating article about M. Fleischmann and S. Pons. The violation of scientific ethics the Times ascribed to M. Fleischmann and S. Pons was that they presented the results of their scientific research at a press conference before the publication of a scientific journal paper. Yet as stated above, the paper was published on the day of the press conference. Furthermore, it was later revealed that the "decision" that the cold nuclear fusion in the experiments of M. Fleischmann and S. Pons are impossible at this meeting of the APS was made by voting by nine members of the society.

There is not a single paper in peer-reviewed journals that scientifically justifies the impossibility of cold nuclear fusion. There is no such thing. There is only information at the interview level, distributed by the media about the statements of scientists who have never been engaged in cold nuclear fusion research, but were engaged in such fundamental and money-intensive areas of physics as plasma fusion, stellar physics, the Big Bang theory, the origin of the Universe, the Large Hadron Collider.

The whole story of attributing cold nuclear fusion to pseudoscience is a story about money, not science. Just look at the schedule of funding for plasma research in the United States 1953-1991.

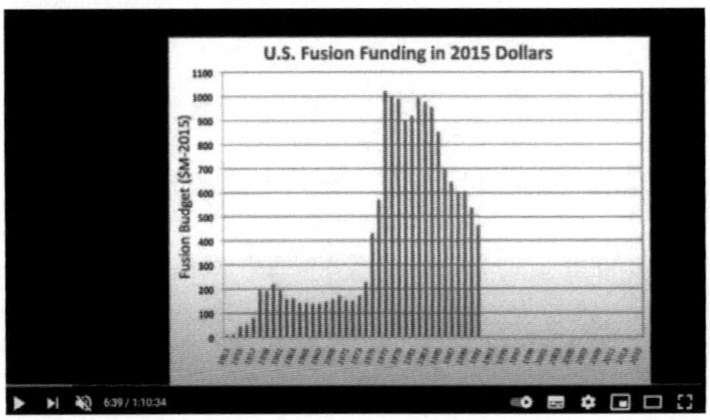


In fact, by 1989, funding for fusion research was halved from one billion dollars per year down to $450 million in 1991. There was something for the plasma fusionists to fight for!

Back at the university, during the course of lectures "Measurement of Physical Parameters", we were taught that the verification of devices for measuring physical quantities must be carried out with a device that has an accuracy class higher than the device is verified. This rule has exactly the same relation to the verification of phenomena! Therefore, the heat tests at MIT and CalTech, which people like to refer to as evidence of the failure of cold fusion, are no tests at all. Compare the accuracy and error in measuring temperature and power with the experimental data of M. Fleischmann and S. Pons, which are given in this report by Melvin Miles[14].

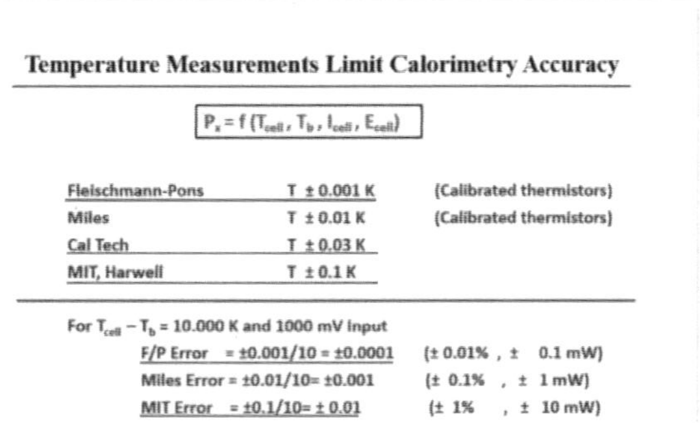


They differ by tens and thousands of times!

Regarding what "... if the main reason for the increased output energy yield is considered to be nuclear reactions involving deuterons, then the neutron yield would be significantly higher (by 11–14 orders of magnitude)."[9] Here the calculation is simple: if 4 MJ/cm$^3$ of excess heat is released from the cathode, then at least $4.29 \times 10^{18}$ neutrons should be formed. If at least, one neutron leaves the reaction zone and does not give up its energy inside the cell from 2.45 MeV to room temperature, then there is no way to register so much excess heat. And if the emitted neutrons are registered, then the number of fusion reactions occurring, in this case, should be much greater than the minimum of neutrons, and more tritium will be formed. Plus, knowing that the cross-section of the interaction of neutrons and helium-3 is incommensurably large compared to the cross-sections of other possible reactions of the products of d+d fusion reactions (by about two orders of magnitude),

$$^{3}\text{He} + n \rightarrow T + p + 0,764\,\text{MeV}$$

then it becomes clear that no one will be irradiated with neutrons, and it is clear that such a ratio of the amount of registered tritium to the number of registered neutrons appears and where helium-4 is subsequently taken from. It appears as a result of a cascade of fusion reactions of d+d reaction products, but this was already apparent from the experiments of other researchers who measured helium-4. M. Fleischmann and S. Pons do not say a word about this.

The "experts" are playing tricks with neutron irradiation. With such amounts of excess heat released, neutron kinetic energy should all turn into heat, transfer the energy to the materials and water of the electrolyte in the cell, and not carry away 75% of the energy from the reaction zone outside the reactor (thus irradiating the experimenters).

Therefore, M. Fleischmann and S. Pons recorded only a small part of the neutrons – heavy water, as is known, is a good moderator of neutrons.

From a scientific point of view, there is only one error in this article – it is the reduction of the amount of excess energy released normalized to the volume of the palladium electrode used. In this case, the consumable component and energy source is deuterium. It would be logical to attribute the released excess amount of energy to the amount of deuterium absorbed by palladium and compare it with the assumed heat in nuclear fusion as a result of the d+d reaction, but, as mentioned above, the energy balance of this process should not be limited to the products of this reaction.

## 8. Conclusion

The magic terms: the Coulomb barrier, fusion, plasma, which sound from the lips of plasma physicists are fascinating. But I want to ask them: What does the relation of the temperature below 1000°C and the fourth aggregate state of matter – plasma – have to do with the electrolysis process of Martin Fleischmann and Stanley Pons? Plasma is an ionized gas. Hydrogen ionization begins at 3,000 degrees Kelvin and by 10,000 degrees Kelvin, the hydrogen is completely ionized, that is, it is approximately 2,727°C - the beginning of ionization, and by 9,727°C - fully ionized hydrogen - plasma.

Question: How can the description of the fourth aggregate state of a substance be applied to an ordinary gas? It's like comparing warm and transparent. You can, of course, try to measure the distance to the Moon by determining the amount of dew that fell in the Sahara Desert at night, but what will be the result? Similarly, the results of cold nuclear fusion cannot be described and considered from the point of view of plasma fusion. By this method, one can only deny the possibility of cold nuclear fusion!

The critics present only one doubt about the possibility of conducting nuclear fusion reactions at such thermodynamic parameters. But nuclear physics does not say a word about the zero probability of nuclear reactions occurring at temperatures close to room temperature. And this means only that these probabilities begin to increase when the room temperature rises to 1000°C.

## 9. CUI Prodest?

And here comes the age-old question: Cui prodest? - Who benefits from this? Of course, the one who first starts shouting: "Stop the thief!" I don't want to hint at anyone, but they were the first to shout: "It can't be!" the physicists involved in plasma fusion, and immediately began to tell their fairy tales and horror stories about plasma, neutrons, and how it's all incomprehensible to the simple mind. And here they are, spending another couple of decades and a little more than a few tens of billions of dollars, in the end, almost fulfilling the long-standing dream of humanity about endless and such "free", "clean" energy for all time. Only this reminds me of the bright communist future that we were so fervently promised in the USSR, and it turned out that this is just a horizon line that recedes when you strive for it, moving on the surface of the ball called planet Earth – an incentive, in one word.

The biggest mistake of cold nuclear fusion, which the plasma fusionists "slipped" to us, is the inability to overcome the Coulomb barrier with equally charged hydrogen nuclei at low temperatures. I must disappoint them and all the "theorists" who have come running into cold nuclear fusion with their "astrolabes" and are trying to come up with something exotic to overcome this Coulomb barrier in the form of the hydrino, dineutrino, dineutronium, etc. There are quite physical laws and phenomena that explain all the logic of the registered products of cold nuclear fusion, if you think a little and remember your university courses in physics.

We must understand that cold nuclear fusion is a natural process that created, i.e. synthesized, the entire world around us, and this process occurs both in the bowels of the Sun and inside the Earth. It can't be any other way. And we'll all be absolute idiots if we can't take advantage of this discovery by two electrochemists!

Cold fusion is not pseudoscience, it is only assumed to be pseudoscience for the sake of and in defense of the misguided and fearful responsibility of plasma and large collider scientists, who have turned modern physics into a profitable business for a narrow circle of people who only call themselves scientists.

The discovery of M. Fleischmann and S. Pons has done modern physicists a disservice, which are firmly on the cutting edge of science. They once famously skipped a small area of the possibility of nuclear fusion reactions at low temperatures and costs, and now they are at a great loss and will never admit this to anyone.

Therefore: Long live cold fusion! Oh, come on your plasma fusion! It's time to stop flushing a lot of money down the plasma fusion toilet bowl. You should give at least a drop of these funds for normal research of cold nuclear fusion and the creation of working power plants based on it! Someone, on whom it depends, makes these overeager plasma physicists work, as they should, with their heads, and not with their tongues. How much more time does it take to understand that plasma fusion is a dead end, and the Sun - is not a plasma reactor? How long can you calm everyone with the unfulfilled (unrealizable) dream of plasma energy?

So, long live cold fusion!

We must write this discovery in all school textbooks.

And Stanley Pons should be awarded the Nobel Prize!

## Bibliography

[1] https://elementy.ru/nauchno-populyarnaya_biblioteka/431405.

[2] https://brights-russia.org/blog/odyssey/creeping-cold-fusion.html.

[3] https://www.youtube.com/watch?v=DoZ7RHgnZw8.

[4] https://yandex.ru/video/preview/?text=??%20?%20?????%2C%20?????%20?%20??%3A%20???????????%20 ????????????%20???????»%2026%20????????%202016%20????&path=yandex_search&parent-reqid=1655625 950401277-1495366787l695061455-vla1-0276-vla-l7-balancer-8080-BAL-6225&from_type=vast&filmId=9880512705904578593

[5] https://naked-science.ru/article/sci/termoyadernyy-sintez-chudo-kotoroe.

[6] http://www.mathnet.ru/php/archive.phtml?wshow=paper&jrnid=ufn&paperid=1626&option_lang=eng.

[7] K.A. Gladkov. Atom from A to Z. 2nd Edition, Moscow, Atomizdat, 1974, pages 13–15.

[8] http://www.cleanhme.eu/\#home.

[9] Electrochemically induced nuclear fusion of deuterium. / Fleischmann Martin, Pons Stanley // J. of Elecroanal. Chem. -1989. - Vol. 261. - No. 2a. - pp. 301–308. (The entire paper is here: http://lenr-canr.org/acrobat/Fleischmanelectroche.pdf)

[10] Abstract Journal 18B Nuclear Physics. -1989. -6. - ref. 6B1.

[11] Fleischmann, M., et al., Calorimetry of the palladium-deuterium-heavy water system. J. Electroanal. Chem., 1990, 287, p. 293.

[12] D. Gozzi, F. Cellucci, P. L. Cignini, G. Gigli, M. Tomellini, E. Cisbani, S. Frullani, G. M. Urciuoli, J. Electroanalyt. Chem. 1998, 452, p. 254.

[13] T. Roulette, J. Roulette, and S. Pons. Results of ICARUS 9 Experiments Run at IMRA Europe. IMRA Europe, S. A., Centre Scientifique Sophia Antipolis, 06560 Valbonne, FRANCE (http://lenr-canr.org/acrobat/RouletteTresultsofi.pdf)

[14] M.H. Miles. The Fleischmann–Pons Calorimetric Methods, Equations and New Applications. Journal of Condensed Matter Nuclear Science 24 (2017) 1–14.

Research Article

# Electromagnetic Excitation of Coaxially-Coiled Constantan Wires by High-Power, High-Voltage, Microsecond Pulses

Francesco Celani* and A. Spallone

*ISCMNS_L1, International Society of Condensed Matter Nuclear Science, Via Cavour 26, 03013 Ferentino-IT*
*EU Project H2020: CleanHME*
*Istituto Nazionale di Fisica Nucleare, Via E. Fermi 56, 00044 Frascati (RM)-IT*

C. Lorenzetti, E. Purchi, S. Fiorilla, S. Cupellini, M. Nakamura, P. Cerreoni, R. Burri, P. Boccanera, and E. F. Marano

*ISCMNS_L1, International Society of Condensed Matter Nuclear Science, Via Cavour 26, 03013 Ferentino-IT*

G. Vassallo

*ISCMNS_L1, International Society of Condensed Matter Nuclear Science, Via Cavour 26, 03013 Ferentino-IT*
*DIDI, Univ. of Palermo, 90128 Palermo (PA)-IT*

L. Gamberale

*LEDA srl, Univ. Milano-Bicocca, 20126 Milano-Italy*

**Abstract**

This paper is mostly based on our presentation given at the 23$^{rd}$ International Conference on Condensed Matter Nuclear Science (ICCF23) held at Xiamen University (XMU)-PRC on June 9-11, 2021. Upon the suggestion of the Conference Organizers, we added a detailed explanation of our procedures; moreover, we included the latest results and data from our recently published papers. In the framework of LENR-AHE (Low Energy Nuclear Reaction - Anomalous Heat Effects) studies, since 2021, we have been focused on innovative, low-cost materials instead of the usual, precious metal palladium. We found that the Cu-Ni alloy, known as Constantan (also used for J-type thermocouple construction), has the peculiarity of being able to easily dissociate $H_2$ (or $D_2$) from molecular to an atomic state. While the catalytic feature appears at a rather low temperature (150 °C), Constantan is also able to store atomic hydrogen within its lattice and/or at surfaces, up to temperature of 700-800 °C, and in a wide range of pressures. Thanks to our long experience (since 1994) with wires, we designed various experiments that take advantage of such specific shapes, namely in terms of electromigration of absorbed and adsorbed H, under a suitable longitudinal electric field (0.4-2.5 V/cm), DC and/or pulsed. In 1995 we got noticeable results using Pd wires in electrolytic environments ($D_2O$) at mild temperatures (40-60 °C). Later on, in some experiments, we used even gaseous environments at high temperatures (up to 700-800 °C). The main problem of Pd was its

*Email: franzcelani@libero.it and francesco.celani@lnf.infn.it.

large brittleness after H or D absorption. Moreover, we experimentally reconfirmed that one of the key conditions to induce AHE is the "*flux*" of H moving inside its lattice (longitudinal) or through the surface (transversal). Pioneers of transversal flux were G. C. Fralick-NASA [1]; M. McKubre-SRI [2]; Y. Iwamura-Mitsubishi [3], [4], [5]; Y. Arata-Osaka Univ [6]; F. Celani-INFN [7]. We focused mainly on *longitudinal flux* (following also the theoretical models developed by G. Preparata-Milan Univ. [8], although some of our unconventional electrolytic experiments (1995-1998) had both. In any case, apart from the initial state (the situation of intrinsic hydrogen concentration gradients), the flux needs external energy to be continuously activated. Based on our own experiences since April 1989, AHE are due to *non-equilibrium conditions*, i.e., they need continuous stimulation, usually energy consuming, apart from some specific (but delicate) geometric set-ups (such as the *Capuchin knot*, which we began testing in 2018, in some of our geometric arrangements where local thermal gradients were extremely large: $> 100\ ^{\circ}$C/mm). Recently, we developed an unconventional geometry of the electrode we aimed to use, and at almost the same time, longitudinal and transversal flux at high temperatures in gaseous environments: our goal was to minimize extra energy added, to maximize the AHE and keep it operative for as long as possible. We report here on procedures to enable flux by high power electric pulsing with both related problematics and peculiar advantages, some unexpected. The core of the reactor is arranged as a *reversed coaxial coil* with inner electrode made by a Fe tube, which we described beginning in 2019 at ICCF22.


*Keywords:* $H_2$-induced LENR-AHE, $H_2$ electromigration, High power electric pulses, Hydrogen flux, Memory effect

## 1. Introduction

This paper summarizes the presentation given at the 23rd International Conference on Condensed Matter Nuclear Science (ICCF23 at the Xiamen University (XMU)-PRC on June 9-11, 2021). This presentation was given via web-conference due to Covid-19 limitations. We are obliged to highlight that this paper is rather complex due to the exploitation of techniques coming from different and distant fields of science.

We report some preliminary results on the electromagnetic field stimulation of LENR phenomena in coated Constantan wires in a hydrogen atmosphere. In particular, the use of microsecond longitudinal current/voltage pulses triggers the occurrence of modest anomalous heat effects (AHE). Speculatively, the phenomenon is mediated by large non-equilibrium conditions and the electro-migration of active species in the Constantan wires. This stimulation technique seems to induce an *activation* of the material, and its effects on AHE remain noticeable for long times, even after pulse stimuli are ceased. The preliminary findings reported in this presentation confirm earlier experiments with the same technique already studied, by us during the nineties, in electrolytic experiments using Pd wires [7].

## 2. TOPICS

- Introduction to the current-voltage pulse approach[1] in the study of anomalies in hydrogen and/or deuterium-metal systems. From the initial electrolytic near room temperature experiments with LiOD-$D_2O$ and palladium [7] to (present) gas phase experiments with Constantan up to 900 °C [9].
- Overview of the techniques explored from 2018 to induce AHE (Anomalous Heat Effects) at different temperatures, pressures, voltages, and the exploitation of electron emission as a tool for non-equilibrium conditions[2].
- Stimuli based on the Richardson effect and space charge modulation (Child-Langmuir law [10]).
- Use of Dielectric Barrier (DBD) or Paschen discharges.
- The correlation between AHE occurrence and electron emission mediated by: Richardson effect, Child-Langmuir effect, Paschen discharges.

*Email: franzcelani@libero.it and francesco.celani@lnf.infn.it.
[1]The use of pulse had been pioneered at INFN in 1994, other authors have explored similar approaches since then [36] [38].
[2]The authors have clear evidence that LENR phenomena tend to occur only in presence of non-equilibrium conditions.

- Modified Constantan wires heated with direct current (DC) in the presence of a transversal mild excitation (a few milliamps, ±600 V, 50 Hz sinusoidal): recent advances, on the reactor design based on coaxial cathode-anode wire geometry[3], and related operating conditions [9]. Further progress was as follows:
  - A) Description of the latest high power, fast rise-time, pulse excitation approach to induce a large instantaneous flux of active species across interfaces[4] of the materials:
    - A1. longitudinal electromigration and surface high current density (skin effect included but with its effective role to be fully quantified).
    - A2. some preliminary evidence of the surface activation of the wires by a plasma of hydrogen-noble gas mixtures.
  - B) Some experimental results and next steps: advantages and drawbacks of nickel alloy field excitation.

## 3. Introduction to Current-Voltage Pulse Approach [9]

The use of longitudinal/transversal current/voltage pulses as LENR stimuli was pioneered by our group at INFN beginning in 1994. The original implementation was conducted in electrolytic experiments comprised of palladium-yttrium alloys (specially developed for us by the precious metal company, Tanaka K. K. in Japan). The use of pulses aimed, mainly, at enhancing the electromigration of deuterium or protium. Moreover, pulses were already considered as the enabler of strong non-equilibrium conditions. The whole program, since 1995, started with the study of electromigration of protium[5] isotopes and its effects on LENR occurrence, and with the investigation of the potential occurrence of "*Coherence*" phenomena as per Giuliano Preparata's model [8]. See **Annex A.**

In short, the so-called electromigration effect is controlled by the amplitude of the voltage drop divided by the length (V/cm). Usual values are on the order of 0.05-1 V/cm. In the case of Preparata's effect, in some specific conditions of "coherence", the effect depends on the TOTAL voltage along the wire. His advantage is evident, mainly by a large reduction of power dissipated. The main drawback is the extreme difficulty of achieving such conditions. As a practical matter, it is quite difficult to obtain Coherence for wire lengths over 10-20 cm, at least in our experimental tests since about 1994. Moreover, we observed that over such length some unusual phenomena, like increased conductivity, almost disappear or were difficult to detect. A possible reason is some unpredictable over-accumulation of impurities in specific points along the wire, due to electromigration itself. In other words, this could be a self-limiting phenomenon.

The electromigration effect was experimentally discovered by M. Gerardin in 1861 [11] and studied later by Alfred Cohen *et al.* in 1929 [12], among others.

After an initial study of the effect of direct current, our focus shifted to the use of powerful microsecond pulses.

This program enabled us to compare the effect of direct current and pulsed stimuli and resulted in various publications.

Already at that time it became clear that palladium did not have much future in *practical applications*, due to both its high cost and brittleness upon H or D absorption. The lack of robustness had a negative impact on every experiment with thin wires[6] which were prone to break very easily. Afterward, and beginning in the early 2000's, the work shifted from the investigation of a palladium-based systems to simpler, and cheaper, DC experiments with nickel (2002) and later with its alloys (since 2011). In 2011 we started the investigation of Constantan ($Cu_{55}Ni_{44}Mn_1$) due to its low cost and the capability to withstand temperatures as high as 900°C. Moreover, this alloy shows AHE activity using both

[3]Presented at ICCF22, 8-13 Sept. 2019, Assisi-Italy.

[4]Surface or bulk interfaces.

[5]The study of proton electromigration was pioneered by Alfred Cohen [12]-Germany.

[6]The best geometry to maximize and observe electromigration, or coherence, phenomena.

deuterium and hydrogen gases, although frequently we found "transition time" of AHE when changing gas (typically $H_2$, $D_2$) and the AHE vanished and/or was even negative in respect to the calibrations (usually by He gas).

In other words, the unique set of properties of this alloy can be summarized as follows:

- Low cost.
- Its noteworthy capability to dissociate $H_2$, $D_2$ molecules to the atomic state H, D, and their absorption from 150 °C up to about 700 °C [13].
- High robustness in harsh experimental conditions.
- Once "activated", a Constantan wire heated by a large direct current shows the occurrence of AHE, provided that the temperature is sufficiently high. However, the wire "activation" remains the most difficult task to achieve.

We have been able to detect a modest AHE, with satisfactory reproducibility (i.e., of academic interest only, at least at the moment), when the following conditions are met:

- Sufficiently high temperatures and large thermal gradients along the wire. We tested such effects developing the so-called "Capuchin knot" geometry. Anyway, although this is a very promising approach to minimizing added external energy, it was difficult to perform because of several practical geometric constraints. In the end, we were forced to quit, even though it is in principle an excellent approach.
- The surface is rough, being micro-structured, or better, nano-structured, and coated with Low Work Function (LWF) stable oxides [14], [15], [16];
- A flux of atomic or ionized deuterium or protium is induced in the wire.

In short, *non-equilibrium conditions*, enabled by various physical-chemical techniques, are compulsory to produce excess heat and possibly other effects, similarly to the better/deeper studied Pd-based systems.

## 4. Overview of the Techniques Explored

The critical parameters for a practical application of Anomalous Heat Effects (AHE), are:

(1) Establishing a reliable and simple procedure to "activate" and keep stable over time the metal (or alloy) where AHE occurs.
(2) Finding a process to "renew-refresh" in-situ the material, counteracting the "aging and deactivation" that almost always occurs. By the way, such "negative" behavior is like the well-known deactivation of Fe, Ni, Cu, Mn, K based catalysts used in chemical processes.
(3) Minimizing the extra power used to stimulate the "non-equilibrium" conditions needed for AHE occurrence.

In Fig. 1 we summarize the technical solutions adopted over time to obtain conditions for AHE occurrence with surface-modified Constantan wires.

The main degrees of freedom of the experimental set-up wires are:

a) core geometry.
b) wire diameter.
c) pressure and gas mixture recipe ($H_2$, $D_2$ or their mixtures with noble gases, mainly Ar).
d) maximum peak voltage allowable between the Constantan wire cathode and the tubular anode.
e) temperature.

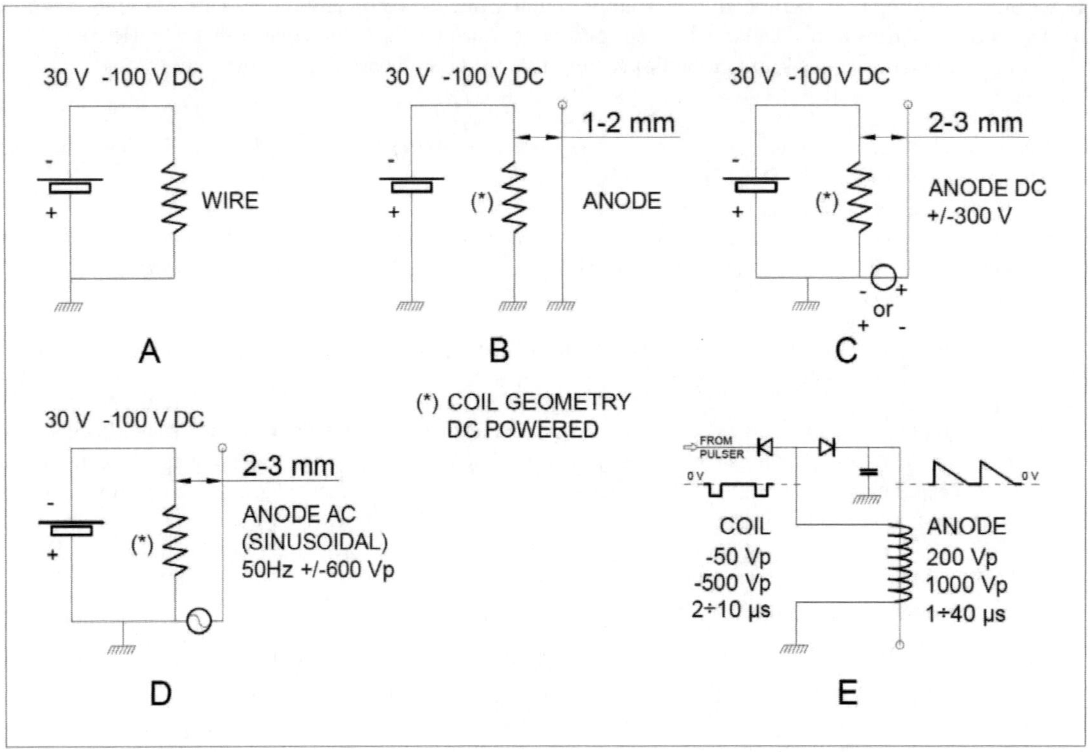


**Figure 1.** Different wire arrangements used to enable an increasingly more effective field stimulation in the last decade (2011–2021), starting from the simplest A in 2011, to the more elaborated E (2020–2021).

Background: reactor core geometry:

a) On the basis of the strong experimental evidence that non-equilibrium conditions are essential to obtain AHE, a new geometry of the reactor core was designed.
b) The wire is arranged to fully exploit the power of pulses and their electromagnetic effects, as follows:
   b.1) First, high-power longitudinal negative pulses over-heat mainly the wire surface. This is due to combined effects to get a large skin effect with: very fast-rise time and short pulse duration; magnetic material at the surface. The skin depth is given by the well-known simplified formula:

$$\delta = (\rho / (\pi * \nu * \mu_r * \mu_o))^{0.5}.$$

   In this formula: $\delta$ is the skin-depth of the current injected; $\nu$ is the frequency in Hz; $\mu_r$ is relative magnetic permeability of the conductor; $\mu_o$ is the magnetic permeability of free space ($4\pi * 10^{-7}$ H/m). Although the virgin Constantan ($Cu_{55}Ni_{44}Mn_1$) we usually used (IsabellenHutte-Germany) is not magnetic, we very often *measured* at room temperature magnetic behavior after prolonged current treatments at high temperatures ($> 800°C$) needed for the oxidation, to get spongy surfaces, and the addition of relatively large value of Fe. In fact, the LWF mixture is made by: Sr=10, Fe=10; K=1, Mn=1. This liquid mixture is "painted" several times over the Constantan wire surface by intermediated high temperature

cycles. We recall the unique ability of Constantan to dissociate $H_2$ to H is quite constant even in the large relative range of composition of Cu/Ni from 80/20 to 20/80. Regarding the possibility of demagnetizing Fe (because the temperature was close to the Curie temperature of 768 °C), we recall that the stationary temperature of the wire, before pulsing, is on the order of 500-700 °C: measured by the K type thermometer put inside the Fe counter electrode.
*In fact, although very promising, the effectiveness of the skin effect is difficult to assess precisely with the setup used: this should be further investigated in the future.*

This results in the "boiling-off" of the electrons emitted by Low Work Function material covering the sponge Constantan wire due to the Richardson effect: apart from the "help" by the instantaneous skin effect, at least when the repetition rate of pulses (also considering gas composition-pressure) is large enough to keep the mean wire temperature suitable for such electron emissions.

b.2) Second, self-generated High Voltage-Low Current positive pulse. Such pulse is sent to the Fe counter-electrode tube (Fig. 2A) on which the Constantan wire is wound in the shape of a compact coil, having a high value of inductance L. By elementary considerations of electrodynamics (e.g., Faraday-Neumann law and very fast rise-time, $<100$ ns at the end of the input pulse) a high voltage with opposite polarity to the main is generated (Fig. 8, red color curve). The value of peak voltage is related to the variation of current (I) over time (t) as: $V=-L*dI/dt$. It is used, depending on the pressures, both to accelerate the electrons toward the anode (Child-Langmuir effect) and/or to induce the breakdown of the active gas (Paschen/Dielectric Barrier Discharge effects). The diameter of the Constantan-Fe counter electrode is typically $<1$ mm because the glassy sheath is thin. The glass insulation sheath covering Fe is micrometric porous. It fills the volume between the cathode (Constantan wire) and the anode (Fe tube on which the insulated Constantan wire is coiled).

c) The "in-situ" pulse generation leverages on the short distance between the negative Constantan coil (that generates the positive pulse at the end of the input pulse) and the Fe core anode. This approach enables an effective energy transfer from extremely fast pulses, thus reducing the unwanted problematics of: impedance mismatching, parasitic capacitive effects. The complex circuit is a suitable compromise among values of inductance and storage capacitor (C2 in Fig. 3). Anyway, because of large variations of operating points and "aging effects" no attempts were made to tune by C2. Its value was selected mainly in respect to maximum voltage allowed before uncontrolled short circuits could happen, although circuitry (based on Zener diodes and other ancillary circuitry, not shown in Fig. 3 for simplicity) was used to prevent further damage arising from overvoltage situations.
d) In addition, this allows us to have a sequence of thermal pulses able to increase electron emission when the temperature is appropriate for the Richardson effect of the Low Work Function (LWF) materials (multilayered mixture of Sr, Fe, K, Mn) coating the wire. The multilayer structure in LENR was pioneered by Yasuhiro Iwamura-Japan [3], [4], [5], since 1998 (at that time at Mitsubishi Heavy Industries, Yokohama). Since 1998, although the layers used by Iwamura were much thinner, we used simple chemical paint procedures. We tried to have some control of the coating thickness by varying the concentration of the solution (which was usually quite low). Obviously, by our simple procedure, the control of the thickness is less accurate and reproducible compared to the sophisticated one adopted by Iwamura.
e) Together, these are the starting points for longitudinal impulsive electromigration, with even diffusion speed of hydrogen increased because of higher temperatures.
f) Moreover, if suitable conditions of gas pressure and voltage are reached, as shown in Fig. 5, it is possible to enter the Paschen regime that can allow both increased non-equilibrium condition and renewing surface of

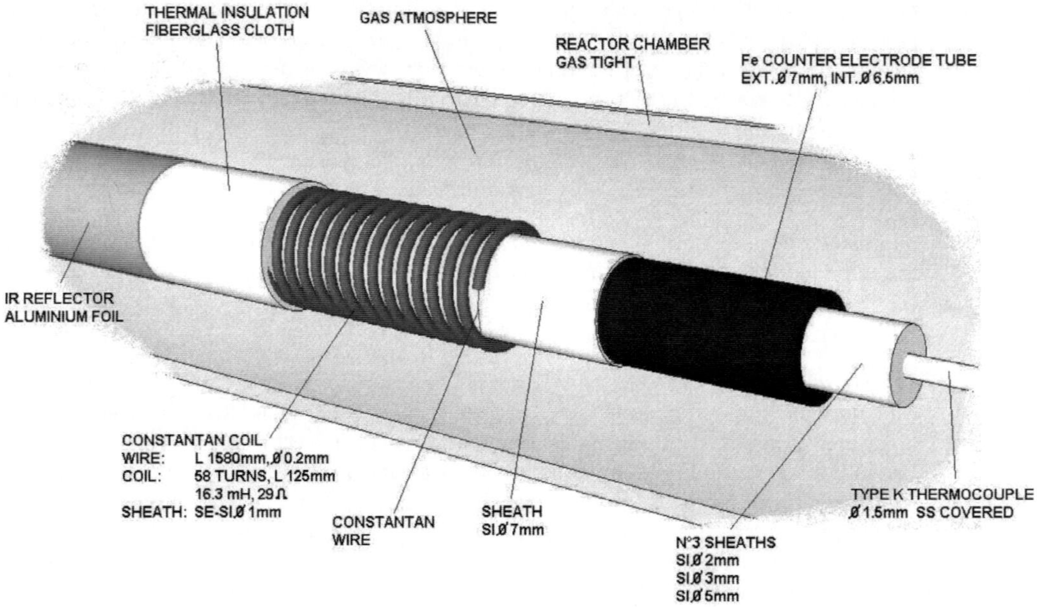


**Figure 2.** Schematic of the coaxial coil with its inner Fe counter-electrode. The wire, length of 158 cm, is arranged as a coil with a typical length of 12-14 cm. The coil usually had 75 turns; recently reduced to about 50 (i.e., with a larger diameter) because of problems with HV insulation. The sheath on Constantan wire is a fabric made of two kinds of main materials: a) SE, a glass similar to the composition of a tungsten light bulb, with a maximum operating temperature of 700 °C, and unusual atomic hydrogen surface adsorption; b) SI, a mixture of $Al_2O_3$ and $SiO_2$, with a maximum operating temperature of 1200 °C; c) SE-SI, a mesh of SE and SI (produced under our specific request by SIGI-Favier Company).

constantan wires (if sintered). We think that the Paschen regime, because it acts each time on different localized points of the wire, is able to promote "fresh" surfaces just by electric erosion. The main drawback is that if the erosion is excessively strong in localized points, the wire will break. *This is still now an open problem.* DBD regime is possible in condition far from the optimal Paschen one, as qualitatively shown in Fig. 6.

g) In conclusion, the reactor is based on a coiled coaxial structure (made of sponge Constantan wire with surface coated by LWF materials) with the inner core used as a counter electrode. Its schematic and key details are shown in Fig. 2A, 2B.

We recall that the unusual property of SE glass was discovered in 1928, by chance, by Nobel Laureate Irving Langmuir. In that case, the atomic hydrogen was produced by tungsten wires at low pressure and very high temperature (2700–3200 °C): research related to the development/optimization of incandescent light bulbs [37]. In our case, the operating temperature is only 200–800 °C, thanks to the unusual property of hydrogen dissociation by Ni-Cu alloys (from 20/80 up-to 80/20 relative composition).

The system was organized to be quite versatile, allowing it to enter several chemical-physics regimes: Richardson, Child-Langmuir, Paschen, DBD. Such regimes are activated just by changing gas pressure and anode-cathode voltage. The cathode, put along the geometric center of the coil, is just the iron tube as shown in Fig. 2A.

1) Richardson operation mode: electrons are emitted by LWF materials covering the Constantan wire by a simple thermionic effect. The anode is almost disconnected (or grounded by high impedance): low voltage in respect to the cathode, also movement of electrons at surface are at low values, just "boiling off" at the surface.

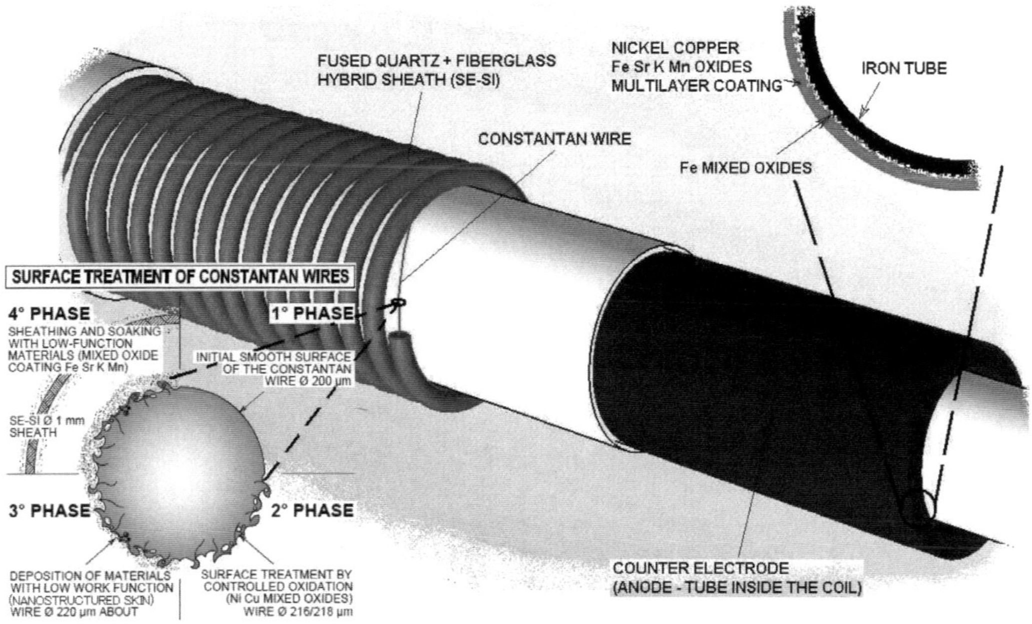


**Figure 2B.** Overview of the coil assembly: multilayer coating of LWF oxides; high temperature insulating hybrid glass ($Al_2O_3$-$SiO_2$) porous sheaths; external IR reflector. The structure of the wire has 3 main multilayer sections: a) inner with pure Constantan ($Cu_{55}Ni_{44}Mn_1$), "true bulk" about 190 μm diameter after oxidation procedures; b) external with mixed $Cu_x$-$Ni_y$ composition oxidized to create sponge structures; c) sponge structures filled by Sr, Fe, K, Mn mixed oxides. Because of uncontrolled leakage during the coil conditioning with sheaths inserted, some amounts of mixed oxides (even Cu-Ni) could cover the Fe surface. The several steps for wire constructions, and related details, are described in the insert named: phase 1°, 2°, 3°, 4°.

2) When gas pressure is low enough ($\ll$ 1 mbar) it is possible to activate a "Child-Langmuir" regime that allows boiling electrons to flow toward the anode with current density that depends on voltage as $V^{1.5}$, according to the literature [17] in parallel plate geometry.
3) Paschen/Dielectric Barrier Discharge modes: if the voltage of applied pulses and pressure are suitable, a gas breakdown occurs. This regime, we believe, is the most suitable to generate AHE because the hydrogen atoms stored inside the sponge surface of Constantan are forced to go through the surface. Previous experiments, shown at ICCF22, provided evidence that in these conditions (2&3) the highest and longest lasting AHE can be obtained. Anyway, as pointed out several times, we were not able to "control" the intrinsic spark regime for a long enough time (days or weeks) in a reliable way, suitable to develop practical applications.

## 5. Latest High Power, Fast Rise-Time, Pulse Excitation Approach

A) To continue, for each pulse injected, impulses cause the following effects:
- As clearly shown in Fig. 2A, there is a K type stainless-steel (SS) covered thermocouple at the inner core of the reactor. As a consequence, the temperature reported is lower than the real wire temperature. In respect to instantaneous overheating of the thin-skin wire's surface it is *estimated*, by considering the conditions of the largest injected power for each pulse (4.5 kW peak, about 4 kW mean), with a rise

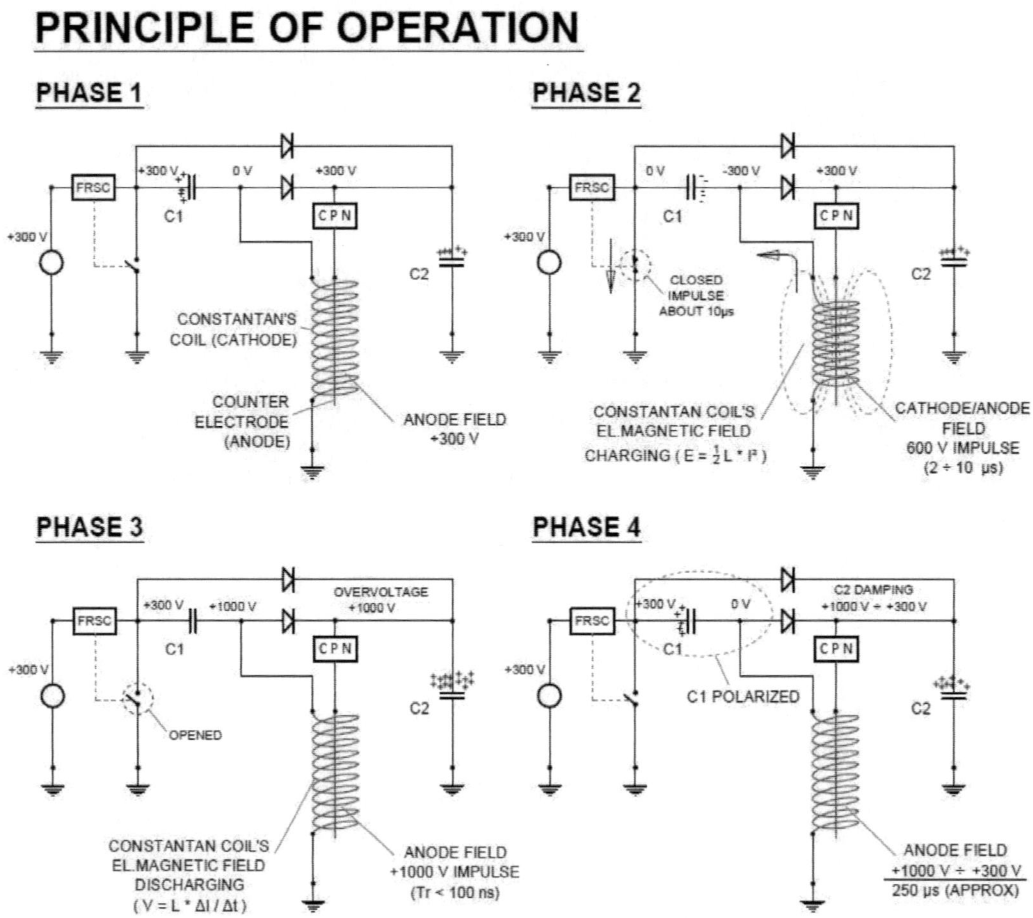


**Figure 3.** Schematic of the pulser with main auxiliary circuits, depicted as a "black box".Pulse duration/fall at Fe counter-electrode is determined by the discharge mode along the electrode itself (i.e., DBD, Paschen): it is dependent on the kind of gas, pressure, and inter-electrode distance. The term **FRSC** means: Fast Recovering Synchronized Charging. The (unusual) circuitry was developed by us since 1994. The aim is to operate at quite high repetition rate (up to 30 kHz in some cases) at high energetic efficiency, without using the usual inductance-resistance network. It decouples the main switch from the power supply, in the figure at +300 V, during the time of power switching conditions. The term **CPN** means: Controlled Peak power Network. It was introduced to reduce excessive power injected to the Fe tube (the counter electrode) that is connected, by surrounding gas, to the main coil. Anyway, up to now, the practical results by using the present CPN circuitry were not satisfactory enough for long term (days or weeks) operation: the thin Constantan wire ($\Phi$=200 μm) frequently failed catastrophically.

time of a few microseconds and skin depth of a few micrometers, that the maximum peak temperature increasing could reach large values (several tens of degree Celsius) in respect to the stationary one (500-700 °C), depending on the operating conditions of pulses and insulation. Although the temperature increase apparently is not very large (some tens of degrees Celsius), it could be quite effective in electron emission phenomena recalling the exponential nature of it toward temperature (Richardson law, Fig. 4 and formula therein).

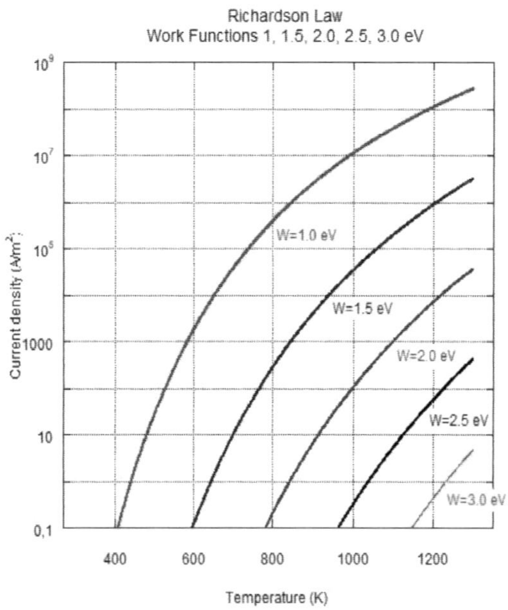


Electrons are emitted following the **Richardson law** for the thermionic emission at reduced pressure

$$J = AT^2 e^{-\frac{W}{kT}}$$

Where:

J is the current density emitted (A/m$^2$)

T is the emitter temperature (K)

k is Boltzmann's constant

W is the work function (eV)

A is a constant (in the simplest form of the law)

**Figure 4.** Calculated dependence of density of electron emission J (A/m$^2$) on Temperature (273–1500 K) and Work Function (1.0–3.0 eV), according to the Richardson law. Because of specific treatments at the wire surface, which is sponge-like, the effective surface is about 100 times larger (based on SEM observations): in the case of our 200 μm diameter, 158 cm long wire, it increases from $10^{-3}$ to $10^{-1}$ m$^2$. With W = 2 eV, J is $10^1$ A/m$^2$ at 730 °C and jumps to $10^3$ A/m$^2$ at 980 °C. J depends largely on both temperature and work function.

- Extremely large electro-migration "driving force" values, on the order of 1.5-2.5 V/cm, as measured by a 100 MHz digital oscilloscope during pulse conditions. Sometimes an analog oscilloscope was also used, 200 MHz BW, for cross-check purposes. The data were consistent.

B) At the start of a voltage pulse only the Richardson effect is present. It takes less than 100 ns to reach a voltage level sufficient to start the Paschen/DBD regime.

## 6. Operating Principle

The reactor operating procedure is based on the application of various concurrent stimuli that were associated with AHE occurrence in previous experiments (like the ones presented at ICCF22).

The "new" pulsed power supply and reactor assembly allows us in particular to optimize the timing of the stimuli producing the non-equilibrium conditions. As anticipated, these are consequences of the following effects:

### 6.1. Richardson Effect

The Richardson effect and its formulation (Fig. 4) occurs at high temperatures. It allows a continuous emission of electrons from the surface of any material. In our case, the emission of electrons is large even at relatively low temperature thanks to a coating of Low Work Function oxides (1.9-2.3 eV).

From the Richardson law equation [18] (shown in Fig. 4), it is evident that values of *W as low as possible* are needed to avoid operating temperatures excessively large (over 1000°C). In our experiments we used mainly SrO, K doped (at 10% concentration), to have a value of W close to 2 eV. All the wire pretreatments processes (i.e., repeated surface oxidation <–> reduction steps to get submicrometric sponge-like surfaces, i.e., large effective area) and multiple chemical coatings (multilayers procedure), are homemade.

### 6.2. Child-Langmuir Law

The negatively charged electrons that "boil-off" at the surface of hot material, at *low gas pressures*, can be expelled/moved when a counter electrode with a certain positive voltage is positioned at a sufficiently close distance to the material surface. The current intensity depends as $V^{1.5}$.

In detail:

$$\mathrm{J} = \boldsymbol{B} * \boldsymbol{S} * \frac{\boldsymbol{V}^{1.5}}{\boldsymbol{d}^2}$$

Where:

- J is the Current density,
- B is a constant,
- S is Anode surface,
- V is Voltage among Anode and Cathode,
- d is Anode-Cathode distance.

We point out that the Child-Langmuir regime was used only in some specific conditioning procedures of the wire: sometimes we verified, under vacuum, the relationship between voltage applied and measured current density. Anyway, in our experimental conditions (which are not conventional!), the measured value of the exponent was 1.3-1.4, not 1.5. The reduction is due to two main effects: coaxial geometry (instead of flat parallel plates), and the presence of a porous sheath instead of a void gap.

### 6.3. Dielectric Barrier Discharges (DBD) or Paschen Gas Breakdown

The current of the discharge depends on the distance between the electrodes, the pressure, and the type of gas. Figure 5 shows this dependence for a typical Paschen discharge, while Fig. 6 shows, schematically, the occurrence of a DBD discharge on the surface of the wires, and possible effects on electron density (due to the Maxwell-Wagner-Sillars effect at the interface of materials featuring different dielectric constants).

With reference to the Paschen discharge curves (Fig. 5) we would highlight that the use of the LWF oxides coating on the wires, and of thoriated tungsten ($WTh_{5\%}$, bundle of 2 mm diameter long rods, inside a SS airtight thin-wall tube) very close to the glass reactor external wall, allows some reduction of the breakdown voltage because of $\gamma$ emissions. In the experiments of 2019, we collected evidence confirming that the thoriated tungsten is a useful, safe and simple, enabler of Paschen discharges also in our experimental set-up, especially in the optimal Paschen regime. The experiment with $WTh_{5\%}$ were based on previous, specific tests where we observed the behavior of high voltage induced current by oscilloscope. Tests made mainly in DC conditions up to +2 kV at Fe counter-electrode. The main result was that using the $WTh_{5\%}$ the number of the discharges increased and were more regular (i.e., less "violent").

## 7. Recent Results Using Pulsed Power

- Since March 2021 the main improvements made to the pulser, to "match" the unusual needs of coiled coaxial core (magnetic materials both at the wire surface ($Fe_xO_y$) and at the inner section (Fe tube)), were satisfactory for the planned heavy-duty test. The pulser was designed according to the principles of "capacitive discharge"

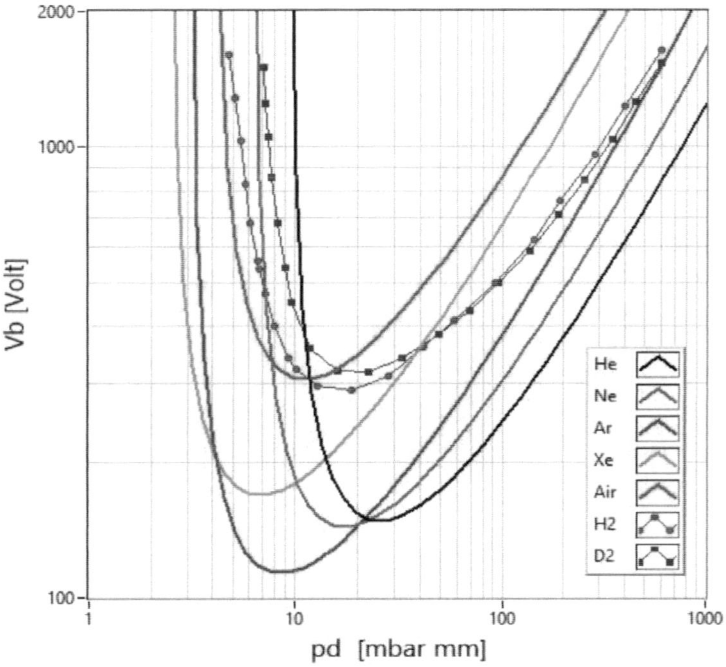


**Figure 5.** Direct current breakdown tension (Vb) of several gases as a function of the product of pressure and distance between electrodes (p*d). The addition of Ar (pure Ar has the lowest value of Vb, about 120 V) to $H_2$ or $D_2$ clearly enables discharges at lower voltages. Moreover, He could be very useful at higher pressures.

and is characterized by an unusual (proprietary) high energy-transfer efficiency, even at quite large repetition rates (up to 30 kHz; specific circuitry labelled as FRSC in Fig. 3), as shortly described in the explanations of the figure.

- In particular, the new unconventional procedure to generate useful (i.e., under control) high voltage at the end of the main power pulse was tested for long times without failure of electronics, even in the most severe operating conditions *($\Delta V/\Delta t$ on the order of* 5-10 *kV/μs*, up to 2500 Hz of the present repetition rate, typical pulse duration of 10 $\mu$s).
- Moreover, thanks to the performance at very high frequency of the main high-power switch adopted (based on SiC technology), the commutation losses were kept low, thus increasing the *overall efficiency of the pulser (close to 90%), and its long-term robustness.*
- Due to some experimental observations, we were prompted to increase the highest operating repetition-rate from 1 kHz up to 2.5 kHz. Even in this case, except the time spent to modify the pulser design, we did not observe damage to the circuitry and/or to the Constantan wire. Upcoming tests are planned to investigate higher pulse rates and reduced pulse duration (currently 10 $\mu$s) thus reaching the overall mean input power up to 100-130 W.
- Up to now, the "useful" maximum peak power applied to the Constantan wire (l=158 cm, $\Phi$=200 μm, weight about 450 mg) was over 4500 VA, i.e., over 10 kVA/g of material. The overall time duration of the pulse was kept, for now, at 10 μs. We have indications, based on previous short-duration tests, that it could be possible to

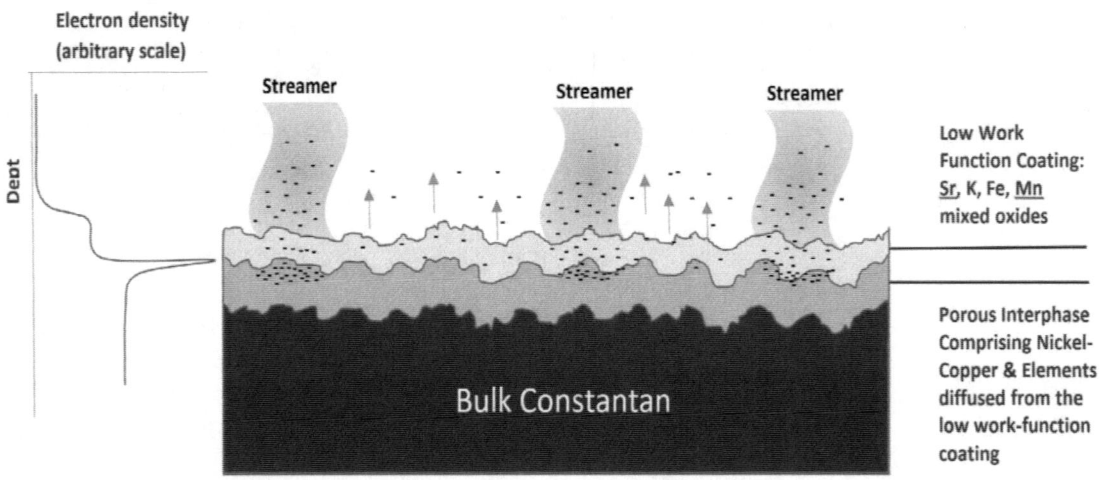


**Figure 6.** Schematic drawing of electron behaviour during a negative unipolar pulse. The occurrence of destructive discharges is prevented by the low conductivity coating on the Constantan wires as well as by the porous dielectric sheaths. The range of voltage-frequency used during the experiments enables the production of a mild filamentary discharge [19].

increase by 30% the peak power currently used by increasing the operating voltage of the main power supply (now at 400 V).

- It is important to highlight that the effective mean power is about 80% of the peak power due to inductive effects of the coil, that comprises a magnetic core (rise-time of the pulse is limited by the value of L/R). Also, we recall that the extra voltage of the opening switch that we used to generate high voltage has the following expression $V=-L*di/dt$. Therefore, a sufficiently high value of inductance is needed due to the intrinsic limits of the speed of the main switch adopted (SiC technology), featuring a fall time on the order 50-100 ns at the maximum peak current used (on the order of 12 A in the present experiments).
- The electromigration effect by itself, and local heating of the wire by longitudinal current flowing, to provide mean power up to about 95 W, was successful: no damage even after 60 h of continuous operations at 2.5 kHz repetition rate. The measured mean inner coil temperature was close to 600 °C: temperature as measured by the axial thermocouple. The temperature of the Constantan must be higher. To rule out the possibility of RFI during the pulse, we abruptly disconnected the pulse and observed a temperature decrease. It was regular (i.e., exponential-like decreasing, with no jumps). When we adopted $H_2$ gas largely enriched by Ar (i.e., $Ar/H_2=84/16$), we reduced the gasses thermal conductivity to the cold side (i.e., the room temperature), just thanks to the lower thermal conductivity of Ar (about 7 times lower compared to $H_2$). As a consequence, the internal temperatures were much higher: it increased by 50-150 °C.
- The *SIMPLE* glass reactor, in the present work just the dissipation type (thermal equilibrium time <30 minutes), worked well for our purposes and it was even possible to make sufficiently accurate thermometry using several "blanks" (sometimes quite complete: Ar, He, Xe, air, at different pressures, usually: vacuum; He, Ar at about 0.1 and 3 bar). We used only 2 K-type thermocouples (max temperature 1200 °C), SS covered (external body grounded to minimize possible RF interference) and electrically insulated. The first thermocouple, Fig. 2A, was put into the center of the coil (T_int.) inside the Fe counter-electrode (tube geometry); the second at the external wall (T_ext.) of the borosilicate glass reactor (3 mm thick; tested for high temperature

and pressure operations): the same type used since 2011 in our experiments. We note that, thanks to several blanks and using the T_ext values as the key parameter, the thermometry is almost equivalent to calorimetry for the purposes of the present experiments. Fully blank tests (i.e., with several different gases) were made from time-to-time, for a cross check of the whole system. The usual standard blanks were made using He gas because its thermal conductivity is the closest to $H_2$. *In the case of pulse conditions, the blank test are the values of temperatures measured under $H_2$ in DC conditions*. So, the values of AHE could be under-estimated. The power applied were 5, 10, 15, 20, ... i.e., steps of 5 W, up to 60 W; and later steps 10 W up to 120 W.

- The reference temperature is the room temperature, air conditioned and continuously measured. The thermocouple, at the 3 mm thick glass (type 3.3 Schott-Duran) surface, is attached by a quite complex procedure as follows: A) at first, the glass is covered by twin layers of 40 $\mu$m thick Al foil, 2 cm $\times$ 3 cm. It is used because it has the local mean value of temperature, and to avoid potential IR interference from (possibly) the localized hottest area of the inner core of the reactor toward the tip of thermocouple (positioned at the center of 2 x 3 cm foil); B) the thermocouple is covered by 4 layers of the same kind of Al described before, with the same dimensions; C) the side of Al toward ambient is covered by a high thermal insulation material, 1 cm thick, to avoid abrupt interference from localized ambient temperatures; D) the whole assembly is elastically pushed toward the glass tube by 2 independent springs that circumvent the tube. In future experiments we plan to make further improvements related to temperature measurements.
- At present we are not able to evaluate potential "aging" due to material degradation. In fact, the experiments ends when the wire breaks, therefore it is not possible to perform a control run at the end of the experiment (i.e., with helium). In any case, due to the fact that various regimes of operation are associated with different values of AHE, we can make a self-check of the system even after several weeks of operation: the operative operating points that do not give AHE results, or give poor results, are all quite similar, (and all are similar to a "blank" run).
- In the near future, once we obtain enough satisfactory results as a whole, we will move to a usual air-flow calorimeter (equilibrium time about 4–6 hours), operating in the Frascati Laboratory (e.g., our ICCF22 paper, published by JCMNS 33, 2020, pg. 1-28). So, we will be able to evaluate the *absolute value* of AHE, if any, and not only relative to blank experiments (which are supposed to be the "zero").
- All the tests at sufficiently *high pressure (several bars, i.e., far from the Paschen regime*) of gas (He, $H_2$, Ar-$H_2$) used were successful. Also, tests under dynamic vacuum were successful. In short, there were no catastrophic failures of the wire, and we detected measurable values of AHE as described below. In other words, the 200 μm thick wire is able to withstand *longitudinal* current as large as 12 A for pulse duration of 10 μs and repetition rate up to 2.5 kHz. As reference, in free-air, the maximum DC current recommended by the Constantan's manufacturer (IsabellenHütte-Germany) is 2.34 A: at that current the surface overheating (in respect to NTP conditions) is 600 °C with a related surface power load of 13.6 W/cm$^2$. We experienced that, at such wire's surface temperature (620°C), no oxidation occurred during a 24 h test in free air, using a virgin wire (as provided by the producer).
- When we tried to enter into the Paschen/DBD regime, i.e., lowering the pressure step-by step (from 4-5 bar down to 0.1 bar, or lower), under high voltage pulsed situations at the Fe thick counter-electrode (about 1000 V), we experienced several catastrophic failures of the coils. Anyway, such unwanted effect was expected: Paschen/DBD discharges are always highly localized. Breakdown occurs first at a particular point and all the energy is dissipated on that point. It is destructive for small wires, which is not surprising. We made some counter-measures to reduce such problems, but they are not yet sufficient, as detailed later.
- As anticipated, we did not observe catastrophic failures when operating in a full dynamic vacuum (up to mean power on the order of 50-60 W, measured inner temperatures close to 900 °C). We adopted the procedure of first making a vacuum (dynamic) and after giving High Pulsed Power. About pumping, a usual 2-stage rotary

is enough: a TMP type is not necessary. We just needed to pay attention to minimize contamination by oil vapour.

- The most probable reason for breaking, according to our visual observations, were localized high voltage (about 1000 V peak) discharges (i.e., like a hot-spot). Calculated peak power, supposing 1 $\mu$s of time to transfer the energy, was on the order of $2.5*10^3$ W/pulse.
- The simplest explanation is that the Paschen regime is excessively "severe" to the thin wire used ($\Phi$=200 μm) in our present experimental conditions. We are aware that sparks (arcs) are locally destructive to all metallic surfaces: because of such intrinsic effects, innovative solutions are needed. Work is in progress.
- Up to now we changed 5 coils. We even made changes to the geometric construction details, but they were, so far, not enough to solve the problems.
- Even the addition of some current protection network, shown as the CPN black box in Fig. 3, to limit the "long duration" (>>1 μs) current flowing between the electrodes, was not fully effective. The CPN was simply based on a RC network in parallel: at short times (i.e., high frequency), the impedance of the capacitor is low, so enough current is allowed to be transferred to obtain the high positive voltage planned. Anyway, the procedure was not enough to solve the problem for a long-enough time of operation (at least 30 minutes, i.e., the equilibrium time of the system). In some experiments it was really a big problem. Deeper analysis of damaged sheaths is necessary: future specific work is planned.
- Waiting for some new idea/devices/circuitry to make the electric field more homogeneous among the electrodes (i.e., to avoid a localized hot-spot), we deeply studied the effect of electromigration by itself, neglecting, for now, the other very important effects we were looking for: Paschen/DBD; Richardson with related Child-Langmuir NOT under usual high vacuum but even at 10-50 mbar of pressure. At such intermediate pressures there is some coexistence with the Paschen curve, scientifically interesting but at present out of our control. Anyway, the amount of data collected are not enough to allow the building of a proper plot.
- We began our in-depth analysis because in our older experiments (1994-1998), when we used only pulsed electromigration, with plates at the beginning, and later wires, in an electrolytic environment at about 1 bar of pressure, Pd-Pt or Pd-Ni electrodes. In these experiments we observed some AHE in specific operating conditions of: electrolytic conductivity/composition, pulse duration, repetition-rate, power injected. Some information about this research is scattered among several Conference/Workshop/Meetings papers of that time period, although no systematic final study, was performed.
- In the recent tests some of the operating conditions are even more "intense" in comparison with old ones, with Pd based material, so the probabilities to observe something "strange/useful" are increased.
- Figure 7 shows the pulse regime typical values of Voltage, Current, Power applied in experiments at high mean power (>90 W). Voltage, (100 V/div., in red), Current (2 A/div., in blue) Power (800 V*A/div., in green) applied to the Cathode, time scale 1 μs/div. Gas pressure, mixture Ar/$H_2$ (84/16): about 4 bar at RT, 5.2 bar at HT. The maximum temperature measured inside the wire coil (by the axial thermocouple T_in) was about 580°C.
- Figure 8 is similar to Fig. 7 for operating conditions. It shows the high voltage (in red, 200 V/div) applied to the anode, maximum value about +1060 V. As reference, the current (in blue) but with time scale of 2 μs/div, produces some evolution of peak voltage versus time. The reference value of voltage did not start from 0 V but from a higher value, about +400 V (i.e., the value of main power supply at the test): deliberately *applied,* by suitable auxiliary circuitry (shown in Fig. 3)*, to get pre-ionization of the gas mixture (see Paschen curve, Fig. 5)* in the case of future suitable gas pressure reduction.

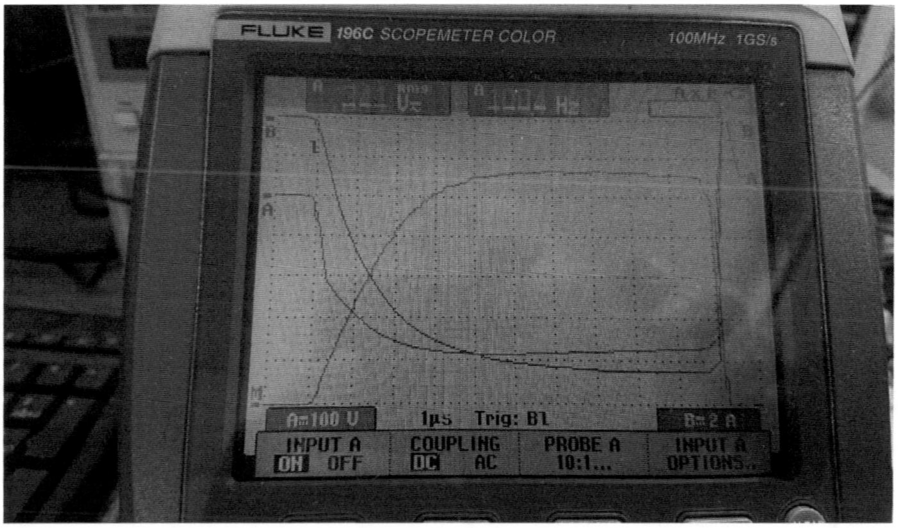


**Figure 7.** Snapshot of Voltage (red color, 100 V/div, maximum value −380 V), Current (blue color, 2 A/div, maximum value −12.4 A), Power (green color, scale factor 4, 800 W/div, maximum value 4500 W) of a typical high-power pulse. Time scale is 1 $\mu$s/div. The repetition rate was the highest up to now used (2.5 kHz). As previously reported, the progressive buildup of the negative current at the beginning of the pulse is due to the L/R ratio of the wire (added with auxiliary coil, not shown for simplicity in Fig. 1 or 3), in contrast to the very rapid extinction made possible by the SiC circuitry shown in Fig. 3.

## 8. Main Results

The main details are shown under the figure captions, as shown in Fig. 9, Fig. 9A, Fig. 10, Fig. 11, Fig. 12.

Figures 9 and 9A show the difference between the external glass temperature (T_ext) and the reference temperature (T_ref=room temperature) as a function of the input power. In red we have the DC calibration curve obtained by $4^{th}$ order polynomial fitting, based on 20 measurements (i.e., starting from 0, each power step is as low as 5 W up to 60 W, and later on 10 W up to 120 W with an added intermediate measurement at 105 W). In the case of pulsing, made 8 measurements. Power varied changing the main power supply, shown (as an example) at +300 V in Fig. 3, from +50 V up to +400 V, step 50, for each repetition rate adopted (1.0, 1.5, 2.0, 2.5 kHz). In total we made 32 measurements for each combination of gas composition. Anyway, for simplicity we reported only measurements using concentrated $H_2$ at a few bars of pressure.

Fig. 9A is just the same figure at low power, to make observation of details easier.

Figures 9 and 9A together show the measurements in the DC and pulse regime, with pulse peak power varying from 64 W to 4640 W, changing the main power supply from +50 V up to +400 V. The *repetition rate,* from 1 kHz to 2.5 kHz, affects the *mean powe*r. The power is displayed on the X-axis as equivalent DC heating power.

In other words, in Fig. 9 and 9A we have the same core, the same atmosphere ($H_2$) and similar pressures as shown in both the DC calibration (data fitted with equation shown) and several temperatures/equivalent power by pulsing procedures. Fitting for pulsing conditions are performed but not reported for simplicity. Anyway, the R quality-factor of fitting was always >0.9999.

Observing the green color curve in Fig. 7, we have to consider that the mean power is lower than peak power shown because of the time needed to reach the flat regime (to be specific, about 3.5 $\mu$s). Moreover, the nominal duty-cycle

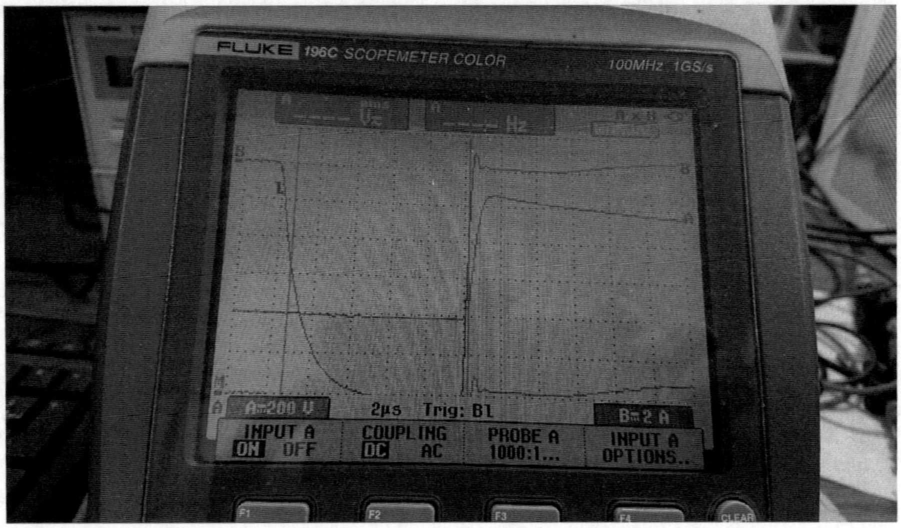


**Figure 8.** Snapshot of the peak voltage at the counter electrode (red colour, 200 V/div, maximum value +1060 V). Shown also, as reference, the current applied (blue colour, 2 A/div). The large positive voltage is due to the "opening circuit phase" of the pulse. It is extremely fast because combined effects of both overall electronic arrangements and intrinsic fall time performance of SiC switch used at our operating point (<<100 ns).

(supposing ideal rectangular pulses with rise/fall times close to zero ns), with the present pulse width (10 $\mu$s), were: 1%, 1.5%, 2%, 2.5%.

The power input is just what is observed for each pulse, using the 100 MHz oscilloscope (Fluke Mod. 96 C), with in-phase power (green color) multiplied by the repetition rate. We made several snapshots of the pulses, noted them, and determined their mean value.

The accuracy of measurements in the procedure was ensured as follows:

a) First we calibrated the system using an array of 8 non-inductive resistors (Wishay model RCH25, each 220 Ohm, TEMPCO 150 ppm/°C), with high power (25 W each) connected in parallel. Nominal maximum power is 200 W: used at 100 W. Moreover, the array was fan cooled to keep temperature variations as low as possible. The value of final resistance (nominal 27.5 Ohm) is close to the value of the wire used, i.e., l=158 cm, $\Phi$=200 $\mu$m. The value of resistance was measured, in DC conditions, by V-I procedures (accuracy better than 0.1%) up to the maximum power of 200 W.
b) The measurements were repeated, under pulsed conditions, at a mean power of 100 W by the scope. Each measurement was mediated by 64 individual measurements (one of the mathematical functions of the scope). No detectable differences happened between DC and pulsed conditions. We recall that, for the kind of resistor used, under 10 $\mu$s of single pulse duration, the safety overload factor is as large as 30 kW in comparison to nominal 25 W in DC condition.
c) The evaluations of measurements, by scope, i.e., power injected as shown in Fig. 7, were made by 3 independent researchers.
d) Considering the several cross checks performed, and routinely repeated, we estimated the overall accuracy to be on the order of 1%.

Anyway, the input power, apart from initial times (about 5 minutes) from changing power, is quite constant.

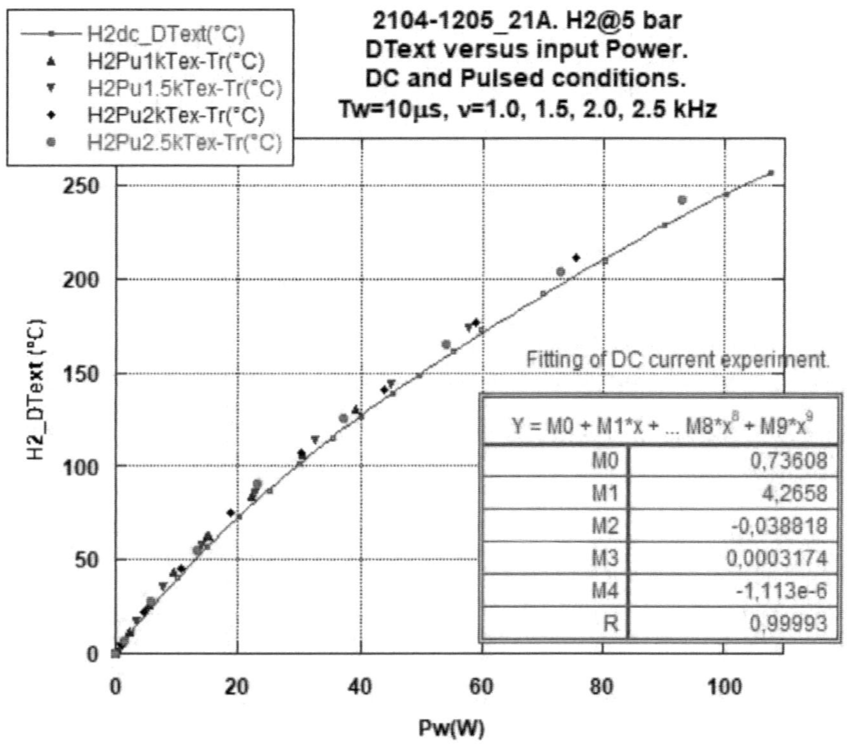


**Figure 9.** Behaviour of temperature difference, at reactor glass wall, changing the Pw from DC to pulsed. In the case of DC (20 measurements in total, including values at 0, 105, 110, 120 W) is shown. The curve is a 4° order polynomial. Part of the full range of measurements performed is shown, up to about 105 W. The fitting of data has a R value very close to 1 (→0.99993).

We suppose that AHE in the DC regime is negligible and therefore the greater temperature increment in the pulse regime is due to only AHE triggered by the pulse activation mode. We have obtained graphs in Fig. 10 and Fig. 11 that show directly the AHE variation versus the equivalent DC heating power, for 4 different pulse repetition rate values and 8 values of peak power pulse.

To calculate AHE we used the formula AHE=((ΔText–ΔTdc)/ΔTdc)*Pw, valid only for low AHE values. Then, it is possible to consider a thermal exchange coefficient K constant such that ΔTdc=K*Pw and ΔText≈K*(Pw+AHE). We specify that ΔText means ΔT under pulse regime. Actually, K decreases as temperature increases because of thermal losses, resulting in some *underestimation* of AHE value.

The obtained measurements are very conservative, as partially anticipated, for the following reasons:

1. We suppose that AHE obtained in DC regime has a value of zero, as referred to similar curves obtained with He as thermal exchange gas, although we have often observed AHE>0 even in such conditions, i.e., AHE_H2>AHE_He.
2. We recall that, after suitable activation cycles, the AHE measured in DC conditions under hydrogen is larger in respect to helium gas, as reported in our ICCF22 paper.

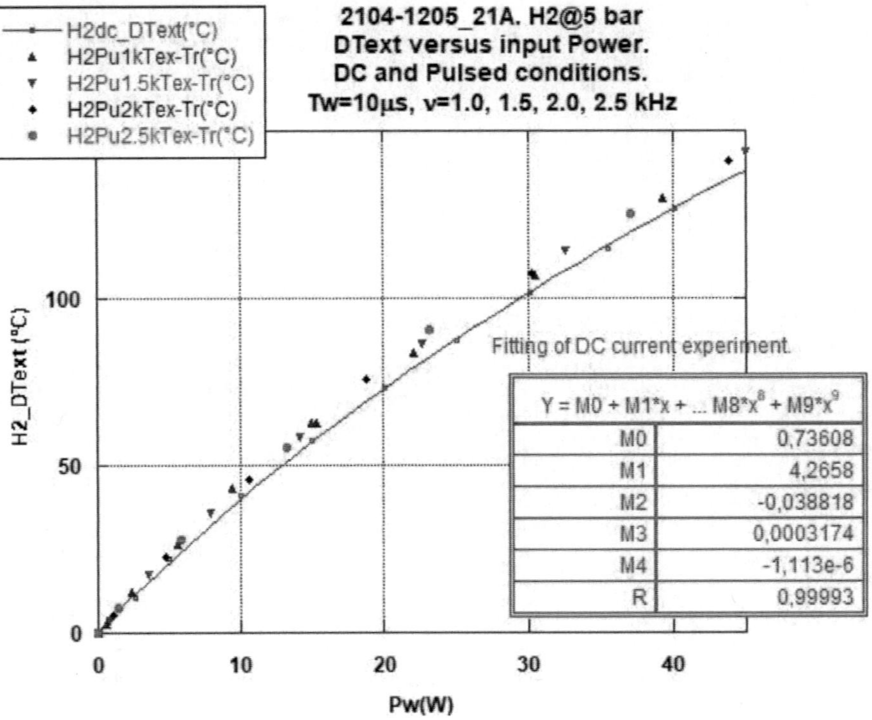


**Figure 9A.** Same measurement as Fig. 9 but at lower power (0-45 W) for an easier evaluation at low power regions. In the plots the terms DText and Tex-Tr have the same meaning, i.e., temperature difference in respect to room temperature. Different names from previous experiments are used in the plot just for historical reasons. In the case of DC measurements, the reactor wall temperature is shown as DText (i.e., T_wall reactor-T_room). In the case of pulsing it is explicitly quoted as T_reactor-T_room.

3. The formula used inherently gives AHE values less than real values because we assume there is no increment of thermal losses as temperature increases. For example, in Fig. 9 the AHE seen by graphical evaluation is over 6.7 W, while the same using the formula it is just 3.7 W. Moreover, at the maximum power of pulsing, it is about 9 W.

- At the end of Fig. 12 the unexpected beneficial effect of HPPP (High Peak Power Pulse) is shown. HPPP is discussed in our publications since 1993 experiments [20], [21], [22], [7]). The design of current pulser shares several features with the original one, nevertheless it has been improved in some specific respects. In fact, the original device was conceived for an electrolytic system where a very high current was needed (for instance in the case of plate electrode geometry, a current up to 150 A and a peak voltage up to 200 V; typical duration was 1-3 $\mu$s). Now, using thin wires in a gas system, we need instead a maximum current of about $-20$ A, a maximum voltage of $-500$ V, and an extra positive voltage up to $+1200$ V, $+1$ A. Anyway, in our paper, for simplicity, we reported in detail only results with maximum current of $-12$ A and voltage of $-400$ V. The sequence of events in the case of the present experiment are shown in Fig. 3, phases 1→4.

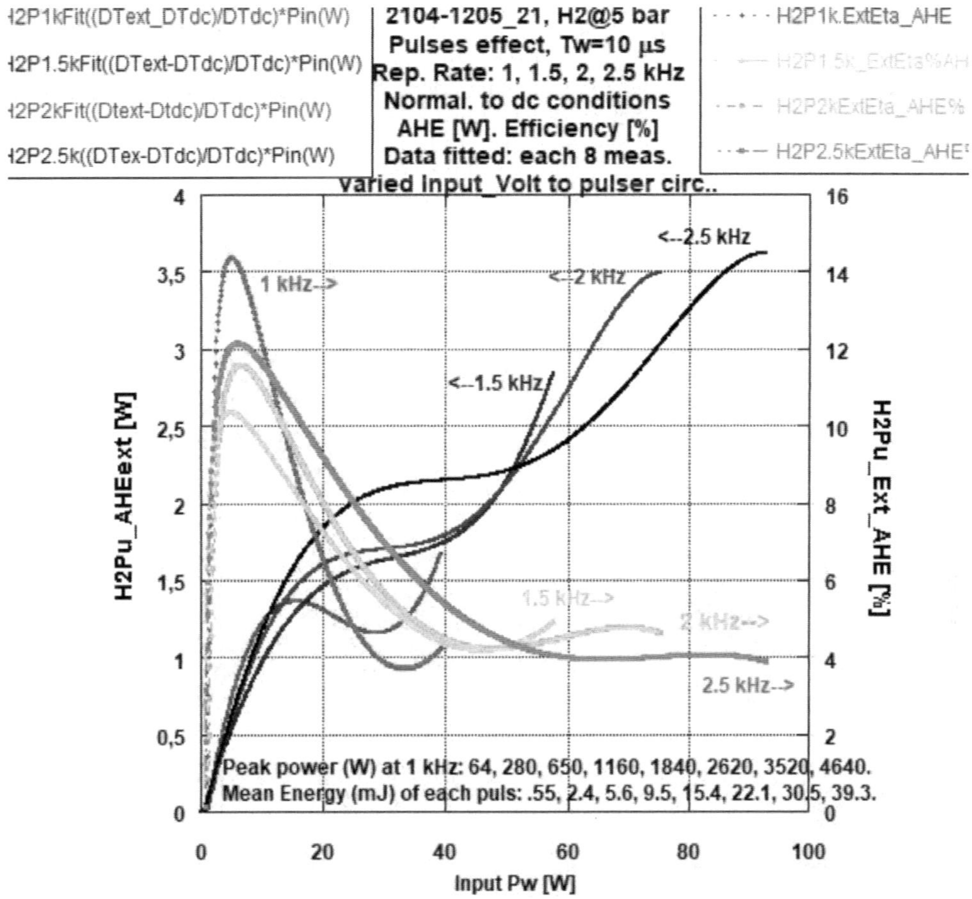


**Figure 10.** Only the electromigration effect is shown; $H_2$ at 100%, 5 bars. Plot of percent of AHE changing the peak power pulses (explored 8 values, 64→4640 W, detailed at the foot of the figure) and repetition rate (4 values: 1, 1.5, 2, 2.5 kHz). Data is normalized at the same input power in DC conditions. AHE values on the left scale (W) and relative values on the right scale (%). We observed that while the absolute input power increases the efficiency tends to a flat value of about 4-5%. In the data shown it is not possible to show each data points because calculations are based on *fitting* of the calibrations (i.e., under DC measurements) and the values of exact input power can be different (up to few watts) from the effective one.
*Regarding the value of the mean energy of each pulse, for the sake of comparison, we recall that the typical energy needed to ignite the gasoline-air mixture in an Otto-Benz engine is on the order of 20-160 mJ, depending on the gasoline/air ratio and the compression ratio. Moreover, the total time of ignition is on the order of some milliseconds. In our case, we have energy up to 40 mJ and time of pulse of only 10 μs. In other words, the energy involved is over 100 times more concentrated in respect to time. The drawback is large stress to the thin wire.

- To prove such beneficial effectiveness, we made specific tests by changing the gas mixture composition ($H_2$→$H_2$/Ar 60/40→$H_2$/Ar 16/84). The typical pressures were in the range of 3-5 bars, apart from specific, explorative tests at different (low) pressures. We even de-loaded the wire from the hydrogen stored in the bulk, as much as possible, by vacuum treatments.

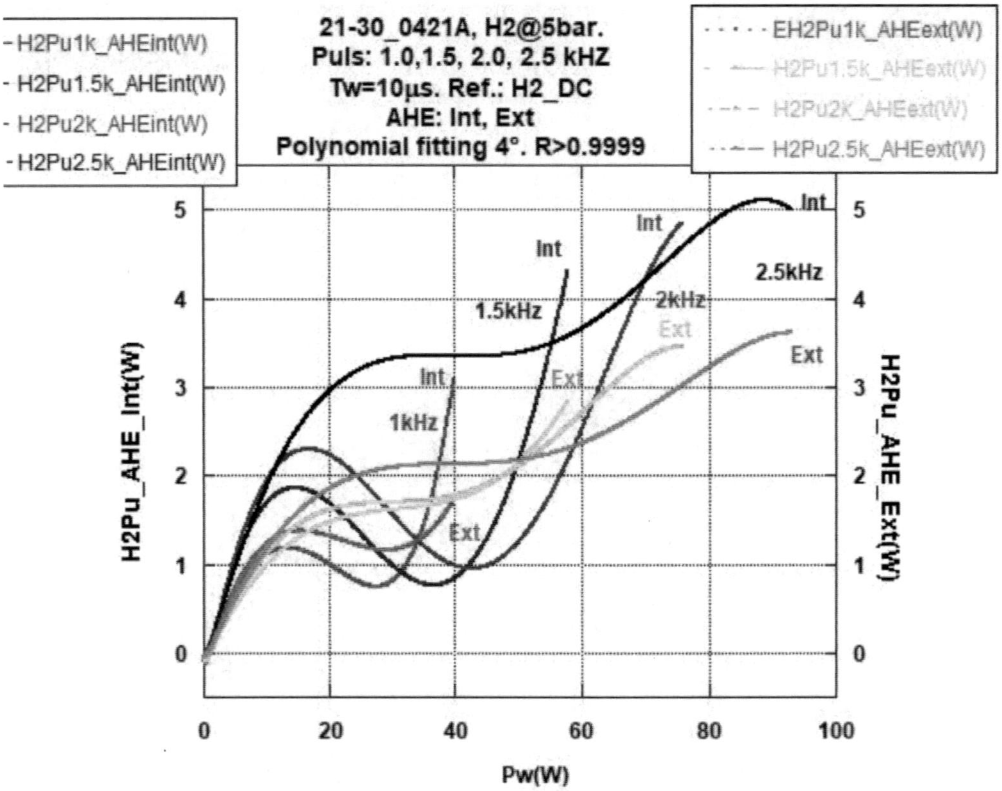


**Figure 11.** Only Electromigration effect; $H_2$ at 100%, 5 bars. Plot of absolute AHE. The values of AHE external are the same of Fig. 10, left side. Comparison between internal and external AHE, calculated with the same procedures and considering the DC condition as blank. We observed that the internal one is always higher than the external one. We guessed that some of the energy coming out from the coil is not fully converted to temperature increase through the gas and thick glass wall. We observed by oscilloscope quite large HF signal coming out to the environment. The shapes of HF signals are quite irregular and chaotic, so they can be used only as starting points for future improvements to the apparatus. In conclusion, such observation suggest that in the next experiments we should modify the reactor geometry to recover energy that is otherwise not included in the computation and lost.

- Restoring of the initial "good conditions", at the moment has only been partially achieved. We just guess that, in the de-loading procedures (mainly by vacuum at high temperatures and long times) severe sintering occurred at the surfaces of the submicrometric, useful, "active" location that are the main places where AHE is generated. The "*active Site*" model (but with different explanations) was pointed out by Edmund Storms and Mitchell Swartz, both in the USA, among others. Further experiments are needed to validate these statements more deeply.
- We recall that in our project, one of the roles of the mixed *plasma* of Ar and $H_2$, meant for the Paschen/DBD regimes, was to "reactivate" surfaces that went inactive because of spontaneous, severe "aging effects". This work is in progress.

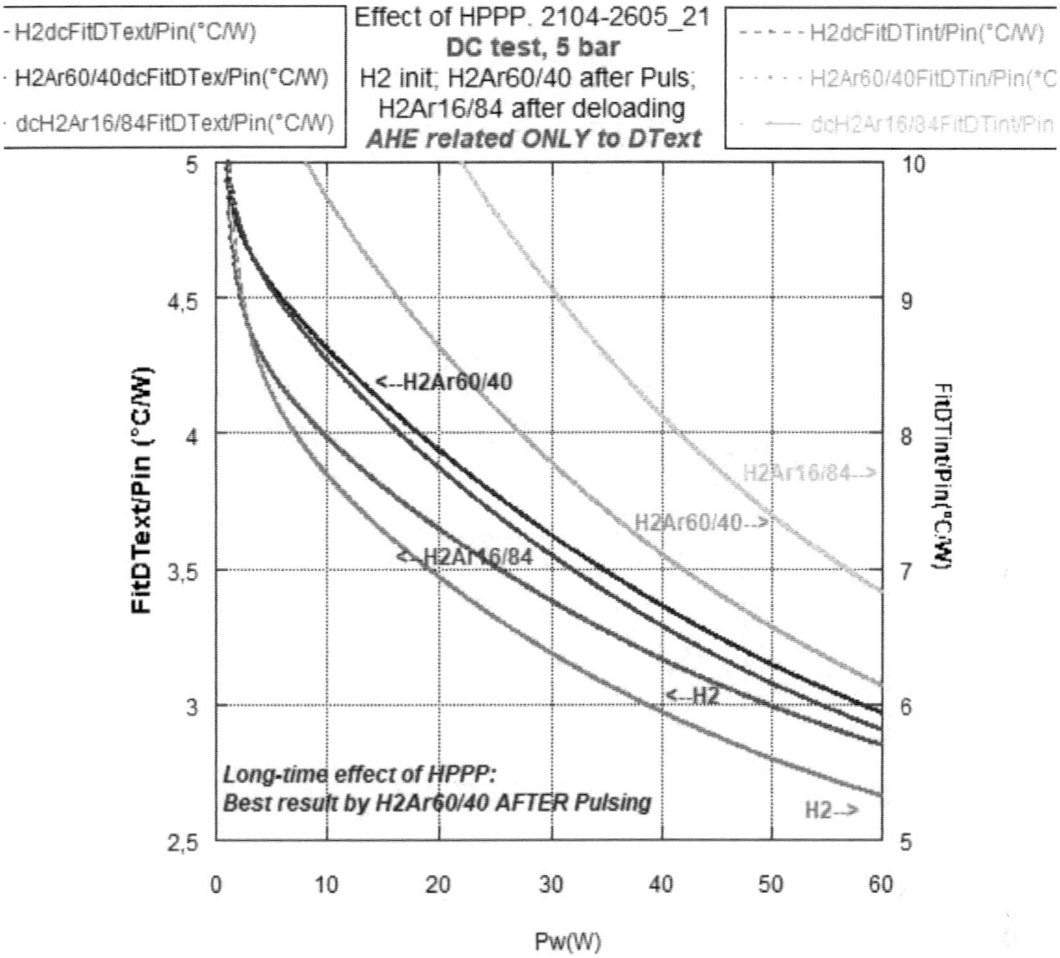


**Figure 12.** An unexpected, useful effect after HPPP (High Peak Power Pulse) treatments are applied, over 2*$10^7$ pulses at 10 kVA/g. Shown is the ratio of DT/Pwin (°C/W) versus input power. Vertical left-side: DT_ext; Vertical right-side: DT_int.

- We recall also that in our old experiments, with Pd (1994-1998), we observed some "useful" *memory effects* by long-time pulsing but we have never analyzed the phenomena deeply enough.
- Several similarities among the behaviors of Pd-D and Constantan-H are observed. We guess that other pure materials or alloys may exist (with suitable amounts of hydrogen or deuterium stored at the surface or into the bulk) that can produce AHE in the right conditions, hopefully at higher levels than the present one.

All the measurements reported are in DC conditions: the calibration curve is the usual in $H_2$ in DC at the beginning of the specific experiment (26 April 2021), before pulsing.

Regarding the values of DT_int (right side of the curve), it is necessary to consider that, *because of the lower thermal conductivity of Ar in respect to $H_2$, the mixture $H_2$/Ar=16/84 has the largest values of DT_Int/Pin but lower in respect to the DT_Ext/Pin at H2/Ar=60/40.* ***Anyway, results are always better than with pure $H_2$ at the beginning of the experiment.***

Over time we made 3 long main tests, detailed as follows:

- $1^{st}$ test with $H_2$ only in DC conditions (**reference**, pink and red colors). The tests, with power up to 61 W, were made *BEFORE* applying the pulse mode, i.e., at the beginning of the experiment (April 21-26, 2021). The plot is the ratio of DT/Pin, so it is not necessary to have the zero values shown but only their trend for comparison purposes.
- $2^{nd}$ test after pulses on $H_2$. R/Ro decreased from 1.000 to 0.820 (i.e., H absorbed into the bulk of the Constantan); R/Ro measurement performed by V/I procedures with I at low intensities, <200 mA, to avoid self-heating. Later on, the gas composition was changed from $H_2$ at 100% concentration to $H_2$/Ar=60/40. Increased both external (blue color) and internal (sky-blue color) temperatures, the last one expected because lower thermal conductivity of Ar/$H_2$ mixtures.
- $3^{rd}$ test: We de-loaded the previously adsorbed $H_2$ as much as possible (R/Ro, went from 0.820 back to 0.969). After further vacuum cleaning, the gas composition was changed to $H_2$/Ar=16/84. Results: the internal temperature (light green) increased largely because of the lower gas conductivity (as expected); the external temperature (green curve) DECREASED in respect to 60/40 composition. Anyway, the results were still *better than the initial reference*. This is a clear demonstration of HPPP: it "activated" the material, most probably in the inner bulk of Constantan. We have evidence that with our procedure it was not possible to FULLY de-load the wire from the most inner locations of the lattice. In other words, the activation seems quite stable for long times (weeks): ***the target was achieved.***
- Again, we remind the reader that all the experiments in pulsed condition are normalized in DC under $H_2$. As a consequence, the AHE values could be underestimated.

We guess that, at least in these specific experiments, devoted to study bulk properties, that the "movement", or more accurately termed the flux of atomic hydrogen, even at low concentrations, has a key role in AHE production. Further work is necessary to reinforce this hypothesis.

- For this specific experiment we were able to observe the beneficial effect, after pulsing, for only a few days because the wire was destroyed some hours after last experiment: a heavy-duty test (which was planned long before it was done). Anyway, after also reconsidering previous experiments, we *estimate* that the time of beneficial effect of pulsing to be on the order of 1 week.

## 9. Conclusions

(1) It was experimentally demonstrated that the "old" beneficial effects of HPPP, discovered by us starting in 1994 with the very costly Pd system (Pd has recently cost about 75 €/g), *can be "transferred" also to the low cost Constantan alloy (about 20 €/kg).*

(2) The pulsing procedure, by itself, can induce some AHE, which is stable over time. *The effect seems to behave positive feedback over the temperature range we have explored up to now (maximum about 600 °C, pure $H_2$). We expect further advantages at temperatures over 700 °C (by using Ar-$H_2$ mixtures at low pressures), where the Richardson regime is further enhanced.*

(3) Moreover, we have found, just by chance, a procedure to "*activate*" our specific Constantan wires, allowing them to produce AHE when other conditions (gas, high temperature, sponge surface covered by LWF materials, thermal gradients, ...) are fulfilled.

(4) The activation was found to be stable enough to be beneficial also in DC conditions, even after several days after ending the pulsing.
(5) From a theoretical point of view, one of the most interesting hypotheses on the origin of anomalous heat is based on the active role of Nickel-Copper alloys in the decomposition of molecular hydrogen and in the catalysis of Ultra-Dense Hydrogen (UDH) formation, according to Leif Holmlid's (Univ. Goteborg-SE) models and experiments [23]. In the case of Holmlid's experiments, non-equilibrium conditions were applied, mainly, by intense Laser pulses (see also: [24] for the interpretation of the results from a very innovative although complex, point of view). We hope to soon overcome the problems of catastrophic coil destruction and to can use at fully the possibilities given by the combined effects of Richardson, Child-Langmuir, Paschen/DBD, Skin, from the point of view of increasing the values of AHE measured up to now.
(6) In conclusions, only a few of the possibilities we explored to induce and increase the AHE were realized because of several technical constraints. We hope that the "indications" given by us can be used by other researchers active in the field.

## Annex A.

### Brief theoretical description of the Coherence processes underlying LENRs and the Preparata's effect.

In this annex we give a short description of the fundamental concepts of the theory of the *electrodynamic coherence* developed in the 90's by Giuliano Preparata [8], which describes the coherence processes in the context of quantum electrodynamics in condensed matter.

According to this theoretical view, in a metal hydride a reorganization of the quantum vacuum of QED takes place caused by the self-interaction of the charges of the deuterons through the radiative electromagnetic field when they overcome a *concentration threshold*, causing the spontaneous formation inside the metal of a large number of *coherence domains* (CD) each consisting of roughly $10^{11}$ deuteron and described by a single quantum state. These coherence domains have typical dimensions on the order of 1-10 $\mu$m and their thermodynamic stability is guaranteed by the fact that the deuteron belonging to the domain acquire an energy gap on the order of 1 eV per deuton, thus guaranteeing the existence of such states for temperatures up to the melting temperature of metals.

The existence of such coherence domains has recently been theoretically studied also by other authors [25], who independently confirmed the existence of the states.

The interest in these states stems from the fact that:

1. In the scattering processes between coherent states the transition cross sections are proportional to the square of the number of particles that constitute them, differently from what happens for the incoherent processes which are proportional to the number of particles. Consequently, processes that would be totally negligible in the incoherent case are amplified by a factor $10^{11}$ and may become predominant.
2. The kinematics of such processes is substantially different from that of processes involving a small number of particles (single particle processes).
3. The involvement of coherent states in LENR processes is supported by the experimental evidence of the suppression of highly energetic particles (T, p, $^3$He, n), typical of the well-known nuclear processes occurring in a vacuum.
4. The dynamics of interaction of the coherent processes allows the distribution of the energy associated with a nuclear-type reaction to a very large number of deuterons ($\sim 10^{11}$), explaining in a natural way the absence of highly energetic reaction products.

5. Since the coherent states are in thermodynamic equilibrium with the incoherent fraction of the deuterons, the excess energy acquired by the nuclear reactions by the final coherent states can be efficiently transferred to the incoherent fraction in the form of heat.

We note in passing that the described features of the coherent interactions provide a natural explanation of the long-standing Huizenga three miracles [26].

The coherent nature of the deuteron plasma confined inside the metal produces the so-called *Preparata effect* that can be advantageously exploited to obtain enhanced deuterium loading of thin hydride wires when subjected to a longitudinal electric potential.

A quantum system, whose dynamics is described by a unique wave function, is able to "see" the electrostatic potential applied to a palladium wire as a chemical potential that confines the D+s in the lattice, thus increasing the achievable concentration (Preparata's effect) [27]. The theory applies equally well for Constantan. This effect is analogous to the well-known Böhm-Aharonov effect [28] where the dynamics of a quantum system is affected by a change in the e.m. potential through a modification of the phase of the wave function. The chemical potential μ of H+ (D+) in a metal is thus shifted by the applied electric potential V(*r*) multiplied by the screened charge Z*e of the H+ (D+) in the metal

$$\mu[V(r)] = \mu[0] + Z * eV(r) \tag{1}$$

In eq. (1) μ[0] is the chemical potential of the point 0 of the cathode where the electric potential is highest, so that V is the (negative) relative potential of the point *r* with respect to 0. The profile of the chemical potential μ is changed in such a way that the chemical potential in some regions of the cathode can fall below: a) the chemical potential μ_sol of the ions in solution, in the case of electrolytic loading; b) below μ_gas, in the case of gas loading. Consequently, an inflow of ions occurs in those regions.

The Preparata effect is most effective when it is possible to apply large electric potential differences across the cathode without inducing sizeable Joule heating that might spoil the pre-condition of the effect, i.e., the H+(D+) coherence. The optimal effect is expected when very thin cathodes whose resistance

$$R = \rho l/S \tag{2}$$

is increased as much as possible by taking a large length (l) divided by a very small cross-section S; $\rho$ is its resistivity. Since the expected height of the chemical potential barrier is a fraction of 1 eV and Z* has been estimated [29] as about 0.1, a voltage of about 10 V applied along the one-dimensional cathode should be sufficient to induce a massive inflow of ions increasing the loading by a factor 1.3-1.4 compared to a two or three dimensional cathode (plate or rod) exposed to the same electrolytic current density.

The intermediate voltages along the wire will decrease the chemical potential of the coherent state so that in the regions where the electric potential is lower, the loading will be favoured, resulting in a sort of confinement of the hydrogen inside the metal and the activation of AHEs in the more negative part of the wire.

The increasing factor of 1.3-1.4 can have huge effects regarding AHE production. In some experiments, made by Preparata's Team around 1999-2000 at ENEA-Frascati in collaboration with the LEDA Company, the effect was experimentally verified and more deeply analysed [30].

## 10. Acknowledgements

We are indebted to a Metallurgical Company in the **North-Eastern part of Italy (NEMC)**, which since 2011 provided some financial support and performed key experiments in their own laboratories (by their scientists and technicians): a fully independent cross-check of our most critical experiments was useful to increase our confidence on reported results.

Since 2017 we initiated collaborations with **NEMC** and **SIGI-Favier** (Italy-France), to design an original hybrid sheath obtained by crossing glass and Alumina–Quartz fibers. The sheaths are used for the electric insulation of the wires. These original sheaths can continuously operate up to 1200°C and, thanks to a tailored geometry, can adsorb significant amounts of Atomic Hydrogen. Moreover, the sheaths are porous, with holes of micrometric dimension: *one of the key aspects of our experimental set-up.*

**I**nstitute **F**luid **A**ssociation, Rome-Italy, provided us some economic support for consumables in the laboratory, and the expense of trips to attend Conferences/Workshop/Meetings.

We thank the company **"Franco Corradi S.A.S."** (Rho, Milan) for providing high performances thin alumina tubes used for heavy-duty test up to 1100°C.

We thank **Jed Rothwell** who, as usual, improved not only the English Language aspects of our presentations, but also gave us useful advice.

This time the work performed by the **Referee** (unknown) and the Editor of the Journal (**Jean Paul Biberian**) were the key factors to improve, largely, our original paper: not enough detailed to be understood by the typical reader, even expert in the (complex) field of LENR.

*Special thanks to the Scientists involved in the CleanHME European project, chaired by Konrad Czerski (Szczecin University, Poland). In particular, for the fruitful collaboration with Prof. Bo Hoistad and Collaborators (Uppsala University, Sweden) as well as with Dr. Andras Kowacs from the Broadbit Company.*

**This work has received funding from the European Union's Horizon2020 Framework Programme, under grant agreement #951974.**

## Disclaimer

This work was written under the sole responsibility of the Authors and does not contain nor represent necessarily the opinion nor the position of the whole CleanHME European project.

## References

[1] G. C. Fralick, A. J. Decker and J. W. Blue, "Results of an attempt to measure increased rates of the reaction D-2 + D-2 yields He-3 + n in a nonelectrochemical cold fusion experiment," NASA-TM-102430, E-5198, NAS 1.15:102430 , Cleveland, 1989.

[2] M. C. H. McKubre, S. Crouch-Baker, A. M. Riley, S. I. Smedley and F. L. Tanzella, "Excess Power Observations in Electrochemical Studies of the D/Pd System; the Influence of Loading," in *Frontier of Cold Fusion, Proc. ICCF3 Nagoya*, 1992.

[3] Y. Iwamura, T. Itoh, M. Sakano, S. Sakai and S. Kuribayashi, "Low Energy Nuclear Transmutation in Condensed Matter induced by D2 Gas Permeation through Pd Complexes: Correlation between Deuterium Flux and Nuclear Products," in *Proc. of ICCF10, Condensed Matter Nuclear Science*, Cambridge, MA, 2006.

[4] Y. Iwamura, T. Itoh and N. Gotoh, "Characteristic X-Ray and Neutron Emission from Electrochemically Deuterated Palladium," in *Proc. 5th Int. Conf. Cold Fusion, Monte Carlo*, Monaco, 1995.

[5] Y. Iwamura, T. Itoh and M. Sakano, "Nuclear products and their time dependence induced by continuous diffusion of deuterium through multi-layer palladium containing low work function material," in *8th International Conference on Cold Fusion*, Lerici (La Spezia), 2000.

[6] Y. Arata and Y. C. Zhang, "Reproducible "Cold" Fusion Reaction Using a Complex Cathode," *Fusion Technology*, vol. 22, no. 2, pp. 287–295, 1992.

[7] F. Celani, A. Spallone, P. Tripodi, A. Petrocchi, D. d. Gioacchino, M. Boutet, P. Marini, V. D. Stefano, M. Diociaiuti and W. Collis, "Reproducible D/Pd Ratio >1 and Excess Heat Correlation by 1-$\mu$s-Pulse, High-Current Electrolysis," *Fusion Technology*, vol. 29, no. 3, pp. 398–404, 1996.

[8] G. Preparata, "QED Coherence in Matter," 1995.

[9] F. Celani, C. Lorenzetti, G. Vassallo, E. Purchi, S. Fiorilla, S. Cupellini, M. Nakamura, P. Cerreoni, R. Burri, P. Boccanera, A. Spallone and E. F. Marano, "Stimulation of LENR-AHE by high power electric pulses on coiled coaxial Constantan wires at high voltage and temperature," ResearchGate, 2020.

[10] K. Jensen, Introduction to the Physics of Electron Emission. Chapter 16: Child–Langmuir law, John Wiley & Sons Ltd , 2017.

[11] M. Gerardin, *Comptes rendus hebdomadaires des séances de l'Académie des sciences*, vol. 53, p. 727, 1861.

[12] A. Cohen, "Nachweis von Protonen in Metallen," *Naturwissenschaften*, vol. 16, pp. 183–184, 1928.

[13] S. Romanowski, W. M. Bartczak and R. Wesołkowski, "Density Functional Calculations of the Hydrogen Adsorption on Transition Metals and Their Alloys. An Application to Catalysis," *Langmuir*, vol. 15, no. 18, p. 5773, 1999.

[14] F. Celani, A. Spallone, B. Ortenzi, P. S, E. Purchi, F. Santandrea, S. Fiorilla, A. Nuvoli, M. Nakamura, P. Cirilli, P. Boccanera and L. Notargiacomo, "Observation of Macroscopic Current and Thermal Anomalies, at High Temperatures, by Hetero-structures in Thin and Long Constantan Wires Under H2 Gas," *J. Condensed Matter Nucl. Sci.*, vol. 19, pp. 29–45, 2016.

[15] F. Celani, E. F. Marano, B. Ortenzi, S. Pella, S. Bartalucci and S. B. F. Micciulla, "Cu–Ni–Mn Alloy Wires, with Improved Sub-micrometric Surfaces, Used as LENR Device by New Transparent, Dissipation-type Calorimeter," J. Condensed Matter Nucl. Sci., vol. 13, pp. 56–67, 2014.

[16] F. Celani, C. Lorenzetti, G. Vassallo, E. Purchi, S. Fiorilla, S. Cupellini, M. Nakamura, P. Cerreoni, P. Boccanera, R. Burri and A. Spallone, "Unexpected effects due to voltage waveform at anode, in gaseous LENR experiments based on sub-micrometric surface coated Constantan wires.," in *Terzo Convegno "Assisi nel Vento", Domus Laetitiae*, Assisi (PG)-Italy, May 17-19, 2019.

[17] I. Langmuir, "The Effect of Space Charge and Residual Gases on Thermionic Currents in High Vacuum," *Phys. Rev.*, vol. 2, no. 450, 1913.

[18] O. W. Richardson, Thermionic Phenomena and the Laws which Govern Them. Nobel lecture, 1929.

[19] M. A. Lieberman and A. J. Lichtenberg, in *Principles of Plasma Discharges and Materials*, vol. 1, Wiley Blackwell, 2005, 2nd Edition, p. 700.

[20] F. Celani and A. Spallone , "High power us pulsed electrolysis for high deuterium loading on Pd plates," in *Selected papers following ICCF4, Dec. 6-9, 1993*, Maui-Hawaii, 1994.

[21] F. Celani and A. Spallone, "High power us pulsed electrolysis using palladium wires: evidence for a possible "phase" transition under deuterium overloaded conditions and related excess heat," in *Proceedings from ICCF5, 9-13 April 1995*, Montecarlo.

[22] F. Celani, A. Spallone, P. Tripodi, A. Petrocchi, D. Di Gioacchino, P. Marini, V. Di Stefano, S. Pace and A. Mancini, "Deuterium overloading of palladium wires by means of high power $\mu$s pulsed electrolysis and electromigration: suggestions of a "phase transition" and related excess heat," *Physics Letters A*, vol. 214, no. 1–2, pp. 1–13, 6 May 1996.

[23] L. Holmlid , P. U. Andersson and S. Badiei, "Fusion reactions in high-density hydrogen: A fast route to small-scale fusion?," *International Journal of hydrogen energy*, vol. 34, no. 1, 2009.

[24] A. Kovacs, Unified Field Theory and Occam's Razor: Simple Answers to Deep Questions, ISBN 978-1-80061-129-0: World Scientific Publishing , 2022.

[25] L. Gamberale and G. Modanese, "Coherent Plasma in a Lattice," *Submitted to PRB*, 2022.

[26] J. R. Huizenga, Cold Fusion: The Scientific Fiasco of the Century, Oxford University Press, 1994.

[27] M. Cola, E. Ded Giudice, A. De Ninno and G. Preparata, M. Cola, E. Del Giudice, A. De Ninno, G. Preparata, Proceedings of 8th International Conference on Cold Fusion (ICCF-8), Bologna: SIF, 2000.

[28] Y. Aharonov and D. Bohm, *Phys. Rev.*, vol. 11, p. 485, 1959.

[29] G. Alefeld and J. Volkl, Hydrogen In Metals, Vol. I, II, Springer-Verlag, 1978.

[30] A. De Ninno, A. Del Giudice and A. Frattolillo, "Excess heat and calorimetric calculation: evidence of coherent nuclear reactions in condensed matter at room temperature," in *Low-Energy Nuclear Reactions Sourcebook. ACS Symposium Series Vol. 998*, American Chemical Society, 2008, pp. 127–152.

[31] M. Marić, D. Savić and S. J, "Gas breakdown and secondary electron yields," *The European Physical Journal D*, vol. 68, no. 155, 2014.

[32] F. Paschen, "Ueber die zum Funkenübergang in Luft, Wasserstoff und Kohlensäure bei verschiedenen Drucken erforderliche Potentialdifferenz," *Annalen der Physik*, vol. 273, no. 5, pp. 69–75, 1889.

[33] L. F. Berzak, S. E. Dorfman and J. Smith, "Paschen's Law in Air and Noble Gases," Berkley-USA, 2006.

[34] E. M. Bazelyan and Y. P. Raizer, "Spark Discharge," Boca Raton, FL-USA, CRC Press 1998, 1998.

[35] D. Fisher and J. d. Bitetto, "Townsend Ionization Coefficients and Uniform Field Breakdown," *Physical Review*, vol. 104, p. 1213, 1956.

[36] R. Ahern, "Amplification of Enegetic Reactions". USA Patent US 2011/0233061, 29 9 2011.

[37] I. Langmuir, "Atomic Hydrogen as an Aid to Industrial Research," *Science*, vol. 67, no. 1730, pp. 201–208, 1928.

[38] J. J. J. Dufour, J.-C. X. Dufour, J. H. Foos and D. M. L. Murat, "Method and device for producing energy from a methal type hydride". Europe Patent EP1072041A1, 19 4 1999.

**International Society for Condensed Matter Nuclear Science (ISCMNS)**
The Willows, Hobro,
Wolverley, Kidderminster
Worcs DY11 5ST
England

ISBN 978-1-7138-6328-1

9 781713 863281

CURRAN ASSOCIATES INC.
proceedings.com

# MFPT 2018

## Intelligent Technologies for Equipment and Human Performance Monitoring

Virginia Beach, Virginia, USA
15-17 May 2018

MFPT